FOUNDATIONS

OF

PHYSICS

FOUNDATIONS

OF

PHYSICS

Steve Adams

MERCURY LEARNING AND INFORMATION

Dulles, Virginia
Boston, Massachusetts
New Delhi

Publisher: David Pallai
MERCURY LEARNING AND INFORMATION
22841 Quicksilver Drive
Dulles, VA 20166
info@merclearning.com
www.merclearning.com
800-232-0223

Steve Adams. *Foundations of Physics.*
ISBN: 9781683921448

192021321   This book is printed on acid-free paper in the United States of America.

Our titles are available for adoption, license, or bulk purchase by institutions, corporations, etc. For additional information, please contact the Customer Service Dept. at 800-232-0223(toll free).

For Alison

CONTENTS

PREFACE

The aim of this book is to draw on the essential physical principles that typical physics courses use to provide a strong conceptual base for the further study of more advanced topics. As such this book provides support for both introductory courses (calculus-based) and for readers interested in a basic review of key topics in physics. It will also be a useful reference work for instructors.

The focus is on physical principles. Applications are used to exemplify the physics but do not divert attention from the underlying concepts. Mathematics is the language of physics and a mathematical approach is taken throughout, drawing mathematical techniques including basic calculus. The approach here acknowledges this and helps to secure a foundation of relevant mathematical skills in the context of real physical problems.

Practical techniques, including the collection, presentation, analysis, and evaluation of data, are discussed in the context of key experiments linked to the theoretical spine of the work. There are also sections on testing mathematical relationships, the analysis of uncertainties, and how to approach, carry out, and write-up experimental investigations.

Every chapter concludes with a set of exercises and an appendix on Fermi problems provides an open-ended challenge that allows the reader to practice their skills in unfamiliar contexts.

How to use the book

Although the order of topics in this book mirrors that of most physics courses, it is not intended to be read in order or from cover to cover, and most chapters can be read in isolation. The book is there to be consulted, as and when a topic is studied or revised. The early sections on the language of physics and representing and analyzing data can be referred back to from any of the other sections. The appendices contain summary lists of units and useful data, equations, solutions to exercises, and a guide to planning, carrying out, and writing up experiments. There is also an extensive glossary of terms.

CHAPTER 1

THE LANGUAGE OF PHYSICS

1.0   Introduction

NASA's Mars climate orbiter was launched in 1998 and should have gone into orbit around Mars 286 days later. Instead it fell too close to the planet and broke up in the atmosphere. The mission had cost upwards of $100,000,000. Why did this happen? Because Lockheed Martin, who was calculating the thrust to maneuver the spacecraft, used English units (pound-seconds), while NASA, which controlled the thrusters, was expecting metric units (newton-seconds). Units matter!

1.1   The S.I. System of Units

The international system of units is based on seven base units:

Quantity	Name	Symbol
	S.I. base unit	
Length	Meter	M
Mass	Kilogram	kg
Time	Second	s
Electric current	Ampere	A
Temperature	Kelvin	K
Amount of substance	Mole	mol
Luminous intensity	Candela	cd

The definition of each base unit is related to the experimental method that is used to establish the unit in the laboratory. All mechanical quantities can be expressed in terms of just three base units—length, mass, and time—and the definitions of each of these is given below.

Base unit	Definition
Meter	The distance traveled by light in a vacuum in a time of 1/299,792,458 seconds
Kilogram	The mass of the international prototype of the kilogram, a platinum–iridium block kept under controlled conditions at the International Bureau of Weights and Measures
Second	The duration equal to 9,192,631,770 periods of the radiation corresponding to the transition between the two hyperfine levels of the ground state of a cesium 133 atom

You will notice that the meter is actually defined in terms of the speed of light. This has been the case since 1983 when the speed of light, which had been measured with ever-increasing precision, was defined to have the value 299,792,458 ms^{-1}.

Derived units

Many physical quantities are measured in derived units. These are combinations of base units such as m^3 (for volume) or kgm^{-3} (for density). Some common derived units that are given their own names are shown in the table.

Derived quantity	Name	Symbol	
		S.I.-derived units	**S.I. base units**
Force	Newton	N	kgms^{-2}
Pressure	Pascal	Pa	$\text{kgm}^{-1}\text{s}^{-2}$
Energy	Joule	J	$\text{kgm}^2\text{s}^{-2}$
Power	Watt	W	$\text{kgm}^2\text{s}^{-3}$

When you come across a new physical quantity that is unfamiliar, you can work out its S.I. units by making sure that an equation containing the new quantity is balanced. All equations in physics must balance in terms of units. Here are two examples.

Energy

Any equation involving energy will suffice, e.g., $E = \frac{1}{2} mv^2$. The units of the right-hand side are $\text{kgm}^2\text{s}^{-2}$, so these must be the units of energy. The name "joule" is a convenient alternative for such a common physical unit, so $1 J = 1 \text{ kgm}^2\text{s}^{-2}$.

Viscosity

There is an equation for viscous drag on a sphere moving through a fluid called "Stoke's law." This has the form $F = 6\pi\eta rv$ where r and v are the radius and velocity of the sphere, respectively. η is the viscosity of the fluid. What are the S.I. units of viscosity? First of all, rearrange the equation to give $\eta = F/6\pi rv$ and then balance the units (6π has no units; it is simply a number). The right-hand side has units of $Nm^{-2}s$ or Pas (this is because $1\ Nm^{-2} = 1\ Pa$). This can be reduced to base units by substituting for N ($= kgms^{-2}$). The base units for viscosity are, therefore, $kgm^{-1}s^{-1}$.

1.2   Dimensions

All mechanical quantities depend on mass (M), length (L), and time (T). These are the fundamental dimensions of mechanics and are measured in terms of the base units of kg, m, and s. All equations in physics must balance in terms of numerical value, units, and dimensions. When they do, they are said to be **homogeneous**. This can be very useful for checking your working during a derivation (if you made a mistake in the algebra, then the dimensions might not balance) and for testing proposed equations to see if they are viable. It can also be used to construct possible equations if you have an idea of the relevant parameters.

To indicate that we are dealing with dimensions, we use square brackets so that $[E]$ means "the dimensions of" energy and $[v]$ means "the dimensions of" velocity. The fundamental dimensions are related in a similar way to base units, so it is possible to work them out by balancing simple equations. For example, if we want to find the dimensions for energy, we can use the equation $E = mgh$. The dimensions of E will be the same as the dimensions of mgh:

$$[E] = [m][g][h] = MLT^{-2}L = ML^2T^{-2}$$

If the dimensions are now replaced by base units, we see that the S.I. unit of energy is kgm^2s^{-2} as before.

Five basic dimensions can be used to express most quantities in physics.

Dimension	Mass	Length	Time	Current	Temperature
Symbol	M	L	T	I	θ

Dimensionless numbers play an important role in many areas of physics. These are combinations of physical quantities where the dimensions cancel so that the result is a pure number. These often have (or seem to have) great significance. The fine structure constant α in quantum electrodynamics is a good example:

$$\alpha = \frac{e^2}{2_0 hc} = 0.0072973525664$$

where e is the charge on an electron, ε_0 is the permittivity of free space, h is the Planck constant, and c is the speed of light. This dimensionless constant determines the strength of the interactions between electrons and photons. It is also, approximately, the speed of an electron in units of the velocity of light. Since $\alpha \ll 1$ the motion of electrons in atoms does not have to be treated relativistically.

Another interesting dimensionless number is the ratio of the mass of a proton to the mass of an electron.

$$\mu = \frac{m_p}{m_e} = 1836.15267389$$

Physicists think that we should be able to derive numbers such as μ from a fundamental physical theory. The fact that, so far, we have been unable to do this suggests that there is more new physics to be discovered!

1.2.1 Method of Dimensions

The method of dimensions uses the requirement that all physical equations must balance to construct possible equations for physical quantities. The starting point must be some physical intuition about the system. For example, a few experiments with a mass spring oscillator should convince you that the time period of the oscillator depends only on the mass m and spring constant k. If this is correct and if the equation for time period has the form

$$T = \text{constant} \times m^x \times k^y$$

we can find x and y by balancing the dimensions (the constant is assumed to be dimensionless).

$$[T] = [m]^x \times [k]^y$$

Dimensions of k are MT^{-2}

$$T^1 = M^x(MT^{-2})^y = M^{x+y}T^{-2y}$$

Equating powers of each dimension:

For the dimension of time: $1 = -2y$, so $y = -1/2$

For the dimension of mass: $0 = x + y$, so $x = +1/2$

This suggests that a coherent form of the equation could be:

$$T = \text{constant} \times m^{1/2} \times k^{-1/2}$$

or
$$T = \text{constant}\sqrt{\frac{m}{k}}$$

Theory shows that the equation is actually

$$T = 2\sqrt{\frac{m}{k}}$$

The method of dimensions cannot determine the numerical value of the constant (because it is dimensionless), but it can show a possible coherent form for the equation.

1.3 Scientific Notation, Prefixes, and Significant Figures

Physicists have to deal with a huge range of numerical values. For example, the charge on an electron is 0.000000000000000000160 C and the breaking stress of steel is 210000000000 Pa. Writing numbers out in full like this is tedious and makes them hard to manipulate, so physicists use scientific notation that reduces them to a number between 1 and 10 multiplied by a power of 10. In scientific notation,

Charge on an electron $= 1.60 \times 10^{-19}$ C

Young modulus of steel $= 2.10 \times 10^{9}$ Pa

Another way to deal with very large and very small values is to use prefixes. These represent multiplication by a power of 10. For example, the prefix "milli-" represents multiplication by 10^{-3}, so 1 mm $= 10^{-3}$ m $= 0.001$ m. Here is a list of common prefixes and their multiplication factors.

Name	Symbol	Multiplier
milli-	m	10^{-3}
micro-	μ	10^{-6}
nano-	n	10^{-9}
pico-	p	10^{-12}
femto-	f	10^{-15}
atto-	a	10^{-18}
kilo-	k	10^{3}
mega-	M	10^{6}
giga-	G	10^{9}
tera-	T	10^{12}
peta-	P	10^{15}
exa	E	10^{18}

These go up (or down) in multiples of 10^3, but there are some prefixes that are in common use that do not fit this pattern: "centi-" multiples by 10^{-2} and "deci-" multiplies by 10^{-1}. In chemistry it is quite common to state volumes in decimeters-cubed. 1 decimeter is 10 cm or 0.10 m, so this is a volume of 1000 cm³ or 1 liter. In S.I. base units, it is 0.001 m³. Beware!

When presenting data it is important to use an appropriate number of significant figures, even if some of these are zeroes. The number of significant figures used represents the precision of the data, so a length of 1.20 m is more precise than 1.2 m. In principle, 1.2 m could have been rounded from anything between 1.150 m and 1.249 m, with an uncertainty of ±0.05, whereas a length of 1.20 m must really lie between 1.195 m and 1.205 m, with an uncertainty of ± 0.005. Quoting the third significant figure has increased the precision by a factor of 10.

However, it is also important not to quote data to too many significant figures, if this is not justified by the measurements that were used to obtain the data. For example, if you were trying to calculate the density of a block of wood and had measured a mass of 40.5 g and a volume of 24.2 cm³, the value of density is 40.5/24.2 = 1.673553719 gcm⁻³. The result must be rounded off so that it is consistent with the data used to calculate it. In this case, data was given to three significant figures, so an appropriate result is 1.67 gcm⁻³.

* **A useful rule of thumb:** Quote calculated values to the same number of significant figures as the least precise piece of data used in the calculation.

1.4 Uncertainties

In 2012, the discovery of the Higgs boson was announced at the Large Hadron Collider at CERN. Physicists had tuned the collider to the energy range in which the particle was expected to be found and sure enough it turned up! This allowed physicists to measure the mass of the Higgs: 125.09±0.21 *GeV/c²* (the *GeV/c²* is a convenient mass unit used in particle physics). The quoted uncertainty is < 0.2% of the mass. This pins the mass of the Higgs into a small enough range so that its properties can be compared with theoretical predictions.

If the uncertainty in a measured value is too large, it is not very useful. For example, if you were asked to measure the acceleration due to gravity in a laboratory and you got a value of 9.8 ms⁻², you might be quite happy and feel that you had done a good job. However, if the uncertainty associated with that value was ± 1.0 ms^{-2} then the acceleration due to gravity as measured in your experiment could lie anywhere between 8.8 and 10.8 ms^{-2}, so the fact that the actual value turns out to be close to the true value was probably just by chance. If you repeated the experiment in the same way, you would probably get a very different value.

▪ Whenever you calculate a value from experimental data, you should include the estimated uncertainty. This is a measure of the reliability of the measured value.

While a detailed analysis of uncertainties and their effect on calculated values involves a lot of statistics, there are some simple methods that will give a reasonable estimate and that can be used quite easily.

1.4.1 Types of Uncertainty

The table below defines types of uncertainty. In each case the measured quantity is a variable x.

Type		Example
Absolute uncertainty	δx	$g = 9.81 \pm 0.05$ ms^{-2}
Fractional uncertainty	$\dfrac{\delta x}{x} \dfrac{\delta x}{x}$	Fractional uncertainty in g is $0.05/9.81 = 0.0051$
Percentage uncertainty	$\dfrac{\delta x}{x} \times 100\%$	Percent uncertainty in g is $0.0051 \times 100 = 0.51\%$

1.4.2 Combining Uncertainties

It is often important to be able to calculate the uncertainty in a calculated quantity that depends on several other measured quantities, each with its own uncertainties. The simplest way to do this is to calculate the maximum and minimum possible values of the quantity using the known uncertainties in the measured values and then to use these extreme values to find the average and range. For example, in an experiment to measure the acceleration of free fall, the following results are obtained:

Distance fallen from rest, $s = 2.500 \pm 0.005$ m

Time taken, $\qquad t = 0.710 \pm 0.020$ s

From theory, $\qquad g = 2s/t^2$

$$g_{max} = (2 \times 2.505)/(0.690)^2 = 10.52 \text{ ms}^{-2}$$

$$g_{min} = (2 \times 2.495)/(0.730)^2 = 9.36 \text{ ms}^{-2}$$

$$g = 9.93 \pm 0.58 \text{ ms}^{-2}$$

While this method can be used, there are also some simple ways in which uncertainties can be combined mathematically. Here are some simple rules.

Combination	Rule	
Uncertainty in a sum: $y = a + b$	Add absolute uncertainties	$\delta y = \delta a + \delta b$
Uncertainty in a product: $y = ab$	Add fractional uncertainties	$\dfrac{\delta y}{y} = \dfrac{\delta a}{a} + \dfrac{\delta b}{b}$
Uncertainty in a quotient: $y = a/b$	Add fractional uncertainties	$\dfrac{\delta y}{y} = \dfrac{\delta a}{a} + \dfrac{\delta b}{b}$
Uncertainty in a power: $y = a^n$	Multiply fractional uncertainty by power	$\dfrac{\delta y}{y} = n\dfrac{\delta a}{a}$

Using the data from the example above (acceleration of free fall) and applying the rules from the table, we have:

$$g = 2s/t^2 = (2 \times 2.500)/(0.710)^2 = 9.92 \text{ ms}^{-2}$$

$$\frac{\delta g}{g} = \frac{\delta s}{s} + 2\frac{\delta t}{t} = \frac{0.005}{2.500} + 2\frac{0.020}{0.710} = 0.058$$

$$\delta g = 0.58 \text{ ms}^{-2}$$

This gives a result of **g = 9.92 ± 0.58 ms^{-2}** (almost identical to the first method).

1.5   Dealing with Random and Systematic Experimental Errors

In 2011, results from CERN suggested that neutrinos might be traveling faster than the speed of light. This made it to the front page of newspapers around the world. However, the scientists whose data had led to the claim realized that, if it was true, it would turn physics upside down, so they invited other scientists to check their work and to repeat the measurements. A year later it was confirmed that the original experiment had introduced measurement errors, which had not been accounted for. When they were included, it was clear that neutrinos had not traveled faster than the speed of light. It is always important to consider possible sources of error in your experiments and to try to reduce them as much as possible. Of course, it will be impossible to reduce them to zero, so you must try to estimate their size in order to calculate the uncertainty in your results.

Random errors

If a quantity is measured repeatedly, the results are likely to vary as a result of measurement errors. If these errors are random, the results will be scattered above and below the actual value of the quantity being measured.

- The best way to reduce the impact of random errors is to use a large number of repeats and to take an average.

Repeated measurements are also helpful for estimating experimental uncertainties. A useful estimate is half the range of the data (but it is important to eliminate any anomalous results before calculating the range).

Systematic errors

A systematic error is one that introduces a consistent bias to all of the measured data, usually making it all too big or too small. Taking an average will not help to reduce the impact of a systematic error. However, if a systematic error is known, then it can be corrected for by adjusting the measurements. For example, if there is a zero error, this must be subtracted from all subsequent measurements.

If a systematic error is present but unidentified, the final result might seem very precise and yet disagree with an accepted value. It is always important to compare your results with accepted values (where these exist). If you have carried out a good experiment, your value with its uncertainties should overlap the accepted value for the quantity measured.

1.6   Differential Calculus

1.6.1   Derivatives and Rates of Change

Calculus was invented by Isaac Newton in order to solve physics problems. Newton realized that he needed to be able to deal with quantities that were continuously changing, so he needed a technique that could cope with changes that occurred in vanishingly small (infinitesimal) intervals of time. Differential calculus deals with instantaneous rates of change. For example, to find the velocity of a moving object we need to measure its change in displacement δs during a time δt, but if velocity is changing continuously, the value of $\delta s/\delta t$ will be an average during the interval δt and the instantaneous velocity will vary during the time interval. In order to find the instantaneous velocity at a particular moment, we would need to use an infinitesimally short time interval $\delta t \to 0$. This is not possible in an experiment, but mathematically the derivative of displacement with respect to time (ds/dt) represents this instantaneous value. It is a mathematical limit of the ratio ds/dt as $\delta t \rightarrow 0$.

$$v = \frac{ds}{dt} = \lim_{\delta t \to 0} \frac{\delta s}{\delta t}$$

This approach is used for rates of change throughout physics:

Velocity $= ds/dt$

Acceleration $= dv/dt$

Electric current $= dQ/dt$

Rate of change of flux-linkage $= d(N\Phi/dt)$

…

The "operator" (d/dt) … can be read as "rate of change of…."

All of the above examples are rates of change with time, but rates of change with respect to other variables are also common. For example, dV/dx can represent a potential gradient in an electric field.

The process of deriving a rate of change is called "differentiation" and the result is a "derivative." Derivatives can be represented graphically as the gradient of a graph. The graph below shows how the charge builds up on a capacitor as it is charged. The gradient is dQ/dt, which is equal to the current that flows at a particular instant as the capacitor charges.

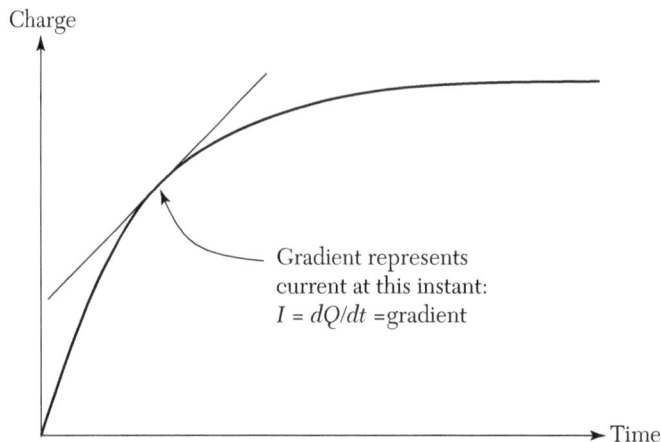

Some common derivatives (rates of change) that often occur in physics are listed in the table below.

Function (y)	Derivative (rate of change) $\left(\dfrac{dy}{dx}\right)$
Constant value, e.g., $y = 8$	0
Power law, e.g., $y = Ax^n$	$\dfrac{dy}{dx} = nAx^{n-1}$
Exponential function, $y = e^x$	$\dfrac{dy}{dx} = e^x$
Exponential relationship, e.g., $y = Ae^{bx}$	$\dfrac{dy}{dx} = bAe^{bx}$
Sine function, $y = \sin x$	$\dfrac{dy}{dx} = \cos x$
Cosine function, $y = \cos x$	$\dfrac{dy}{dx} = -\sin x$
Sinusoidal variation, e.g., $y = A \sin (bx)$	$\dfrac{dy}{dx} = bA\cos(bx)$
Cosinusoidal variation, $y = A \cos (bx)$	$\dfrac{dy}{dx} = -bA\sin(bx)$

Second derivatives

A derivative is a rate of change, so the derivative of displacement with respect to time is velocity and the derivative of velocity with respect to time is acceleration. This means that acceleration is the rate of change of the rate of change of displacement.

$$a = \frac{dv}{dt} = \frac{d}{dt}\left(\frac{ds}{dt}\right)$$

This is called the **second derivative** of displacement and is written as:

$$a = \frac{d^2 s}{dt^2}$$

Whereas first derivatives are equal to gradients, second derivatives are related to the sharpness of curvature of a graph (the rate of change of the gradient). Acceleration is related to the sharpness of curvature of a graph of displacement against time.

1.6.2 Maximum and Minimum Values

Calculus can be used to find where something has its maximum or minimum value. Look at the graph below, which shows how the power transferred from a supply to a resistor in an electric circuit depends on the value of the load resistor.

   The maximum power P_{max} occurs at some resistance R_{peak}. This is where the curve has a gradient of zero. However, gradients are equal to derivatives, so we can find the maximum value if we can find where the gradient is zero. To do this we need an equation for power in terms of the resistance, $P(R)$. It turns out that this is easy to find (see Section 18.5.1). Once we have the equation, we differentiate it, set it equal to zero, and solve the resulting equation to find the value of R.

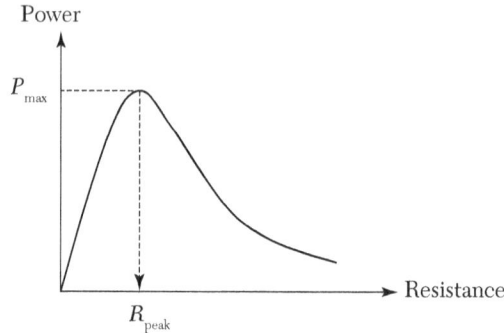

$$\frac{dP(R)}{dR} = 0$$

To find a maximum or minimum of some function $y(x)$:

▪ Differentiate to find the derivative $\frac{dy}{dx}$

▪ Set the derivative equal to zero, $\frac{dy}{dx} = 0$

▪ Solve for x

Calculus can be used to show the nature of the stationary point:

▪ Find $\frac{d^2y}{dx^2}$

▪ If $\frac{d^2y}{dx^2}$ is positive, the gradient is increasing: minimum position

▪ If $\frac{d^2y}{dx^2}$ is negative, the gradient is decreasing: maximum position

▪ If $\frac{d^2y}{dx^2}$ is zero, the position is a saddle point, neither a maximum nor a minimum

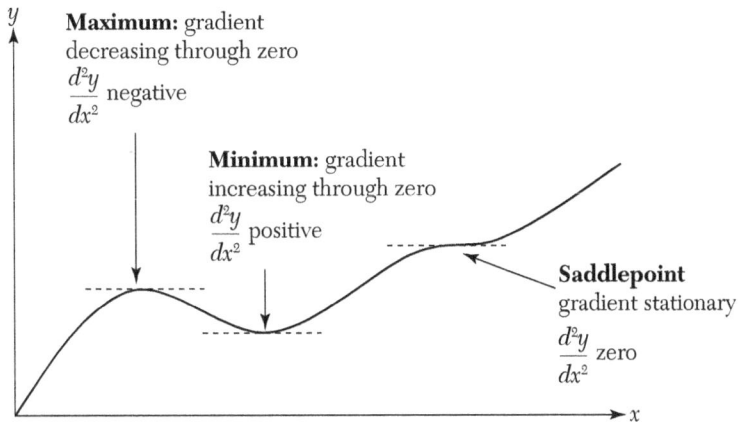

1.7   Differential Equations

An algebraic equation such as $s = ut + \frac{1}{2}at^2$ can be solved to find the *value* of one of the variables if all of the others are known. For example, if we know the initial velocity, acceleration, and time for which an object accelerates, we can use the equation to find its displacement. Differential equations include derivatives and their solution is not a value but a *function*.

For example, the simple equation for constant velocity v,

$$\frac{ds}{dt} = v,$$

is an example of a first-order differential equation. It is called first order because it only involves first derivatives; there are no second- or higher-order derivatives in the equation. Its solution is an expression for displacement as a function of time $s(t)$ that satisfies the equation. It turns out that there are an infinite number of such expressions. One of them is

$$s = vt$$

where k is a constant (representing a constant velocity).

We can show that this is a solution by differentiating it:

$$\frac{ds}{dt} = \frac{d}{dt}(vt) = v$$

However, the expression

$$s = vt + 10$$

is also a solution.

$$\frac{ds}{dt} = \frac{d}{dt}(vt + 10) = v$$

The general solution to this differential equation is

$$s = vt + s_0$$

where s_0 is another constant (equal to the initial displacement at $t = 0$).

Differential equations are used to express many of the fundamental relationships in physics, including Newton's laws of motion, Maxwell's equations of electromagnetism, and Schrodinger's equation in quantum theory. If you study physics at a higher level, you will need to get to grips with a range of methods for solving differential equations. The table below lists some examples of differential equations that will feature later in this book along with some useful solutions.

Topic	Differential equation	Solution	Conditions
Radioactive decay	$\dfrac{dN}{dt} = -\lambda t$	$N = N_0 e^{-\lambda t}$	$N = N_0$ at $t = 0$
Capacitor discharge	$\dfrac{dQ}{dt} = -\dfrac{Q}{RC}$	$Q = Q_0 e^{-\frac{t}{RC}}$	$Q = Q_0$ at $t = 0$
Newton's second law (constant acceleration)*	$\dfrac{d^2 x}{dt^2} = \dfrac{F}{m}$	$x = ut + \dfrac{1}{2}\left(\dfrac{F}{m}\right)t^2$	F, m, u constants $x = 0$ at $t = 0$
Simple harmonic motion°	$\dfrac{d^2 x}{dt^2} = -\omega^2 x$	$x = A \cos(\omega t + \phi)$	A, ω, ϕ constants $(\omega = 2\pi f)$

*The equations for Newton's second law and for simple harmonic motion are second-order differential equations because they involve second derivatives. Whereas first-order differential equations generate one arbitrary constant, second-order differential equations generate two.

1.8   Integral Calculus

Integration is the inverse process to differentiation and is related to the limit of a sum. The graph below shows how velocity varies with time for a particular object. The area under the graph represents displacement and can be approximated by adding up a large number of thin rectangular strips, each of width δt.

The area of the small shaded strip, $v(t)\delta t$, approximates to the extra displacement during a short time δt at time t. The area between any two times t_1 and t_2 is equal to the displacement during that time interval and is given, approximately, by the sum of the areas of all such strips between those two times:

$$s = \sum_{t_1}^{t_2} v(t)\delta t$$

Velocity

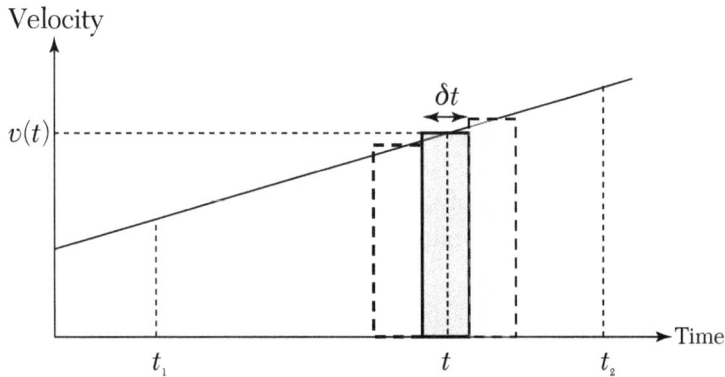

This becomes a better approximation to the actual area if we take thinner and thinner strips by making δt smaller and smaller. It would be a precise value in the limit that δt approached zero: $\delta t \rightarrow 0$. In this limit the sum becomes a continuous process called an integral:

$$s = \lim_{\delta t \to 0} \sum_{t_1}^{t_2} v(t)\delta t = \int_{t_1}^{t_2} v(t)\,\mathrm{d}t$$

The table below gives some derivatives and related integrals that are used in this book.

Context	Differential form	Integral form
Dynamics	$v = \dfrac{ds}{dt}$	$s = \int v\,dt$
Dynamics	$a = \dfrac{dv}{dt}$	$v = \int a\,dt$
Newton's laws	$F = \dfrac{d(mv)}{dt}$	$mv - mu = \int F\,dt$
Electric circuits	$I = \dfrac{dQ}{dt}$	$Q = \int I\,dt$
Radioactivity	$\dfrac{dN}{dt} = -N$	$\int \dfrac{dN}{N} = -\int dt$
Capacitors	$\dfrac{dQ}{dt} = -\dfrac{Q}{RC}$	$\int \dfrac{dQ}{Q} = -\int \dfrac{dt}{RC}$

1.9   Vectors and Scalars

Scalars are physical quantities that have magnitude but no direction, e.g., distance, speed, mass, energy, power, and temperature. Scalar quantities simply add together, so if 3.0 kg is added to a body of mass 7.0 kg, the resulting body has a mass of 10 kg. Scalars do not have a direction but can have a sign, and this must be taken into account when they are added.

Vectors are physical quantities that have magnitude and direction, e.g., displacement, velocity, force, momentum. Vectors can be represented by arrows; the length of the arrow represents magnitude, and the direction of the arrow is the direction of the vector. They are often distinguished from scalar quantities by underlining them, e.g., $\underline{v}$ is a velocity vector and v is its magnitude.

1.9.1 Adding Vectors

When vectors are combined we need to consider their direction as well as their magnitude, so we cannot simply add the magnitudes. For example, if I walk 10 m north and then 10 m east I have walked a distance of 20 m (scalar), but my displacement is 14 m NE (vector). This can be shown on a vector diagram.

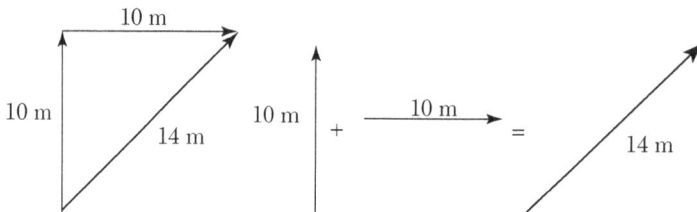

A displacement vector 10 m north has been added to a displacement vector 10 m east to obtain a resultant displacement vector of 14 m northeast. The resultant vector is found by placing all the vectors to be added end to end and then connecting the start of the first vector to the end of the last vector.

▪ This diagrammatic method of adding vectors can be used to solve problems if all the vectors are drawn to the same scale.

▪ Pythagoras' theorem and trigonometry can be used to find the resultant vector from the vector diagram.

▪ Vector subtraction is carried out by adding the negative of the vector (i.e., reversing its direction).

1.9.2 Resolving Vectors into Components

If a projectile is fired at some angle to the ground, its velocity will be partly vertical and partly horizontal. The velocity has a horizontal **component** and a vertical **component**.

Any vector can be resolved into two perpendicular components using trigonometry.

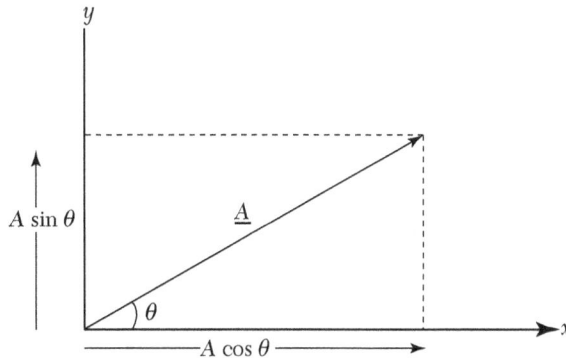

The vector $\underline{A}$ has been resolved into two components:

- x-component has magnitude $A_x = A \cos \theta$
- y-component has magnitude $A_y = A \sin \theta$

The components can also be used to reconstruct the original vector:

- The magnitude of $\underline{A}$ is, by Pythagoras's theorem, $A^2 = A_x^2 + A_y^2$
- The angle θ is found from $\tan \theta = \dfrac{A_y}{A_x}$

Vector addition can also be carried out by adding components along a common set of perpendicular axes:

- If $\underline{C} = \underline{A} + \underline{B}$ then $C_x = A_x + B_x$ and $C_y = A_y + B_y$

1.9.3 Multiplying Vectors

There are two different ways to multiply two vectors: the scalar product and the vector product.

Scalar product

The work done by a force is the product of the force and the displacement of the point of action of the force parallel to the force. Since both force and displacement are vectors but work is a scalar, this is an example where the product of two vectors results in a scalar. For this reason it is called a scalar product (sometimes referred to as a "dot product").

The magnitude of a scalar product is equal to the product of the magnitudes of the two vectors multiplied by the cosine of the angle between them.

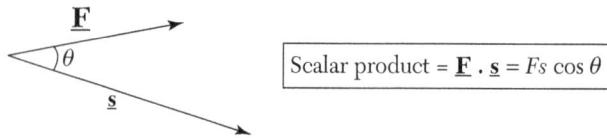

Scalar product = $\underline{\mathbf{F}} \cdot \underline{\mathbf{s}} = Fs \cos \theta$

The scalar product is zero when the vectors are perpendicular.

Vector product

The force on a moving charged particle in a magnetic field has a magnitude given by $F = qvB \sin\theta$ where B is the magnetic field strength, v is the velocity of the particle, and θ is the angle between $\underline{\mathbf{v}}$ and $\underline{\mathbf{B}}$. Since force, magnetic field strength, and velocity are all vectors, this is an example where the product of two vectors results in another vector. For this reason it is called the vector product (sometimes referred to as the "cross-product"). The direction of the resulting force is found by using a right-hand rule and is perpendicular to both of the vectors $\underline{\mathbf{v}}$ and $\underline{\mathbf{B}}$.

A vector product is written as: vector product of $\underline{\mathbf{v}}$ and $\underline{\mathbf{B}} = \underline{\mathbf{v}} {}^{\wedge} \underline{\mathbf{B}}$

The force on a moving charge in a magnetic field is therefore $\underline{\mathbf{F}} = q \, (\underline{\mathbf{v}} {}^{\wedge} \underline{\mathbf{B}})$

The magnitude of the vector product is equal to the product of the magnitudes of the two vectors multiplied by the sine of the angle between them.

The direction of the vector product is perpendicular to the plane defined by the two vectors being multiplied together. To work out the direction of the product vector, imagine rotating from the first vector to the second vector and then the resultant is in the direction of movement of a right-handed screw! The vector product is zero if the vectors are parallel ($\theta = 0$).

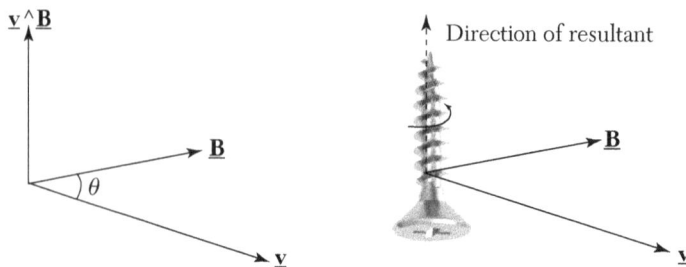

1.10   Symmetry Principles

Symmetry in nature is often linked to beauty. Apparently human beings with more symmetrical faces are more attractive to others. But in physics, symmetry is a powerful tool that can be used as a guide to the underlying laws of nature and that makes the equations of physics both powerful and beautiful.

A geometrical symmetry leaves a shape unchanged under certain rotations or reflections. For example, if an equilateral triangle is rotated through 60° or reflected about a line passing through a vertex and perpendicularly through the opposite side, it is unchanged. The diagram shows three symmetry axes of an equilateral triangle. A circle would have an infinite number of these, all parallel to diameters.

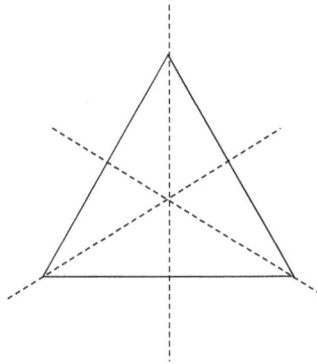

The shape of the triangle is **invariant** under these rotations and reflections. In Newtonian mechanics the laws of physics are the same in all uniformly moving (inertial) reference frames, so Newton's laws are invariant under a change of velocity. Einstein's theory of special relativity goes further and includes all of the laws of physics (see Chapter 24). Hermann Minkowski realized that Einstein's equations for relativity were similar to those for a geometrical rotation and identified physical quantities that are the same for all inertial observers—these are four-dimensional quantities called invariants and are constructed from space and time components.

Emmy Noether showed that symmetry principles are linked to conservation laws. This is not really surprising because a conservation law identifies some quantity that stays the same (is invariant) when other things change. For example, the total linear momentum of a collection of colliding bodies is the same before and after the collisions, and the total energy of the universe is the same before and after an explosion. Noether showed that conservation of momentum is linked to the laws of physics remaining the same under translation; conservation of angular momentum is linked to the laws staying the same under rotation; and the law of conservation of energy is linked to the laws staying the same at all times. This link between mathematical symmetries and conservation laws is a powerful idea in theoretical physics.

In particle physics, the search for symmetry in mathematical equations can lead to the prediction and discovery of new types of particle. Paul Dirac realized that the equation for the rest energy of an electron is a quadratic equation, one that has both positive and negative solutions:

$$E^2 = p^2 c^2 + m_0^2 c^4$$

When $p = 0$ we can write

$$E^2 = m_0^2 c^4$$

which has the solutions

$$E = \pm m_0 c^2$$

The positive solutions correspond to normal electrons. By taking the negative energy solutions seriously, he predicted the existence of anti-electrons or positrons. The symmetry between matter and antimatter is now built into the Standard model and is linked to other fundamental symmetries such as time reversal. There is even a sense in which anti-matter can be regarded as ordinary matter traveling backward in time!

The time reversal symmetry in particle physics is one of three symmetries that are fundamentally linked:

T: Time reversal symmetry—changing the sign of the time coordinate

P: Parity symmetry—changing the sign of all three spatial coordinates

C: Charge (charge conjugation) symmetry—changing the signs of all of the charges

It turns out that none of these symmetries are observed in all interactions, but if C, P, and T are all applied, the symmetry is observed. No violation of CPT symmetry has ever been seen.

The Standard model also contains some apparent symmetries that have not been explained. There are three pairs of leptons and three pairs of quarks (see Section 26.4.5) but no known physical process that links them or that will allow the transformation of quarks into leptons or vice versa. Attempts to construct a theory that would allow this has led physicists to predict the existence of additional "super-symmetric" particles for all the existing particles. So far none have been discovered.

Symmetry principles have also been used to investigate alternative approaches to physics. Richard Feynman and John Wheeler, for example, realized that Maxwell's equations for electromagnetism could be solved both forwards and backwards in time. This sounds odd, but by including both solutions they were able to construct a version of the theory that worked just as well as the original one and did not neglect half of the solutions. However, the interpretation of the so-called "absorber theory" requires us to take the idea of future events influencing the present seriously. Perhaps we should.

1.11 Exercises

1. Express:
   (a) 267 g in kg,   (b) 25 km in mm,   (c) 5.0 m³ in cm³,
   (d) 80 km/h in m/s,   (e) 45 cm² in m².

2. Light travels at 3.0×10^8 m/s and it takes about 8 min for light to travel from the Sun to the Earth.

   (a) How far away is the Sun in km?

   (b) How far is a light year? Give your answer in meters.

   (c) The distance from the Earth to the Moon is 380,000 km. How far is this in light-seconds?

3. Round off the following to three significant figures:
   (a) 2.000009   (b) 0.0020900
   (c) 0.009502   (d) π

4. Newton's equation for gravitational forces is $F = Gm_1m_2/r^2$. Use the method of dimensions to find correct S.I. units for the universal gravitational constant G.

5. The lengths of the sides of a rectangular box are measured using a ruler marked with a millimeter scale. The measurements give side lengths of 85, 62, and 20 mm respectively. The uncertainty in each measurement is ±1 mm.

   (a) Calculate the fractional uncertainty in the length of each side.

(b) Calculate the percentage uncertainty in the length of each side.

(c) What is the area of the largest face of the box? Include its absolute uncertainty.

(d) What is the volume of the box? Include its absolute uncertainty.

6. The time period of a mass spring oscillator is given by

$$T = 2\pi\sqrt{\frac{m}{k}}$$

where m is the mass oscillating and k is the spring constant (Nm^{-1}).

In an experiment the time period is measured to be 0.68 ± 0.04 s and the spring constant is 20 ± 2 Nm^{-1}.

(a) Calculate the mass of the oscillator including its absolute uncertainty.

(b) Which value (k or T) contributed most to the uncertainty in m?

7. (a) Explain the difference between a systematic error and a random error.

(b) How would you reduce the size of random errors when carrying out an experiment?

(c) A micrometer screw gauge reads 0.02 mm when it should read 0. The diameter of a wire is measured to be 0.34 mm using the same micrometer screw gauge. What value should be recorded?

8. Convert these numbers to scientific notation:

(a) 5500   (b) 0.0000000007

(c) 120000000000000000000000000

9. Evaluate the expressions below (do not use a calculator):

(a) $2.0 \times 10^6 \times 4.5 \times 10^9$; (b) $2.0 \times 10^6 \times 4.5 \times 10^{-9}$;

(c) $6.0 \times 10^{12}/3.0 \times 10^4$; (d) $6.0 \times 10^{12}/3.0 \times 10^{-4}$

10. The Apollo spacecraft traveled to the Moon at an average speed of 1.5 km/s. How long would it take at this speed to get to:

(a) Mars,   (b) Pluto,

(c) the nearest star,   (d) to cross our galaxy?

Speed of light = 300000000 m/s = 3.0×10^8 m/s

Distance to Mars (varies) = 50000000000 m = 5.0×10^{10} m

Distance to Pluto (varies) = 6000000000000 m = 6.0×10^{12} m

Distance to the nearest star (about) = 4.5 light-years

Diameter of our galaxy (about) = 100000 light years = 1×10^5 light-years

REPRESENTING AND ANALYZING DATA

2.0   Introduction

Experiments generate data and it is important to know how to select, record, and process this data in order to find out what the experiment has revealed. This is a huge problem for large experiments. According to the CERN website, the four experimental stations on the Large Hadron Collider (LHC) each generate between 750 Mbs^{-1} and 4 Gbs^{-1} of data when the accelerator is operating! The task of selecting and analyzing relevant data is carried out by the Worldwide LHC Computing Grid, which might reject the majority of results in order to focus on those that could reveal interesting new physics. Writing the algorithms for this system is extremely challenging.

When you carry out experiments in a laboratory, you will not face such a daunting task. However, presenting your data clearly and to an appropriate precision—rejecting anomalous results, deciding how to process it, and then extracting information from the processed data—is an essential part of physics.

2.1   Experimental Variables

In most physics experiments you will vary one parameter and measure another. The variable you change is called the **independent variable** and the one that responds to this is called the **dependent variable**. For the experiment to be a fair test, you must keep all other parameters that might affect the dependent variable constant; these are called **control variables**.

For example, imagine you are investigating how the acceleration of a trolley along a runway depends on the force applied to it. The dependent variable is the acceleration, which might be measured using light gates. The independent variable is the applied force, which might be varied using falling masses attached to the trolley via a pulley. There are several control variables, all of which must be kept constant. The obvious one is the mass of the trolley. If this was not constant, then changes in acceleration might be because of changes in mass rather than force, so the results of the experiment would not be clear. Another control variable is the angle of the runway; it should be horizontal in each trial. It would also be important to control frictional forces (e.g., by using the same trolley and runway and measuring the acceleration over the same distance on the same part of the runway).

- *Independent variable*: the one you deliberately vary

- *Dependent variable*: the one that changes in response to the change in the independent variable

- *Control variables*: the ones that you have to keep constant so they do not also affect the dependent variable

When you display results on a graph it is conventional for the dependent variable to go on the y-axis and the independent variable to go on the x-axis.

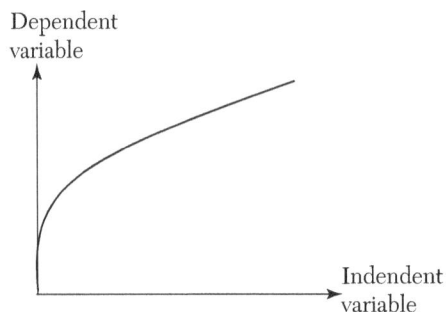

2.2   Recording Data

Here are some guidelines for recording experimental data:

- Record all of the raw data—if you repeat measurements and calculate an average, record the repeats, not just the average.

- Record data to the precision of the instrument—if you are measuring time periods using a clock that reads to 0.01 s, then all times should be recorded to 0.01s (e.g., 3.54 s, 3.57 s, etc.).

- Record calculated values to an appropriate number of significant figures; usually the smallest number of significant figures used in the calculation. If velocity v is calculated from a displacement of 4.26 m and a time of 0.45 s, a calculation gives $v = 9.466666…$ ms^{-1}. This should be rounded to two significant figures: 9.5 ms^{-1}.

- Include units at the head of the table and not with the numerical values. The correct way to label a column is quantity/unit. This means quantity divided by the unit, and that is why just numbers appear in the column. For example, displacement/m labels a column of displacements measured in meters.

Here is an example (showing just two rows in a table).

A student carried out an experiment to find out how the time period of a mass oscillating on a spring depends on the mass. In order to reduce the effect of timing errors at the start and end of the timing period, he measured 20 oscillations. In order to reduce the effect of random errors, he repeated the experiment three times. The table includes all of his raw data, averages, and calculated values.

Mass/kg	Times for 20 oscillations/s				Time period/s
	Trial 1	Trial 2	Trial 3	Average	
0.100	15.78	15.57	15.69	15.68	0.7840
0.150	X	X	X	X	X
Etc.	X	X	X	X	X

Not that, while the time period has been given to four significant figures (consistent with the raw data), the uncertainty is likely to make one or more of the final figures meaningless.

Spreadsheets (such as Excel) are often used to record and analyze data. They have some strong advantages over simple written tables. The main one is that the data, once it is recorded, can be processed within the spreadsheet. For example, if you have calculated displacement and time values, it is simple to use the spreadsheet to calculate velocities or accelerations. Mean values and standard deviations can also be calculated. Once the data has been processed, it can be selected and displayed graphically in a wide variety of ways (scatter graphs, bar charts, pie charts, etc.). It is also possible to insert error bars to extrapolate a line backwards or forwards, and to fit a line or curve to plotted data and display an equation for the fit.

Much of the analysis described above is done "behind the scenes," but when you need to present your data in a report, it is important to make sure that you include the correct table headings (and units) and that you round data to an appropriate number of significant figures. *You* should be in charge of the format; don't let the spreadsheet present its default style!

2.3 Straight Line Graphs

A linear relationship means that the rate of change of the dependent variable with respect to the independent variable is constant. Linear graphs are important in physics, and it is also important to interpret them correctly.

2.3.1 Interpreting Straight Line Graphs

Here are three linear graphs each representing a different relationship between the variables.

Beware—a common mistake is to think that a linear graph implies direct proportion. This is only the case if the graph passes through the origin.

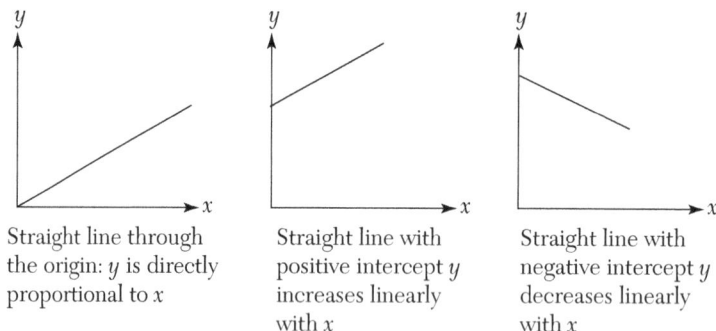

Straight line through the origin: y is directly proportional to x

Straight line with positive intercept y increases linearly with x

Straight line with negative intercept y decreases linearly with x

It is also important to realize that a linear graph with negative gradient does *not* imply inverse proportion.

2.3.2 Analyzing Straight Line Graphs

A straight line graph can be represented by the equation $y = mx + c$ where:

- y is the dependent variable
- x is the independent variable
- m is the gradient
- c is the intercept on the y-axis when $x = 0$

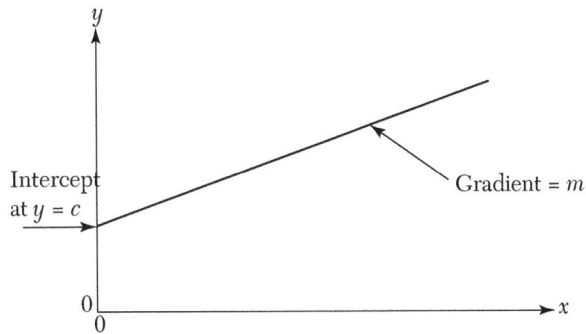

If there is a linear relationship between two physical variables x and y, then a graph of y against x can be used to find m and c and hence determine an equation for this relationship. If $c = 0$ then y is directly proportional to x.

2.4   Plotting Graphs and Using Error Bars

2.4.1   Plotting Graphs by Hand

Graphs are used to display data, to test mathematical relationships, and to calculate physical values from intercepts and gradients. They can be plotted by hand or using a spreadsheet program such as Excel; but however they are produced, it is important to present them properly and in such a way that relationships are shown clearly and any significant values can be extracted accurately.

Here are some guidelines for plotting graphs by hand:

▪ Give the graph a heading.

▪ Select scales such that the plotted data extends over at least half of the width and height of the graph paper. This might require the use of a "false origin," that is, a graph origin that is not at $(0, 0)$.

The two graphs below display the same data. However, the range of y-values is from 92 to 98. The use of a false origin makes the relationship clearer and would make calculation of a gradient from the graph more accurate.

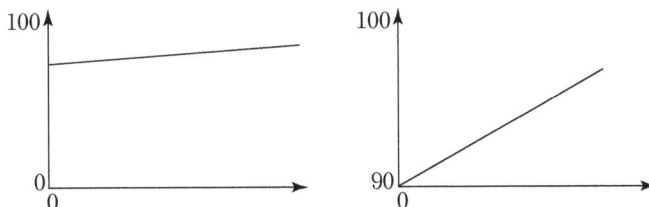

- If the origin (0, 0) is included on the graph, then mark it by a zero on the scale on each axis.

- Mark values onto each axis in equal simple intervals, e.g., 0.0, 10.0, 20.0, 30.0 … or 0.02, 0.04, 0.06… Avoid unusual or awkward intervals such as 3, 5, 7….

- Label both axes with the physical quantity and unit in the same way as if you were putting in a table heading, e.g., velocity/ms^{-1}.

- Mark each data point carefully with a cross or a point surrounded by a circle using a sharp pencil. If you are plotting on millimeter graph paper, the points need to be within half a millimeter of the correct value.

- Look carefully at the pattern of the data and decide whether any of the points are anomalies. This means that they seem "out of place" compared to the others. If possible these data points should be checked by repeating the experiment. If not they should be labeled and ignored when drawing a best-fit line.

- Look at the pattern of data and decide whether it is best represented by a straight line or a smooth curve. Remember, this is experimental data, so each point has a degree of uncertainty—this means that the line does not have to pass through the points; it is there to represent the relationship revealed by the data.

- If the data is best represented by a straight line, then use a ruler to draw it, trying to balance approximately equal numbers of points on either side of the line. It is also possible to find the gradient and y-intercept of a linear graph directly from the data by using an algebraic method called linear regression (omitting data from anomalous points).

- If the data is best represented by a curve, then draw a smooth curve, again trying to balance the distribution of points on either side of the curve.

2.4.2   Finding a Gradient from a Straight Line Graph

Many physical relationships can be tested by plotting straight line graphs and it is often important to find the gradient of the graph.

Here are some guidelines for finding an accurate gradient from a linear graph.

- Choose two points on the line that are well spaced from one another—at least half the length of the line apart.

- If possible choose points at convenient x or y values so that their coordinates can be read easily from the graph.

- Use a ruler to draw lines across to the y-axis and down to the x-axis and read off coordinates for each point (x_1, y_1) and (x_2, y_2).

- Mark these coordinates on the graph beside the chosen points.

- Calculate the gradient from the equation:

$$\text{gradient} = \frac{(y_2 - y_1)}{(x_2 - x_1)}$$

- Don't forget units! The units for the gradient are the units of y divided by the units of x. For example, a graph of charge/C against time/s would have units of Cs^{-1} or amps.

Here is an example of a gradient calculation using data for the extension of a steel spring.

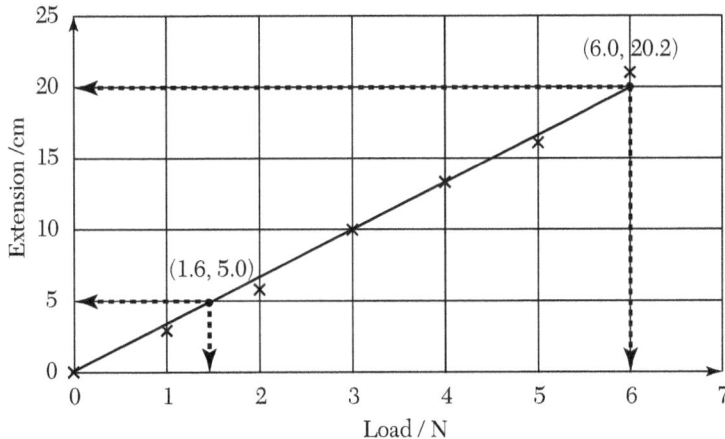

$$\text{gradient of line} = \frac{(20.2 - 5.0)}{(6.0 - 1.6)} = 3.5 \text{ cmN}^{-1}$$

2.4.3 Using a Spreadsheet Program

Whether you are plotting a graph by hand or using computer software, your aims are the same—to represent the relationship between two variables accurately and clearly and to extract any useful physical information or values from the graph as accurately and precisely as possible. The great advantage of a spreadsheet program is that once data has been stored it can be used to generate a graph at the click of a button. However, this can also lead to problems. The default settings of the program will determine how the graph appears, and this may not be the best way to display this particular dataset. The ease and rapidity of creating a graph can also mean you forget to do basic things such as ensuring the axes are labeled (or labeled correctly), having sensible scales with appropriate numbers of significant figures, etc.

Here are some questions to ask yourself when using a spreadsheet to produce a graph and analyze your data:

▦ Could I do a better job by plotting the graph by hand?

▦ Do I want the program to fit a line or curve to this data, or shall I print out the graph once the points have been plotted and then draw the line by hand?

▦ Does the graph have an appropriate heading?

▦ Do I want to include gridlines and if so with what divisions?

▦ Are the axes correctly labeled with quantity and unit?

▦ Are the scales marked correctly and are they easy to read?

▦ Are values shown with the correct number of significant figures?

▦ Do I need to use a false origin?

▦ Should the line be extrapolated back or forward?

▦ Should the line go through the origin?

▦ What kind of fit do I want to use (e.g., linear, exponential, polynomial, etc.)?

▦ Do I want to include error bars?

▦ Do I want the program to display a formula and if so how should it be formatted?

▦ How large should the graph appear in my report (avoid tiny graphs!).

Remember, you—not the computer—should be *in control*!

2.4.4   Using Error Bars

All experimental measurements will have a degree of uncertainty. This can be included on a graph by the use of error bars. These can then be used to work out the uncertainty in an intercept or gradient that is calculated from the graph.

If a data point has values x and y with uncertainties $\pm\delta x$ and $\pm\delta y$, respectively, then the error bar will be a line extending from $x - \delta x$ to $x + \delta x$ parallel to the x-axis and from $y - \delta y$ to $y + \delta y$ parallel to the y-axis.

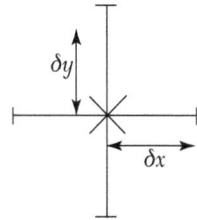

It is often the case that the error bars for one variable are far more significant than for the other. When this is the case, only one set of error bars is drawn. This is usually the dependent variable (y).

Once all of the error bars have been drawn onto the graph, you can add a trend line. If you are plotting a graph that is linear there will be a range of possible lines that can be drawn that pass through all of the error bars. The extreme lines (steepest and shallowest) are called the "worst acceptable" lines. These can be used to find the range of possible gradients and the range of possible intercepts.

The graph below uses the same data as in the previous example (for stretching a steel spring), but error bars (±1 cm) have been added to the data for extension. In addition to the original best fit, a "worst acceptable line"(WAL) has also been drawn. This is slightly steeper than the best fit line so it gives a greater value for the gradient.

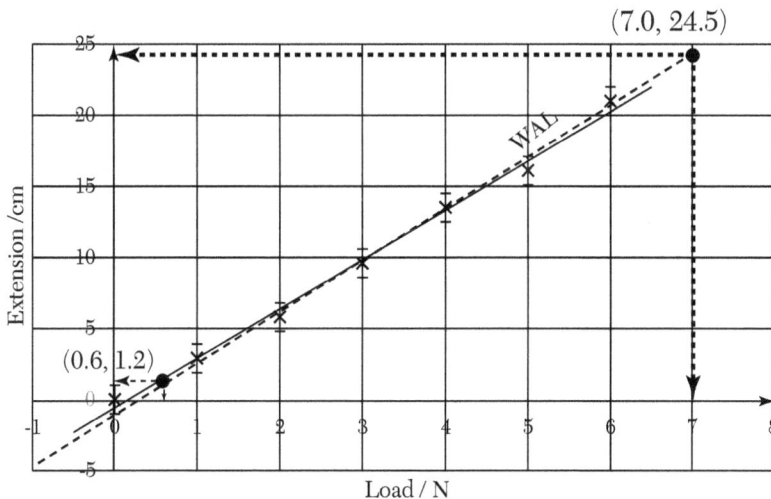

gradient of worst acceptable line $= \dfrac{(24.5 - 1.2)}{(7.0 - 0.6)} = 3.6 \, \text{cmN}^{-1}$

Using this value and the gradient of the best fit line gives

gradient $= 3.5 \pm 0.1 \, \text{cmN}^{-1}$

2.5   Logarithms

2.5.1   Logarithmic Scales and Logarithms

Many quantities in physics have values that spread over an extremely wide range, so it is often convenient to represent them using a scale that increases in multiples rather than equal amounts. Such a scale is called logarithmic, e.g., 1, 10, 100, 1000…; 1, 2, 4, 8, 16…

Each step raises the power of some base quantity by $1 : 10^0$, 10^1, 10^2, 10^3…; 2^0, 2^1, 2^3, 2^4….

The logarithm of any number to a particular base is the power that the base must be raised to in order to get the number. For example,

Base 10: logarithm to base 10 of 1000 = 3   or $\log_{10}(1000) = 3$

Base 2: logarithm to base 2 of 8 = 3   or $\log_2(8) = 3$

Another common base is the number e. Logarithms to base e are called "natural logarithms" and are written using the prefix "ln."

Base e: logarithm to base e of 10 = 2.3026   or $\ln(10) = 2.3026$

The values of logarithms to base 10 and of natural logarithms can be found directly with your calculator.

If you are working with logarithms you will also need to be able find anti-logarithms or inverse-logarithms. For example, if you know that the logarithm to base 10 of some physical quantity is 5, you can find the value of the physical quantity by raising the base to the power 5. For example,

if $\log_{10}(x) = 5$ then $x = 10^{\log_{10}(x)} = 10^5 = 100\,000$

if $\log_{10}(x) = 2.3$ then $x = 10^{\log_{10}(x)} = 10^{2.3} = 199.5$

Antilogarithms (or inverse-logarithms) can be found directly with your calculator.

2.5.2   Using Logarithms

Logarithms behave like powers, reducing multiplication to addition and division to subtraction:

$$\log_{10}(6 \times 7) = \log_{10}(6) + \log_{10}(7)$$

$$\log_{10}(6 \div 7) = \log_{10}(6) - \log_{10}(7)$$

Powers are reduced to multiples:

$$\log_{10}(6^7) = 7\log_{10}(6)$$

This relationships is useful when analyzing data for variables that are linked by a power law (if the base is omitted, it is assumed that everything is to base 10):

If $y = Ax^n$ then $\log(y) = \log(A) + n \log(x)$

A graph of $\log(y)$ against $\log(x)$ (called a "log-log graph")will be linear with gradient n and intercept $\log(A)$.

Exponential relationships can also be analyzed using natural logarithms:

If $y = Ae^{bx}$ then $\ln(y) = \ln(A) + bx$

A graph of $\ln(y)$ against x (called a "log-lin graph") will be linear with gradient b and intercept $\ln(A)$.

2.6 Testing Mathematical Relationships Between Variables

Graphs are often used to identify or test mathematical relationships between two physical variables, and it is important to understand how different relationships can be identified. In most cases it involves plotting a suitable straight line graph and then interpreting the gradient and intercept. The sections that follow give examples of common mathematical relationships that can be analyzed using straight line graphs.

2.6.1 Direct Proportion

If y is directly proportional to x then when x is multiplied by any number y will be multiplied by the same number. For example, y will double when x doubles. To test for this, you can plot a graph of y against x. The variables are directly proportional if the graph is a straight line AND it passes though the origin $(0, 0)$.

The relationship can be represented by $y = kx$ where k is the gradient of the graph of y against x.

2.6.2 Inverse Proportion

If y is inversely proportional to x then when x is multiplied by any number y will be divided by the same number. For example, y will halve when x doubles. To test for this you can plot a graph of y against $1/x$. The variables are inversely proportional if the graph is a straight line AND it passes through the origin $(0, 0)$.

The relationship can be represented by $y = k/x$ where k is the gradient of the graph of y against $1/x$.

2.6.3 Inverse-Square Law

If y is related to x by an inverse-square law then when x is multiplied by any number y will be divided by that same number-squared. For example, y will fall to 1/4 of its initial value when x is doubled. To test for this you can plot a graph of y against $1/x^2$. The relationship is an inverse-square law if the graph is a straight line AND it passes through the origin $(0, 0)$.

The relationship can be represented by $y = k/x^2$ where k is the gradient of the graph of y against $1/x^2$.

You can also test an inverse-square law by plotting a log-log graph as for any power law. The gradient will be −2.

2.6.4 Power Law

If y is related to an unknown power of x by a relationship of the form $y = Ax^n$ then this power can be found by plotting a graph of log (y) against log (x). This can be shown by taking logarithms of both sides of the equation:

$y = Ax^n$

$\log(y) = \log(A) + n \log(x)$

A graph of log (y) against log (x) should be linear with a gradient equal to n (the power) and an intercept equal to log (A).

2.6.5 Exponential Decay or Growth

If y is related exponentially to x by a relationship of the form $y = Ae^{bx}$ then this can be tested by plotting a graph of ln (y) against x. This can be shown by taking natural logarithms (base e) of both sides of the equation:

$y = Ae^{bx}$

$\ln(y) = \ln(A) + bx$

A graph of ln (y) against x will be linear if the relationship is exponential. The gradient of this graph is b and the intercept on the ln (y) axis is A. For exponential growth the gradient b is positive; and for exponential decay (e.g., capacitor discharge or radioactive decay) the gradient b will be negative.

Here is a summary of how to test the relationships described above, all of which are tested using straight line graphs.

Relationship	Equation	Plot	Interpretation/ comments
Direct proportion	$y = kx$	y against x	Gradient = k passes through origin

Relationship	Equation	Plot	Interpretation/comments
Inverse proportion	$y = k/x$	y against x	Gradient $= k$ passes through origin
Inverse-square law	$y = k/x^2$	y against $1/x^2$	Gradient $= k$ passes through origin
Power law	$y = Ax^n$	$\log(y)$ against $\log(x)$	Gradient $= n$ intercept $= \log A$
Exponential	$y = Ae^{bx}$	$\ln(y)$ against x	Gradient $= b$ intercept $= \ln(A)$

2.7 Exercises

1. Find the logarithms to base 10 of the following numbers:
(a) 1000 000, (b) 0.001, (c) 56,
(d) 0.325, (e) 1

2. Find the numbers whose logarithms to base 10 are:
(a) 4, (b) 2.7, (c) 0.05,
(d) –2.7, (e) –0.05

3. When a thin convex lens is used to form an image of a bright object on a screen, the relationship between object distance from the lens u, image distance from the lens v, and the focal length of the lens f is:

$$\frac{1}{u} + \frac{1}{v} = \frac{1}{f}$$

A student carries out an experiment to find the focal length f. He varies the object distance and measures pairs of values for u and v.
(a) Suggest a suitable graph that he can plot in order to find f.
(b) Explain how he would obtain a value for f from his graph.

4. The data below shows how the pressure and volume of a gas changes when it is compressed isothermally (at constant temperature).

Pressure/kPa	100	128	155	180	215	250	280	305	335	345
Volume/cm³	35.5	28.0	22.7	19.6	16.0	13.5	12.0	11.0	10.0	9.5

Assume that the uncertainty in pressure measurements is ±10 kPa and the uncertainty in volume measurements is ±1.0 cm³.

Boyle's law suggests that the pressure is inversely proportional to volume under these conditions.
(a) Use a graphical method to show that Boyle's law applies. Include error bars.

(b) Find an equation to relate pressure to volume and calculate the value of any constants in this equation (stating their units and the associated uncertainty).

5. The table below gives some data for the planets in the solar system.

Planet	Av. distance from the Sun (million km)	Orbit time, days (d)/ years (y)
Mercury	58	88 d
Venus	108	?
Earth	150	365 d
Mars	228	687 d
Jupiter	778	11.9 y
Saturn	1430	29.5 y
Uranus	2870	84 y
Neptune	4500	165 y

Kepler proposed that the orbital period of the planets T is linked to their mean radius of orbit r by an equation of the form: $T = r^n$

(a) Use a graphical method to verify that such a power law is valid and determine the value of n.

(b) Use your graph (or the equation) to predict the orbital period of Venus (check online to see if your value is acceptable).

6. When a capacitor discharges through a resistor, the discharge current I falls exponentially according to an equation of the form:

$$I = I_0 e^{-\frac{t}{\tau}}$$

where I_0 is the initial value of the current and τ is the "time constant" for the decay (in seconds).

The table below shows how current falls with time when a particular capacitor is discharged through a resistor.

Current/mA	48	38	30	24	19	15	12	9	8	6	5
Time/s	0	10	20	30	40	50	60	70	80	90	100

Use a graphical method to find the time constant τ.

CAPTURING, DISPLAYING, AND ANALYZING MOTION

3.0   Introduction

Kinematics is the study of motion. Dynamics is the study of how forces affect motion. In this chapter, we focus on how motion is described in space and time, and how we can capture data from moving objects in order to display their motion graphically. In Chapter 5, we will see how Newton's laws describe how motion is affected by unbalanced or resultant forces.

3.1   Motion Terminology

The table below lists the key terms used in this chapter.

Term	Symbol	S.I. Unit	Comment
Displacement	s	m	Distance moved in a particular direction: a vector
Distance	d	m	How far something has traveled along the path taken: a scalar
Initial Velocity	u	ms^{-1}	Initial rate of change of displacement: a vector
Final velocity	v	ms^{-1}	Final rate of change of displacement: a vector
Speed	v	ms^{-1}	Rate of change of distance: a scalar
Acceleration	a	ms^{-2}	Rate of change of velocity: a vector
Time	t	s	Duration of motion

It is often important to distinguish between instantaneous and average values of velocity.

$$\text{average velocity} = \frac{\Delta s}{\Delta t}$$

Where Δs is the change in displacement during a time interval Δt. If the time interval is small (but finite), this might be written as:

$$\text{average velocity} = \frac{\delta s}{\delta t}$$

To obtain the instantaneous velocity we would need to take the ratio of δs to δt in the limit that $\delta t \to 0$. This is the same as taking the derivative of displacement with respect to time:

$$\text{instantaneous velocity} = \lim_{\delta t \to 0} \frac{\delta s}{\delta t} = \frac{ds}{dt}$$

or

$$v = \frac{ds}{dt}$$

The table below lists symbols used to describe changes and rates of change.

Δx	A change in x
δx	A small change in x
$d(x)/dt$	The rate of change of x

Average and instantaneous acceleration are defined in a similar way:

$$\text{average acceleration} = \frac{\Delta v}{\Delta t}$$

$$\text{instantaneous acceleration} = \frac{dv}{dt}$$

3.2 Graphs of Motion

There are three key variables—displacement, velocity, and acceleration—so it is important to be able to recognize and interpret graphs where any one of these is plotted against time. It is also useful, when any one of these graphs is known, to be able to work out the form of the others.

The most useful graph is usually one of velocity against time because:

▪ The area under a velocity time graph is equal to displacement.

▪ The gradient of a velocity time graph is equal to acceleration.

The relationships between the three graphs are summarized in the diagram below, which shows graphs of motion for an object that starts from rest and accelerates at a constant rate.

Displacement

Differentiate:
$v = ds/dt$
Velocity is **gradient**
of displacement
time graph

Time

Velocity

Differentiate:
$a = dv/dt$
Acceleration is
gradient of velocity
time graph

Integrate:
$s = \int v dt$
Displacement is **area**
under a velocity time
graph

Time

Acceleration

Integrate:
$v = \int a dt$
velocity is **area** under
an acceleration
graph

Time

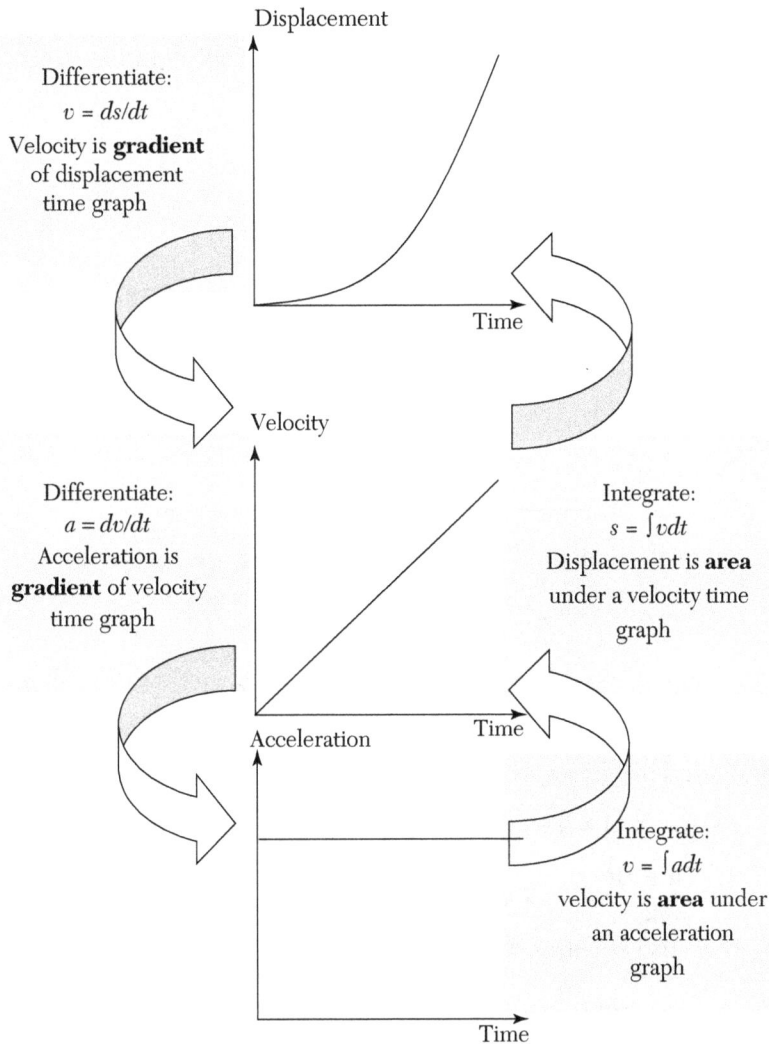

3.3   Equations of Motion for Constant Acceleration: The "*Suvat*" Equations

There are five variables—s, u, v, a, and t—so it is possible to construct five different equations, each of which omits one of these variables. We will do this in two ways—firstly using the graphs of motion and secondly using calculus.

3.3.1 Derivation 1: From Graphs of Motion

The velocity time graph below is for a particle accelerating from an initial velocity u to a final velocity v in a time t.

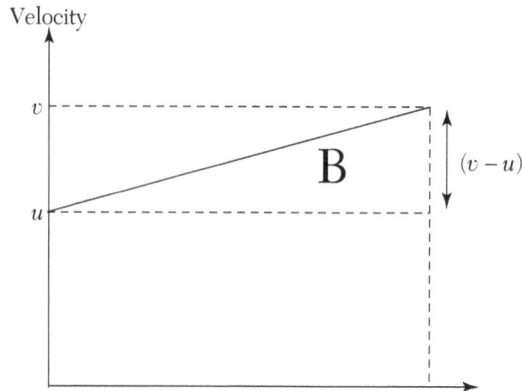

Acceleration is the gradient of a velocity time graph, so

$$a = \frac{(v-u)}{(t-0)} = \frac{(v-u)}{t}$$

$$\boldsymbol{v = u + at} \quad (suvat \text{ equation } 1)$$

Displacement is the area under a velocity time graph, and this can be configured in various ways.

(i) $s = \text{area} = A + B = ut + \frac{1}{2}(v-u)t = \frac{(u+v)t}{2}$

$$\boldsymbol{s = \frac{(u+v)t}{2}} \quad (suvat \text{ equation } 2)$$

(ii) If we use equation 1 we can eliminate v from equation 2:

$$s = \text{area} = ut + \frac{1}{2}(u + at - u)t = ut + \frac{1}{2}at^2$$

$$\boldsymbol{s = ut + \frac{1}{2}at^2} \quad (suvat \text{ equation } 3)$$

(iii) Alternatively we can use equation 1 to eliminate u from equation 2 leading to:

$$\boldsymbol{s = vt - \frac{1}{2}at^2} \quad (suvat \text{ equation } 4)$$

(iv) Finally we can use equation 1 to eliminate t from equation 2 leading to:

$$\boldsymbol{v^2 = u^2 + 2as} \quad (suvat \text{ equation } 5)$$

3.3.2   Derivation 2: Using Calculus

We start with the definition of acceleration: $a = \dfrac{dv}{dt}$

This is equivalent to the integral: $\displaystyle\int_u^v dv = \int_0^t a\,dt$ (a is constant)

which gives $v - u = at$ or $\boldsymbol{v = u + at}$ (*suvat* equation 1)

Now use the definition of velocity: $v = \dfrac{dv}{dt}$

which is equivalent to the integral: $\displaystyle\int_0^s ds = \int_0^t (u + at)dt$ (a is constant)

which gives $\boldsymbol{s = ut + \dfrac{1}{2}at^2}$ (*suvat* equation 3)

The other equations can be derived from these by a series of substitutions.

These equations are valid for motion at constant acceleration.

3.4   Projectile Motion

3.4.1   Independence of Horizontal and Vertical Components of Motion

A projectile is an object moving under the influence of gravity. In the simplest case this is an object moving in a uniform gravitational field when frictional forces can be neglected. The resultant force acting on the object is its weight, and this acts vertically downwards. In order to analyze the motion we consider the horizontal and vertical components of the motion separately. This is because:

- Horizontal and vertical components of projectile motion are independent of one another.
- The horizontal motion is at constant velocity.
- The vertical motion is at constant acceleration.

The graph below shows the trajectory of a particle launched horizontally at 8.0 ms^{-1} from a point 19.6 m above the Earth's surface. Its vertical acceleration is constant and equal to 9.81 ms^{-2}.

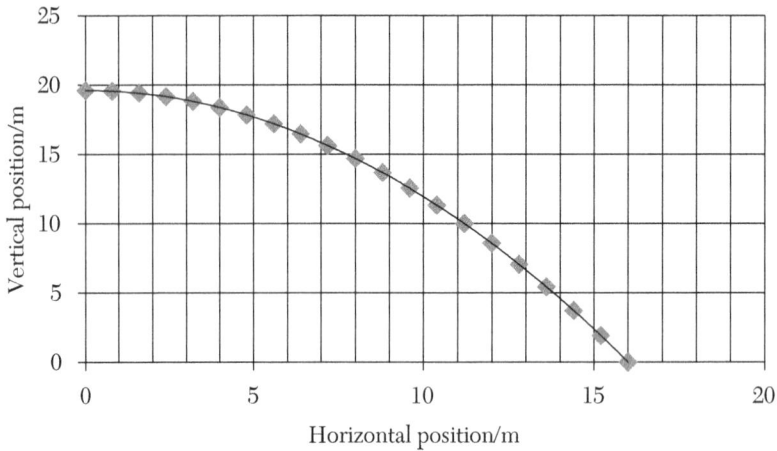

Horizontal position/m

Each data point represents the position of the projectile at 0.10 s intervals. Notice that the horizontal displacement in each 0.10 s interval is always the same (0.80 m).

3.4.2 Parabolic Paths

The path of a projectile in a uniform gravitational field (in the absence of frictional forces) is parabolic. This can be shown by eliminating t from the equations for horizontal and vertical displacement. While this can be done for a projectile launched at any speed or direction, the example below is for the simple case of a projectile launched horizontally from a point at $x = 0$, $y = 0$ at $t = 0$.

Horizontal motion: $x = ut$

where u is the initial horizontal velocity

Vertical motion: $y = vt - \frac{1}{2}gt^2$

where v is the initial vertical velocity

Substituting for t in terms of x in the equation for y gives:

$$y = \left(\frac{-g}{2u^2}\right)x^2 + \left(\frac{v}{u}\right)x$$

which has the form $y = ax^2 + bx + c$ (a parabolic curve with $c = 0$)

The equation above is only valid if g is constant, and there are no frictional forces. This is approximately true for projectile motion over small distances close to the surface of the Earth. However, when considering the motion of rockets, varying g and frictional forces must be considered.

3.4.3   Range of a Projectile

Imagine a projectile launched from a horizontal surface with an initial speed u at an angle θ to the horizontal. What is its horizontal range? In the absence of friction the horizontal range is equal to the constant horizontal velocity $u \cos \theta$ multiplied by the time of flight. To find the time of flight we can analyze the vertical motion.

Projectile motion - no drag.

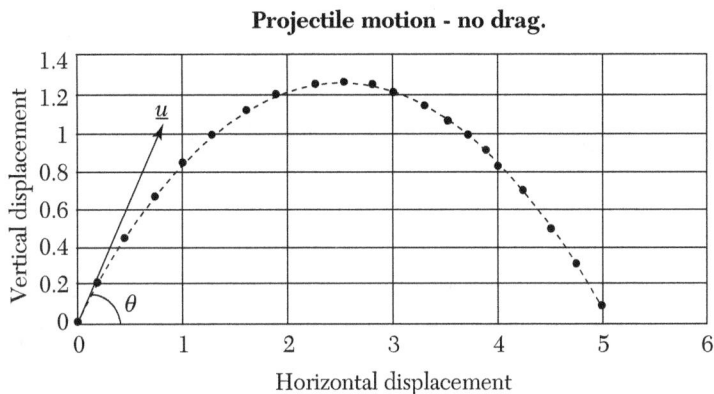

Initial vertical component of velocity $= u \sin \theta$

Final vertical component of velocity $= -u \sin \theta$ (the motion is symmetric up and down)

Vertical acceleration $= -g$

The displacement when the projectile returns to the surface is $s = 0$

Using *suvat* equation 1 from Section 3.2:
$$-u \sin \theta = u \sin - gt$$

So the time of flight is: $t = \left(\dfrac{2u \sin \theta}{g} \right)$

and the range is: $R = \dfrac{2u^2 \sin \theta \cos \theta}{g} = \dfrac{u^2 \sin 2\theta}{g}$

This has a maximum value when $\sin 2\theta = 1$ suggesting that maximum range will be achieved when the projectile is launched at an angle of 45° to the horizontal (in the absence of friction).

3.5   Equation of Motion

The *suvat* equations cannot be used when the acceleration varies. However, it is still possible to analyze the motion if the nature of the variation is known. The general approach to analyze these situations starts with an equation of motion, i.e., an equation for the variation of acceleration.

For example, a simple harmonic oscillator (such as a mass on a spring) has an acceleration a that is directly proportional to its displacement x from an equilibrium position and directed back toward that position:

$$a = -kx$$

where k is a constant (for a mass spring system it is the spring constant).

This is a second-order differential equation:

$$\frac{d^2x}{dt^2} = -kx$$

This can be solved to find equations for the displacement $x(t)$ and velocity $v(t)$ for the oscillator (see Section 11.2.1).

3.6   Methods to Capture and Display Graphs of Motion

There are many different ways to capture motion data. The simplest is to use a stop clock and meter ruler, and this method should not be neglected. The main problem with this method is the uncertainty in time caused by human reactions in starting and stopping the clock. Typical human reaction times for a visual cue, such as watching someone release a ball, is about 0.25 s, so this method is unlikely to be useful unless the measured time is significantly longer than this.

3.6.1   Motion Sensors and Dataloggers

Motion sensors and dataloggers have the advantage that they can take rapid repeated measurements and record the data electronically. This can then be processed by a computer in order to plot graphs of displacement, velocity, and acceleration. Many motion sensors work by sending out ultrasound pulses and detecting their reflection from the moving object. The displacements are calculated automatically using the known speed of ultrasound in air and the time between the pulse being emitted and its reflection being detected.

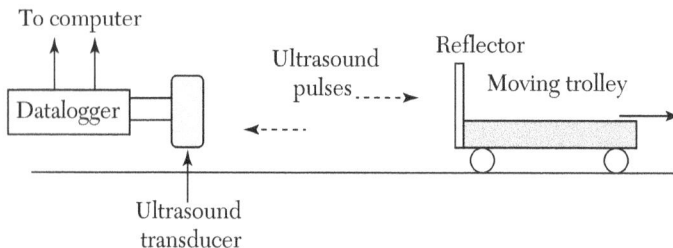

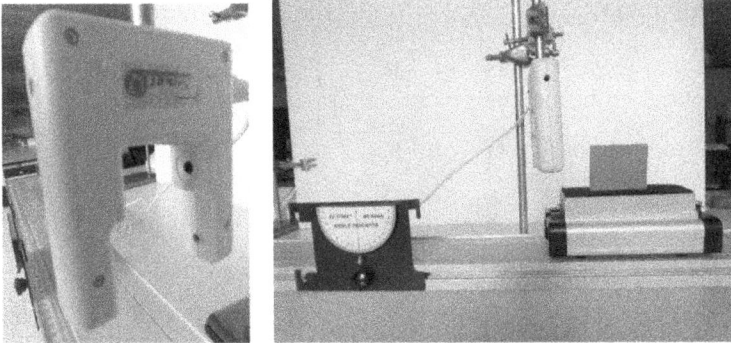

3.6.2 Light Gates

Light gates can also be used. Each light gate contains a light source and a detector. When an object passes between these, the beam is cut by a card and a datalogger measures and records the time at which the beam was cut and how long it is blocked. When an experiment is carried out, the moving object (e.g., a dynamic trolley) usually has a vertical card attached and the card travels between the "jaws" of the light gate. The length of the card must be measured and used as a parameter in the setup to enable the software to calculate velocities.

The diagram below shows an enlarged view of a light gate and a typical experimental setup.

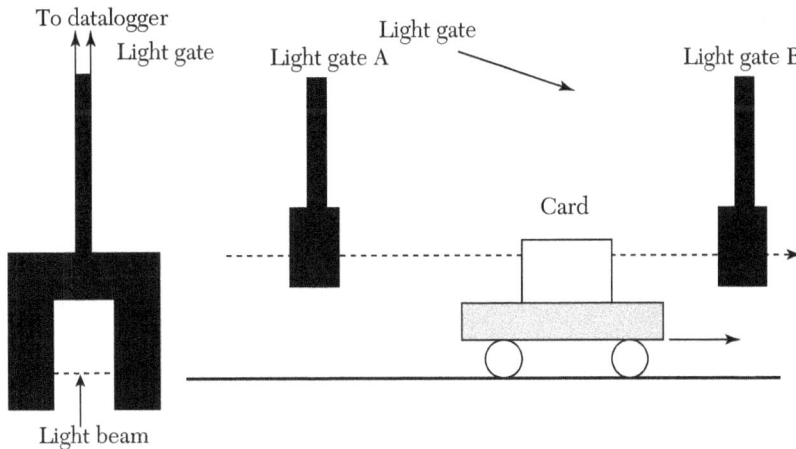

If two light gates (A and B) are used, the following measurements can be made and recorded:

▪ Time at A and time at B (from starting the experiment)

▪ Time to move between A and B

▪ Velocity at A and velocity at B

▪ Acceleration from A to B

Bear in mind that if the object is accelerating, the velocity measurements at each light gate will be average values over the time taken for all of the card to pass through the beam.

3.6.3 Mobile Phones and Tablets

Mobile phones and tablets usually have a built-in camera with both slow-motion and video capabilities as well as a range of sensors, such as a three-axis accelerometer. This makes them ideal for capturing, displaying, and analyzing data, especially since there are a wide range of apps designed for just this purpose.

Accelerometer sensor

This detects the real-time acceleration of the mobile phone along three perpendicular axes. The accelerometer samples the acceleration (a sampling interval of 20 ms is typical), and the captured data can be exported to a spreadsheet program such as Excel. When used with a suitable app, tables of data and graphs for acceleration, velocity, and displacement against time can be displayed.

The acceleration due to gravity can be measured by dropping the phone (but make sure it has a soft landing!), and the acceleration of a dynamic trolley can be measured by simply attaching the phone to the trolley. The portability of the phone makes it ideal for investigating acceleration in real-world situations such as cars or trains.

Video capture

The high frame rates available on some cameras are ideal for investigating short-lived, transient phenomena such as the impact of a water droplet with the surface of water or the formation of a crater when a ball bearing falls into sand. The cameras in mobile phones typically offer 240 frames per second for slow-motion playback, but more sophisticated cameras can record video at several thousand frames per second.

Slow-motion cameras can also be used to make accurate measurements. For example, if you are trying to measure the bounce height of a squash ball, it is difficult to judge this by eye. A slow-motion video of the bounce can be analyzed to locate the maximum height much more accurately.

A great deal of software is available for the analysis of videos of motion. One of the best free packages is called Tracker. This is ideal for the analysis of projectile motion or rotational motion. Once your video file has been loaded into Tracker, you can use the pointer to mark the position of the object you wish to track and then advance the video a few frames at a time to mark subsequent positions. This positional data is used by the software to generate displacement, velocity, and acceleration data (and graphs) in two dimensions. For rotational motion you are able to track angular displacement and angular velocity.

Here is a screen grab from an experiment to capture and analyze the motion of a "magnus glider"—a spinning object projected from top left of the image. You can see the individual tracking points behind the object, and the data table and graph of the motion are on the right.

While these applications enable us to track and display complex motions, they do have limitations, and their accuracy will be limited by the quality of the information put into them. Displacements in a video file are measured by the number of pixels across the image; they are not absolute measurements. In order to measure displacements in meters, the user must calibrate the software by identifying a known distance in the image. However, parallax effects across the field of view can distort results, so it is important for the experimenter to think hard about the measurements that are being taken!

3.7 Exercises

1. A ball is thrown vertically upwards at $4.0\ \mathrm{ms^{-1}}$ from an initial position 1.2 m above the ground on Earth and is allowed to fall down to the ground. Sketch graphs to show:

   (a) displacement versus time
   (b) velocity versus time
   (c) acceleration versus time.

   Take upwards as positive and label your axes with appropriate values.

2. An experiment performed on the Moon finds that a feather falls 20.75 m from rest in 5 s.

   (a) Calculate its speed as it hits the Moon's surface.
   (b) Calculate the acceleration of free fall at the Moon's surface.

3. A sprinter reaches his maximum velocity of $12\ \mathrm{ms^{-1}}$ after running 30 m from rest. How long does this take him and what is his average acceleration (assume acceleration is constant).

4. A car is moving at $50\ \mathrm{kmh^{-1}}$, what is this speed in $\mathrm{ms^{-1}}$?

   (a) When the brakes are applied the car can stop in 4.0 s;what is its braking distance?
   (b) A cat runs out in front of your car when you are traveling at $50\ \mathrm{kmh^{-1}}$ and your reaction time is 0.6 s. What is the total distance the car will travel before stopping (i.e., its "stopping distance")?

5. A stone is dropped into a deep well. It strikes the surface of the water at the bottom of the well 2.2 s after its release. How deep is the well?

6. A boat slows down from 6.0 to $4.0\ \mathrm{ms^{-1}}$. During the deceleration it travels 150 m. Calculate the average deceleration.

7. Two dragsters line up for a 500 m race. Car A accelerates at $4.0\ \mathrm{ms^{-2}}$ and then maintains a constant maximum speed of $50\ \mathrm{ms^{-1}}$. Car B accelerates at $5.0\ \mathrm{ms^{-2}}$ and then maintains its maximum speed of $45\ \mathrm{ms^{-1}}$. Which car wins the race and by how much?

8. A cricket ball is bowled horizontally at a speed of $20\ \mathrm{ms^{-1}}$ from a height of 2.5 m above the ground.

   (a) How far from the bowler does it first hit the ground?
   (b) What is the maximum distance from the cricketer that the ball hits the ground if he throws the ball upwards at the same initial speed at an angle of 45° to the horizontal? Ignore air resistance.

9. Describe an experiment to measure the acceleration of free fall on Earth. You should include a labeled diagram of the apparatus; a list of measurements to be taken including the instruments used to take these measurements; an explanation of how you will maximize precision and accuracy; an explanation of how you will process the data to get an accurate value of the acceleration.

4

FORCES AND EQUILIBRIUM

4.1 Force as a Vector

Forces have direction and magnitude; they are vector quantities and add and subtract as vectors. When several forces act on the same body the vector sum of these forces is called the **resultant force**. If this is zero the forces are said to be in **equilibrium**. If there is a non-zero resultant force the object accelerates in the direction of this force.

4.1.1 Free Body Diagrams

Free body diagrams show all the forces acting on a single object and are useful when starting to solve a problem involving several forces. For example, if a block rests on a rough slope there will be several forces acting on the block *and* on the surface, but what happens to the block is determined *only* by the forces that act *on it*. Here is an example, showing the free body diagram for the block. If the resultant force is zero the block remains in equilibrium. If it is non-zero the block will accelerate down the slope.

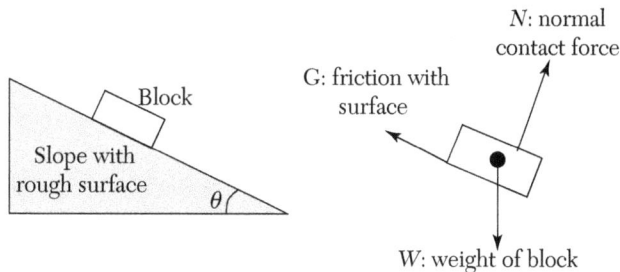

4.1.2   Resolving Forces

It is often helpful to resolve forces along two perpendicular axes. In the example above it might be helpful to consider the components of forces acting parallel (call this the x-direction) and perpendicular (call this the y-direction) to the slope. The reason for doing this is that it is obvious that forces perpendicular to the slope must be in equilibrium because the block cannot accelerate in this direction. It is constrained so that it is only free to move—if it moves at all—parallel to the slope. If we are only considering forces (and not moments) we can simplify the free body diagram by treating the object (the block) as a point when we are resolving forces.

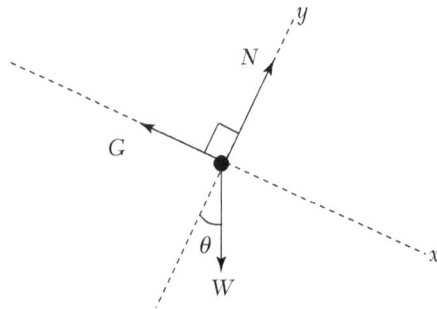

The resultant component F_x in the x-direction is given by:
$$F_x = W \sin \theta - G$$

The resultant component F_y in the y-direction is given by:
$$F_y = N - W \cos \theta = 0$$

4.1.3   Finding a Resultant Force

The resultant force on an object is the vector sum of the forces that act on it. This can be found by:

- Scale drawing

- Constructing the resultant from its components

Here is an example. Three forces act in the same plane on a single object. We need to find the resultant force.

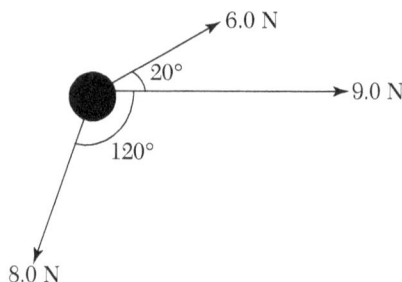

Method 1: Scale Drawing

Draw the vectors to the same scale, e.g., 2.0 mm = 1.0 N, and place them end to end in any order. The resultant vector can then be drawn from the start of the first vector to the end of the final vector. Now measure its length and use the scale to find its magnitude. Its direction relative to any of the other vectors can be measured using a protractor.

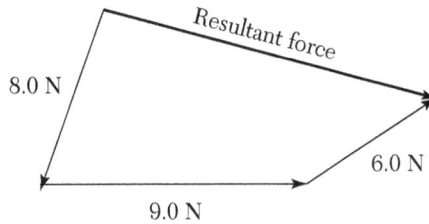

Method 2: Constructing the Resultant from Its Components

Choose two convenient perpendicular axes. At least one of these should be parallel to one of the existing forces because that force will not then need to be resolved.

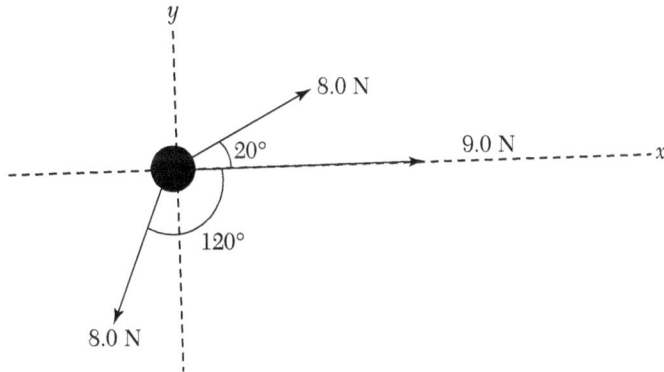

Resolving parallel to the x-axis: $F_x = 9.0 + 6.0 \cos (20°) - 8.0 \sin (30°)$

$$= 10.64 \text{ N}$$

Resolving parallel to the y-axis: $F_y = 6.0 \sin (20°) - 8.0 \cos (30°) = -4.88 \text{ N}$

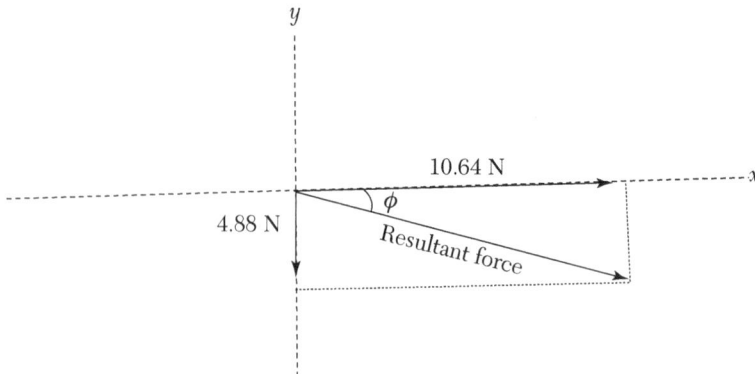

Resultant magnitude is found using Pythagoras' theorem:

$$F = \sqrt{10.64^2 + (-4.88)^2} = 11.7 \text{ N}$$

The angle between the resultant and the x-axis is:

$\tan \phi = 4.88/10.64 = 0.459$, so $\phi = \mathbf{24.6°}$

An analytic approach like the one used here is preferable to scale drawing because it will give a more accurate answer.

4.2 Mass, Weight, and Center of Gravity

4.2.1 Mass

Mass and weight are often confused. Mass is a scalar quantity related to the amount of matter in an object and does not depend on the strength of the gravitational field. A body of mass 1.0 kg has the same mass on Earth, on the Moon, and in deep space. Mass determines the inertia of a body, that is how hard it is to change its motion, and this will be discussed in much more detail later when we look at forces and motion. This property of a body is called its inertial mass.

Mass also determines how strongly an object responds to a gravitational field; this is often called the gravitational mass of a body. The fact that all objects fall with the same acceleration in the same gravitational field (see Section 5.1.4) suggests that inertial and gravitational mass are directly proportional to each other.

$$a = \frac{F}{m_{\text{inertial}}} = \frac{m_{\text{gravitational}}\,g}{m_{\text{inertial}}} = \left(\frac{m_{\text{gravitational}}}{m_{\text{inertial}}}\right)g = g$$

Einstein assumed that inertial and gravitational mass are equivalent. This helped him to construct the general theory of relativity (see Section 23.6).

4.2.2 Weight

Weight is the gravitational force acting on a body and is a vector quantity. The weight of a body depends on the strength of the gravitational field in which it is placed, given by the formula:

$$W = mg$$

where m is the mass in kilograms and g is the gravitational field strength measured in Nkg^{-1}.

The gravitational field strength near the surface of the Earth is on average about 9.81 Nkg^{-1} (standard gravity is 9.80665 Nkg^{-1}) but varies by about 0.7% at different locations (from about 9.76 Nkg^{-1} on a mountain in Peru to about 9.83 Nkg^{-1} in Oslo).

Here are some values for gravitational field strength elsewhere in the solar system.

	Moon	**Sun**	**Venus**	**Mars**	**Jupiter**	**Saturn**
Surface gravity/Nkg^{-1}	1.62	275	8.87	3.69	24.8	10.5
Surface gravity/g_{Earth}	0.165	28.0	0.904	0.376	2.53	1.07

4.2.3 Center of Gravity

The center of gravity is the point through which the resultant gravitational force on a body (its weight) can be considered to act. This point can lie inside or outside the body itself depending on its shape and the distribution of mass within it. For a body of uniform density and shape, the center of gravity is at its geometric center as shown in the diagrams below:

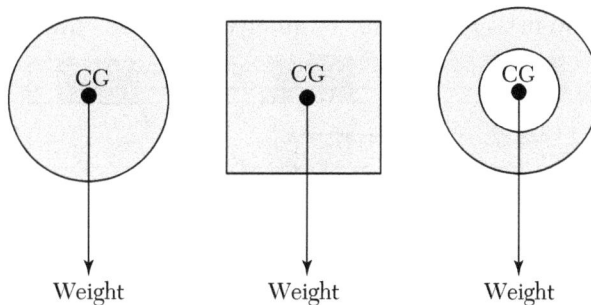

If an object is suspended from a point it will align itself with the center of gravity vertically below the point of suspension. This means that the location of the center of gravity can be found by suspending the object separately from two different points (*A* and *B*) and noting where the two lines of suspension intersect inside the object. This is shown below for a two-dimensional object.

The reason that the center of gravity lies beneath the point of suspension is that its line of action then passes through the point of suspension and so has no moment or turning effect about that point. If the object is rotated slightly so that the center of gravity lies on either side of this line, then there would be a resultant moment causing the body to rotate back toward the equilibrium position. Equilibrium of moments (see Section 4.4.4) can be used to calculate the position of a center of gravity for an extended body. The idea is simple.

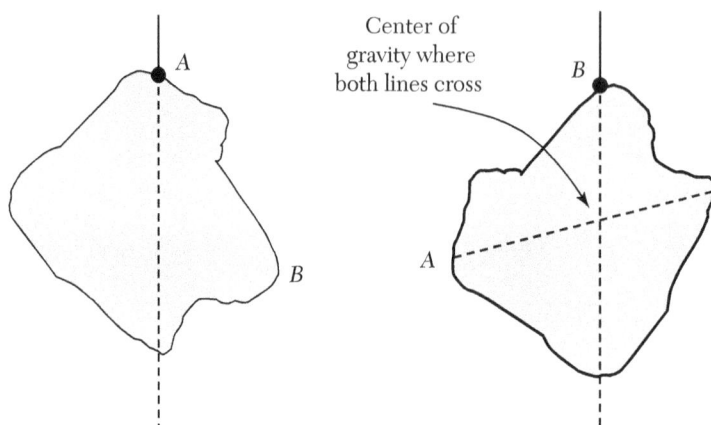

The resultant moment of the body about any point must be equal to the moment produced by its weight acting through the center of gravity. From this equality we can find the distance of the center of gravity from the point about which we are taking moments. Here is an example for a uniform rod of length l and mass m. We take moments about one end of the rod.

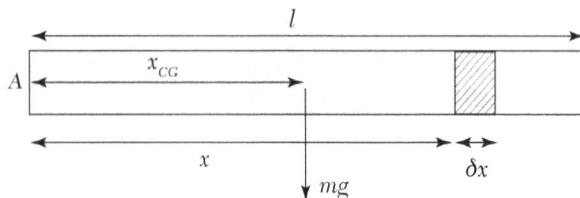

Mass of strip of length, $\delta x = m\left(\dfrac{\delta x}{l}\right)$

Moment about A of strip of length, $\delta x = \dfrac{mgx\delta x}{l}$

Moment of entire rod about $A = \displaystyle\int_{0}^{l} \dfrac{mgx\mathrm{d}x}{l} = \dfrac{mgl}{2}$

This must equal the moment of the weight, mg, acting through the center of gravity, about A.

$$mgx_{CG} = \frac{mgl}{2}$$

so $x_{CG} = l/2$; the center of gravity is at the midpoint of the rod, as expected.

The terms "center of gravity" and "center of mass" are often used interchangeably, but they are only in the same position when the object concerned is placed in a uniform gravitational field. If the field varies significantly across the object, then they will be in different positions. For example, for the rod above, if the value of g increases from left to right along the bar then the contributions to the resultant moment would be greater from strips near the right-hand end. This would move the position of the center of gravity to the right of the center of the bar, i.e., $x_{CG} > l/2$. The center of mass on the other hand would not be affected and would remain at the center of the uniform bar. In practice the difference between the two positions is rarely significant.

4.3 Equilibrium of Coplanar Forces

4.3.1 Using the Triangle of Forces to Solve Equilibrium Problems

Many problems in physics involve coplanar forces, often in a horizontal or vertical plane. The forces will be in equilibrium if they add to give zero resultant force. When this is the case the forces themselves will, when placed end to end in any order, form a closed shape. In the case of three forces this is called a "triangle of forces,"and if two of the forces are known the triangle can be solved to find the third force that puts the system into equilibrium. Here is an example; it shows a picture suspended from two strings that pass over a nail in a wall. The three forces acting on the picture come from the tension in each side of the string and from its weight. The system is in equilibrium, so the forces form a closed triangle of forces.

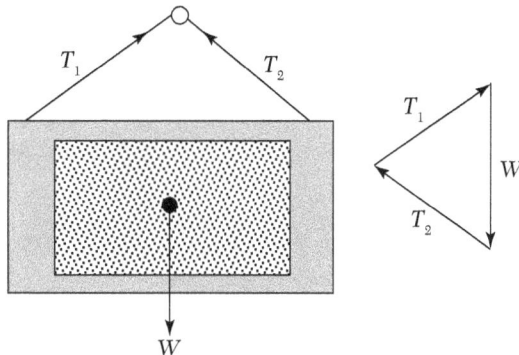

If there are more than three forces in equilibrium then they will form a closed quadrilateral of four or more sides.

Here is an example where the triangle of forces is used to calculate an unknown force keeping the system in equilibrium. The diagram shows a pendulum held in equilibrium by a horizontal force F. The problem is to find the magnitude of F.

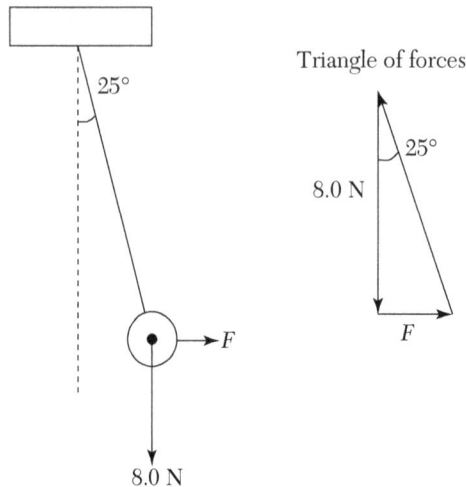

From the triangle of forces, $\tan (25°) = F/8.0$, so $F = 8.0 \tan (25°) = 3.7$ N

Using a triangle of forces is sometimes a convenient method to solve a simple equilibrium problem, especially if two of the forces are perpendicular to one another. However, for more complex problems we need a more general method; this involves resolving forces along perpendicular axes.

4.3.2 Resolving Forces to Solve Equilibrium Problems

If a system is in equilibrium there is no resultant force acting on it. This means that the components of the resultant force along any set of axes will be zero. We can use this fact to solve equilibrium problems—resolve all forces along a common set of axes, and set each component separately equal to zero. This method can be illustrated using the example from Section 4.3.1. We will resolve along horizontal and vertical axes since two of the forces act in these directions (taking right and up as the positive directions). Let the tension in the supporting string be T.

Resolving horizontall: $F - T \sin (25°) = 0$ or $T \sin (25°) = F$ (1)

Resolving vertically: $T \cos (25°) - 8.0 = 0$ or $T \cos (25°) = 8.0$ (2)

Dividing (1) by (2), tan (25°) = F/8.0, so F = **8.0 tan (25°)** = **3.7 N** as before.

In this particular case the method of resolving is a little more involved than using the triangle of forces, but this method is much simpler if there are a large number of forces and if none of them are perpendicular.

4.4 Turning Effects of a Force: Moments, Torques, and Couples

4.4.1 Moments and Torques

The moment of a force about a point P (regarded as the pivot) is defined as the magnitude of the force multiplied by the perpendicular distance of its line of action from that point, as shown below:

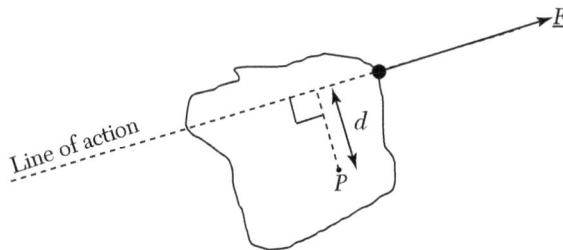

Moment about $P = Fd$

The S.I. unit of moment is Nm and the direction of the moment is usually described as clockwise or counterclockwise about the pivot. In this example it is clockwise.

You might be tempted to think that this unit is equivalent to the joule, since 1 J = 1 Nm. However, when calculating a moment, the force and displacement are perpendicular, whereas for work done, they must be parallel.

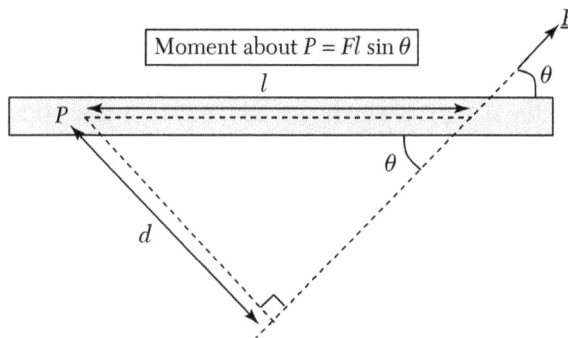

Moment about $P = Fl \sin \theta$

Notice that d is the perpendicular distance from the pivot, not the distance between the point where the force acts on the body and the pivot. This means that we often have to resolve the force to calculate the moment. The example below shows how to calculate the moment of an inclined force acting on a uniform rod at a distance l from the pivot P.

This can be regarded as the magnitude of the force multiplied by the component of l perpendicular to the line of action of the force $(F \times l \sin \theta)$, OR the component of force perpendicular to l multiplied by l $(F \sin \theta \times l)$.

The term torque is often used in engineering or when studying rotational dynamics. This is simply another name for a moment and is calculated in the same way and measured with the same units.

If the line of action of a force acts *through* P its moment about P is zero.

4.4.2   Resultant Moment

If several coplanar forces act on an object, the resultant moment (or torque) is the sum of the moments created by each force taking into account their direction (clockwise or counterclockwise). Here is an example in which several blocks rest on a uniform beam of weight 3.N:

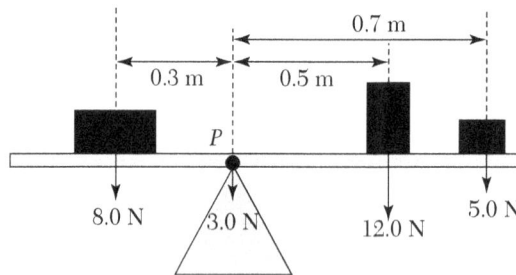

Resultant moment about P (taking clockwise is positive) = $(0.5 \times 12.0) + (5.0 \times 0.7) - (0.3 \times 8.0) = 7.1$ Nm clockwise. This would result in the beam rotating clockwise with an angular acceleration.

Note that the weight of the beam has zero moment about P because the center of gravity of the beam is directly above P, so the line of action of the weight acts through P.

4.4.3   Couples

When the line of action of a resultant force on a body acts through its center of mass (CM), it causes a change in translational motion (e.g., an acceleration). If it acts through another point in the body, it will change its translational *and* rotational motions.

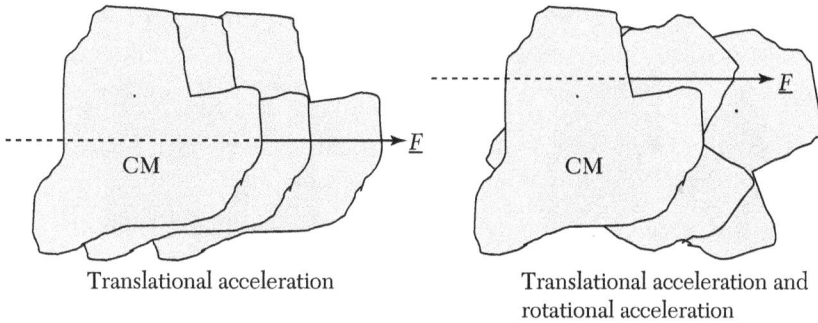

Translational acceleration

Translational acceleration and rotational acceleration

A **couple** is a pair of parallel forces of equal magnitude acting in opposite directions on the same body. They produce a resultant moment but no resultant force.

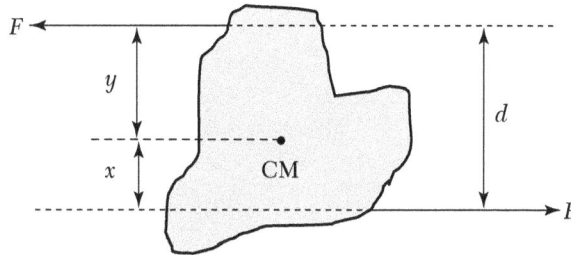

The perpendicular distance of each line of action from the center of mass is x and y respectively.

Moment of couple = $Fx + Fy = F(x = y) = Fd$

Notice that the moment of the couple is independent of the individual values of x and y and depends only on their sum—in other words, on the *separation* of the two lines of action. This means that the moment of a couple has the same value about any point in the body (or outside it).

Moment of a couple = magnitude of one force × perpendicular separation of lines of action

Moment of a couple = Fd

4.4.4 Principle of Moments

For an extended object to be in equilibrium there are two conditions:

▪ No resultant force

▪ No resultant moment

We have already seen how to determine if there is a resultant force. The principle of moments states that there will be no resultant moment if, for a particular object,

■ The sum of clockwise moments about any point is equal to the sum of counterclockwise moments about that same point.

Since this applies to *any* point we are free to take moments about a *convenient* point—for example, one through which one or more of the applied forces acts. This will simplify calculations. Here is an example.

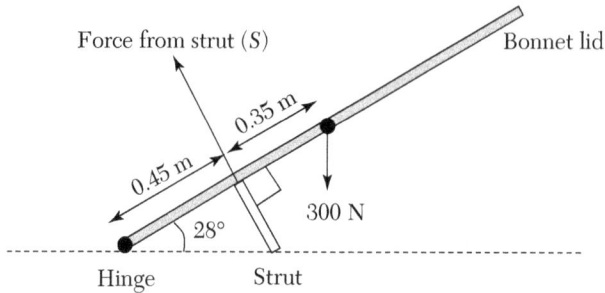

The hood of a car is held open in equilibrium by a strut as shown above. The problem is to find the force from the strut. However, there is also an unknown force from the hinge. The easiest way to solve this problem is to take moments about the hinge position because this immediately eliminates the force that acts through the hinge (because it will have no moment about this position).

Taking moments about the hinge position,

Clockwise moment from weight of bonnet = $300 \cos (28°) \times 0.80$

$$= 212 \text{ Nm}$$

Counterclockwise moment from strut force = $S \times 0.45$ Nm

Applying the principle of moments, $0.45 \, S = 212$, therefore $S = 471$ N

We could continue to solve for the unknown force H from the hinge by considering the equilibrium of forces. This gives us two more equations—one for the horizontal components and one for the vertical components:

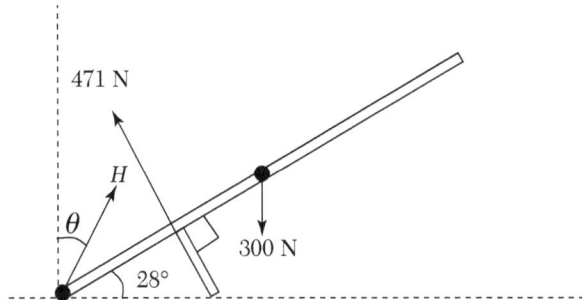

Resolving horizotally,

$$H \sin \theta = 471 \sin (28°) \quad \text{or} \quad H \sin \theta = 221 \tag{1}$$

Resolving vertically, $H \cos \theta + 471 \cos (28°) = 300 \quad \text{or} \quad \cos \theta = 116 \tag{2}$

Dividing (1) by (2) givs $\tan \theta = 221/116 = 1.91 \quad \text{and} \quad \theta = 62.3°$

Substituting back into (1) gives $\quad H = 250$ N

When coplanar forces act on an extended structure (as in the example above), the principle of moments gives one equation and the equilibrium of forces gives two more, so with three independent simultaneous equations, it is possible to solve for up to three unknowns.

4.5 Stability

4.5.1 Types of Mechanical Equilibrium

When a system is in equilibrium the forces and moments acting on it are balanced. However, if the system is disturbed, it might return to equilibrium or depart from it. The behavior of the system when disturbed is determined by its stability. Consider the three objects below, all of which are in equilibrium and all of which are given a small clockwise displacement from equilibrium.

Another way to look at this is in terms of potential energy. In *A* any disturbance tends to raise the center of mass and increase GPE, so forces act toward the equilibrium position because this is the local minimum of potential energy. In *B* the disturbance lowers the center of mass and the GPE, so forces continue act in the direction of the disturbance to lower the potential energy of the system. In *C* the disturbance has no effect on the

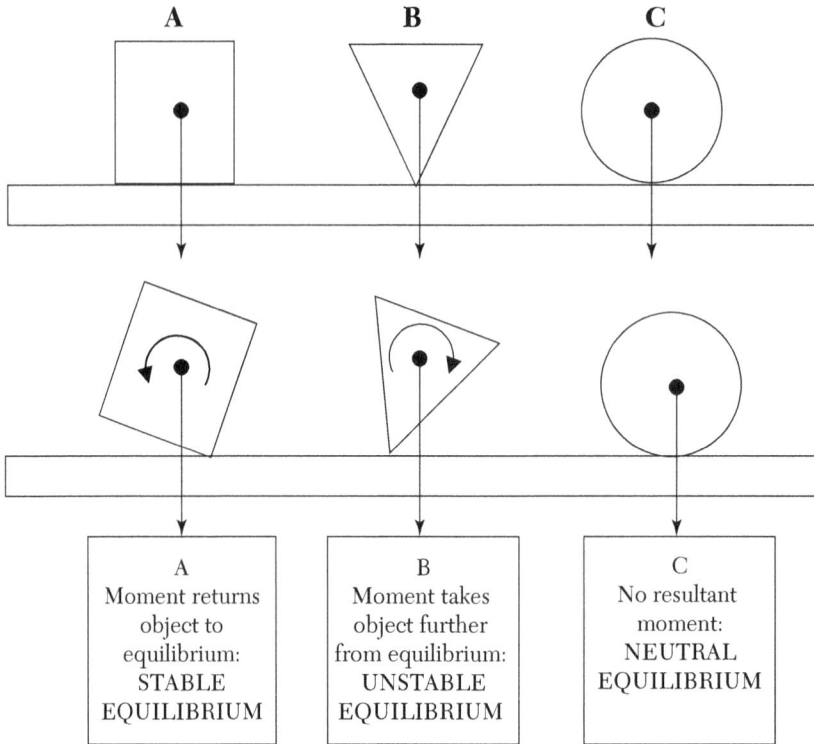

A	B	C
A Moment returns object to equilibrium: STABLE EQUILIBRIUM	B Moment takes object further from equilibrium: UNSTABLE EQUILIBRIUM	C No resultant moment: NEUTRAL EQUILIBRIUM

height of the center of mass, so the potential energy is unchanged. We can show this graphically:

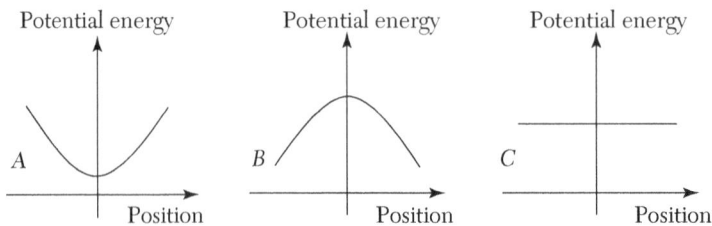

In terms of potential energy,

- Equilibrium occurs when the gradient of potential energy with position s zero: $\dfrac{dPE}{dx} = 0$.

- Stable equilibrium occurs at a local minimum of potentialenergy: $\dfrac{d^2PE}{dx^2}$ is positive.

◾ Unstable equilibrium occurs at a local maximum of potential energy: $\dfrac{d^2PE}{dx^2}$ is negative.

◾ Neutral equilibrium occurs when $\dfrac{dPE}{dx} = 0$ and $\dfrac{d^2\mathrm{PE}}{dx^2} = 0$.

4.5.2 Degrees of Stability

If an object in stable equilibrium is displaced a small amount and released, it will return to its equilibrium position. However, a large displacement might cause it to become unstable.

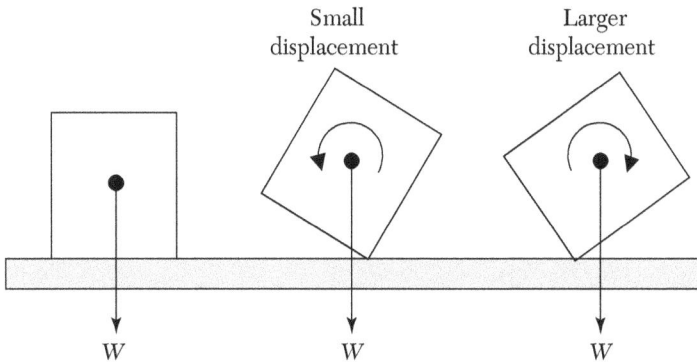

The limit of stability is reached when the line of action of the weight passes beyond the corner of the base, which acts as a pivot. The resultant moment then changes from counterclockwise to clockwise, and when the object is released, it continues to rotate and falls over. The potential energy curve looks something like this:

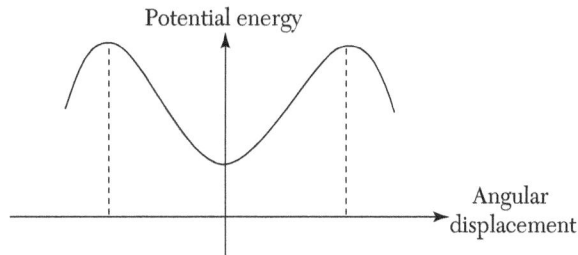

The dotted lines represent the limits of stability. These will move farther apart if the object has a wide base and low center of gray.

4.6   Frictional Forces

Frictional forces can arise in a variety of ways, for example, surfaces rubbing against one another, air resistance, or the drag on an object moving through a liquid. However, all frictional forces oppose motion or the tendency to move (e.g., when a car is parked on a hill, frictional forces act up the hill). We will discuss fluid friction when we consider viscosity (see Section 6.3), but here we will concentrate on the frictional force between two surfaces in contact with one another.

4.6.1   Origin of Frictional ForcesBetween Surfaces in Contact

On a microscopic scale all surfaces are uneven. This means that when they are put together they only actually make contact at points, and these are under considerable pressure. The frictional force between the surfaces arises because temporary bonds form at these points and these must be broken in order for one surface to slide over the other. Work must be done to break these bonds, and this transfers energy to heat as they slide past one another.

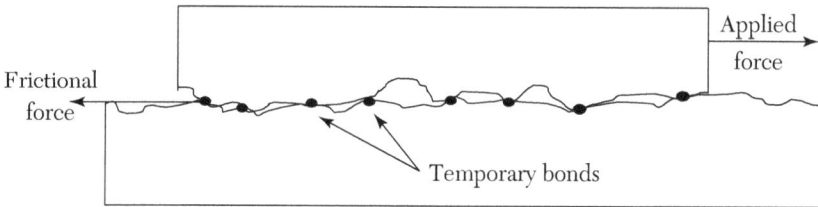

The work done by the frictional force is $W = Fs$, where s is the displacement of one surface relative to the other.

In most cases, when two surfaces are in contact the frictional force depends on:

▪ The nature of the surfaces

▪ The normal reaction force between the surfaces; the frictional force F between two surfaces is directly proportional to the normal reaction force, N: $F \propto N$

▪ But is independent of:

▪ The area of contact

These simple rules work very well in many cases, but there are exceptions. For example, a pointed object might dig into a surface. Another common exception is for car tires in snow. Having wider tires does not affect the normal force, but it does increase the friction because the snow does not pack so much. Another way of thinking about this is to say that the coefficient of friction is dependent on the normal force.

4.6.2   Static and Dynamic (Kinetic) Friction

The frictional force between two surfaces is not constant; it depends on the applied forces. For example, if a block rests on a rough surface and the horizontal force applied to the block is gradually increased, the block remains at rest until the applied force reaches a limiting value and then it suddenly starts to accelerate. This means that the frictional force must have been equal to the applied force up until the point of slipping and must then have decreased for the block to have an initial acceleration. This is shown in the graph below:

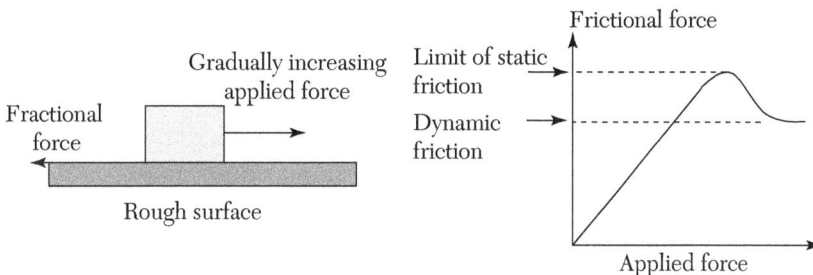

The **limit of static friction** is the maximum frictional force acting between the two surfaces when they are at rest. To move the block, an applied force greater than this limit must be used. Once the block moves and the surfaces are sliding over one another, the friction drops to a lower dynamic value.

4.6.3   Coefficients of Friction

The limiting frictional force between two surfaces is directly proportional to the normal reaction between them.

$$F_{\text{limit}} = \text{constant} \times N \quad \text{or} \quad F_{\text{limit}} = \mu N$$

where m is the "coefficient of friction," a dimensionless constant dependent upon the nature of the surfaces.

We have already seen that there is a difference between the limit of static friction and dynamic friction, so there are two coefficients of friction between a pair of surfaces:

- μ_S = coefficient of static friction

- μ_K = coefficient of dynamic (kinetic) friction

4.6.4 Measuring the Coefficient of Static Friction

There are two simple methods that can be used to measure μ_S.

Method 1: By Measuring Minimum Force Needed to Make the Block Slip

Here is a diagram of the apparatus:

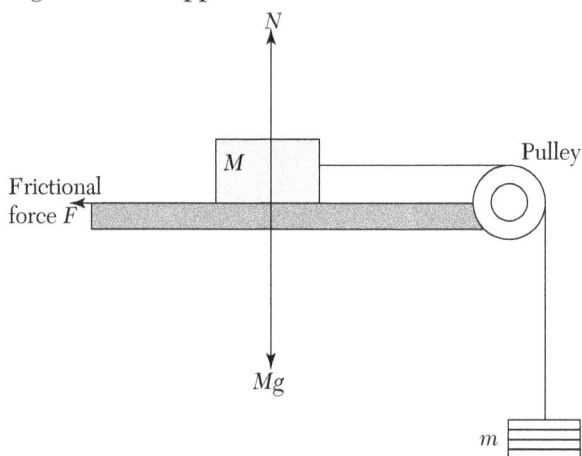

Masses are added to the hanger until the block *just* begins to slip. The limit of static friction is then mg.

The normal reaction is $N = Mg$.

Therefore $mg = \mu_S Mg$, so $\mu_S = m/M$

Method 2: By Tilting the Surface Until the Block Just Slips

Here is a free body diagram for the block when limiting angle after which the block just begins to slip is reached.

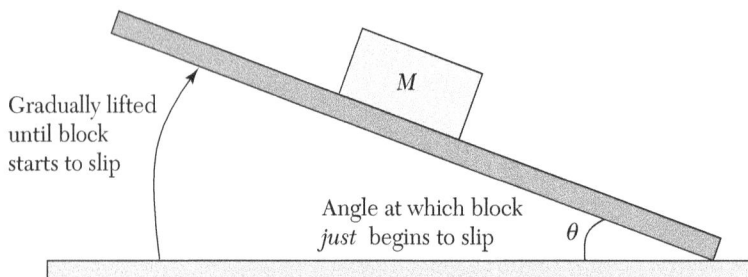

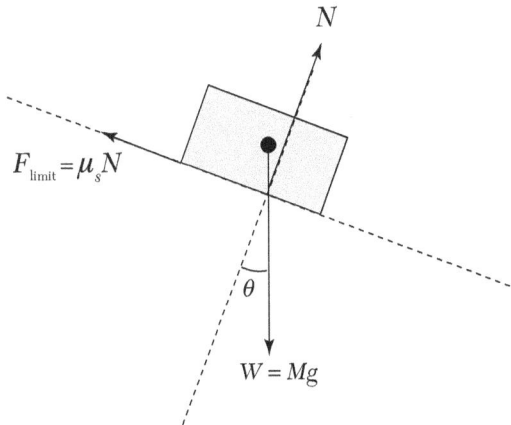

At the limiting angle, the block is in equilibrium, so we can resolve forces parallel and perpendicular to the surface:

Resolving parallel to the surface, $\mu_s N = Mg \sin \theta$ (1)

Resolving perpendicular to the surface, $N = Mg \cos \theta$ (2)

Dividing (1) by (2) gives $\mu_s = \tan \theta$

▨ The coefficient of static friction is equal to the tangent of the limiting angle.

4.6.5 Measuring the Coefficient of Dynamic (Kinetic) Friction

A method similar to method 2 above can be used. However, in this case the angle must be increased in small increments, and each time the block should be given a small push. When you reach the angle fat which the block, once pushed, continues to move down the plane at a small constant velocity, ten $m_K \tan \phi$.

4.7 Exercises

1. In which of the following situations (if any) are the forces acting on a man in equilibrium?

▨ Lying still in his bed

▨ Sitting in a car seat when the car is traveling at constant velocity along a motorway

▨ Standing in a lift that is moving upwards at constant velocity

▨ Floating apparently weightless inside an orbiting spacecraft

▨ Floating in a swimming pool

2. Calculate the magnitude and direction of the resultant force acting on each block in the free body diagrams below:

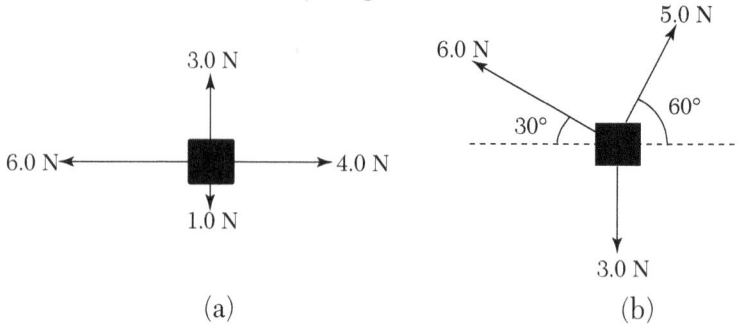

3.0 N

6.0 N←   →4.0 N

1.0 N

(a)

5.0 N

6.0 N

30°   60°

3.0 N

(b)

3. A frame of mass 1.2 kg is suspended from a rigid support by two wires as shown below.

Calculate the tension in each wire.

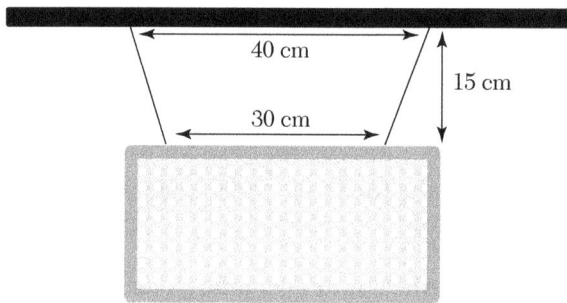

40 cm

15 cm

30 cm

4. It is possible to balance the weight of a 1-meter ruler by placing the pivot off-center as shown below. Assume that the center of gravity of the ruler is at its geometric center.

Calculate the weight the ruler.

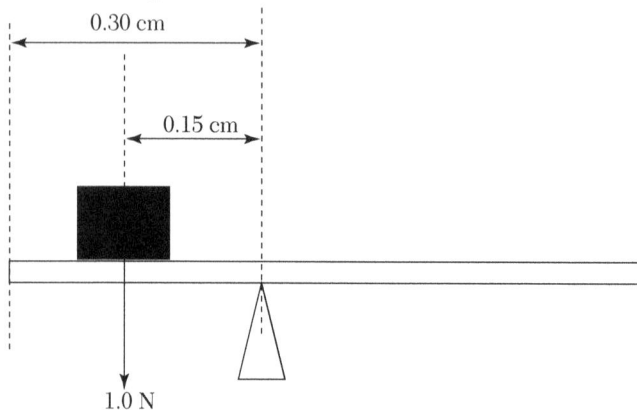

0.30 cm

0.15 cm

1.0 N

5. The diagram below shows a paving slab of mass 300 kg resting on two wooden supports, A and B. Assume that the center of gravity of the slab is at its geometric center and that the forces between the supports act at the centers of their areas of contact with the slab.

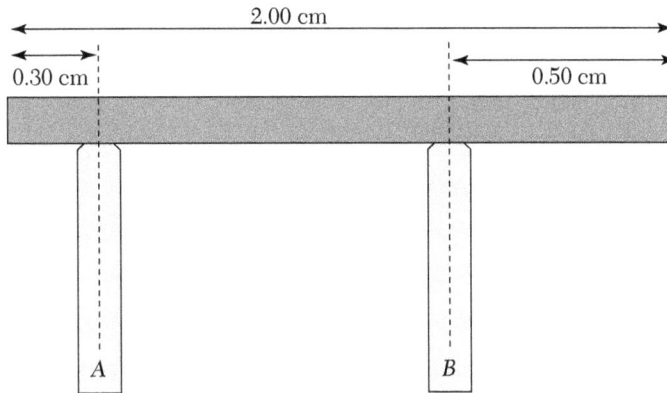

(a) Calculate the support forces from A and from B.

(b) Calculate the minimum downward force required at the right-hand end of the slab to cause it to tip.

6. A wooden block of mass 0.65 kg is at rest on a rough plank. One end of the plank is slowly lifted and the block just begins to slip down the slope when the angle to the horizontal is 25°. The plank is held at that angle and the block moves with a constant acceleration of 0.20 ms^{-2} until it reaches the bottom of the slope.

(a) Calculate the coefficient of static friction between the block and the plane.

(b) Calculate the coefficient of kinetic friction between the block and the plane.

(c) Sketch a graph to show how the frictional force between the block and the plank varies from the time the plank is first lifted until the block reaches the bottom of the plank. Include values on the force axis.

CHAPTER 5

NEWTONIAN MECHANICS

5.0 Introduction

Newton's masterwork, *Philosophiæ Naturalis Principia Mathematica* ("Mathematical Principles of Natural Philosophy") is probably the most famous and important book (actually three books) in the history of physics. It sets out the laws of motion and the law of gravitation and then applies these laws to the motion of planets in the solar system. It also marks the beginning of mathematical physics. This is not because mathematical arguments had not been used in physics before Newton but because Newton provided a mathematical framework in which to tackle an enormous range of physical problems. Newton's work had, and still has, a remarkably wide impact on science and philosophy and is even used as a model in other apparently unrelated disciplines (e.g., economics!). If you want to be good at physics, you need to be good at mechanics.

5.1 Newton's Laws of Motion

5.1.1 Newton's First Law of Motion

This is a deceptively simple law, but, like many simple statements in physics, it has deep significance. It describes the natural state of motion of objects when they are not subjected to a resultant force. While we know it as Newton's first law, it had already been stated by Galileo. He used a thought experiment to explain it, and this is described below.

Imagine releasing a small ball from the top of a U-shaped ramp. In the absence of friction we would expect it to rise up to the same height on the far side of the U (A below). Now let the ramp on the far side slope up more gradually. We would expect the ball to travel further along the ramp until, once again, it reached its starting height (B below). Now let the ramp on the far side continue on horizontally; what will happen to the ball? Logically it seems it must continue to move at a constant velocity until such time as the ramp rises back up again (C below). Galileo argued that there is no need for an unbalanced force to keep things moving at constant velocity.

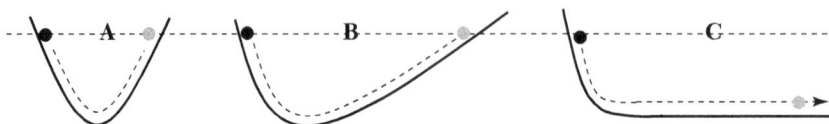

Newton's first law

An object continues to remain at rest or move at constant velocity (in a straight line) unless acted upon by a resultant force.

Comments

- Most moving objects are acted upon by many forces (e.g., thrust, gravity, contact forces, drag). If they are moving at constant velocity, the resultant of all these forces must be zero.

- If we know that the forces acting on a particular object are in equilibrium, we cannot assume it is at rest; it might be moving at constant velocity.

- If an object is accelerating, decelerating, or changing direction, the forces acting on it must be unbalanced; there is a resultant force. An example is an object moving at constant speed along a curved path: since the speed is unchanging, there is no force component parallel to the motion, but there must be a component of force perpendicular to it in order to cause a change of direction.

5.1.2 Galilean Relativity

Galileo thought deeply about the nature and causes of motion. He realized that the law of inertia (effectively Newton's first law) implied that the laws of mechanics will be the same in a moving reference frame as they are in one at rest. He used another thought experiment to show this:

Shut yourself up with some friend in the main cabin below decks on some large ship, and have with you there some flies, butterflies, and other small flying animals. Have a large bowl of water with some fish in it; hang up a bottle that empties drop by drop into a wide vessel beneath it. With the ship standing still, observe carefully how the little animals fly with equal speed to all sides of the cabin . . . When you have observed all these things carefully (though doubtless when the ship is standing still everything must happen in this way), have the ship proceed with any speed you like, so long as the motion is uniform and not fluctuating this way and that. You will discover not the least change in all the effects named, nor could you tell from any of them whether the ship was moving or standing still.[1]

This was a profound observation. Einstein realized that if the laws of physics are the same in stationary and uniformly moving reference frames, then there is no fundamental difference between rest and motion; it just depends on what you choose as your reference frame. It led Einstein to the special theory of relativity (see Chapter 24), but in Galileo's time, it served a different purpose. Galileo was convinced that the Earth orbited the Sun rather than the other way around as was believed by most people at the time. They thought that it was obvious that the Earth was not moving because we cannot feel the motion. Galileo's thought experiment showed that you would not expect to feel the motion—everything would happen on Earth as if it was at rest; so the argument against the Earth's motion was flawed.

Galilean relativity is the idea that the laws of mechanics are the same in all uniformly moving reference frames.

5.1.3 Newton's Second Law of Motion

The second law of motion deals with the effects of a resultant force. When a resultant force acts on something, it causes a change in motion: an acceleration. The definition of acceleration is rate of change of velocity, and since velocity is a vector, this means that accelerations include speeding up, slowing down, and changing direction. The second law can be stated in a number of different ways. Here we state how it applies when a resultant force acts on a body of constant mass. Later we will consider situations in which the mass might not be constant.

---

[1] *Dialogue Concerning the Two Chief World Systems*, translated by Stillman Drake, University of California Press, 1953, pp. 186–187 (second day).

Newton's second law

When a resultant force F acts on a body with constant mass m, it produces acceleration a in the direction of the resultant force that is directly proportional to the resultant force and inversely proportional to the mass:

$$a \propto \frac{F}{m}$$

We can replace proportionality with equality if we define the units in which we measure resultant force in the following way.

Definition of newton

A resultant force of 1 N accelerates a mass of 1 kg at 1 ms^{-2}.

This makes the constant of proportionality 1, and the equation becomes:

$$a = \frac{F}{m}$$

Comments

- This is a vector equation: the resultant force and acceleration vectors are in the same direction.

- The term inertia is given to the property of mass that resists acceleration. The m in the equation above is sometimes called "inertial mass." The greater the mass of an object, the greater its inertia.

- Newton argued that because objects in free fall accelerate toward the Earth, they must be acted upon by a resultant force directed toward the Earth—a gravitational force.

- Newton realized that because the planets follow curved paths, they must also be acted upon by a resultant force. This is a gravitational force toward the Sun.

- Newton thought that the orbital motion of the Moon was caused by a gravitational force toward the Earth. This force has the same origin as the force that makes objects close to the surface fall downwards. By comparing the acceleration of an object near the surface with that of the Moon, he was able to derive the famous inverse-square law of gravitation (see Section 23.1.1).

5.1.4   Free Fall

The ancient Greek philosopher Aristotle thought that more massive objects should fall faster than less massive objects. Galileo proposed a thought experiment that showed that both should fall at the same rate. It is simple but compelling, and it goes something like this.

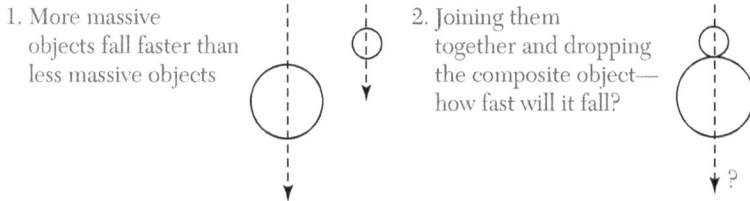

1. More massive objects fall faster than less massive objects

2. Joining them together and dropping the composite object—how fast will it fall?

According to Aristotle the combined object should fall faster than either of the original objects because it is more massive. However, shouldn't attaching the smaller mass to the faster mass slow the larger mass down and shouldn't attaching the larger mass to the smaller mass speed the smaller mass up? This argument suggests that the composite body should have a speed intermediate between the speeds of the small mass and large mass alone. Aristotle's idea leads to a contradiction. On the other hand, if all objects fall at the same rate then there is no problem.

Galileo is said to have tested this idea by dropping two cannon balls of different sizes from the top of the Tower of Pisa and showing that they landed at the same time. Most historians doubt whether he actually did this, but in 1971, during the Apollo 15 mission to the Moon, a hammer and a feather were dropped to the Moon's surface; both landed at the same time.

Newton's second law can explain this. The resultant force on an object of mass m in a gravitational field of strength g is its weight, mg. Using the equation above,

$$a = \frac{F}{m} = \frac{mg}{m} = g$$

gravitational field strength is equal to the acceleration of free fall in that field.

Comment

▨ It seems obvious that we can cancel the two m's in the equation above. However, if we think about this a little more deeply, it is not so obvious that they are the same thing. The m on the bottom of the equation is the **inertial** mass; the m on the top is related to how strongly the mass responds to a gravitational field, sometimes called the **gravitational mass**. They are two different properties of mass, so the fact that all objects fall with the same acceleration in the same gravitational field shows that they are at least proportional to one another and possibly identical. This subtle point was one of the clues that helped Einstein toward his general theory of relativity, a new theory of gravity.

5.1.5 Newton's Third Law of Motion

The third law is the one that is most often misunderstood. It is usually stated as: *For every action there is an equal and opposite reaction.*

However, to really understand this, we must unpack it. What are "actions" and "reactions"? Here is another statement of the third law.

Newton's third law

When *A* exerts a force on *B*, *B* exerts a force of equal magnitude on *A*. The two forces are of the same type and act in opposite directions.

Comments

- Forces never arise by themselves; they always come in pairs as opposite ends of an interaction.

- The "action" and "reaction" forces act on *different* bodies (otherwise they would always cancel out and it would be impossible to change the motion of anything).

- The "action" and "reaction" forces are always of the same type—e.g., both are contact forces, or both are gravitational forces.

- The fact that forces arise from interactions means that changing the motion of one body must cause an opposite change in the motion of another—this leads to the law of conservation of momentum (see Section 5.3.4).

EXAMPLES

1. The gravitational forces on the Earth and Moon are an "action–reaction" pair.

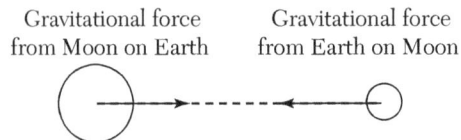

Gravitational force from Moon on Earth   Gravitational force from Earth on Moon

2. When a ball rests on the ground, there are two "action–reaction" pairs.

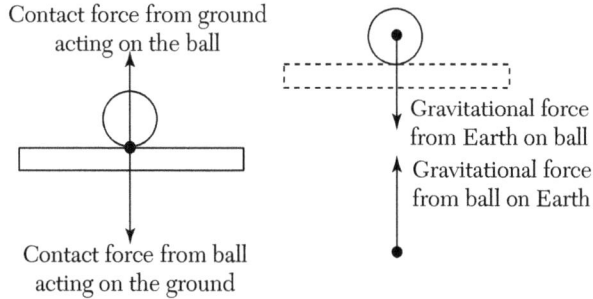

Contact force from ground
acting on the ball

Contact force from ball
acting on the ground

Gravitational force
from Earth on ball

Gravitational force
from ball on Earth

Notice that while it is true that the weight of the ball and the upward contact force from the ground have equal magnitudes and act in opposite directions, they do not form an action–reaction pair. They fail in three respects: they are not the same type of force; they do not act on different bodies; and they are not part of the same interaction.

3. When a ball rests on an accelerating surface, the contact force and the weight are not equal. While each action–reaction pair remains balanced, there is now a resultant upward force on the ball, as shown by the free body diagram on the right in the figure below.

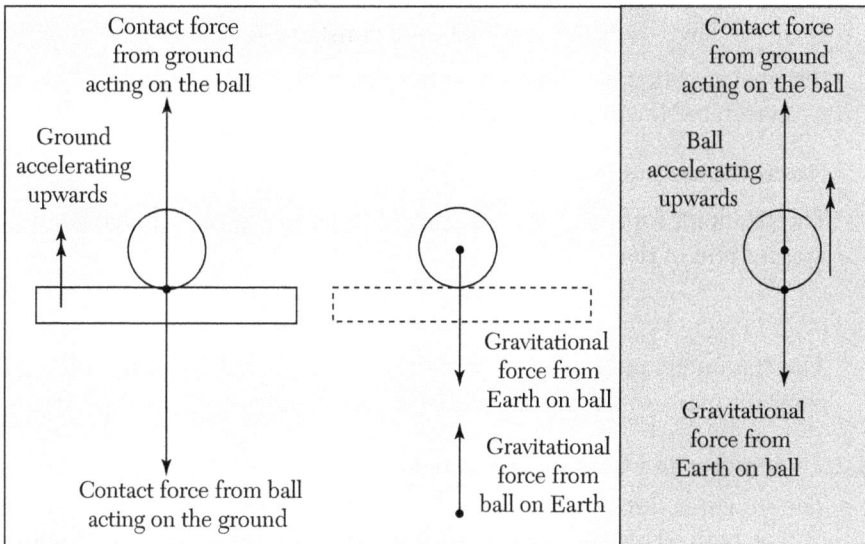

Contact force
from ground
acting on the ball

Ground
accelerating
upwards

Contact force from ball
acting on the ground

Gravitational
force from
Earth on ball

Gravitational
force from
ball on Earth

Contact force
from ground
acting on the ball

Ball
accelerating
upwards

Gravitational
force from
Earth on ball

5.2   Linear Momentum

Linear momentum is one of the most important physical quantities. It is conserved in all closed systems and by all interactions.

Linear momentum p is defined as the product of mass and velocity:

$$\underline{p} = m\underline{v}$$

It is a vector quantity, and its S.I. units are kgms^{-1} (equivalent to Ns).

5.2.1   Newton's Second Law in Terms of Linear Momentum

The statement of Newton's second law in Section 5.1.3 leads to the equation $F = ma$. We can rewrite this in terms of the linear momentum of moving mass.

For a constant mass and force,

$$F = ma = m\frac{(v - u)}{t} = \frac{(mv - mu)}{t} = \frac{\text{change in momentum}}{\text{time}}$$

Or *resultant force = rate of change of momentum*

We can derive the same result using calculus:

$$F = m\frac{dv}{dt} = \frac{d}{dt}(mv) = \frac{dp}{dt}$$

Here we have used the fact that m is constant so that $m\dfrac{dv}{dt} = \dfrac{d(mv)}{dt}$, but the final equation is valid more generally, even if m is changing. This allows us to restate Newton's second law of motion.

Newton's second law

The resultant force acting on a body is equal to the rate of change of linear momentum of that body.

$$F = \frac{dp}{dt}$$

The statement and equation $(F = ma)$ in Section 5.1.3 is a special case, for constant mass, of the general equation given here.

5.2.2   Impulse and Change of Momentum

The longer a force acts on a body, the greater the change of momentum of that body. This suggests that the product of force and time might be a useful physical quantity. It is called "impulse." When the force and mass are constant, we can derive an equation for impulse:

$$F = \frac{(mv - mu)}{t}$$

so $\quad Ft = (mv - mu)$

Or $\quad$ impulse = change of momentum

The S.I. unit for impulse is newton-second (Ns), which is equivalent to the unit for momentum, kgms^{-1}.

We can derive a more general equation for impulse by integrating Newton's second law $F = \dfrac{dp}{dt}$:

$$\int F dt = \int \mathrm{d}p = p$$

The term on the left-hand side is equal to the area under the graph of force against time, and this is equal to the change of momentum. For example, the force exerted on a football when it is kicked might vary as shown in the graph below. The area under the graph will be equal to the momentum transferred to the ball as it is kicked.

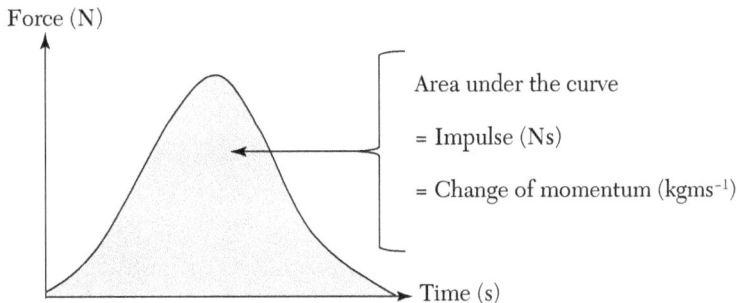

Force (N)

Area under the curve

= Impulse (Ns)

= Change of momentum (kgms^{-1})

Time (s)

5.2.3 Conservation of Linear Momentum

Consider an interaction between two bodies; this might be an attraction, a repulsion or collision of some kind. From Newton's third law, each body will exert an equal but opposite force on the other, and these forces will act in opposite directions along the same line and will act for the same time.

$F_{B \text{ on } A}$ $\quad$ A $\quad$ B $\quad$ $F_{A \text{ on } B}$ $\quad$ x

Taking positive values in the x-direction, the impulse on each body during a short time δt is given by:

Impulse on $A = F_{B \text{ on } A}\,\delta t$

Impulse on $B = F_{A \text{ on } B}\,\delta t = -F_{B \text{ on } A}\,\delta t$

During the interaction, the impulse given to each body, calculated from the integral $\int F dt$, is also of equal magnitude but opposite in direction. Since impulse is equal to change in momentum, the change in momentum of A is equal and opposite to the change in momentum of B. The momentum change of the complete system is zero.

$$\text{change of momentum of system} = \int_0^t F_{A\,on\,B}dt + \int_0^t F_{B\,on\,A}dt = 0$$

Linear momentum is conserved in an interaction between two bodies: Newton's third law tells us that *all* forces arise as a result of interactions, so the above argument will apply many times over for a complex system of interacting bodies, as long as we include all the pairs of forces. This can be stated as the *law of conservation of linear momentum*.

The linear momentum of a closed system is constant.

Comments

◾ A "closed system" means that we include all pairs of forces within the system. Another way of saying this is that the linear momentum of a system is constant if no external resultant force acts on the system.

◾ Linear momentum is a vector quantity, so the conservation of linear momentum implies a separate conservation of each component of linear momentum.

◾ An object moving along a curved path has a continuously changing momentum (it is changing direction even if its magnitude is constant).

To solve problems involving conservation of momentum, it is useful to consider the total momentum of the objects before and after they interact and then to equate them. Here is a simple one-dimensional example from nuclear physics.

Alpha decay: An unstable nucleus of mass M is initially at rest and decays by emitting an alpha particle (this consists of 2 neutrons and 2 protons) of mass m. The new nucleus recoils with velocity u. What is the velocity, v, of the emitted alpha particle?

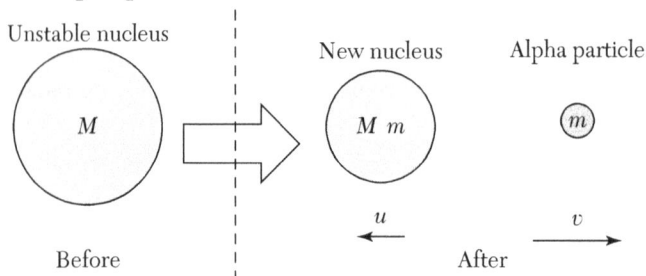

Taking velocities to the right to be positive:

Momentum before $= 0$

Momentum after $= mv - Mu$

Therefore, $mv - Mu = 0$ and $v = -\dfrac{M}{m}u$

In alpha decay, the new nucleus is usually much more massive than the alpha particle, so the alpha particle has a much larger velocity and takes away most of the energy released in the decay (kinetic energy is proportional to velocity-squared).

Here is an example of a two-dimensional collision between two balls of masses m_1 and m_2.

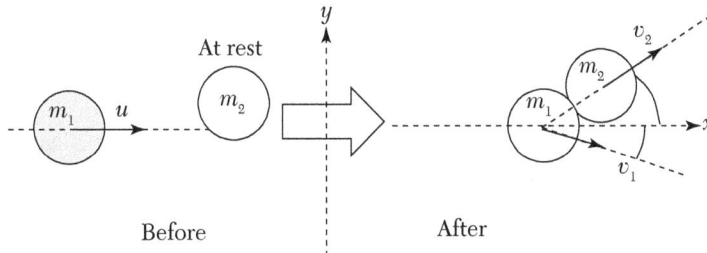

The x- and y-components of momentum must be conserved independently of one another:

x-momentum before $= m_1u + 0 = m_1u$

x-momentum after $= m_1v_1 \cos \theta + m_2v_2 \cos \varphi$

Therefore, $m_1u = m_1v_1 \cos \theta + m_2v_2 \cos \varphi$ (1)

y-momentum before $= 0$

y-momentum after $= m_2v_2 \sin \varphi - m_1v_1 \sin \theta$

Therefore, $m_2v_2 \sin \varphi = m_1v_1 \sin \theta$ (2)

Equations (1) and (2) can then be used to determine the values of up to two unknown quantities.

Conservation of momentum can also be shown by drawing vectors to represent the momenta of the bodies before and after the collision.

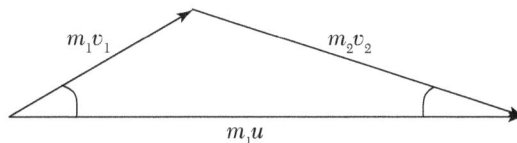

The sum of the vector momenta after the collision must equal the total vector momentum before the collision. Solving the triangle using trigonometry is equivalent to taking components and deriving the equations (1) and (2) above.

5.3   Work Energy and Power

When a stone is kicked, it moves for a while and then comes to rest. It appears that the movement energy has disappeared. James Joule realized that the energy of motion in the moving stone has not disappeared but has been transferred into other forms of energy to the objects with which it has interacted. If you slide a book over the surface of a table, it eventually stops moving, but the motion energy it had has been transferred to the particles on the surfaces of the table top and the book, and when the book stops moving, these particles are vibrating more vigorously than before. Some thermal energy has been generated, and there has been an increase in temperature. Count Rumford noticed a similar effect when cannons were bored in the arsenal in Munich; he even carried out an experiment to show that if the cannons were bored under water, the water could be brought to the boil. This suggests that the motion involved in boring the cannons had produced thermal energy.

Joule realized that there is a mechanical equivalent of heat; in other words, the book does work as it comes to rest, and this transfers its kinetic energy into thermal energy to the surroundings. Joule's work set the stage for the idea that energy is conserved. However, energy comes in a variety of different forms, and while it cannot be created or destroyed, it can be transferred. When the book stops moving, its kinetic energy has been transferred into thermal energy, which spreads into the surroundings.

5.3.1   Work

Work is the transfer of energy when the point of application of a force moves in the direction of the force:

Work done (J) = force applied (N) × displacement parallel to force (m)

The S.I. unit of energy is joule (J).

1 joule of energy is transferred when a force of 1 N moves through 1 m (1 J = 1 Nm).

If the force is at an angle to the displacement, then the component of force parallel to the displacement must be used.

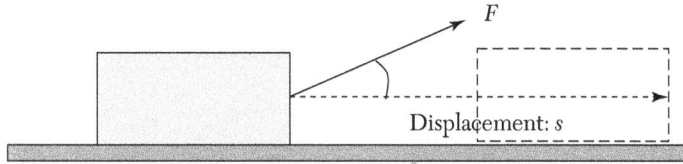

Work done to move a block through displacement s is: $W = Fs \cos \theta$

If $\underline{\mathbf{F}}$ and s are expressed as vectors, the work done is the **scalar product** of the vectors: $W = \underline{\mathbf{F}} \cdot \underline{\mathbf{s}}$

In most cases the force varies with time or position, so the work done can be found graphically or—if there is a formula for the variation of force—by using calculus.

1. Work calculated from a graph of force against displacement (parallel to the force)

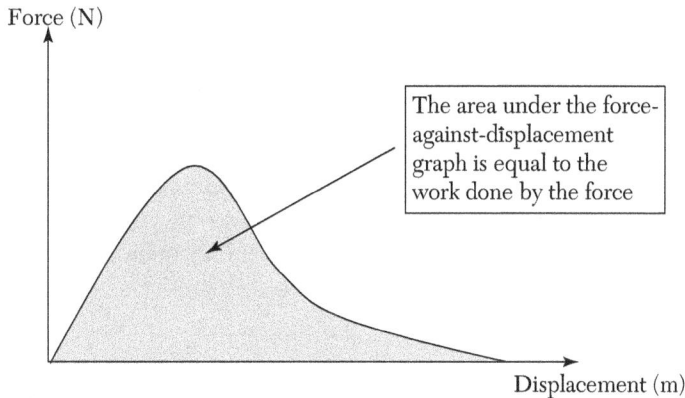

The area under the force-against-displacement graph is equal to the work done by the force

2. Work done calculated by integration

Consider a varying force $F(x)$ acting along the x-axis. The work done δ Was the force moves from x to $(x + \delta x)$ is given by:

$$\delta W = F(x)\delta x$$

The total work done W_{ab} as the force moves from $x = a$ to $x = b$ is given by:

$$W_{ab} = \sum_{x=a}^{x=b} F(x)\delta x$$

And if we let the small increments in x tend to zero $(\delta x \to 0)$, this sum becomes a continuous integral:

$$W_{ab} = \int_{x=a}^{x=b} F(x)dx$$

5.3.2   Gravitational Potential Energy Changes (Uniform Field)

The strength and direction of the gravitational field g is almost constant close to the Earth's surface. g is given by:

$$g = F_{grav}/m$$

and has a value of about 9.81 Nkg^{-1} close to the Earth's surface.

The fact that the field is almost uniform makes it very easy to calculate the work needed to lift a mass near the Earth's surface.

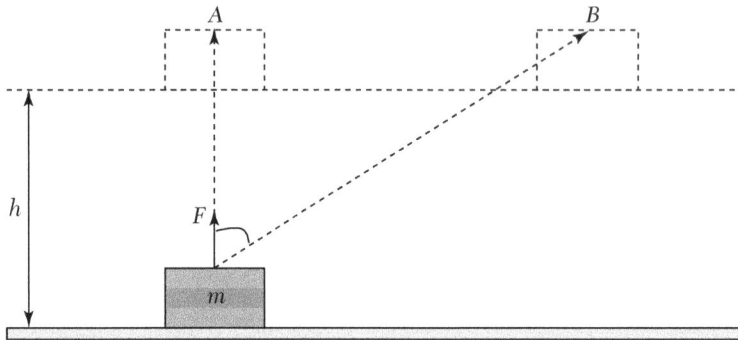

The diagram above shows a mass m being lifted through a height h by two different routes: vertically to A and along an inclined path to B. The vertical force needed to lift the mass is always equal to mg since the applied force must balance the weight of the mass. In the absence of frictional forces, there is no horizontal component of force to consider.

Work done lifting mass vertically: $W = mgh$

Work done along the inclined path: $W = mg\,(h/\cos\theta)\cos\theta = mgh$

The work done is *independent of the path* taken and depends only on the height through which the mass is lifted. If the mass is allowed to fall back to the ground, it will gain kinetic energy, which could be used to do some useful work (e.g., generating electricity). Lifting it in a gravitational field has given it the potential to do work.

- Gravitational potential energy (GPE) is the potential energy an object has due to its position in a gravitational field.

In a uniform gravitational field, change in GPE is given by:

$$\Delta GPE = mgh$$

If a mass is moved around a closed loop inside a gravitational field (any gravitational field, not just a uniform field), the work done by an external agent in lifting it is equal to the work done by the gravitational field as it comes back down. If it returns to its original position, the work done in the loop is zero. This is an example of what is called a "conservative field." Gravitational fields are conservative fields.

5.3.3 Kinetic Energy

Kinetic energy is the energy a body has due to its motion. We can use the definition of work to derive an expression for kinetic energy by considering the work that must be done to accelerate a mass m from velocity u to velocity v.

Consider a mass that is moving at velocity v and acted upon by a resultant force F.

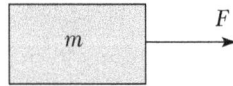

In a short time δt, the work done on the mass is given by: $\delta W = F\delta s = Fv\delta t$

Using $\qquad F = ma = m\dfrac{\delta v}{\delta t}, \delta W = m\dfrac{\delta v}{\delta t}v\delta t = mv\delta v$

Taking the limit of $\delta t \to 0$ and integrating between initial and final velocities, we get:

$$W = \int_{u}^{v} mv\,dv = \frac{1}{2}mv^2 - \frac{1}{2}mu^2$$

The work done is equal to the change in kinetic energy, so $KE = \frac{1}{2}mv^2$

It is also possible to derive this relationship by considering a constant resultant force acting on a constant mass to produce a constant acceleration. The displacement during acceleration is given by:

$$s = \frac{(v^2 - u^2)}{2a}$$

Multiplying throughout by mass m and acceleration a gives:

$$mas = \frac{m(v^2 - u^2)}{2}$$

Using $F = ma$, we obtain:

$$W = Fs = \frac{1}{2}mv^2 - \frac{1}{2}mu^2 \quad \text{as before}$$

5.3.4 The Law of Conservation of Energy

The law of conservation of energy can be stated in several different ways. Here are two of them:

- Energy is never created or destroyed, but one form can be transferred to another.

- The total energy of a closed system is constant.

Energy is a scalar quantity, so working out how much energy is present in a system is simply a matter of adding the energies of each part of the system. Alpha decay is a good example. We have already seen that the total momentum of the system is zero before and after the decay. However, nuclear energy has been transferred to kinetic energy of the new nucleus formed in the decay and the emitted alpha particle.

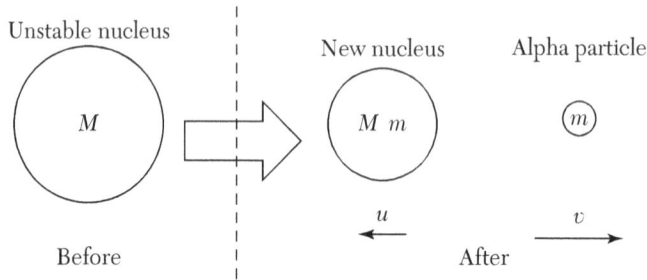

Total momentum = $mv - (M - m) u = 0$

Total energy = $\frac{1}{2} mu^2 + \frac{1}{2} mv^2 \neq 0$ (direction does not the affect sign of KE)

Notice that, in this example, kinetic energy is not conserved. It is important to realize that the law of conservation of energy is about total energy and not any one kind of energy. In some interactions and collisions (usually on the atomic scale), kinetic energy *is* conserved. When this is the case, the interaction is described as an **elastic** interaction. In avast majority of interactions, some of the initial kinetic energy is transferred to other forms (such as thermal energy), and these are described as inelastic interactions.

Type of interaction	Momentum	Kinetic energy
Elastic	conserved	conserved
Inelastic	conserved	not conserved

5.3.5   Energy and Momentum in a 2D Collision

We can use a 2-dimensional collision to illustrate how the laws of conservation of momentum and energy are used in problems. In this example, a ball of mass m_1 coming in from the left strikes another stationary ball of mass m_2, and then the two balls move off in different directions.

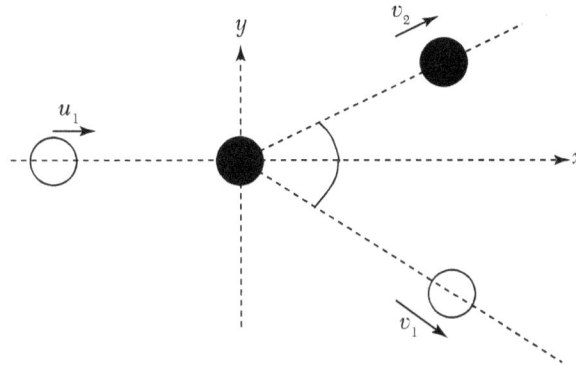

We can apply the conservation of momentum along each axis:

x-axis: $m_1 u_1 + 0 = m_1 v_1 \cos\theta + m_2 v_2 \cos\varphi$ $\qquad$ (1)

y-axis: $0 = m_2 v_2 \sin\varphi + m_1 v_1 \sin\theta$ $\qquad$ (2)

Kinetic energy before the collision: $E_K \text{(before)} = \frac{1}{2} m_1 u_1^2$ $\qquad$ (3)

Kinetic energy after the collision: $E_K \text{(after)} = \frac{1}{2} m_1 v_1^2 + \frac{1}{2} m_1 v_2^2$ $\qquad$ (4)

If the collision is elastic, then $E_K \text{(before)} = E_K \text{(after)}$.

If the collision is inelastic, then $E_K \text{(before)} - E_K \text{(after)} = \text{KE transferred}$ in to other forms

5.3.6 Energy Transfers

There are different types of energy:

- Kinetic energy: the energy of a body as a result of its motion

- Potential energy: the energy stored as a result of the position of a body in a field, e.g.:

    Gravitational potential energy

    Electrical potential energy

    Nuclear potential energy

There are other forms of potential energy that are related to electrical potential energy because they relate to the bonds between particles inside materials:

    Chemical potential energy

    Elastic potential energy (strain energy)

- Thermal energy: the sum of random kinetic energies of all particles in a body; it differs from kinetic energy in that these motions average to zero at the center of mass frame of the body

- Radiant energy: the energy of electromagnetic waves (or photons)

- Rest energy: the energy associated with mass through Einstein's equation, $E = mc^2$

Heat and work are not forms of energy. These are two important ways in which energy is transferred:

- Work is the energy transferred by a force when the point of action of the force is displaced in the direction of the force.

- Heat is the energy transferred due to temperature differences between two systems.

Devices that transfer energy from one form to another are called **transducers**. Energy transfers are often displayed using a flow diagram or Sankey diagram. For example, the diagram below is for a compact fluorescent lamp that converts 60% of the electrical energy supplied to it into visible radiant energy (light).

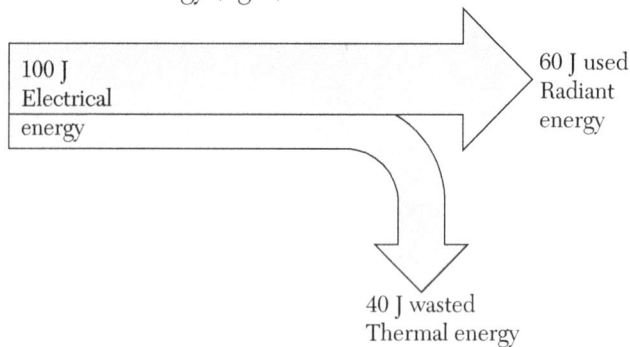

100 J
Electrical
energy

60 J used
Radiant
energy

40 J wasted
Thermal energy

The efficiency of a transducer is the ratio of useful output energy to total input energy, usually expressed as a percentage. In the case of our compact fluorescent light, this would be 0.40 or 40%.

$$\text{Efficiency} = \frac{\text{Useful output energy}}{\text{Total input energy}} \times 100\%$$

5.3.7 Power

Power is the rate of transfer of energy.

$$\text{Power} = \frac{\text{Energy transferred}}{\text{Time}}$$

For continuous energy transfers, $P = \dfrac{dE}{dt}$

The S.I. unit of work is watt (W), and 1 W = 1 Js^{-1}.

Be careful not to confuse W (symbol of watt) with W (work, a physical quantity)!

If the energy is transferred as work this is the rate of doing work. If work W is done in time t, we derive the power as $P = \dfrac{dW}{dt} = \dfrac{d(Fs)}{dt}$

When the force remained constant, this becomes: $P = F\dfrac{ds}{dt} = Fv$

This is the scalar product of force and velocity: $P = \textbf{F.v}$

Efficiency can also be expressed in terms of power:

$$\text{Efficiency} = \frac{\text{Useful output power}}{\text{Total input power}} \times 100\%$$

5.4 Energy Resources

If energy is a conservable quantity, why we say we may face a global energy crisis? The problem is that when energy-dense fuels are used to provide useful work or heating, the energy itself becomes spread out among an ever-increasing number of particles and usually ends up as thermal energy at a low temperature (see Chapter 10).As a result, the availability of energy to do useful work decreases. There is a limit to our ability to extract work from heat generated per the laws of thermodynamics (see Section 10.2.3); so we continue to look for new sources of fuel (chemical or nuclear) and to improve technologies that can extract energy from our environment. The table below gives information about some primary energy sources.

Primary source	Energy type	Primary energy transfer	
Fossil: oil, coal, gas	Chemical	Chemical → thermal	Non-renewable
Uranium, plutonium	Nuclear fission	Nuclear → thermal	
Isotopes of hydrogen	Nuclear fusion	Nuclear → thermal	
Solar	Radiant	Radiant → electrical	Renewable
Wind	Kinetic	Kinetic → electrical	
Geothermal	Thermal	Thermal → kinetic	
Hydroelectric	GPE	GPE → kinetic	
Tidal	GPE	GPE → kinetic	
Biomass	Chemical	Chemical → thermal	
Waves	Kinetic	Kinetic → electrical	

Renewable energy resources are those that are naturally regenerated. Fossil and nuclear fuels are **non-renewable energy sources**. They can get depleted over time. However, some non-renewable fuels, such as those used for nuclear fusion, would be capable of providing energy for millennia. Unfortunately, we are yet to build an effective reactor working based on nuclear fusion. However, a huge research reactor, ITER, is being constructed in France as a major step toward a commercial fusion reactor, and it is hoped that fusion power is achievable by the mid-2020s.

5.5 Propulsion Systems

A propulsion system is one that changes the way an object moves or maintains its motion in the presence of opposing forces. In order to generate a driving force that acts on the moving object, the system must exert an equal force in the opposite direction on something else, according to the Newton's third law. Newton's second law tells us that the magnitude of the driving force will be equal to the magnitude of the rate of change of momentum of the material that the system pushes against. Wheels, propellers, and jet engines exert forces on their surroundings. Rockets eject burnt fuel and exert a force on that.

5.5.1 Jet Propulsion

In a jet engine, a system accelerates a gas backwards and, in so doing, generates a forward force on the jet engine. The jet pushes back on the atmosphere, and the atmosphere pushes forward on it. Here is a schematic of a jet engine:

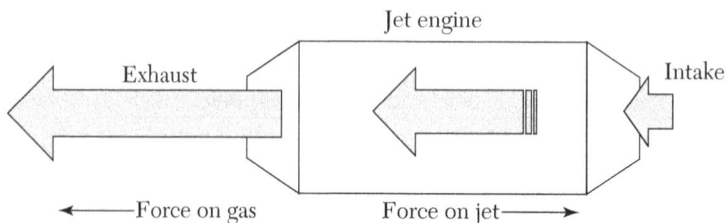

The force exerted on the jet engine will be equal in magnitude to the rate of change of momentum of the gas passing through the engine. While it is true that the combustion of fuel inside the engine adds mass to the air passing through it, this is in practice a very small contribution; so we can simplify the analysis by assuming that the mass flow in and out of the system is solely due to the air. This means that the mass flow rate in and out of the jet is the same.

If u is the air speed of the aircraft and v is the speed of exhaust gases, then the thrust is given by:

$$F = \frac{dm}{dt}(v-u)$$

Clearly v must be greater than u for the jet to produce a forward thrust.

There are two ways to increase the thrust:

▨ Increase the mass flow rate (dm/dt)

▨ Increase the exit velocity of gases $(v-u)$

Different types of jets use different methods.

5.5.2 Rockets

A jet cannot work in a vacuum, because there would be no external material against which to push. Rockets get around this problem by ejecting a large amount of matter (in the form of burnt fuel) at a very high velocity. The change in momentum of the ejected matter is equal and opposite to the change in momentum of the rocket. Once again, we can derive an equation for the thrust of a rocket using Newton's second law, but this time we are dealing with an object of changing mass.

$$F = \frac{d(mv)}{dt}$$

If the rocket is in space with no external forces acting upon it, then we use the conservation of momentum to show that the change in velocity of the rocket depends on the proportion of its mass expelled as burnt fuel. Consider a short time δt during which the rocket expels a mass δm at a velocity u relative to the rocket. At this time, the rocket has a forward velocity v and a mass m.

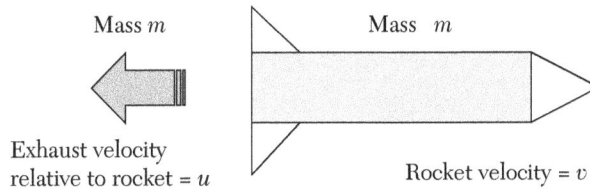

Mass m

Mass m

Exhaust velocity
relative to rocket = u

Rocket velocity = v

Taking motion to the right to be positive and using conservation of momentum, we have

Change of momentum of exhaust gases = $-u\delta m$ (1)

Change of momentum of the rocket = $(m-\delta m)\,\delta v = m\delta v - \delta m\delta v$ (2)

The final term in Equation (2) can be neglected because it is a product of two second-order terms. Now we use conservation of momentum before and after this mass of fuel is ejected:

$$-u\delta m + m\delta v = 0$$

Now separate variables and integrate from an initial velocity v_0 to a final velocity v_f during which time the mass of the rocket falls from m_0 to m_f.

$$\int_{m_0}^{m_f} \frac{dm}{m} = \int_{v_0}^{v_f} \frac{dv}{u}$$

leading to:

$$v_f = v_0 + u \ln\left(\frac{m_0}{m_f}\right)$$

The final velocity of the rocket can be increased by:

- Increasing the velocity of exhaust gases

- Increasing the ratio of m_0/m_f.

The second condition requires that the final mass of the rocket be small compared to its initial mass, so the majority of the rocket's mass at launch is in its fuel.

5.5.3 Radiation Pressure

Einstein's photon theory assumes that electromagnetic radiation can only transfer energy to and from matter in discrete quanta or photons of energy $E = hf$, where h is the Planck's constant. Einstein also showed that energy and mass are equivalent or $E = mc^2$, where c is the speed of light. Combining these two equations suggests that the absorption or emission of a photon of frequency f also involves a momentum change of size:

$$p = \frac{E}{c}$$

If light is absorbed by a surface of area A, the rate of change of momentum of the photons is:

$$F = \frac{dp}{dt} = \frac{IA}{c}$$

where I is the intensity (number of photons per second × photon energy). This will be equal and opposite to the force exerted on the surface (by Newton's third law).

When radiation is reflected from a surface (e.g., light from a mirror), the force is doubled because the momentum change is doubled for each photon (from a positive value to an equal negative value). While the radiation pressure from ordinary light sources on human-sized mirrors might be tiny (< 10^{-6} Pa for 100 W of radiation falling onto a mirror of area 1 m^2), it has been suggested that high-intensity laser beams directed from the Earth could accelerate a reflective micro-spacecraft up to very high speeds so that they can make trips to the nearest stars within human lifetimes.

5.6   Frames of Reference

In order to analyze the motion of an object, we need to measure it against a fixed set of axes (a coordinate system) using a reliable clock. Often the reference frame we choose may beat rest with respect to the Earth, but the motions would be very different if they were measured with respect to the Sun or the Moon. Galileo used a thought experiment to show that the laws of mechanics are the same in all uniformly moving reference frames.

Shut yourself up with some friend in the main cabin below decks on some large ship, and have with you there some flies, butterflies, and other small flying animals. Have a large bowl of water with some fish in it; hang up a bottle that empties drop by drop into a wide vessel beneath it. With the ship standing still, observe carefully how the little animals fly with equal speed to all sides of the cabin. The fish swim indifferently in all directions; the drops fall into the vessel beneath; and, in throwing something to your friend, you need throw it no more strongly in one direction than another, the distances being equal; jumping with your feet together, you pass equal spaces in every direction. When you have observed all these things carefully (though doubtless when the ship is standing still everything must happen in this way), have the ship proceed with any speed you like, so long as the motion is uniform and not fluctuating this way and that. You will discover not the least change in all the effects named, nor could you tell from any of them whether the ship was moving or standing still. In jumping, you will pass on the floor the same spaces as before, nor will you make larger jumps toward the stern than toward the prow even though the ship is moving quite rapidly, despite the fact that during the time that you are in the air the floor under you will be going in a direction opposite to your jump. In throwing something to your companion, you will need no more force to get it to him whether he is in the direction of the bow or the stern, with yourself situated opposite. The droplets will fall as before into the vessel beneath without dropping toward the stern, although while the drops are in the air the ship runs many spans. The fish in their water will swim toward the front of their bowl with no more effort than toward the back, and will go with equal ease to bait placed anywhere around the edges of the bowl. Finally the butterflies and flies will continue their flights indifferently toward every side, nor will it ever happen that they are concentrated toward the stern, as if tired out from keeping up with the course of the ship, from which they will have been separated during long intervals by keeping themselves in the air. And if smoke is made by burning some incense, it will be seen going up in the form of a little cloud, remaining still and moving no more toward one side than the other. The cause of all these correspondences of effects is the fact that the ship's motion is common to all the things contained in it, and to the air also. That is why I said you should be below decks; for if this took place above in the open air, which would not follow the course of the ship, more or less noticeable differences would be seen in some of the effects noted.[2]

This is an important observation. This means that there is no way to distinguish rest from uniform motion, so there is no way to be sure that any particular reference frame will beat rest in space. All uniformly moving reference frames are called **inertial reference frames**. It is often helpful when selecting a particular inertial reference frame to simplify a problem, and often the best one to choose is the center of mass frame.

5.6.1 The Center of Mass Frame

The center of mass reference frame is thought to be at rest with respect to the center of mass of the system. This is also the reference frame in which the sum of the momenta of all particles in the system may be zero. To transform the laboratory reference frame to the center of mass frame, we simply subtract the velocity of the center of mass relative to the laboratory from all of the velocities of the particles in the system.

Here is an example in which switching to the center of mass system makes it much easier to understand the physics of an interaction. Since matter and antimatter annihilate to form gamma-rays, it seems reasonable to think that a single electron might annihilate with a positron to form a single gamma-ray photon as shown below:

---

2. *Dialogue Concerning the Two Chief World Systems*, translated by Still man Drake, University of California Press, 1953, pp. 186–187 (second day).

If this could occur, the photon would take away the energy and momentum of the electron–positron pair. However, this is in fact impossible, as can be seen if we transform to the center of mass frame by subtracting $u/2$ from the electron and photon.

The momentum before annihilation may be zero since the electron and positron have equal mass but opposite velocities. This means that a single photon would have to carry away energy but have no momentum. We have seen earlier (Section 5.5.3) that photons have momentum $p = E/c$, so this is impossible. In reality, such an annihilation results in a pair of photons emitted in opposite directions in the center of mass frame.

This allows energy to be carried away and the total momentum to remain at zero. The creation of a pair of identical photons when an electron and a positron annihilate is utilized in PET scanners (see Section 29.5).

5.6.2 Galilean Transformation

A transformation between two inertial reference frames is called a Galilean transformation. This is a set of equations that transforms the coordinates of an object in one inertial reference frame to those in another. For simplicity, consider an object at rest at point P with coordinates (x, y, z) in a particular inertial reference frame. Now consider a second inertial reference frame that coincides with the first one at time $t = 0$ but which is moving in the positive x-direction at a constant velocity v. What are the coordinates x', y', z' of P in this reference frame?

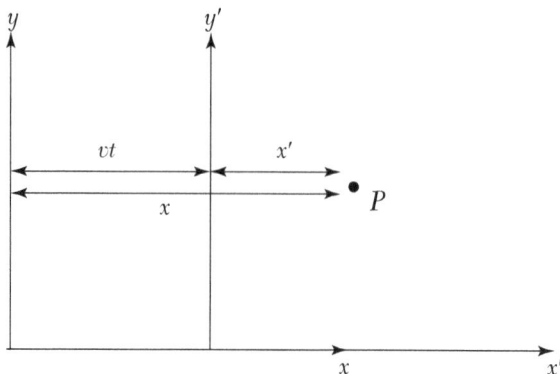

The equations of this transformation are:

$$x' = x - vt$$

$$y' = y$$

$$z' = z$$

$$t' = t$$

The laws of Newtonian mechanics are invariant (do not change) under Galilean transformations. These transformations assume that physics takes place against a background of absolute space and absolute time, which are the same for all observers. However, early in the 20th century, Einstein realized that the laws of electromagnetism are *not* invariant under Galilean transformations. This realization ultimately led him to the special theory of relativity, which postulates that the laws of physics *should* be the same in all inertial reference frames. The only way that this could be true was if measurements in space and time were all relative and not absolute. This is explored further in Chapter 24.

5.7 Theoretical Mechanics

In the century following the publication of Newton's laws of motion, physicists and mathematicians developed alternative ways of solving problems in Newtonian mechanics. These new methods were equivalent to the use of equations such as $F = ma$ but often simplified the solutions of complex problems. They also provided new ways to think about physical processes. One of the most significant approaches was developed by the Italian-French mathematician Joseph-Louis Lagrange in 1788. A great advantage of the Lagrangian method is that it helps to show the links between Newtonian mechanics and quantum mechanics.

5.7.1 Force and Energy

It is often possible to solve a problem by a variety of different methods. For example, consider how you might calculate the final velocity of an object dropped vertically from rest through a height h in the Earth's gravitational field.

Method 1: Using Forces

The downward acceleration of the object is $a = F/m = mg/m = g$

The displacement is h

Using a "suvat" equation, $v^2 = u^2 + 2gh$ so $v = \sqrt{2gh}$

Method 2: Using Energy

The change in GPE as the object falls is mgh.

The gain in KE is $\frac{1}{2}mv^2$.

Energy is conserved, so $\frac{1}{2}mv^2 = mgh$, so $v = \sqrt{2gh}$ as before

Although both methods are equally simple, there is a fundamental difference in what we have done. Method 1, using forces, is based on vector equations. Method 2, using energy, is based on scalar equations. The Lagrangian method is based on scalars and is related to method 2. It works best for solving problems involving conservative forces (i.e., where we can neglect frictional forces so that all of the energies in the system are either kinetic or potential).

5.7.2 Lagrangian Mechanics

The Lagrangian method does not introduce any new physics, but it does provide a different way to look at Newtonian mechanics. The Lagrangian function L is defined as:

$$L = T - V$$

where T is kinetic energy and V potential energy.

The Lagrangian is a scalar quantity. In general, the kinetic energy will depend on velocity (v_x, v_y, v_z) and the potential energy will depend on position (x, y, z). One of the beauties of the Lagrangian method is that it can work equally well with polar or spherical coordinates (this is often difficult when we start from $F = ma$), so we can set up the theory with an arbitrary set of "generalized" coordinates:

Positions: q_1, q_2, q_3 (x, y, z) in Cartesian coordinates

Velocities: $\dot{q}_1, \dot{q}_2, \dot{q}_3$ (v_x, v_y, v_z) in Cartesian coordinates

The dot above the symbol represents differentiation with respect to time (e.g., $\dot{q} = dq/dt$).

Once the Lagrangian for the system is known, we can derive the equation of motion using the Euler-Lagrange equation:

$$\frac{\partial}{\partial t}\left(\frac{\partial L}{\partial \dot{q}}\right) = \frac{\partial L}{\partial q}$$

The method itself is quite simple:

1. Write down the Lagrangian for the system under consideration.

2. Use the Euler-Lagrange equation to derive the equation of motion.

Consider a simple example where a particle of mass m is released from rest in a uniform gravitational field of strength g. We set $x = 0$ at the surface and consider only the x-direction (i.e., solve the problem in one dimension).

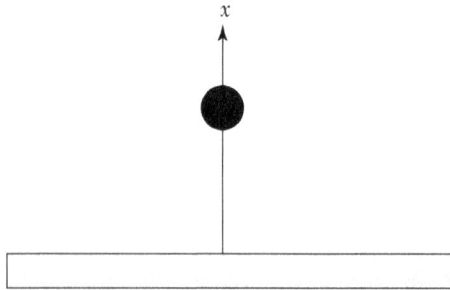

1. $T = \frac{1}{2}m\dot{x}^2$ and $V = mgx$; so $L = \frac{1}{2}m\dot{x}^2 - mgx$

2. L.H.S. of Euler-Lagrange equation: $\frac{\partial}{\partial t}\left(\frac{\partial L}{\partial \dot{x}}\right) = \frac{\partial}{\partial t}(m\dot{x})$

R.H.S. of Euler-Lagrange equation: $\frac{\partial L}{\partial x} = -mg$

These are equal: $\frac{\partial}{\partial t}(m\dot{x}) = -mg$

leading to: $\ddot{x} = -g$

This is hardly surprising. It shows that the object accelerates in the $-x$-direction with an acceleration of magnitude g and that this acceleration is independent of the mass. This is exactly the same result as if we had started with the forces and used $F = ma$. In fact, as you can see, the Euler-Lagrange equation generates this equation in the second line of point (2).

It seems to be a complicated way to solve a trivial problem, and in this case it is! However, in many complex problems the Lagrangian method is far simpler than any attempt to use $F = ma$ directly.

5.8 Exercises

1. It is possible to lift a 1 kg mass slowly using a cotton thread. However, if the thread jerks, it might snap. Why?

2. Use Newton's laws of motion to explain why:
   (a) passengers on a moving bus feel as if they are being thrown forwards when the driver suddenly applies a brake,
   (b) a gun recoils when it is fired,
   (c) you feel "heavier" when you are standing inside a lift that is accelerating upwards,
   (d) the gravitational force pulling the Earth toward the Sun is equal to the gravitational force pulling the Sun toward the Earth.

3. The below diagrams show forces acting on an object with a mass of 1600 kg. Calculate the resultant acceleration in each case.

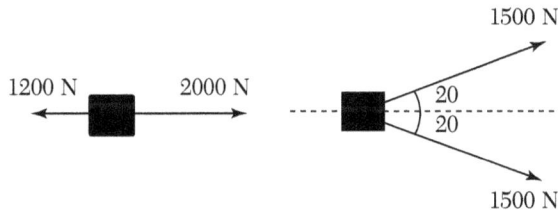

4. Work out the linear momentum of each of the following:
   (a) a 1000 kg car travelling at 30 m/s
   (b) a 200 g stone moving at 80 cm/s

5. A 65 kg rugby player moving due east at 8 m/s collides head-on with a 60 kg rugby player moving due west at 6 m/s. After the collision the two players are initially locked together.
   Calculate the velocity (magnitude and direction) of the two players immediately after the collision.

6. The diagram below shows a 2D collision between two identical pucks on an air table.

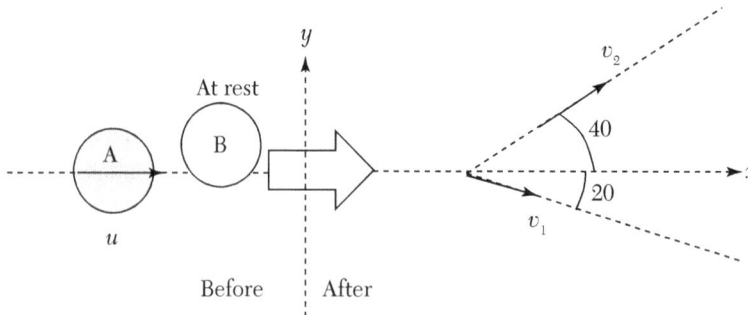

The initial velocity, u, is 0.80 ms^{-1}.

(a) Calculate the magnitudes of the final velocities v_1 and v_2.

(b) Show that this is an inelastic collision and calculate the fraction of the initial energy that is transferred to other forms.

7. When a car of mass 1400 kg is traveling along an horizontal highway at a constant speed of 30 ms^{-1}, its output power is 75 kW.

(a) Calculate the total drag force on the car (assume all the output power does work against drag).

(b) The road begins to climb upwards at a constant angle of 5.0° to the horizontal. The driver maintains the same constant speed of 30 ms^{-1} and the drag on the car is unchanged.

(i) Explain why the power output of the car must increase.

(ii) Calculate the new power output.

8. A car of mass m can apply a maximum braking force B. The driver of the car has a minimum reaction time T.

(a) Derive a formula for the minimum stopping distance for this car when the car is traveling at speed u.

(b) Sketch a graph of the variation of stopping distance with u.

(c) Explain why even a small decrease in the speed limit could significantly reduce the number of pedestrians being hit by cars.

CHAPTER 6

FLUIDS

6.0   Introduction

The particles inside a solid vibrate about fixed positions unless the material is placed under extreme stress. Particles inside a fluid, however, can move past each other and change position; this allows them to flow when stresses are applied to the fluid. Liquids and gases are fluids and their behavior can be modeled using Newton's laws.

Here are some key ideas to describe the behavior of fluids:

Density (kgm^{-3}): $\rho = m/V$ where m is the mass of the fluid (kg) and V is the volume occupied by the fluid (m^3).

Pressure (Pa): $p = F/A$ where F is the force (N) exerted by the fluid perpendicular to an area A (m^2). 1 Pa = 1 Nm^{-2}. Pressure in a fluid acts in all directions, and the pressure at the same level in a static fluid is constant.

Shear stress (Pa):   When two parallel layers of area A are pulled in opposite directions by a force F, the shear stress acting on the layers is $\sigma = F/A$.

Incompressible fluids: Liquids such as water do not compress easily, so a useful model assumes that they have constant volume and so their density is constant.

Viscosity:	When one layer of a fluid moves over another nearby layer, a frictional force between the layers opposes the flow. The greater the resistance to flow, the greater the viscosity of the fluid.
Inviscid fluid:	For situations in which the viscous forces are negligible, or for fluids with very low viscosity, a useful model assumes that the viscosity is zero. Such a fluid is said to be "inviscid."
Ideal fluid:	The simplest model of a fluid is one which is incompressible and has zero viscosity; this is called an ideal fluid. Water can often be treated as an ideal fluid.
Ideal gas:	An ideal gas is one that obeys the equation of state, $pV = nRT$, where p is pressure, V is volume, n is the number of moles, T is temperature in kelvin, and R is the molar gas constant.

6.1 Hydrostatic Pressure

6.1.1 Excess Pressure Caused by a Column of Fluid

The increase in pressure that you feel when you go underwater is caused by the weight of the water above you. This is called **hydrostatic pressure**. A useful equation for hydrostatic pressure can be derived by thinking about the forces that support a horizontal layer of fluid of thickness δz and density ρ at height z in a column of cross-sectional area A.

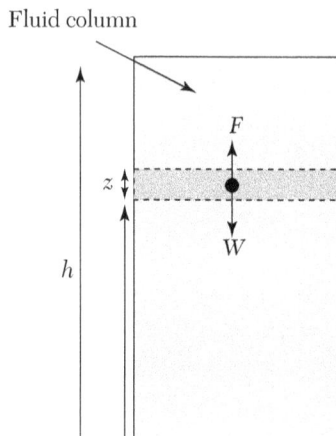

Fluid column

The layer has weight $W = \rho Ag\delta z$, but it is in equilibrium, so there must be a force F of equal magnitude supporting it. This means that the pressure under the layer must be higher than the pressure above it by an amount δp, so that:

$$F = - A\delta p = \rho Ag\delta z$$

The negative sign is because the pressure is greater lower down (p decreases as z increases).

For thin layers, this gives a pressure gradient:

$$\frac{dp}{dz} = -\rho g$$

If the fluid is considered incompressible, its density might be constant and the expression above can be integrated to give the excess pressure at the base of a column of height h caused by the weight of the column.

$$\text{Excess pressure} = p_0 - p_h = \int_{p_h}^{p_0} p = \int_{h}^{0} - \rho g dz = gh$$

This can also be derived by simply calculating the total weight of the column and dividing by the area of the base: $p_0 = \rho Agh/A = \rho gh$. However, the approach above can be used when the density of the fluid changes with depth, for example, to derive an expression of the atmospheric pressure at altitude h.

6.1.2 Atmospheric Pressure

Atmospheric pressure is caused by the weight of the atmosphere. However, air is highly compressible, so the density of air will vary with height—being high near the surface and lower higher up. In order to model this, we make two assumptions:

- Air acts like an ideal gas.

- The temperature of air does not vary with height.

If air is an ideal gas, then its density can be related to its pressure. Considering a mass M with n moles of molecules each of mass m,

$pV = nRT$ (ideal gas equation)

$\rho = M/V = Mp/nRT$ (substitution from ideal gas equation)

Using molar mass ($M = nN_Am$, where N_A is the Avogadro number, i.e., 6.02×10^{23}), we can express density in terms of pressure:

$\rho = N_Amp/RT = mp/kT$, where k is the Boltzmann constant ($k = R/N_A$).

We can now substitute this expression for density into the equation of hydrostatic pressure gradient in a fluid:

$$\frac{dp}{dz} = -\rho g = -\frac{mgp}{kT}$$

The expression on the R.H.S. depends on p, so we have to separate variables before we can integrate between the ground $(z = 0)$ and some height $(z = h)$:

$$\int_{p_0}^{p_x} \frac{dp}{p} = -\left(\frac{mg}{kT}\right) \int_0^x dz$$

$$ln\left(\frac{p_x}{p_0}\right) = -\left(\frac{mg}{kT}\right)x$$

$$p_x = p_0 e^{-\left(\frac{mg}{kT}\right)x}$$

Atmospheric pressure (and therefore density) falls exponentially with height x above the surface.

In reality, the temperature of atmosphere is not independent of altitude and falls significantly (from about 290 K to 220 K) in the first 10 km. These variations can be built into the model to provide a better fit.

6.1.3 Measuring Pressure Differences

A manometer or U-tube can be used to measure pressure differences by connecting one side to the pressure source to be measured and leaving the other side open to the atmosphere. The difference in liquid height can be used as a direct measure of the pressure difference (e.g., mm of water or mm of mercury, or in pascals using $\Delta p = \rho g \Delta h$).

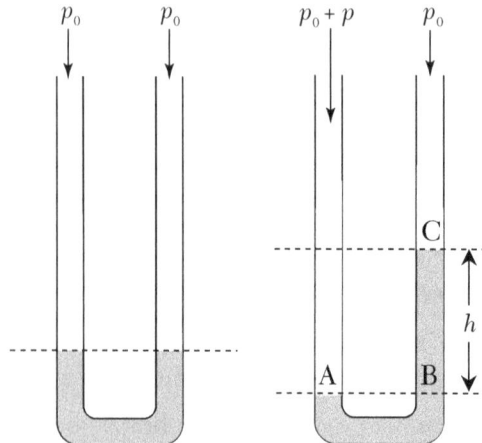

In the manometer shown on the left, both tubes are open to the atmosphere at pressure p_0, so the levels are equal. In the manometer shown on the right, an excess pressure Δp is connected to one tube. The pressures at A and B inside this manometer must be equal because these points are at the same level in the same fluid. However, the pressure at B is also equal to the pressure p_0 plus the pressure caused by the column of fluid BC of height Δh.

$$p_A = p_0 + \Delta p = p_B = p_0 + \rho g \Delta h$$
$$\Delta p = \rho g \Delta h$$

The lower the density of the liquid used, the greater the sensitivity of the manometer (i.e., the greater the change in height per unit change in pressure).

6.1.4 Barometers

Imagine a manometer with one end open to the atmosphere and the other end attached to an effective vacuum pump. The pressure difference is equal to atmospheric pressure p_{At}, so the height of the column can be used to measure atmospheric pressure. This is the principle of the barometer. However, instead of connecting one end of a manometer to a pump, one end is sealed and the fluid is allowed to fall away from the sealed end so that a vacuum forms above it.

The pressure at A is equal to the pressure at B because both points are at the same level in the same liquid. Since A is on the surface of the liquid exposed to the atmosphere, the pressure at both points must be atmospheric pressure.

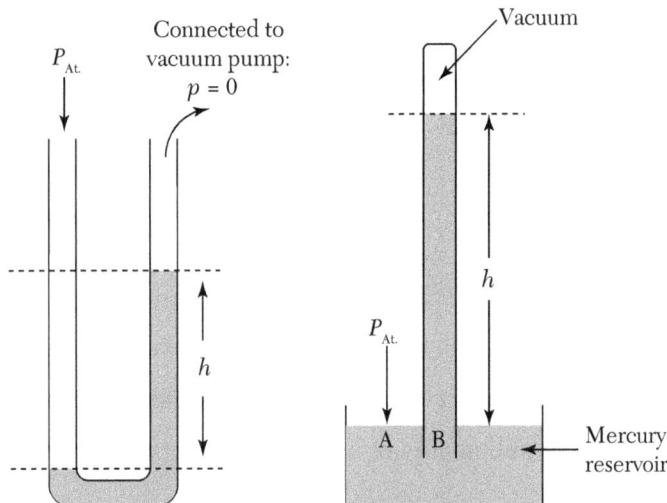

Atmospheric pressure $= p_A = p_B = 0 + \rho g h = \rho g h$

In practice, the vacuum above the mercury in a barometer (called the Torricelli vacuum) may not be perfect. This is because some mercury atoms will leave the surface of the mercury. However, the vapor pressure of mercury at room temperature is only about 1 Pa (0.001% of atmospheric pressure), so it makes no significant difference to the measurement of atmospheric pressure.

Another type of pressure gauge is called a Bourdon gauge (see the image on the right). This contains a hollow curved tube that is closed at one end. When the pressure inside is changed, the tube bends. This small motion is amplified mechanically and used to move a needle against a calibrated scale.

6.1.5 Dams

The design of a dam must take into account all the forces that act on the structure. The most important of these is hydrostatic pressure from the trapped water. In a simple case, this could be the only force acting on the dam assuming that the containing wall of the dam is vertical.

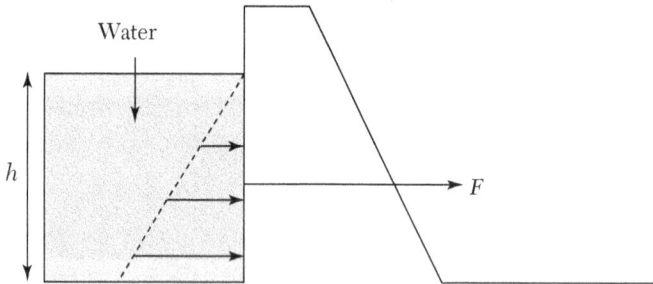

Pressure increases linearly with depth, so the total horizontal force on the containing wall will be equal to the average excess pressure ($\rho g h/2$) multiplied by the area in contact with the water ($A = hl$, where l is the horizontal length of the dam wall):

$$F = \tfrac{1}{2} \rho g l h^2$$

The line of action of this force is at a height $h/3$ from the base of the dam.

In reality, the situation could be more complicated than this.

- There maybe a depth of water on the downstream side.
- The containing wall might not be vertical.

▪ There maybe a hydrostatic pressure gradient under the dam as water penetrates the soil and rocks.

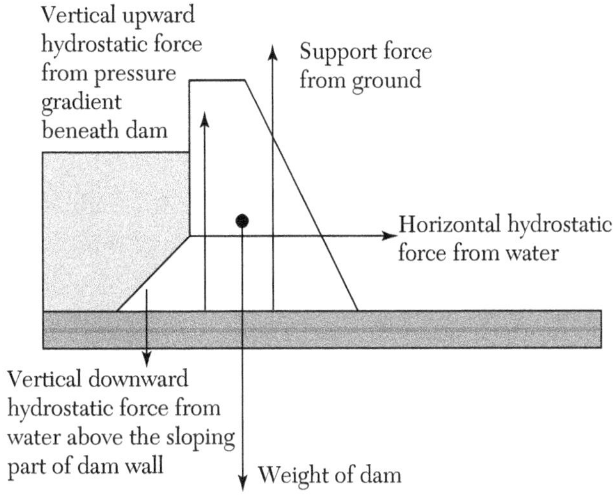

Vertical upward hydrostatic force from pressure gradient beneath dam

Support force from ground

Horizontal hydrostatic force from water

Vertical downward hydrostatic force from water above the sloping part of dam wall

Weight of dam

In addition to hydrostatic forces, engineers must also consider the forces from wind, seismic activity, and ice (if the water freezes). The dam must remain in equilibrium under all harsh conditions.

6.2 Buoyancy and Archimedes Principle

6.2.1 Buoyancy Forces

When an object is wholly or partially submerged in a fluid, there is an upward force on the object called the "upthrust" or "buoyancy" force B. This force arises because the pressure beneath the object is greater than the pressure above, as shown in the diagrams below for a specimen of a rectangular block.

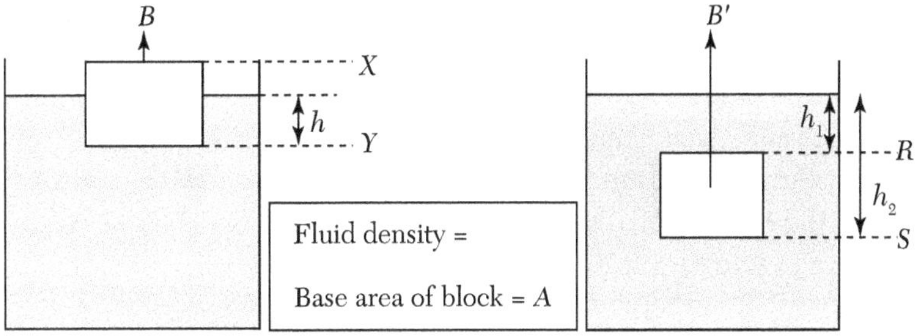

Fluid density =

Base area of block = A

Partial submersion: pressure at X = atmospheric pressure
pressure at Y = atmospheric pressure $+ \rho g h$

$$\text{Force } B = A\Delta p = \rho g h A$$

Total submersion: pressure at R = atmospheric pressure + $\rho g h_1$
pressure at S = atmospheric pressure + $\rho g h_2$
Force $B' = A\Delta p = \rho g\ (h_2 - h_1)\ A$

6.2.2 Archimedes' Principle

Note that the expressions of buoyancy force derived in Section 6.2.1 are both equal to the weight of the fluid that has been displaced by the object:

Partial submersion: hA = volume of displaced fluid, so ρhA = mass of displaced fluid and $\rho g h A$ = weight of displaced fluid.

Total submersion: $(h_2 - h_1)A$ = volume of displaced fluid, so $\rho(h_2 - h_1)A$ = mass of displaced fluid and $\rho g(h_2 - h_1)A$ = weight of displaced fluid.

This is an example of Archimedes' principle: *The buoyancy force is equal to the weight of the fluid displaced.*

This result was derived using a rectangular object but is valid for an object of any shape. This can be understood by considering the buoyancy force on each small vertical column of material inside the object. For each column we can apply exactly the same reasoning as used for the rectangular block, so the total buoyancy force is always equal to the weight of the fluid displaced regardless of the shape of the block.

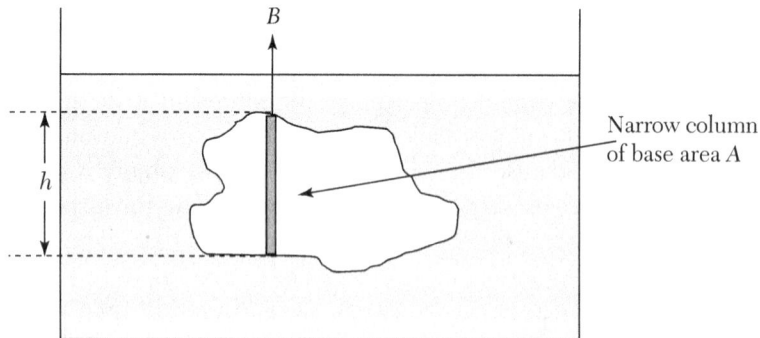

Narrow column of base area A

Using the same reasoning as in Section 6.2.1, the contribution to the buoyancy force from one narrow column will be $\Delta B = \rho g h \Delta A$. This is equal to the weight of fluid displaced by the volume of the column. The total buoyancy force on the object will be the sum of forces on all vertical columns:

$$B = \sum_{\text{all columns}} \rho g h \delta A$$

This is the weight of fluid displaced by the total volume of the block as stated in Archimedes' principle.

6.2.3 Flotation

An object will float if the buoyancy force can support its weight. The maximum buoyancy force is when the object is completely submerged; so for an object of volume V and average density ρ_{obj} to float in a fluid of density ρ_{fluid},

$$\text{weight of object} < \text{buoyancy force}$$

$$\rho_{object} Vg < \rho_{fluid} Vg$$

$$\rho_{object} < \rho_{fluid}$$

An object will float if its density is less than the density of the fluid in which it is placed.

If an object's density is equal to the density of the fluid in which it is submerged, it is said to have "neutral buoyancy." Divers and submarines use neutral buoyancy to remain at the same depth under water.

6.3 Viscosity

6.3.1 Coefficient of Viscosity

A fluid with high viscosity is very resistant to shear. This means that a relatively large shear stress is required to move one layer over another. This can be understood by thinking about the flow of a fluid close to a boundary. Particles in contact with the boundary are assumed to be at rest because of interactions with the surface, whereas those far from the boundary will be flowing with the same speed as the body of the fluid. There is a velocity gradient close to and perpendicular to the boundary, as shown in the diagram.

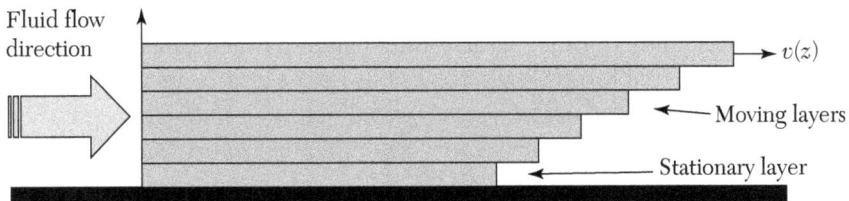

The greater the viscosity, the smaller the velocity gradient dv/dz for the same shear stress.

The coefficient of viscosity η is defined as the ratio of shear stress to velocity gradient:

$$\eta = \frac{\left(\dfrac{F}{A}\right)}{\left(\dfrac{dv}{dz}\right)}$$

The S.I. unit is therefore Nm^{-2}s or pascal-second, Pas. Another common unit of viscosity is poise (P), and 1 P = 0.1 Pas. The coefficient of viscosity of water at room temperature is approximately 1 m Pas or 1 cP (centi-poise). The coefficient of viscosity decreases with increasing temperature.

6.4 Fluid Flow

6.4.1 Laminar and Turbulent Flow

When a fluid flows, the path followed by a particle in the fluid is called a "streamline." At low flow velocities these streamlines are uniform and parallel to one another; this is called laminar flow. However, above a certain critical flow velocity, laminar flow breaks down, the streamlines begin to form eddies, and the flow becomes turbulent.

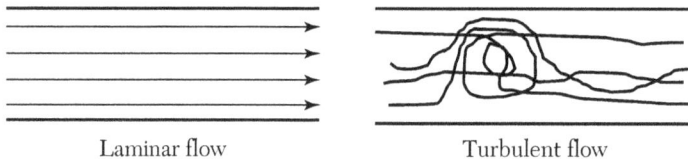

Laminar flow   Turbulent flow

Two regimes exist because there are two competing effects in the moving fluid: inertial forces related to the density and speed of motion of the fluid, and viscous forces related to the viscosity of the fluid and inversely to the physical size of the channel in which the fluid is flowing (e.g., the diameter of the pipe). If viscous forces dominate, eddies cannot form and the flow is laminar. If inertial forces dominate, the flow will be turbulent.

The Reynolds number R_e is a dimensionless constant that represents the ratio of inertial forces ($\propto \rho v$) to viscous forces ($\propto \eta/L$) in a particular flow situation. It is defined as:

$$R_e = \frac{vL}{\eta}$$

where v is the flow velocity, ρ is the fluid density, η is the coefficient of viscosity, and L is the characteristic length. For flow in a pipe, L would be the diameter of the pipe; for flow between two parallel plates, it would be the separation of the plates.

As a very approximate rule, the flow will be turbulent if $R_e > 1000$.

6.4.2 Equation of Continuity

When a fluid flows in a confined channel such as a pipe, the mass of fluid passing each point will be constant. The diagram below shows a fluid passing along a pipe that narrows from a cross-sectional area A_1 to cross-sectional area A_2.

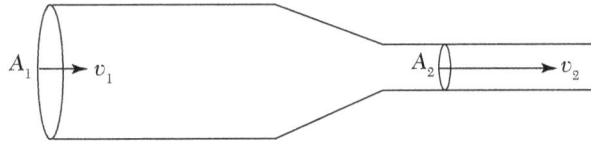

The flow velocity in the wider part of the pipe is v_1 and in the narrower part is v_2. In a short time Δt, the mass flow through area A_1 is $\rho_1 A_1 v_1 \Delta t$ and that through area A_2 is $\rho_2 A_2 v_2 \Delta t$. These must be equal, so in general:

$$\rho_1 A_1 v_1 = \rho_2 A_2 v_2$$

This is called the equation of continuity.

For an incompressible fluid, the density is constant, so:

$$A_1 v_1 = A_2 v_2$$

In this case, the flow velocity is inversely proportional to the area of the pipe:

$$\frac{v_1}{v_2} = \frac{A_2}{A_1}$$

6.4.3 Drag Forces in a Fluid

When an object moves through a fluid, it exerts a force on the fluid to make it flow around the moving object. By Newton's third law, the fluid exerts an equal but opposite force on the moving object; this is the origin of the drag force.

The drag force depends on the nature of the flow around the object (e.g., laminar or turbulent) and is determined by the properties of the fluid, such as density and viscosity, and on the velocity, size, and shape of the moving object. It may also be affected by nearby boundaries (e.g., if the fluid is inside a container).

There are two particular situations that give simple expressions for the drag force.

- Viscous forces dominate ($R_e << 1000$): drag force is directly proportional to velocity.

The drag force arises as a reaction to shearing the layers of the fluid as they flow around the object. These forces depend on the velocity gradient and therefore on the velocity of the moving object.

▪ Inertial forces dominate ($R_e \gg 1000$): drag force is directly proportional to velocity-squared.

The inertial force arises as a reaction to the force needed to accelerate the fluid in front of the moving object up to the velocity of the object. This is directly proportional to the rate of change of momentum of the fluid in front of the object, which is proportional to the mass of fluid encountered per second multiplied by the velocity of the object. Since the mass encountered per second is also proportional to the velocity, the drag force will be proportional to the velocity-squared.

6.4.4 Stoke's Law

Stoke's law gives the viscous drag on a small spherical particle moving through a fluid when the flow around the particle is laminar. The diagram below shows streamlines around an object moving through a viscous fluid. Streamlines show the paths of particles in the fluid as the object passes.

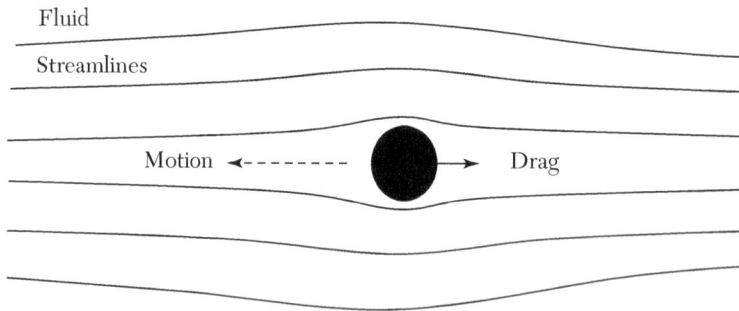

In these circumstances the drag will be directly proportional to v, and we would expect it to depend on the radius r of the sphere and the viscosity η of the fluid. Dimensional analysis can then be used to find an expression for the drag.

$$F = \left(\eta\right)^x \left(v\right)^y \left(r\right)^z$$

$$MLT^{-2} = (ML^{-1}T^{-1})^x \left(LT^{-1}\right)^y \left(L\right)^z$$

For the dimension of mass: $1 = x$

For the dimension of time: $-2 = -x - y$, so $y = 1$

For the dimension of length: $1 = -x + y + z$, so $z = 1$

The drag is therefore $F = \text{constant} \times \eta r v$

Stoke showed that the constant is 6π.

Stoke's law for viscous drag, $F = 6\pi\eta r v$

Conditions under which Stoke's law can be used:

- $R_e \ll 1000$ (flow is laminar): small object, low speed, high viscosity.

- Fluid is homogeneous and of uniform density.

- The radius of the particle is much smaller than the dimensions of the fluid container.

Stoke's law can be used to measure the viscosity η of viscous liquids such as glycerol. If a small ball bearing of radius r is released from the top of a column of viscous fluid, it will soon reach terminal velocity v_t. Once at terminal velocity the sum of viscous drag (given by Stoke's law) and buoyancy must equal the weight of the ball bearing. The viscosity is then given by:

$$\eta = \frac{2gr^2}{v_t}\left(\rho_{bb} - \rho_f\right)$$

where ρ_{bb} is the density of the ball bearing and ρ_f is the density of the fluid.

6.4.5 Turbulent Drag

We can also construct an equation for the drag on an object moving through a fluid at speeds such that the fluid motion becomes turbulent ($R_e \gg 1000$). This is useful when considering large objects moving rapidly through fluids of low viscosity (e.g., air resistance on cars and planes). Here inertial forces will dominate, so the drag is proportional to v^2 and we would also expect it to depend on the cross-sectional area A of the object (in the plane perpendicular to motion) and the density ρ of the fluid. Dimensional analysis can again be used to find an expression for the drag.

$$F = (\rho)^x (v)^y (A)^z$$

$$MLT^{-2} = (ML^{-3})^x \left(LT^{-1}\right)^y \left(L^2\right)^z$$

For the dimension of mass: $\quad 1 = x$

For the dimension of time: $\quad -2 = -y$, so $y = 2$

For the dimension of length: $\quad 1 = -3x + y + 2z$, so $z = 1$

The drag is therefore $F = \text{constant} \times \rho A v^2$

This is usually used in the form $F = \frac{1}{2} C_D \rho A v^2$

where C_D is the "drag coefficient," a dimensionless number that depends on the shape of the moving object. Streamlined shapes will have lower values of C_D.

6.4.6 Bernoulli Equation

The static pressure in a fluid was discussed in Section 6.1 and is calculated from terms of the form ρgh. However, when a fluid is flowing, there is an additional dynamic pressure that arises from the forces needed to stop the flow. The dynamical pressure is related to the potential energy (per unit volume) of the fluid, and the dynamic pressure is related to the kinetic energy (per unit volume) of the fluid. Dynamic pressure is calculated from terms of the form $\frac{1}{2}\rho v^2$. For an inviscid fluid (viscosity is negligible), energy is conserved along a streamline, so the sum of pressure terms is constant. This gives the Bernoulli equation, shown below the diagram.

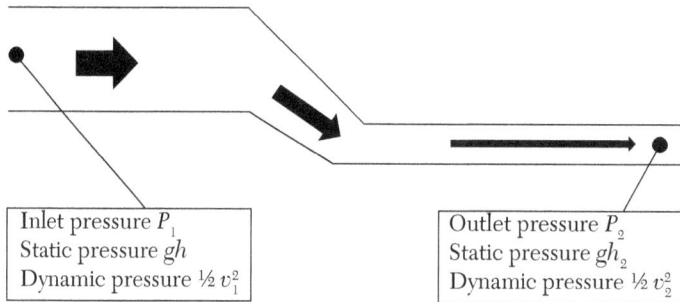

Inlet pressure P_1	Outlet pressure P_2
Static pressure gh	Static pressure gh_2
Dynamic pressure $\frac{1}{2}v_1^2$	Dynamic pressure $\frac{1}{2}v_2^2$

$$P_1 + \frac{1}{2}\rho v_1^2 + \rho gh_1 = P_2 + \frac{1}{2}\rho v_2^2 + \rho gh_2$$

where h_1 and h_2 are heights above some reference level.

The Bernoulli equation only applies when several conditions are met:

- The flow is steady.

- The flow is laminar and not turbulent.

- The fluid is inviscid.

6.4.7 Bernoulli Effect

The "Bernoulli effect" describes a fall in static pressure when the fluid flow velocity increases. This is easily explained using the Bernoulli equation. The diagram below shows a fluid flowing through a horizontal pipe with a central restriction. Since h does not vary, we set $h = 0$ along the streamline.

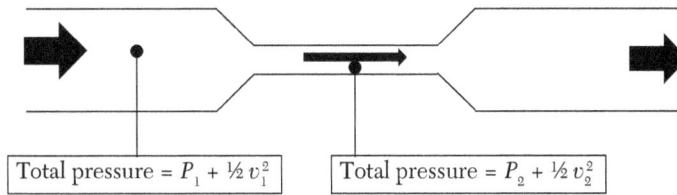

$$\boxed{\text{Total pressure} = P_1 + \tfrac{1}{2}v_1^2}$$ $$\boxed{\text{Total pressure} = P_2 + \tfrac{1}{2}v_2^2}$$

Using the Bernoulli equation, $P_1 + \dfrac{1}{2}\rho v_1^2 = P_2 + \dfrac{1}{2}\rho v_2^2$

The equation of continuity shows that $v_2 > v_1$, so $P_2 < P_1$, and the static pressure falls as the flow velocity increases. The kinetic energy of the fluid has increased, so its potential energy has decreased.

Another way to think about this is by considering Newton's second law. The fluid must accelerate as it enters the constriction, so there must be a resultant force from the wider part of the pipe. This comes from the greater static pressure.

6.4.8 Viscous Flow Through a Horizontal Pipe: The Poiseuille Equation

For an inviscid fluid in laminar flow through a horizontal pipe, there is no pressure difference along the pipe. This is because none of the terms in the total pressure $P + \tfrac{1}{2}\rho v^2 + \rho gh$ changes. This can be considered as an example of Newton's first law of motion—there are no resultant forces acting on the particles of fluid, so they continue at constant velocity. However, if the fluid is viscous, there will be frictional forces opposing the flow, and these must be balanced by a pressure gradient in the tube.

Poiseuille derived a formula for the volume rate of flow of a fluid through a pipe when a constant pressure difference is maintained across its ends. The diagram below shows a horizontal pipe of length l and radius a with pressure difference p across its ends.

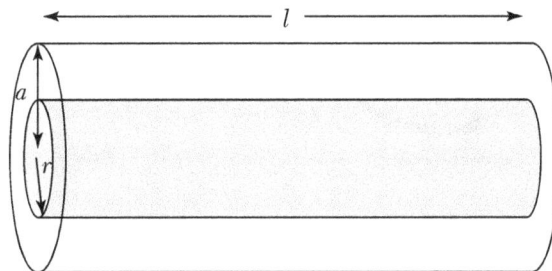

The derivation has two parts: (i) we use the equation for viscosity to work out an expression for the velocity of flow at radius r from the center. Then

(ii) we use this to work out the total rate of volume flow inside the pipe by integrating over cylindrical shells.

(i) The applied force created by the pressure difference must balance the viscous forces along its surface:

$$F = \pi r^2 p$$

$$\text{shear stress along surface} = \frac{F}{(\text{area of surface})} = \frac{\pi r^2 p}{2\pi r l} = \frac{rp}{2l}$$

This must balance viscous forces, so $\dfrac{rp}{2l} = -\dfrac{dv}{dr}$

The negative sign is because the velocity decreases from $r = 0$ to $r = a$.

$$\int_v^0 dv = \int_0^a \left(\frac{pr}{2l}\right) dr$$

which gives $v = \dfrac{p}{4\eta l}\left(a^2 - r^2\right)$

This shows that the velocity profile in the pipe is parabolic.

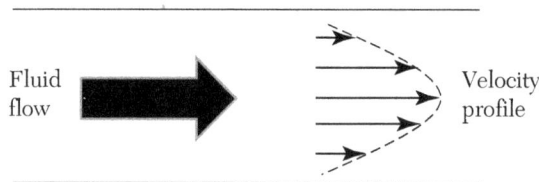

Fluid flow   Velocity profile

(ii) Since different layers flow at different velocities, the total flow can be found by integrating the volume flow for all thin cylindrical shells inside the pipe.

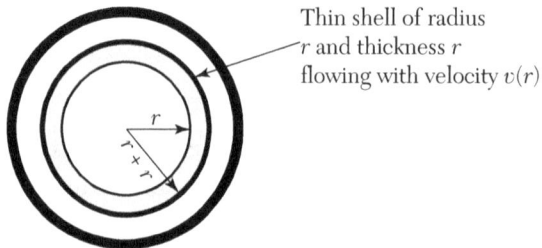

Thin shell of radius r and thickness r flowing with velocity $v(r)$

The contribution δQ to the total flow rate Q (m^{3}s^{-1}) is:

$$\delta Q = 2\pi r v \delta r$$

$$Q = \int_0^a 2\pi r v dr = \int_0^a \frac{\pi r p}{2\eta l}\left(a^2 - r^2\right) dr$$

$$Q = \frac{\pi p a^4}{8\eta l}$$

This is Poiseuille equation. The volume flow rate depends on the fourth power of the pipe radius a and is directly proportional to the pressure gradient p/l.

6.4.9 Measuring the Coefficient of Viscosity

The coefficient of viscosity can be measured by maintaining a constant pressure gradient across a horizontal tube, measuring the volume flow rate and substituting into Poiseuille equation:

$$\eta = \frac{\pi a^4}{8Q}\left(\frac{p}{l}\right)$$

Here is a suitable experimental arrangement.

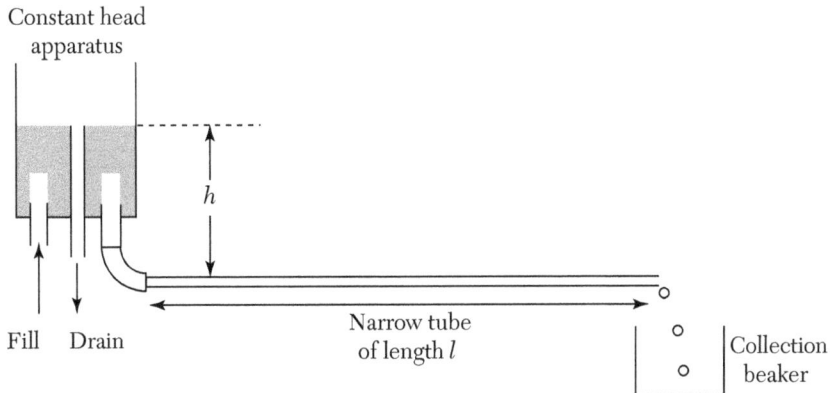

Constant head apparatus

h

Fill   Drain

Narrow tube of length l

Collection beaker

- Constant head apparatus: this is continually filled to maintain a constant depth of fluid. Excess fluid flows away through the drain. The open end of the horizontal tube is at atmospheric pressure, so the pressure gradient in the tube is $\rho gh/l$.

- The horizontal tube must be narrow enough to ensure laminar flow (otherwise the Poiseuille equation is not valid).

- The beaker is used to collect fluid over a set time t. The volume of fluid collected in this time, V, is measured using a measuring cylinder. The volume flow rate $Q = V/t$.

- The average radius of the capillary tube, a, can be determined by filling the tube with water and measuring the volume contained: volume of water $= \pi a^2 l$. As an alternative, the radius can be measured using an average of three diameters measured using a traveling microscope.

Since the coefficient of viscosity depends on temperature, the temperature of the fluid used should also be recorded.

6.5 Measuring Fluid Flow Rates

6.5.1 A Venturi Meter

The change in pressure of a fluid when its flow velocity changes can be used to measure the flow rate in a pipe. The meter works by measuring the change in static pressure as the fluid passes through a constriction in the pipe, as shown below. The derivation assumes that the fluid is ideal and the flow is steady.

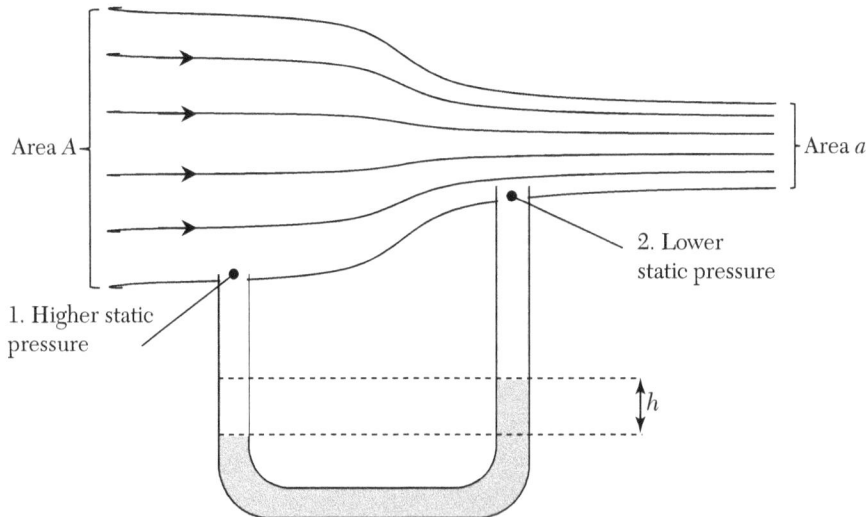

At position 1 the manometer is connected to the fluid in the wider section of pipe (area A), and at position 2 the other side of the manometer is connected to the fluid in the narrower section of the pipe (area a). The pressure difference Δp is equal to $\rho_m g \Delta h$ where ρ_m is the density of fluid in the manometer.

The pressure difference can be related to fluid flow velocity using Bernoulli's equation:

$$\Delta p = \frac{1}{2}\rho\left(v_2^2 - v_1^2\right)$$

The equation of continuity allows v_2 to be written in terms of v_1 and the ratio of pipe areas:

$$v_2 = \left(\frac{A}{a}\right)v_1$$

Combining these two equations, the flow velocity v_1 is given by:

$$v_1 = \sqrt{\frac{2\Delta p}{\rho\left\{\left(\dfrac{A}{a}\right)^2 - 1\right\}}}$$

and the volume flow rate in the tube is $Q = Av_1$

$$Q = A\sqrt{\frac{2\Delta p}{\rho\left\{\left(\dfrac{A}{a}\right)^2 - 1\right\}}}$$

Venturi meters are used in many industrial applications, including water flow.

6.5.2 A Pitot Tube

A Pitot tube is used to measure air speed by comparing the total or "stagnation" pressure with the static pressure.

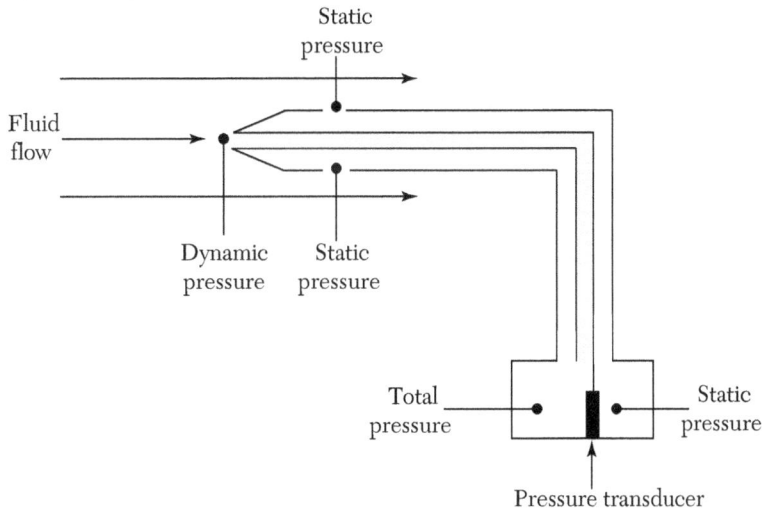

The tube points into the direction of air flow, and pressure sensors are used to measure the difference between the total and static pressure. Since total pressure is equal to the sum of the static and dynamic pressures, the difference between these is just the dynamic pressure $\frac{1}{2}\rho v^2$. This value can then be used to calculate the speed of the fluid relative to the Pitot tube.

Pitot tubes are used to measure the speeds of aircrafts and boats.

$$v = \sqrt{\frac{2\left(p_{tot} - p_{stat}\right)}{\rho}}$$

6.6   Exercises

1. Scuba divers estimate that the excess pressure they experience when they dive increases by an amount equal to the atmospheric pressure for every additional 10 m of depth.
   (a) Show that this is approximately correct.
       Density of seawater = 1030 kgm^{-3}. Atmospheric pressure at sea level = 101 kPa.
   (b) Estimate the total pressure at the bottom of the Mariana Trench in the Pacific Ocean (about 11 km below sea level).
   (c) Seawater is slightly compressible—how will this affect your answer to (b)?

2. A vertical dam wall contains water of density ρ to a depth h. The horizontal length of the dam wall is l.
   (a) Write down an expression for the excess pressure at depth x below the surface of the water.
   (b) Show that the total horizontal force acting on the wall of the dam from the water is given by the expression $F = \frac{1}{2}\rho glh^2$.
   (c) By considering equilibrium of moments show that the line of action of this force is $1/3h$ above the base of the dam.

3. The laminar flow of a viscous fluid through a narrow capillary tube depends only on the radius a, the pressure gradient along the tube (p/l), and the viscosity of the fluid, η.
   (a) Use the method of dimensions to show that the volume flow rate Q is given by an expression of the form:

$$Q = \text{constant}\,\frac{a^4}{\eta}\left(\frac{p}{l}\right)$$

   (b) Explain why the method of dimensions cannot be used to determine the value of the constant in the expression above.
   (c) Explain why this formula is likely to break down for high flow rates or larger diameter pipes.

4. (a) Show that the flow of air around a car is likely to be turbulent. You will need to estimate the relevant quantities.
   (b) A formula used for turbulent drag forces is:

$$F = \frac{1}{2}\,C_D\rho Av^2$$

   where C_D is the "drag coefficient."

(i) Suggest ways in which the drag coefficient might be reduced.

(ii) The drag coefficient for a car is 0.60 and a frontal area of 3.4 m². Calculate the aerodynamic drag force on this car when it is traveling at 25 ms⁻¹ through the air. Air has a density of 1.2 kgm⁻³.

(iii) Suggest, with reasons, how the power required to drive a sports car at 50 ms⁻¹ compares to the power required to drive the same car at 25 ms⁻¹.

(iv) The car above has a maximum power output of 400 kW. Use this to estimate its maximum possible speed.

(v) Suggest a reason why your answer to (iv) is an overestimate.

5. (a) Estimate the volume of your own body (hint: your density is similar to that of water, about 1000 kgm⁻³).

(b) Estimate your weight.

(c) Estimate the buoyancy force on your body from the atmosphere.

(d) Discuss whether or not bathroom scales display your actual mass.

6. A steel ball bearing has a diameter of 1.2 mm. It is released from just below the surface of glycerol inside a wide measuring cylinder. The density of steel is 7700 kgm⁻³ and that of glycerol is 1260 kgm⁻³.

(a) Calculate the weight of the ball bearing.

(b) Calculate the buoyancy force on the ball bearing when it is submerged in glycerol.

(c) Explain why the ball bearing accelerates when it is released but soon reaches a terminal velocity. Your answer should include a free body diagram for the falling ball bearing.

(d) The terminal velocity of the ball bearing is 3.6 mms⁻¹. Use this to calculate a value for the viscosity of glycerol and state any assumptions you use.

(e) In a famous experiment to measure the charge on oil droplets, Millikan used Stoke's law to estimate the viscous drag on tiny oil droplets (with radii of the order of 2 μm) falling through air. He discovered that Stoke's law overestimated the force on the droplets; they actually fell faster than the equation predicted. Suggest a reason for this.

7. A raindrop of radius 0.50 mm is falling through the atmosphere.

(a) Explain why the buoyancy force from the air can be neglected when using Stoke's law to calculate the terminal velocity of the raindrop.

(b) The viscosity of air is 2.0×10^{-5} Pas. Use Stoke's law to calculate the terminal velocity of the raindrop.

    (c) Use your answer to (b) to calculate the Reynold's number for the falling droplet. Comment on your answer.

8. The diagram below shows two pipes, each of length l, but with diameters d and $d/2$ respectively. The volume flow rate into X is Q and the flow is laminar. The pressure at X is p_X and the pressure at Z is p_z. The viscosity of the fluid is η. Find an expression for the pressure at Y assuming the fluid is incompressible.

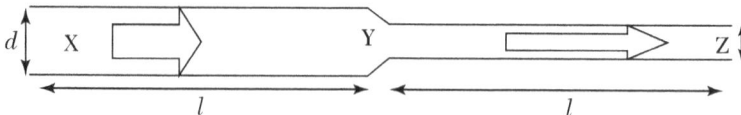

9. (a) Explain how it is possible to drink water from a glass through a straw.

    (b) Discuss whether there is a limit to the length of straw that can be used to drink water. The density of water is 1000 kgm^{-3}.

10. Show that the S.I. base units for viscosity are kgms^{-1}.

11. The pressure difference measured by a Pitot tube on an aircraft's wing is 20 kPa. What is the aircraft's air speed?

12. When an inflated balloon is connected to one side of a water manometer, the height difference between the manometer arms is 14.0 cm. The atmospheric pressure is 102 kPa.

    (a) What is the excess pressure inside the balloon?

    (b) What is the total pressure inside the balloon?

    (c) Suggest one advantage and one disadvantage of using mercury instead of water in a manometer.

13. In a famous experiment, the French physicist Pascal placed one mercury barometer at the base of a mountain and carried a second one to the top of the mountain. Both barometers had mercury columns of equal height when they were together at the base of the mountain. However, Pascal noticed that the mercury column on the barometer he carried with him fell gradually as he climbed the mountain. Explain this effect as carefully as you can.

14. The circulation of blood in the body can be considered as a continuous circuit. Blood is pumped from the heart into the aorta, splits into the arteries, splits again into the capillaries, and then returns to the heart via the veins and finally the vena cava. This can be represented as an electric circuit consisting of series and parallel resistors:

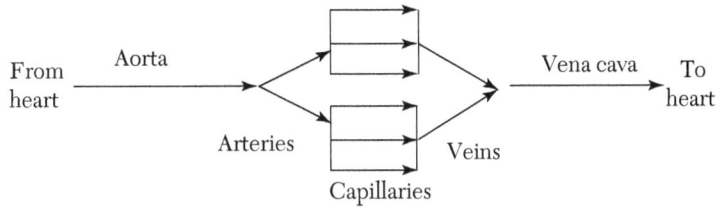

The table below gives the total area of each type of blood vessel along with the average flow speed and volume flow rate.

(a) Complete the table.

	Area (cm^2)	Speed (cms^{-1})	Volume flow rate (cm^{3}s^{-1})
Aorta	3.0	30	90
Arteries	100		
Capillaries	900	0.10	
Veins		0.45	
Vena cava	18		

(b) Blood has a viscosity between 0.003 and 0.004 Pas. Discuss whether blood flow in the human circulation is likely to be laminar or turbulent.

CHAPTER 7

MECHANICAL PROPERTIES

7.1   Density

Density is a property of each material and is independent of the amount of that material. This is in contrast to mass, which depends on the amount of material present. Density is defined as:

$$\text{Density} = \frac{\text{Mass}}{\text{Volume}}$$

The S.I. unit of density is kgm^{-3}, but gcm^{-3} is also in common use. The relation between these is:

$1000 \ kgm^{-3} = 1 \ gcm^{-3}$

$1 \ gcm^{-3} = 0.001 \ kgm^{-3}$

The densities of some common materials are listed below.

	Density/kgm^{-3}	Density/gcm^{-3}
Air (sea level, 15°C)	1.225	0.001225
Water	1000	1.000
Wood	160 (balsa)–1300 (ebony)	0.16–1.3
Concrete	~2400	~ 2.4
Aluminum	2700	2.7
Steel	7750–8050	7.75–8.05
Copper	8960	8.96
Lead	11340	11.34
Mercury	13560	13.56
Uranium	19100	19.10

	Density/kgm^{-3}	Density/gcm^{-3}
Gold	19320	19.32
Osmium (densest naturally occurring element)	22590	22.59

The density of a solid is of the same order of magnitude as the density of an atom because atoms are closely packed together inside a solid material. However, atoms themselves are mainly empty space and the diameter of an atomic nucleus is approximately 20,000 times smaller than the diameter of an atom. This means that the density of nuclear material is around $20,000^3$ (6×10^{12}) times greater than the density of ordinary matter. Typical nuclear densities are $>10^{17}$ kgm^{-3}.

The cores of collapsed stars also have high densities. White dwarf stars have densities of about 10^{10} kgm^3, but neutron stars have densities comparable to that of an atomic nucleus (10^{17} kgm^{-3}) since they are effectively closely packed nucleons (neutrons). On the other hand, "empty space" is not quite empty, having around 1 hydrogen atom per cubic centimeter and a density of the order of 10^{-33} kgm^{-3}.

7.2 Inter-Atomic Forces

A solid material holds itself together because the individual particles from which it is made (atoms or molecules) form bonds. If the particles are pulled further apart they experience an attraction; and if they are pushed closer together they experience a repulsion. While they will have some thermal kinetic energy that causes them to vibrate, they maintain, on average, a constant equilibrium separation. The interatomic forces are electrostatic in

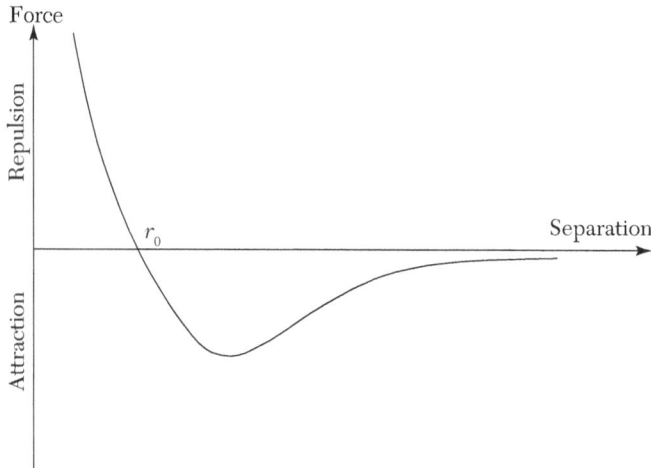

origin and vary with distance as shown below (a positive force represents repulsion and a negative force represents attraction).

The distance r_0 is equal to the equilibrium separation of the particles. For small displacements either side of this position, the graph is approximately linear in many materials, particularly metals. This has two consequences:

- When a force is applied to the material its extension (or compression) will be directly proportional to the force. In other words it will obey Hooke's law in this linear region.

- When a particle is displaced from its equilibrium position it will experience a restoring force directly proportional to its displacement, so that oscillations about the equilibrium will be simple harmonic (see Chapter 11).

The work that must be done to separate two particles from their equilibrium separation is equal to the area between the negative part of the graph and infinity. This is also equal to the energy released when the bond between the particles is formed. Since work must be done to push the particles closer together or to separate them, the potential energy is a minimum when their separation is r_0. They are in a bound state and each particle is in a potential well.

The graph below shows how the potential energy varies with particle separation.

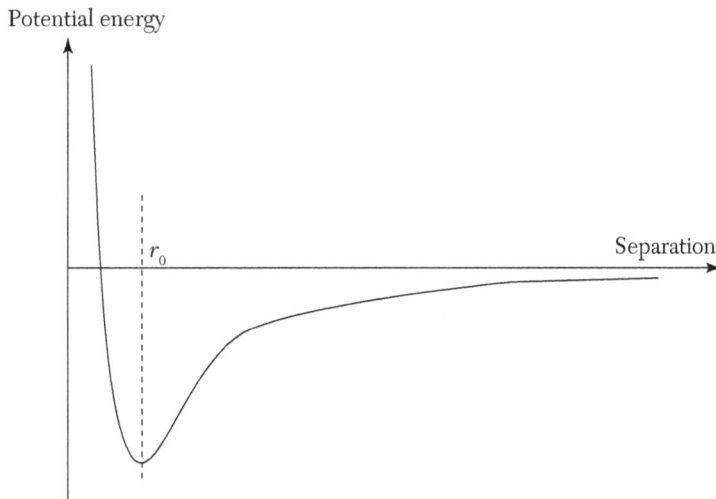

7.3   Stretching Springs

An ideal spring obeys Hooke's law. This states that the extension of the spring is directly proportional to its tension (or to the load it supports). This is a direct consequence of the way the bonds between atoms inside the steel behave. If these bonds obey Hooke's law then so will the material and so will the spring. The simple example of a mass on a spring can serve as a model for atomic and molecular bonds and many other systems in physics and engineering.

Some useful terms

- **Tension**: the force exerted by a spring when it is stretched. If it supports a load in equilibrium this will be equal to the weight of the load.

- **Extension**: the difference between the stretched and unstretched lengths of the spring.

7.3.1   Spring Constant

When increasing loads are suspended from a steel spring the extension increases. A graph of force against extension looks like this:

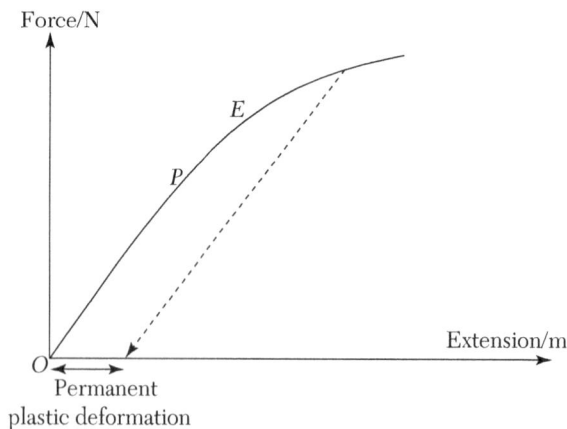

Section OP is (almost) a straight line through the origin, so in this region the extension (x) is directly proportional to the force applied to the spring (or the tension in the spring) (F).

$$F \propto x \quad \text{or} \quad F = kx$$

k is the "**spring constant**" with S.I. unit Nm^{-1}.

The spring constant is a measure of the "stiffness" of the spring.

Point P is the "**limit of proportionality**,"and beyond P the graph becomes non-linear.

In the region OP the spring will return to its original length when the force is removed. This is called **elastic** behavior. Beyond E this is no longer the case, and when the force is removed, the spring does not return to its original length but retains an extension. This is called **plastic** behavior. The dotted line on the graph shows how a spring contracts when a force beyond E is removed.

Point E is the "**elastic limit**" for the spring. Beyond E the spring undergoes permanent plastic deformation.

This graph does not include the fracture point for the spring because once it unravels the applied forces then stretch a steel wire and this will require a large increase in force for a relatively small increase in extension.

7.3.2 Springs in Series and in Parallel

Assume the springs in the examples below have negligible weight and obey Hooke's law with spring constant k. When any one of these springs is extended by an amount e, the tension in the spring is F.

Series combinations

When n springs are connected in series and the system is stretched, the tension in each spring must be the same and equal to F. This means the extension of each spring will be the same as the extension of an individual spring under the same load. The total extension of the system of n springs is therefore ne.

Using Hooke's law for the system of n springs in series gives:
$$k_{\text{SERIES}} = F/ne = k/n$$
Using similar reasoning, a system of different springs with spring constants k_1, k_2, k_3, etc., connected in series has an overall spring constant given by:
$$\frac{1}{K_{\text{series}}} = \frac{1}{k_1} + \frac{1}{k_2} + \frac{1}{k_3} + \cdots$$
Connecting springs in series reduces the stiffness of the system.

Parallel combinations

When n springs are connected in parallel and the system is stretched by a force F, the tension in each spring must be the same and equal to F/n. The total extension of each spring (and the system) will be e/n.

Using Hooke's law for the system of n springs in parallel gives:

$$k_{parallel} = F/(e/n) = nk$$

The spring constant of a system consisting of several springs in parallel is the sum of spring constants of the individual springs. Connecting springs in parallel increases the stiffness of the system.

7.3.3 Elastic Potential Energy (Strain Energy)

Work must be done to stretch a spring. If there are no energy losses in the stretching process, then this work transfers energy to elastic potential energy in the spring that can be released when the spring recompresses. The work done is equal to the area under a graph of force against extension for the spring.

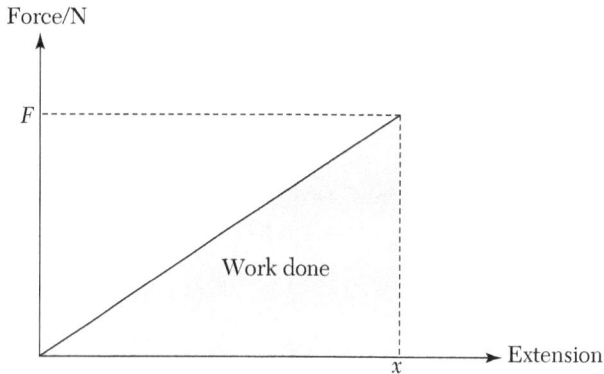

If the spring obeys Hooke's law then the work done to stretch it to an extension x is equal to the shaded area in the graph above:

$$\text{Area} = \tfrac{1}{2}\, Fx$$

or, using Hooke's law,

$$\text{Work done} = \tfrac{1}{2}\, kx^2$$

If we assume that no energy is lost as the spring is stretched, the elastic potential energy in a stretched spring is given by:

$$EPE = \tfrac{1}{2}\, kx^2$$

In general, the work done to stretch something is calculated from the integral:

$$W = \int F(x)dx$$

If the spring obeys Hooke's law this is simply $W = \int kxdx = \dfrac{1}{2}kx^2$. giving the same result as before.

7.4   Stress and Strain

When we are discussing material properties, stress and strain are often preferable to force and extension because the relationship between stress and strain is independent of the dimensions of the sample under test. Stress and strain relate directly to the *type* of material rather than the particular piece of material used in the test.

Some definitions

Tensile stress: σ is the force per unit area when the force is applied along the axis of the sample as shown below.

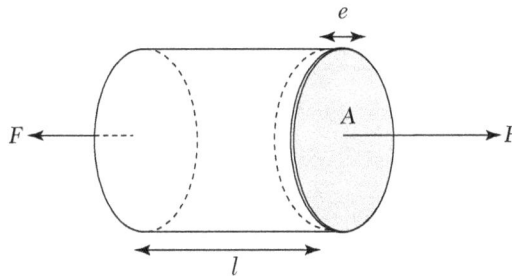

$$\text{Stress}(\text{NM}^{-2}) = \frac{\text{Force (N)}}{\text{Cross-sectional (m}^2)}$$

$$\sigma = \frac{F}{A}$$

The S.I. unit of stress is Nm^{-2} or Pa.

Tensile strain: ϵ is the extension per unit length under a tensile stress.

$$\text{Strain} = \frac{\text{Extension}}{\text{Original length}}$$

$$\epsilon = \frac{e}{l_0}$$

This is a ratio of two lengths, so it is dimensionless.

7.4.1   Young's Modulus

For many materials, particularly metals, tensile strain is directly proportional to tensile stress up to some limiting stress. When this is the case,

$$\text{Young's modulus (Pa)} = \frac{\text{Stress(Pa)}}{\text{Strain}}$$

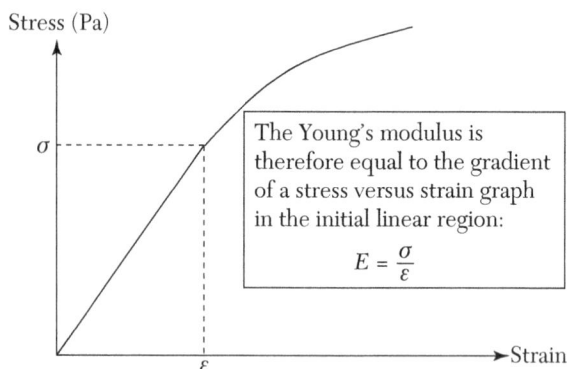

The Young's modulus is a measure of the **stiffness** of a material. A stiff material has small strain for large stress.

Some typical values for the Young's modulus are:

Material	Young's modulus/GPa
Rubber	0.01–0.10
Nylon	2–4
Wood	10
Bone	14
Concrete	30
Copper	120
Steel	280
Diamond	1150

7.4.2 Experimental Measurement of Young's Modulus for a Metal Wire

The Young's modulus for a metal wire of diameter d and original length l_0 can be determined by loading it with masses and plotting a graph of mass m against extension e:

$$E = \frac{(F/A)}{\left(e/l_0\right)} = \frac{Fl_0}{eA} = \frac{4Fl_0}{e\pi d^2} = \frac{4mgl_0}{e\pi d^2}$$

$$m = \left(\frac{\pi d^2 E}{4gl_0}\right)e$$

A graph of m on the y-axis against e on the x-axis has a gradient $\left(\frac{\pi d^2 E}{4gl_0}\right)$. The Young's modulus is given by:

$$E = \text{gradient} \times \left(\frac{4gl_0}{\pi d^2}\right)$$

A suitable experimental arrangement to determine the Young's modulus of a test wire follows.

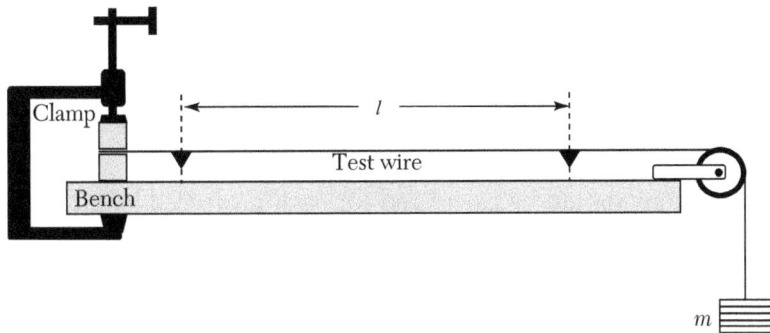

The diameter of the wire must be measured using a micrometer screw gauge. This should be done at least three times in different positions and then an average should be calculated. The images below show how a reading to 0.01 mm is obtained.

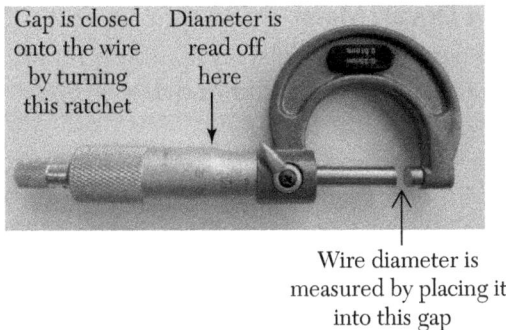

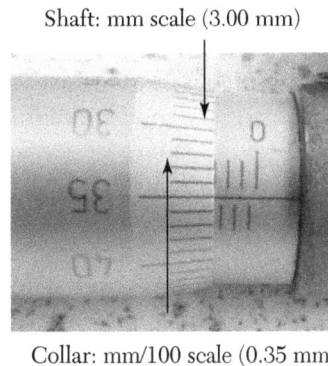

Shaft: mm scale (3.00 mm)

Gap is closed onto the wire by turning this ratchet

Diameter is read off here

Wire diameter is measured by placing it into this gap

Collar: mm/100 scale (0.35 mm)

The reading is taken from the point where the lines on the shaft and collar are aligned. The last reading in millimeters from the shaft is added to the reading from the collar. In the example below the gap is $3.00 + 0.35$ mm = 3.35 mm. Any zero error for the micrometer must be subtracted from this. (You can check for a zero error by closing the micrometer gap completely and checking the reading; it should be 0.00 mm.)

The wire must then be clamped securely at one end.

Two light markers are attached to the wire a distance l_0 apart. This distance should be measured with an unkinked, unloaded but taut wire. Using a larger value of l_0 will increase the extensions and give better results.

The wire is then loaded, adding one mass at a time and recording the mass added and the length of the wire in a suitable results table. This is repeated for at least seven different values of mass, and preferably many more. However, care must be taken not to exceed the elastic limit for the wire. If this does occur the graph will begin to curve, and the Young's modulus must only be calculated using the gradient of the straight part of the graph.

Extensions should be calculated for each measured length and added to the results table:

$$\text{extension, } e = l - l_0$$

Care must be taken throughout the experiment because there is always a danger that the wire could snap. Safety glasses should be worn!

Finally a graph of mass against extension is plotted and the Young's modulus can be determined from the gradient of this graph as shown above (Section 7.4.1).

7.4.3 Stress vs. Strain Graph for a Ductile Metal

A graph of stress against strain is a useful way to display material properties. The stress–strain curve for a metal like copper is shown below.

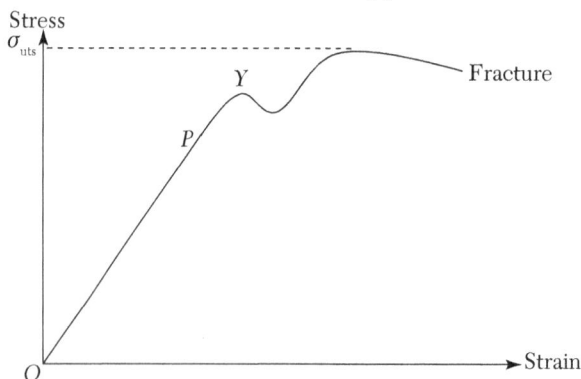

- Region OP: The material obeys Hooke's law in the initial linear region of the graph where stress is directly proportional to strain.

- The Young's modulus of the material is equal to the gradient of line OP.

- Up to Y the material behaves elastically. If the stress is removed it will return to its original dimensions.

▨ *Y* is the "yield point." Beyond this level of stress the material will deform plastically, and when the stress is removed there will be a residual deformation.

▨ The ultimate tensile strength(UTS) is at a stress σ_{uts}.

Some typical values of ultimate tensile strengths are:

Material	UTS/MPa
Carbon fiber	5650
Spider silk	1400
Stainless steel	860
Copper	220
Nylon	75

The dip in the curve is a result of the change in diameter of the sample. As it begins to yield, its diameter becomes smaller, but in this graph the stress has been calculated using the original diameter of the sample, so while the calculated stress decreases the actual stress does not.

The area under a stress–strain graph represents energy per unit volume.

$$\text{area under the curve} = \int \sigma d\varepsilon = \int \left(\frac{F}{A}\right)\left(\frac{dx}{l}\right) \frac{1}{Al} \int F dx$$

(Here we have assumed that the cross-sectional area of the sample has remained constant during strain. This is approximately true for small strains.)

The larger the area up to fracture, the more energy the material absorbs before fracture and the *tougher* the material is said to be.

7.4.4 Rubber Hysteresis

Here is a typical stress–strain graph for rubber showing both loading and unloading. Rubber consists of long-chain hydrocarbon molecules. The bonds within the chain are stiff covalent bonds, but single carbon–carbon bonds allow for rotation, so the long chain can be curled up or stretched out. There are also many cross-links between different chains.

The stress for any particular strain is greater during loading than unloading. This implies that the work done to stretch the rubber is greater than the work done by the rubber as it re-contracts. The area in the loop between the two curves is equal to the work that is transferred to heat internally as the rubber is stretched. You can easily verify that heat is generated by taking an elastic band, rapidly stretching it and immediately touching it to your lip. This curve for loading and unloading is called a

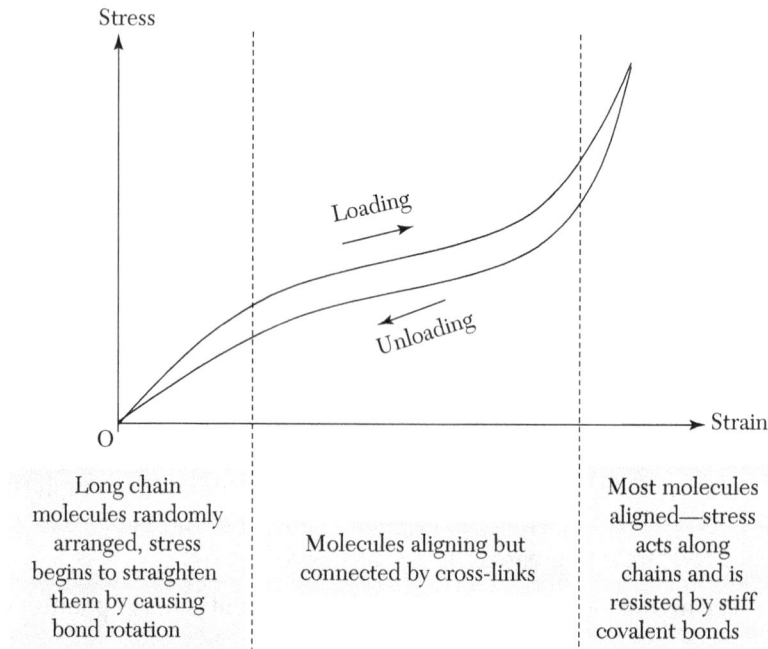

Hysteresis curve. It is interesting to note that if you heat a stretched piece of rubber it will tend to contract as the molecules adopt a more random arrangement.

7.5 Material Terminology

Here is a summary of terms used by materials scientists to describe material properties:

Strong—has a large ultimate breaking stress

Stiff—needs a large stress for a small strain (high Young's modulus)

Elastic—sample returns to its original dimensions when stresses are removed

Plastic—undergoes a permanent deformation when stress is removed

Yield stress—stress needed to cause the onset of plastic deformation

Tough—undergoes significant plastic deformation before breaking, resists crack formation and propagation, absorbs a lot of energy before breaking

Ductile—able to be drawn out into wires (linked to plasticity)

Malleable—able to be beaten into thin sheets (linked to plasticity)

Hard—resists scratching and indentation

Brittle—breaks catastrophically as cracks travel through the material;almost no plastic deformation prior to fracture

Creep—gradual continued strain under a constant stress

7.6   Material Types

Materials fall into a range of different types, which can be identified by their macroscopic properties and explained by their microscopic structures.

Crystalline materials have a microscopic structure in which the particles are arranged in regular geometric patterns. These might form single large crystals (e.g., a diamond) or might consist of a large number of grains (polycrystalline), each of which has a regular structure but which are themselves arranged randomly (many metals are like this although it is possible to grow single metallic crystals). The grain size in a polycrystalline material affects its mechanical properties, and this can be changed by heat treatment. For example, in the process of annealing, a metal is heated up above its recrystallization temperature and then allowed to cool slowly. This allows individual crystals to grow larger and makes the metal less hard, more ductile, and easier to work.

Crystalline structures are rarely perfect, and most contain flaws such as missing atoms (vacancies), atoms of a different material (impurities), and dislocations (where planes of atoms do not join correctly). These have a large effect on properties such as strength and stiffness.

Polymers consist of long-chain hydrocarbon molecules that can be arranged randomly (amorphous polymer) or in a semi-regular or regular (crystalline) way. Their mechanical properties depend on how the polymer chains interact and move past one another. The individual chains form cross-links, and these affect the stiffness and strength of the material. Rubber and polythene are both examples of polymeric materials. When a stress is applied to rubber the molecules initially begin to align, explaining rubber's ability to undergo very large strains. However, if the stress is removed, the cross-links pull the molecules back to their original positions and the rubber behaves elastically. At larger strains most of the molecules have aligned and the rubber becomes very stiff before it finally breaks. Polythene can also undergo a large strain as its molecules align, but in this case the cross-links are unable to pull the molecules back to their original positions, so the polythene undergoes a plastic deformation.

Ceramics are solid non-metallic materials such as brick, tile, pottery, and china that are formed by high-temperature firing. They usually consist of tiny ionic crystals bound together by amorphous glassy regions that formed during firing. Ceramics are usually stiff, hard, and strong and have high melting points and good chemical resistance, but they can be brittle.

Glasses are closely related to ceramics but are characterized by being completely amorphous and result from a rapidly cooled melt. Glasses are distinguished from ceramics by their microscopic structure. For example, silica glass and quartz have identical composition consisting of SiO_4 units, but in the glass they are arranged randomly, while in the ceramic they are arranged regularly.

Composite materials are combinations of two or more different materials designed to take advantage of the desirable properties of each individual component. Examples of composites are fiber glass, concrete, steel, reinforced concrete, etc. Many composites consist of a matrix and a reinforcement. Concrete is strong in compression, but weak in tension and brittle, whereas steel is strong in tension and can prevent crack formation in the concrete, providing a versatile and economical building material.

7.7 Exercises

1. A rectangular wooden block has sides of length 5.0 cm, 8.0 cm, and 12 cm. Its mass is 960 g.
   (a) Calculate its density in kgm^{-3}.
   (b) The absolute uncertainty in each measurement above (i.e., of side lengths and mass) is ±4%. Calculate the absolute and fractional uncertainty in the density.

2. A spring is extended 0.18 m by a force of 30 N.
   (a) Calculate the spring constant of the spring.
   (b) Calculate the work done to stretch the spring.
   (c) A load of 20 kg is supported by 10 of these springs connected in parallel with one another. Calculate the extension of the system under this load.

3. The data below was collected from an experiment to stretch a steel spring. The unstretched length of the spring was 10.0 cm.

Extension/cm	Force/N
0	0
2.8	1.0

Extension/cm	Force/N
6.2	2.0
9.6	3.0
13.2	4.0
16.5	5.0
20.1	6.0
23.2	7.0
26.6	8.0
30.2	9.0
33.0	10
36.3	11
40.2	12
44.4	13
49.4	14
57.1	15
68.5	16
96.5	17
113.2	18

(a) Plot a graph of force (*y*-axis) against extension (*x*-axis).
(b) State the limit of proportionality for this spring.
(c) Calculate the spring constant of the spring.
(d) What was the total length of the spring when it was pulled by a force of 7.0 N?
(e) Calculate the extension of the spring if it supported a load of 850 g.
(f) For forces up to about 10 N the graph is a straight line through the origin. Describe the relationship between force and extension over this range. What law is the spring obeying?
(g) If two springs like the one above were connected in series, how much would they extend when stretched by a force of 14 N?
(h) When the force of 18 N was removed, the spring did not return to its original length. Explain this.
(i) Does the spring become stiffer or less stiff beyond the limit of proportionality? Explain your answer.

4. The steel cable used to moor a ship has a length of 12.0 m and a cross-sectional area of 8.2×10^{-4} m². The force in the cable is 15,000 N.

(a) Calculate the extension of the cable.
(b) Calculate the strain energy (elastic P.E.) stored in the cable.
Density of steel = 8050 kgm^{-3}, Young's modulus of steel = 200 GPa

5. When a rubber band is stretched and released rapidly several times in quick succession, it becomes hot. Use a stress–strain graph for the loading and unloading of rubber to explain this.

6. Here are four stress–strain graphs drawn to the same scale.

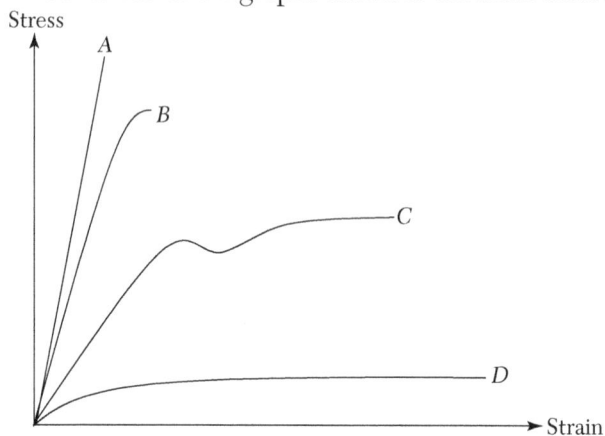

Use the correct terminology to describe the mechanical properties of each material and to compare them.

CHAPTER 8

THERMAL PHYSICS

8.0   Introduction

Thermal energy (often simply referred to as heat energy or heat) is the energy an object has as a result of the random thermal motions of its particles. This is not the same thing as temperature. For example, the Atlantic Ocean has a lower temperature (perhaps 20°C) than a freshly made mug of coffee (perhaps 75°C) but has much more thermal energy because it contains many more particles. While the formal definition of temperature is quite complex and involves an understanding of the concept of entropy, a simple way to think about temperature is to relate it to the mean energy per particle. The mean energy of the water particles in a mug of hot coffee is greater than the mean energy of water molecules in the Atlantic Ocean.

8.1   Thermal Equilibrium

When two systems are placed in thermal contact, heat is able to be transferred between them. On the microscopic level, this is a dynamic process in which energy transfers take place continuously in both directions. However, on the macroscopic level, there might be a net flow in one direction. The macroscopic direction of net heat transfer is always from the system at a higher temperature to the system at a lower temperature, and this continues until the two systems are at the same temperature. This can be stated very simply:

▥ The direction of heat flow is from the system at higher temperature to the system at lower temperature (or even more simply—from hot to cold).

When there is no net heat transfer between systems in thermal contact, they are in thermal equilibrium and are at the same temperature.

The **Zeroth law of thermodynamics** states that, if two systems, A and B, are both in thermal equilibrium with a third system, C, then A and B are also in thermal equilibrium with each other.

This law allows us to use thermometers and to set up formal temperature scales.

8.2 Measuring Temperature

Temperature is measured using a thermometer. Thermometers are based on a physical property that changes with temperature, e.g., volume of a gas, length of a mercury column, resistance of a wire. To get a value from a thermometer it has to be **calibrated**. This means placing a scale on it so that temperatures can be read off. Since different physical properties vary in different ways when temperature changes, it is important to calibrate different thermometers against the same standard—the absolute or thermodynamic temperature scale. In practice, a constant volume gas thermometer is used as the standard to calibrate other types of thermometer. The table below lists some different kinds of thermometer.

Thermometer	Physical property	Useful range
Alcohol	Thermal expansion of a liquid	Depends on type of alcohol used but can measure down to −100°C and up to 78°C
Mercury-in-glass	Thermal expansion of a liquid	−37°C to +356°C
Thermocouple	Thermoelectric effect (See beck effect)	Depends on type but can be used from −270°C to 1250°C
Constant-volume gas thermometer	Thermal expansion of an ideal gas	Used to calibrate other types of thermometer, −200°C to 1600°C
Pyrometer	Spectrum of infra-red radiation emitted by hot object	700–3500°C

8.3 Temperature Scales

Temperature scales are defined using fixed points that are given fixed values. The Celsius scale (previously called the Centigrade scale) uses two fixed points:

▪ Lower fixed point: freezing point of water defined to be 0°C

▪ Upper fixed point: boiling point of water defined to be 100°C

To calibrate an unmarked mercury thermometer the thermometer is put into a water/ice mix and the lower position of the mercury column is marked. It is then put into equilibrium with steam from boiling water, and the upper position of the mercury column is marked. The distance between these two fixed points is then divided into 100 equal divisions or degrees. Temperatures below 0°C and above 100°C can also be marked onto the scale by using the same equal divisions.

A similar method can be used to calibrate other types of thermometer, but in all cases this assumes that the physical property that is changing (e.g., length of column or resistance of a wire) varies linearly with temperature in the range that is being marked. This is usually a reasonably accurate assumption but is not exactly correct, and so thermometers calibrated in this way would need to be corrected (using a constant-volume gas thermometer) if very precise temperature measurements are required.

The thermodynamic scale or Kelvin scale uses two different fixed points:

Lower fixed point: absolute zero of temperature defined to be 0 K (−273.15°C)

Upper fixed point: triple point of water defined to be 273.16 K (0.01°C)

The triple point of water is the unique temperature at which ice, water, and water vapor are in equilibrium. By choosing these values the intervals on the thermodynamic scale T are equal to those on the Celsius scale θ so that the conversion between the two scales is simply

$$T = \theta + 273.15, \text{ or } \theta = T - 273.15$$

	Celsius scale	Thermodynamic scale
Absolute zero	−273.15°C	0 K
Freezing point of water	0°C	273.15 K
Triple point of water	0.01°C	273.16 K
Typical room temperature	About 20°C	About 293 K
Human body core temperature	About 37°C	About 310 K
Boiling point of water	100°C	373.15 K

The best practical way to approximate the thermodynamic scale is to use a constant-volume gas thermometer. This consists of a bulb containing a fixed amount of gas connected to a mercury manometer to measure the gas pressure. When the temperature of the gas in the bulb changes, the pressure also changes. The ideal gas equation, $pV = nRT$, shows that pressure p is proportional to temperature T if volume V and amount of gas n are both constant (which they are). Pressure values can be calibrated to give temperatures.

8.4 Heat Transfer Mechanisms

Heat transfer occurs because of a temperature difference, and the direction of transfer is always from the hotter (higher temperature) to the cooler (lower temperature) system. There are three different mechanisms by which this can occur: conduction, convection, and radiation.

8.4.1 Conduction

Conduction involves the transfer of energy between particles as a result of collisions. It is the most important process for heat transfer inside a solid because particles in a solid remain in fixed positions and so cannot form convection currents. Conduction also occurs in liquids and gases, but here it is often less important than convection since particles in liquids and gases are able to move in convection currents transferring heat energy as they do so. The process of conduction is a dynamic one, but for macroscopic objects the transfer of energy from more energetic particles to less energetic particles has a higher probability than transfer in the opposite direction, so heat flows from higher to lower temperatures. Conduction cannot occur in a vacuum because there are no particles to conduct the heat.

The rate of heat flow $\frac{dQ}{dt}$ (in watts) through an insulated block of material depends on the type of material, its cross-sectional area A, and the temperature gradient between opposite sides of the block.

The temperature gradient in the block is:

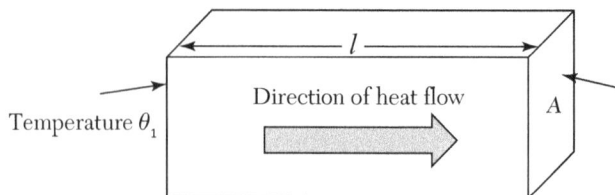

$$\frac{d\theta}{dx} = \frac{(\theta_2 - \theta_1)}{l}$$

where $\theta_1 > \theta_2$.

The rate of heat transfer is given by:

$$\frac{dQ}{dt} \propto -A\frac{d\theta}{dx}$$

The negative sign occurs because of the direction of heat transfer, from higher to lower temperature. A constant of proportionality k can be introduced. This depends on the material of the block. It is called the coefficient of thermal conductivity. The equation of thermal conductivity (Fourier's equation) is:

$$\frac{dQ}{dt} = -kA\frac{d\theta}{dx}$$

The unit of k is $Wm^{-1}K^{-1}$.

The table below lists typical thermal conductivities for common materials at around room temperature.

Material	Coefficient of thermal conductivity/$Wm^{-1}K^{-1}$
Air	0.024
Brickwork	0.6–1.0
Window glass	0.96
Ground or soil	0.33 (dry) to 1.4 (very moist)
Insulation materials	0.035–0.16
Expanded polystyrene	0.03
Dry sand	0.15–0.25
Water	0.58
Rock	2–7
Aluminum	205
Diamond	1000
Gold	310
Iron	80
Stainless steel	16
Timber	0.14

The high values for metals is because they contain large numbers of free electrons that rapidly transfer heat. Diamond has an especially high coefficient of thermal conductivity because its atoms are bonded very tightly together and this couples their motions, so that when one atom is disturbed it has an almost immediate effect on its neighbors and so transfer energy quickly.

8.4.2 Convection

Convection can only occur in fluids, i.e., liquids or gases. This is because convection involves bulk movement of hot matter from one part of the medium to another. Natural convection currents arise as a result of changes of density inside the fluid. In most fluids the material expands as its temperature increases. This increases its volume and decreases its density so that it is then displaced by cooler, denser material from above it. The overall effect is to set up a convection current with warmer material rising and cooler material falling. This is often the dominant mechanism for heat transfer within a fluid.

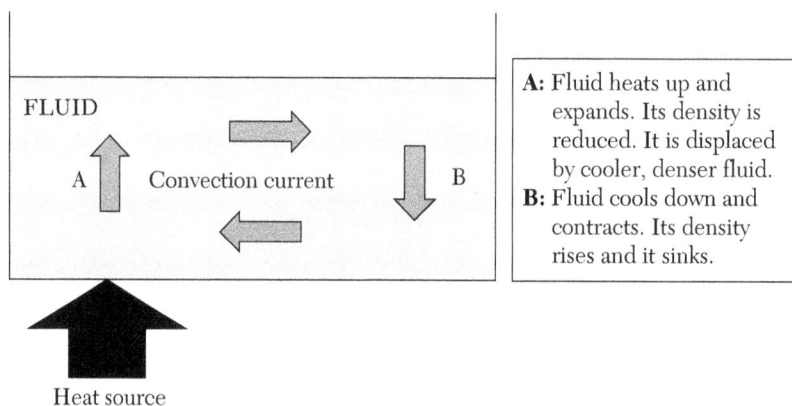

Forced convection is when an external blower or pump is used to force the fluid to move—for example, when you cool a cup of hot tea by blowing across its surface, you are using forced convection. Convection cannot occur in a vacuum because it depends on the movement of particles.

8.4.3 Radiation

All bodies emit and absorb a spectrum of electromagnetic radiation that depends on the nature of their surface and their temperature. For many of the hot objects we encounter in everyday life, the peak of this spectrum is in the infrared region, so thermal radiation is often referred to as infrared radiation although the actual spectrum of thermal radiation is continuous. Thermal radiation is part of the electromagnetic spectrum, so it can be transferred in a vacuum. While the absorption and emission of radiation is a complex subject, it is usually true that matt black surfaces are good absorbers and emitters, whereas light shiny surfaces are poor absorbers and emitters. Thermal radiation, like light, can be reflected from a silvered surface, e.g., on the inside of a thermos flask.

A thermal imaging camera can be used to measure infrared radiation emitted from different objects. The images below show the camera itself and thermal images of a man's face and hand. The image is color-coded (greyscale here) to show different temperatures, and these can be read off from the scale at the bottom of the screen.

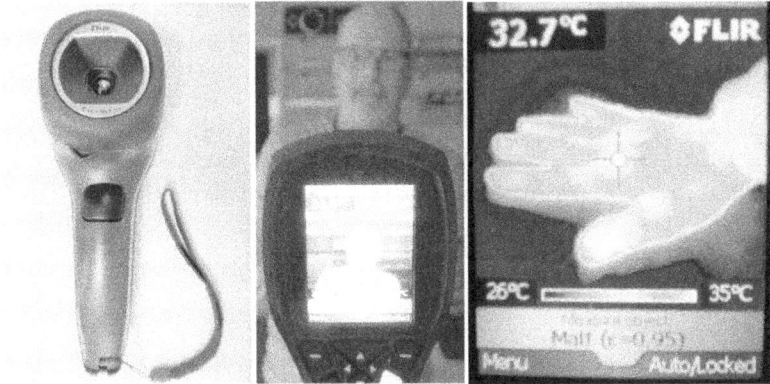

8.5 Black Body Radiation

An ideal radiator and absorber is called a black body. A black body will absorb all the radiation that falls onto it, but in order to stay in thermal equilibrium, it must also emit radiation at the same rate.

The spectrum of radiation it emits at a particular temperature has a characteristic shape that was first measured in the mid-nineteenth century but which could not be explained until Max Planck's quantum theory in 1900 (see Section 27.1.1).

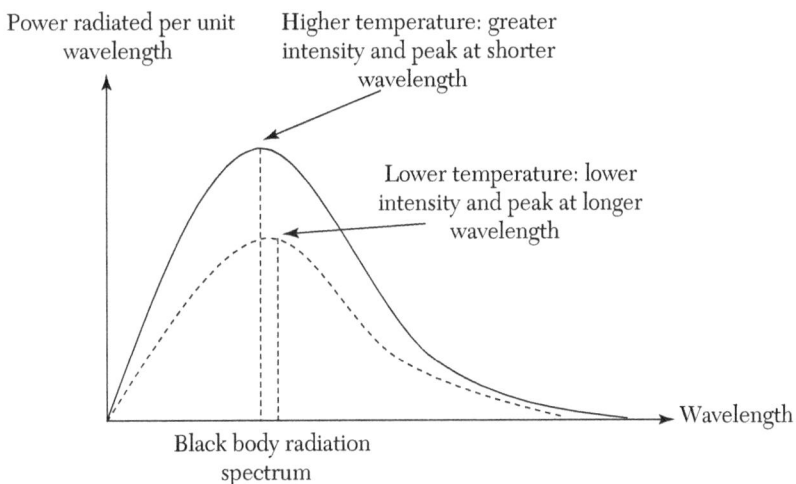

Black body radiation spectrum

As the temperature of a black body increases the total flux of radiation from its surface increases and the wavelength corresponding to the peak of the spectrum becomes shorter. This is apparent when heating a metal. While warm, it emits in the infrared, which we can detect by holding our hand nearby. As its temperature rises, it emits more radiation and begins to glow in the visible part of the spectrum; it becomes "red hot." At even higher temperatures it also emits shorter visible wavelengths and becomes "white hot."

While nineteenth-century physicists were unable to derive the shape of the spectrum from first principles, they did identify two important empirical laws that are very useful when considering thermal radiation. These laws were later derived from Planck's theory.

Wien's displacement law

This law states that the product of the wavelength corresponding to the peak of the spectrum λ_p and the absolute temperature (in kelvin) is a constant.

$$\lambda T = \text{constant} = 2.910^{-3}\,\text{mK}$$

Stefan-Boltzmann law

This law states that the total power P radiated from a black body is proportional to the fourth power of the absolute temperature (in kelvin) T.

$$P = e\sigma AT^4$$

where e is the emissivity ($e = 1$ for an ideal blackbody radiator), σ is Stefan's constant ($\sigma = 5.6703 \times 10^{-8}$ Wm^{-2}K^{-4}), and A is the surface area of the radiator.

8.6 Heat Capacities

8.6.1 Specific Heat Capacity

The specific heat capacity c of a material is the energy needed to raise the temperature of 1 kg of that material by 1°C.

$$c = \frac{\Delta E}{m\Delta\theta}$$

where ΔE is the energy supplied, m is the mass of the sample, and $\Delta\theta$ is the temperature change.

The S.I. unit of specific heat capacity is Jkg^{-1}K^{-1}.

The table below gives values of the specific heat capacity of a range of common substances.

Substance	Specific heat capacity/Jkg^{-1}K^{-1}
Dry air at sea level	1460
Water (pure at 20°C)	4180
Ice (0°C)	2093
Copper	385
Aluminum	897
Iron	449
Mercury	140
Concrete	880

Water has a particularly high specific heat capacity. This makes it an ideal coolant because it can absorb a large amount of heat for a relatively small increase in temperature.

Some terminology

◼ Heat capacity is the energy needed to increase the temperature of a particular sample of a substance by 1°C (or 1K).

◼ Specific heat capacity is the energy needed to increase the temperature of 1 kg of a substance by 1°C (or 1K), i.e., this is the heat capacity per unit mass.

◼ Molar heat capacity is the energy needed to increase the temperature of 1 mole of a substance by 1°C (or 1K), i.e. the heat capacity per unit amount (the S.I. unit of amount is the mole).

8.6.2 Molar Heat Capacities of Gases

The energy needed to raise the temperature of a mole of gas depends on the conditions under which the gas is contained. If the gas is allowed to expand while it is heated, it will do work on its surroundings and so the heat capacity will be greater than if the gas volume is kept constant. For this reason there are two significant heat capacities for a gas:

◼ Heat capacity at constant volume, c_V

◼ Heat capacity at constant pressure, c_P

For the reasons explained above, c_P is greater than c_V.

8.6.3 Measuring Specific Heat Capacity

When measuring the specific heat capacity of a substance it is important to ensure that all of the heat supplied is absorbed by the substance and not transferred to other objects, e.g., the heater itself, the container, or the surroundings. The simple method should reduce most of these losses.

The substance is heated electrically at a constant rate $P = IV$ (where I is the current in the heater circuit and V is the potential difference across the heating element). This power must be kept constant by adjusting the variable resistor if necessary. The initial temperature of the substance under test must be recorded and then regular temperature measurements (e.g., every 30s) are taken after the heater is switched on.

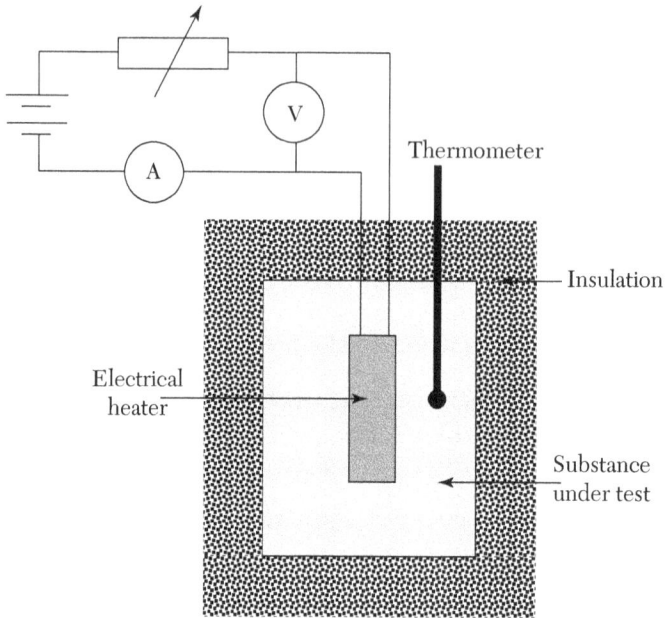

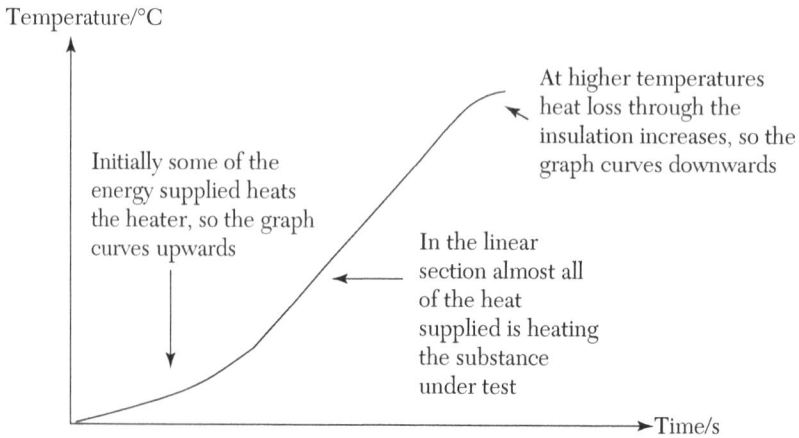

From the equation for specific heat capacity we he $\Delta E = mc\Delta\,\theta$, so we can write:

$$P = \frac{\Delta E}{\Delta t} = mc\frac{\Delta \theta}{\Delta t}$$

Rearranging this equation gives:

$$c = \frac{P}{m\frac{\Delta \theta}{\Delta t}}$$

$\frac{\Delta \theta}{\Delta t}$ is equal to the gradient of the linear part of the graph above.

m can be measured using a top pan balance.

P is calculated from the ammeter and voltmeter readings $(P = IV)$.

8.7 Specific Latent Heat

The specific latent heat L of a substance is equal to the energy that must be supplied to change the state of 1 kg of the substance at its melting or boiling point with no increase in temperature. For example, the specific latent heat of fusion for water is the energy needed to change 1 kg of ice at 0°C to 1 kg of water at 1°C. There are two significant latent heats:

▪ Latent heat of fusion L_f: energy required to change 1 kg of solid to 1 kg of liquid at its melting point (with no change of temperature). This is also equal to the energy released when 1 kg of liquid freezes at its freezing point.

▪ Latent heat of vaporization L_v: energy required to change 1 kg of liquid to 1 kg of gas at its boiling point (with no change of temperature). This is also equal to the energy released when 1 kg of gas condenses to a liquid at its boiling point.

For a change of state,

$$E = mL$$

where E is energy supplied or released, m is the mass of substance that changes state at its freezing or boiling point, and L is the relevant specific latent heat.

The S.I. unit of latent heat is Jkg^{-1}.

The table below gives the values of the specific latent heats of a range of common substances.

Substance	Latent heat of fusion/Jkg^{-1}	Latent heat of vaporization/Jkg^{-1}
Water	3.34×10^5 (0°C)	2.26×10^6 (100°C)
Ethanol	1.09×10^5 ($^-$109°C)	8.38×10^5 (78°C)
Copper	2.07×10^5 (1084°C)	
Lead	2.24×10^4 (327.5°C)	
Tungsten	1.93×10^5 (3400°C)	
Mercury	1.13×10^4 ($^-$ 39°C)	2.94×10^5

8.8 Exercises

1. A brick hut with a flat wooden roof is heated so that the inside temperature is 22°C when the outside temperature is 12°C. The dimensions of the hut's base are 4.5 × 3.5 m and the walls are 2.6 m high and 10 cm thick. There is a glass window of area 1.5 m^2 in one of the walls. The thickness of the glass is 6.0 mm and the thickness of the wooden roof is 3.0 cm.

   (a) Calculate the minimum power of the heater needed to maintain the temperature at 22°C when the outside temperature is 12°C.

   (b) Explain why, in practice, more powerful heating will be needed.

   Thermal conductivities: brick, 0.80 Wm^{-1}K^{-1};wood,0.16 Wm^{-1}K^{-1}; glass, 0.96 Wm^{-1}K^{-1}

2. An 800 g block of copper at 76.0°C is immersed in 2000 cm^3 of water, which is initially at 18.0°C. The water is in a thermally insulated container.

   Calculate the final equilibrium temperature of water and copper assuming there is no heat loss to the surroundings.
   Specific heat capacities: copper, 385 Jkg^{-1}K^{-1};water, 4180 Jkg^{-1}K^{-1}

3. The table below gives the specific heat capacities and molar masses of four metals:

Metal	Specific heat capacity/ Jkg^{-1}K^{-1}	Molar mass/g
Copper	385	63.5
Aluminum	897	27.0
Iron	449	55.8
Mercury	140	200.6

   Calculate the molar heat capacity of each metal. Comment on your results.

4. Calculate the energy needed to change 2.0 liters of water at 20°C into steam at 120°C.

Specific heat capacity of water = 4200 Jkg^{-1} °C^{-1}

Specific heat capacity of steam = 2300 Jkg^{-1} °C^{-1}

Specific latent heat of water = 2.26 MJkg^{-1}

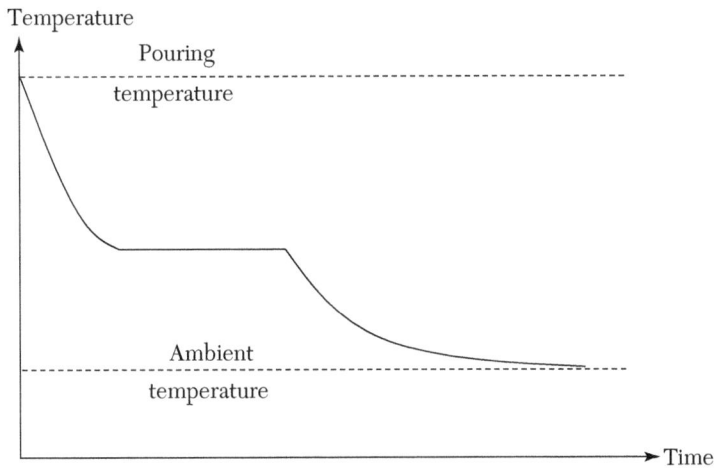

5. When molten metal is poured into a cast it cools down and solidifies. The graph below shows a typical cooling curve for the metal. Explain the shape of the graph and the significance of the three regions.

CHAPTER 9

GASES

9.1 Gas Laws

9.1.0 Introduction

The state of a fixed amount of gas is defined by three parameters: its volume, pressure, and temperature. These are all dependent on one another so that when a gas is compressed its volume and temperature might change. The gas laws are a set of macroscopic empirical laws that describe these relationships and which will be explained later when we consider the microscopic kinetic theory model of a gas (see Section 9.3.2). Given that there are three parameters, there are also three gas laws, each of which holds when one of the parameters is constant.

Boyle's law: dependence of pressure on volume (constant temperature)

Charles's law: dependence of volume on temperature (constant pressure)

Gay-Lussac's law: dependence of pressure on temperature (constant volume)

While many gases under a wide range of physical conditions will approximately obey these laws, an ideal gas is a theoretical gas that obeys them perfectly.

The three gas laws are expressed most clearly when temperature is measured using the thermodynamic or kelvin scale and can be combined into a single equation—the ideal gas equation.

9.1.1 Boyle's Law

The apparatus below can be used to investigate how the volume of a fixed mass of gas varies with pressure. The gas is an air column trapped above some colored water. The experiment must be done slowly so that the gas remains in thermal equilibrium with its surroundings. Changes that take place at constant temperature, like this one, are called **isothermal** changes.

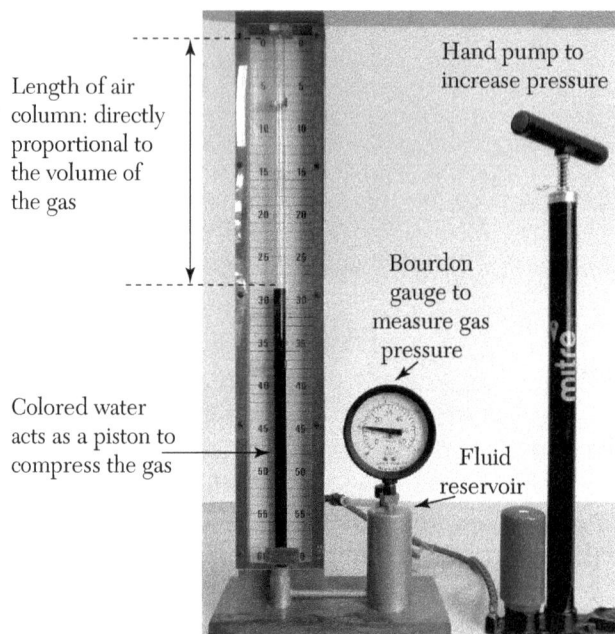

The length of the air column is measured for a range of different pressures. Since the tube containing the air has a constant cross-sectional area, the volume of air is directly proportional to the length of the column. The results of such an experiment show that the pressure p and volume V are inversely proportional to one another. This can be expressed algebraically by $p \propto 1/V$ or pV = constant. This is called Boyle's law:

▣ Boyle's law: pV = constant for a fixed amount of an ideal gas at constant temperature

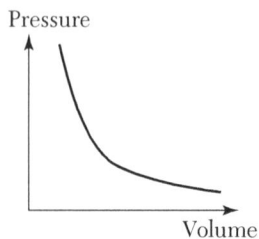

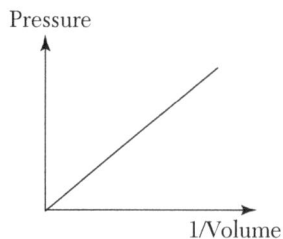

9.1.2 Charles's Law

The apparatus below can be used to investigate how the volume of a fixed amount of a gas varies with temperature when the pressure is kept constant. An air column is trapped inside a capillary tube by a small bead of concentrated sulfuric acid. Concentrated sulfuric acid is used because it absorbs water vapor, keeping the air dry. The capillary tube is then immersed in a water bath and the temperature of the water bath is changed. As temperature increases, the gas expands and the bead moves. The capillary tube has constant cross-sectional area, so the length of the column is directly proportional to the volume of air.

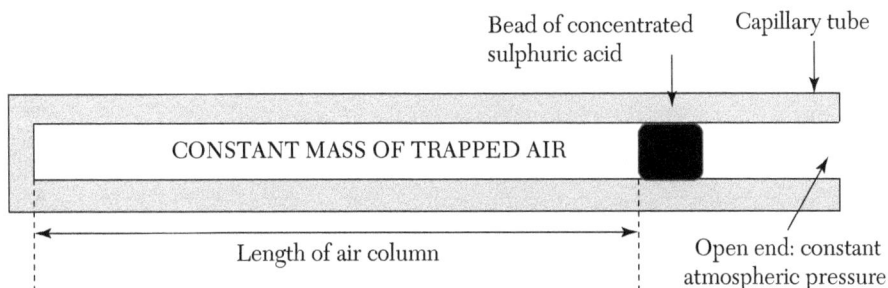

The length of the air column is measured for a range of different temperatures. The tube is open at one end, so the pressure remains constant. The results of such an experiment show that the volume V increases linearly with temperature θ when temperature is measured in Celsius.

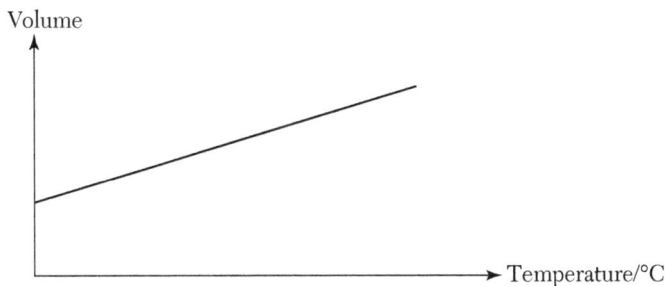

Similar experiments using different amounts of gas also produce linear results but with different gradients. However, if the graphs are extrapolated back, the intercept on the temperature axis is always the same ($-273.15°C$) for all (ideal) gases and all amounts of gas. This is clearly a physically significant temperature and is used as the zero for the thermodynamic or kelvin scale. Using this temperature scale, volume and temperature are directly proportional.

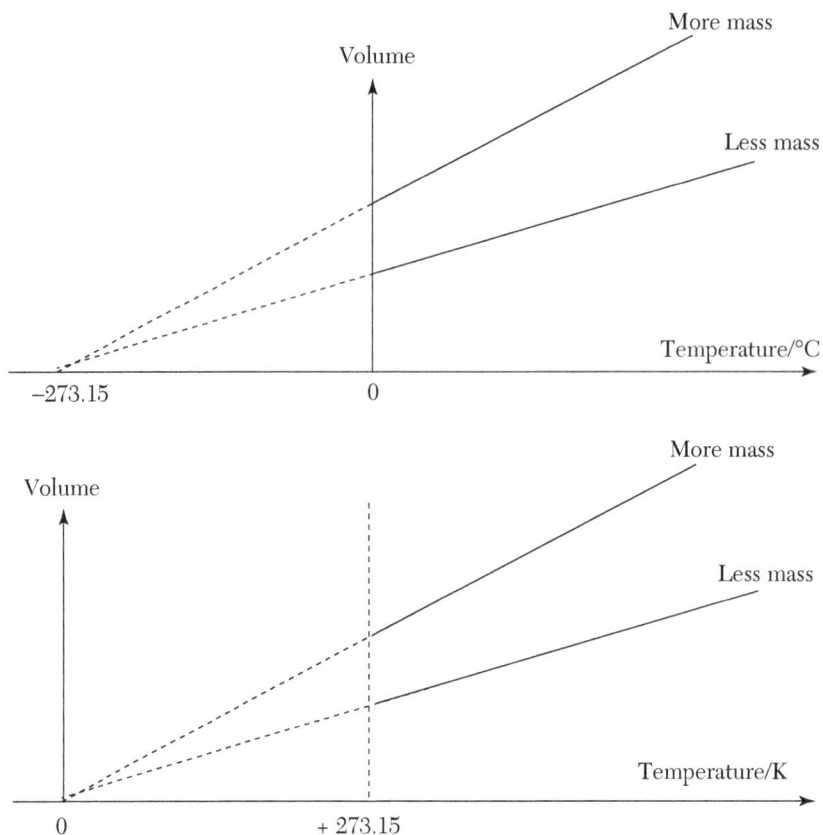

This can be expressed as $V \propto T$ (temperature in kelvin) or V/T = constant. This is Charles's law.

- Charles's law: V/T = constant for a fixed amount of gas at constant pressure (temperature in kelvin)

9.1.3 Gay Lussac's Law (The Pressure Law)

The apparatus below can be used to investigate how pressure of a fixed amount of gas varies with temperature when volume is kept constant.

The pressure is measured for a range of different temperatures. Once again the results show that the pressure varies linearly with temperature in Celsius and that there is a common intercept on the temperature axis of −273.15°C. Using the Kelvin scale, pressure is directly proportional to temperature. This can be expressed algebraically as $p \propto T$ or p/T = constant. This is Gay Lussac's law (the pressure law):

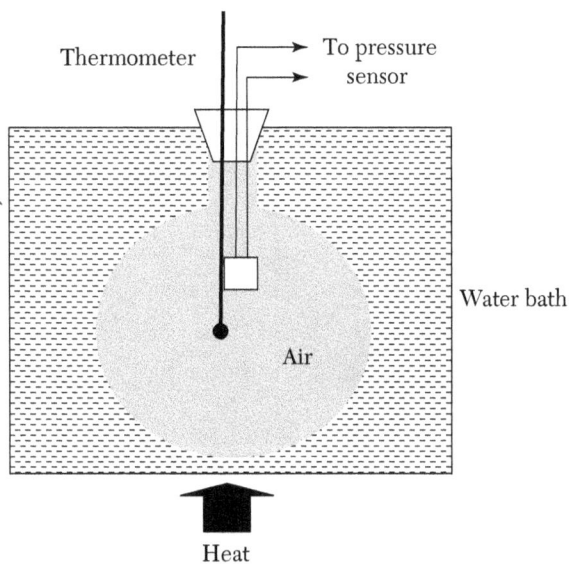

■ Gay Lussac's law: p/T = constant a constant volume of a fixed amount of an ideal gas (temperature measured in kelvin)

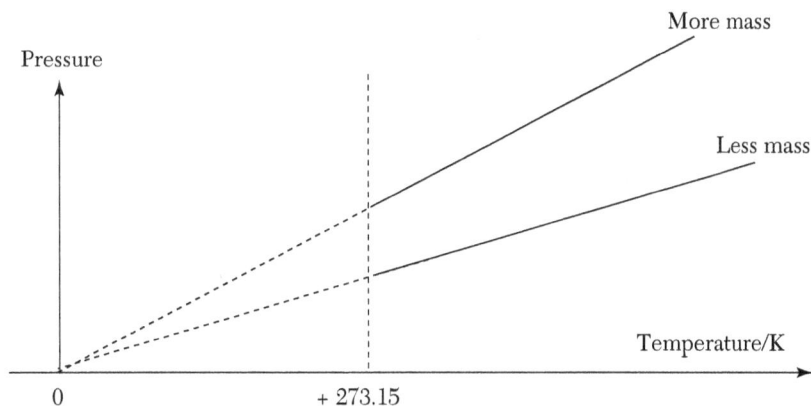

9.2 Ideal Gas Equation

The three gas laws can be combined into a single equation:

$$
\left.
\begin{array}{ll}
\text{Boyle's law:} & pV = \text{constant} \\
\text{Charles's law:} & V/T = \text{constant} \\
\text{Gay Lussac's law :} & p/T = \text{constant}
\end{array}
\right\}
\quad
\frac{pV}{T} = \text{constant}
$$

This is called the ideal gas equation or equation of state for an ideal gas. Holding any one of the three parameters constant reduces it to one of the three gas laws, but it also applies if all three parameters change together.

The constant is directly proportional to the amount of gas. The value of the constant for one mole of an ideal gas is $R = 8.314$ JK^{-1}. The ideal gas equation is often written in the form:

$$pV = nRT$$

where n is the number of moles.

9.3 Kinetic Theory of Gases

9.3.1 Assumptions of the Kinetic Theory

The gas laws are based on experiments but can be explained using Newton's laws if we make a few assumptions about the microscopic nature of a gas. The key assumptions of kinetic theory are:

- Gases consist of a large number of tiny, massive particles in rapid random motion.

- The particles have no long-range interactions.

- All collisions are elastic.

- The volume occupied by the particles themselves is negligible.

Kinetic theory was developed by Boltzmann and others in the nineteenth century, and its great success provided strong support for the idea that matter really is composed of tiny massive particles (the atomic theory). However, even as late as 1900, some physicists and philosophers opposed this theory and suggested that it was simply a convenient model and did not correspond to physical reality. The really convincing evidence that atomic theory was correct came from Einstein's analysis of Brownian motion in 1905.

9.3.2 Explaining Gas Pressure

Gas molecules inside a container are continually colliding with the walls of the container and rebounding. In each collision the wall exerts a force on the molecule that changes its direction, so, by Newton's third law, the molecule exerts an equal but opposite force on the wall. The net effect of billions of collisions creates an almost constant outward force on the wall. Pressure is force per unit area, and force is equal to the rate of change of momentum, so the pressure exerted by a gas is equal to the average rate of change of molecular momentum per unit area.

We will build up an expression for gas pressure by considering first a single molecule bouncing backwards and forwards in a box, and then adding more molecules and allowing them to have random speeds and directions. Let's start with one molecule of mass m moving in the positive x-direction and colliding with the wall of a cubic container of side a:

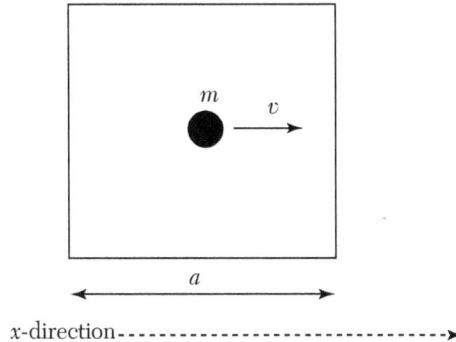

x-direction ------------------------------➤

Change of momentum at wall (momentum is a vector): $\Delta p = -2\,mv_x$

Time between collisions at wall: $\Delta t = 2a/v_x$

Average rate of change of momentum at wall: $\dfrac{\Delta p}{\Delta t} = -\dfrac{mv_x^2}{a}$

Average force on molecule at wall: $F = -\dfrac{mv_x^2}{a}$

Average force on wall (by Newton's third law): $F = \dfrac{mv_x^2}{a}$

Now consider N molecules with x-components of velocity $v_{x1}, v_{x2}, v_{x3}\ldots v_{xN}$.

Average force from N molecules,

$$F = \frac{mv_{x1}^2}{a} + \frac{mv_{x2}^2}{a} + \ldots \frac{mv_{xN}^2}{a} = \frac{m}{a}\sum_{i=1}^{i=N} v_{xi}^2 = \frac{Nm}{a}\,\overline{v_x^2} \tag{1}$$

where $\overline{v_x^2}$ is the **mean-squared x-component of velocity**:

$$\overline{v_x^2} = \frac{v_{x1}^2 + v_{x2}^2 + \ldots v_{xi}^2}{N}$$

Now let's consider a three-dimensional gas in which the molecules move in *all* directions. The derivation above will still be valid for the x-components of velocity, so we need to relate these to the overall velocity. Any particular molecule will have velocity components, v_x, v_y, and v_z, and the magnitude of velocity will be given by:

$$v^2 = v_x^2 + v_y^2 + v_z^2$$

The molecules are moving randomly, so the mean-squared values of v_x, v_y, and v_z will all be equal:

$$v_x^2 = v_y^2 = v_z^2$$

Therefore, $v^2 = v_x^2 + v_y^2 + v_z^2 = 3v_x^2$ (2)

Using equations (1) and (2) above we can find an expression for the average force on the wall of the container from a gas consisting of N particles in rapid random motion:

$$F = \frac{Nm}{a}v_x^2 = \frac{Nm}{3a}v^2$$

where $\langle v^2 \rangle$ is the **mean-squared speed** of the molecules.

We can now derive an expression for the pressure:

$$p = \frac{F}{A} = \frac{Nm}{3a^3}v^2 = \frac{Nmv^2}{3V}$$

This is the **kinetic theory equation** for gas pressure, usually written as:

$$pV = \frac{1}{3}Nmv^2$$

Since density $\rho = nm/V$ we can also write this equation in the form:

$$p = \frac{1}{3}\rho(v)^2$$

Mean-squared and root mean-squared (rms) speeds

- Mean-squared speed, $v^2 = \dfrac{v_1^2 + v_2^2 + v_3^2 + \cdots + v_N^2}{N}$

- rms speed, $v_{rms} = \sqrt{v^2}$

Note that the rms speed is not equal to the mean speed and that the mean-squared speed is not the same as the mean speed-squared!

9.3.3 Molecular Kinetic Energy and Temperature

Compare the kinetic theory equation with the ideal gas equation. Note that N is the number of molecules and n is the number of moles. It will be helpful to introduce N_A, the Avogadro number (number of particles in 1 mole, $N_A = 6.022\ 140\ 857 \times 10^{23}\ \text{mol}^{-1}$) so that N can be replaced by nN_A:

Kinetic theory equation: $pV = \dfrac{1}{3}nN_A mv^2$

Ideal gas equation: $pV = nRT$

The right-hand sides of these two equations must be equal, so gas temperature is directly proportional to molecular kinetic energy:

$$\frac{1}{3}nN_A mv^2 = nRT$$

$$\text{Total KE} = \frac{1}{2}N_A mv^2 = \frac{3}{2}RT$$

Dividing by N_A we get:

$$\text{Mean molecular KE} = \frac{1}{2}mv^2 = \frac{3}{2}\frac{R}{N_A}T = \frac{3}{2}kT$$

The constant $k = \dfrac{R}{N_A}$ is called **Boltzmann's constant**. This is effectively the gas constant per particle, and it is a very important constant in thermodynamics. $k = 1.38064852 \times 10^{-23}$ JK^{-1}. The mean thermal energy per particle in any thermodynamic system is of the order of kT; for an ideal gas, it is $3/2\ kT$.

Notice that, for an ideal gas, the mean kinetic energy depends *only* on temperature (in kelvin). This implies that molecules of different gases, at the same temperature, have the same mean kinetic energy but different mean speeds. For example, air consists of about 80% oxygen and 20% nitrogen. Oxygen molecules are more massive than nitrogen molecules, so they have lower mean-squared speeds. Hydrogen molecules have much lower mass and, therefore move on average at much greater speeds than oxygen or nitrogen molecules. This results in them escaping from the atmosphere much more rapidly, so there is very little hydrogen in the Earth's atmosphere.

The rms speed of a molecule in an ideal gas at temperature T is given by:

$$v_{\text{rms}} = \sqrt{v^2} = \sqrt{\frac{3kT}{m}}$$

The table below shows the rms speeds of oxygen, nitrogen, and hydrogen molecules at 20°C (293 K).

Gas	Mass of molecule/kg	rms speed at 20°C (293 K)/ms^{-1}
Oxygen	5.32×10^{-26}	478
Nitrogen	4.66×10^{-26}	510
Hydrogen	3.34×10^{-27}	1910

9.3.4   Molar Heat Capacities of an Ideal Monatomic Gas

The internal energy U of an ideal gas is equal to its total kinetic energy. For one mole of an ideal gas,

$$U = \text{Total KE} = \frac{1}{2} N_A m v^2 = \frac{3}{2} RT$$

The molar heat capacity at constant volume c_V is the energy required to raise the temperature of 1 mole of the gas by 1 K. Since all of the heat Q supplied increases,

$$\text{Internal energy, } c_V = \frac{Q}{\Delta T} = \frac{\Delta U}{\Delta T} = \frac{3}{2} R$$

Note that this is the molar heat capacity at constant volume. If the gas is instead heated at constant pressure, it will expand and do external work, increasing the heat needed to raise its temperature and therefore increasing its heat capacity.

The first law of thermodynamics (see Section 9.6.2) states that the change in internal energy of a system is equal to the sum of the heat supplied to the system and the work done on the system. The heat supplied to the system at constant pressure is $Q = c_p \Delta T = \Delta U + p \Delta V$ where c_p is the molar heat capacity at constant pressure and $p \Delta V$ is the work done by the gas as it expands (see Section 9.6.3). Using the ideal gas equation we can replace ΔV with $R \Delta T / p$ (because p is constant) to give:

$$Q = \frac{3}{2} R \Delta T + R \Delta T = c_p \Delta T$$

Therefore, $c_p = \frac{5}{2} R = c_V + R$

The ratio of heat capacities is called the adiabatic gas constant γ:

$$\gamma = \frac{c_p}{c_V}$$

For an ideal monatomic gas, $\gamma = 5/3$; for air, it is about 1.4.

9.3.5   Equipartition of Energy

Molecules in an ideal gas are treated as particles with no internal degrees of freedom. This means that the only motion they can have is translation in three dimensions, so they have three degrees of freedom. The derivation above shows that motion along the x-, y-, and z-axes contribute equally to the mean kinetic energy ($3/2\ kT$) of the molecules so that the mean energy per degree of freedom is ½ kT. The classical theorem of equipartition of energy states that when molecules are in thermal equilibrium each independent degree of freedom has a mean energy equal to ½ kT. This applies to translational and rotational degrees of freedom as well as potential energies (which are not present in an ideal gas because there are no long-range forces).

A diatomic molecule is free to rotate about each of three axes. These three rotational degrees of freedom add another $3 \times$ (½ kT) to the overall mean kinetic energy per molecule. A diatomic molecule can also vibrate along the bond between the two atoms. These vibrational modes have both kinetic energy and potential energy, so these two degrees of freedom add $2 \times$ (½ kT) to the overall mean energy per particle. These additional degrees of freedom increase the heat capacity of the gas, and in this case, if the additional 5 degrees of freedom were all excited, the heat capacity should be $4\,kT$.

Classically all of these degrees of freedom should be excited by thermal motion, and physicists expected the molar heat capacities of gases to be in accord with this model. However, this is not the case, especially at low temperatures. Some degrees of freedom (e.g., vibrational modes) do not contribute to the heat capacity and only "switch on" at higher temperatures, resulting in lower heat capacities than those calculated using the equipartition theory. This apparent anomaly was only explained by quantum theory, which resulted in the "quenching" of some degrees of freedom at lower temperatures (see Section 27.1.2).

9.3.6 Law of Dulong and Petit

The focus of this chapter is on gases, but the equipartition theory can explain an empirical law derived by Dulong and Petit. This law states that the molar heat capacity of a crystalline solid has a value of $3R$, and this works pretty well for a large number of solid metallic elements at room temperature.

The explanation is quite straightforward. Imagine an atom in a cubic crystalline lattice with bonds parallel to the x-, y-, and z-axes.

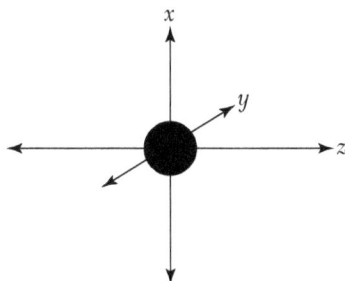

This atom is free to vibrate along three independent directions. Since each vibrational mode has two degrees of freedom, the total mean internal energy is $3 \times (2 \times \frac{1}{2} kT) = 3\,kT$. For one mole of particles, this is $c = 3N_A kT = 3RT$.

9.3.7 Graham's Law of Diffusion

Graham's law of diffusion states that the rate of diffusion is inversely proportional to the square root of gas density:

$$\text{Rate of diffusion} \propto \frac{1}{\sqrt{\rho}}$$

This follows from the kinetic theory equation in the form:

$$p = \frac{1}{3}\rho v^2$$

Diffusion occurs as a result of particles moving randomly and colliding with one another, so it is reasonable to assume that the rate at which this occurs will depend on the rms speed:

$$v_{\text{rms}} = \sqrt{\overline{v^2}}$$

$$\text{Rate} \propto v_{\text{rms}} = \sqrt{\frac{3p}{\rho}} \propto \frac{1}{\sqrt{\rho}}$$

9.3.8 Speed of Sound in a Gas

Sound consists of longitudinal vibrations in a material medium. In a gas the speed at which these disturbances move through the material is directly related to the speed at which the molecules themselves move, so it is not surprising to find that the speed of sound in a gas is directly proportional to the rms speed of the molecules. In fact the speed of sound c in an ideal gas is given by:

$$c = \sqrt{\frac{\gamma p}{\rho}}$$

where g is the adiabatic constant (= 1.4 for a monatomic gas and 1.67 for a diatomic gas). The adiabatic constant is the ratio of the specific heat of a gas at constant pressure to the specific heat at constant volume.

9.4 Maxwell Distribution

While v_{rms} gives a good idea of typical molecular speeds in a gas, the constant random motion and collisions results in a distribution of speeds from zero up to very high values. Maxwell derived an expression for the fraction of molecules DN/N lying within a small range of speeds from v to $v + \Delta v$:

▨ Maxwell distribution, $\frac{\Delta N}{N} = Av^2 e^{-\frac{E}{kT}} \Delta v$

where A is a constant for a particular gas and E is the kinetic energy ½ mv^2 of a molecule moving with speed v.

The derivation of this result is beyond the scope of this book, but we will be concerned with the qualitative implications of the distribution, which are useful to understand a wide range of thermodynamic phenomena.

Consider the following graph illustrating the main features of the Maxwell speed distribution.

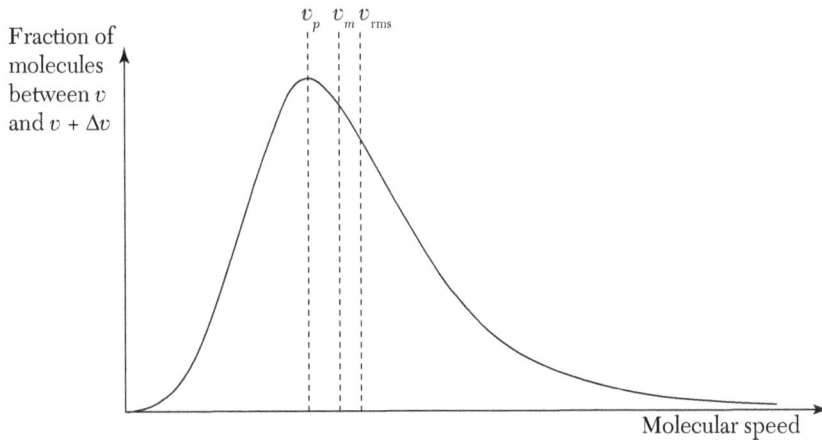

▨ v_p is the most probable speed. This means that there are more molecules with speeds between v_p and $v_p + \Delta v$ than in any other similar-sized range.

▨ v_m is the mean speed ($v_m = 1.13\ v_p$).

▨ v_{rms} is the rms speed ($v_{rms} = \sqrt{v^2} = 1.23\ v_p$).

As the temperature of the gas increases the peak moves to the right and falls. The number of molecules with higher speeds increases.

The area under each graph is the same because the total number of molecules has not changed.

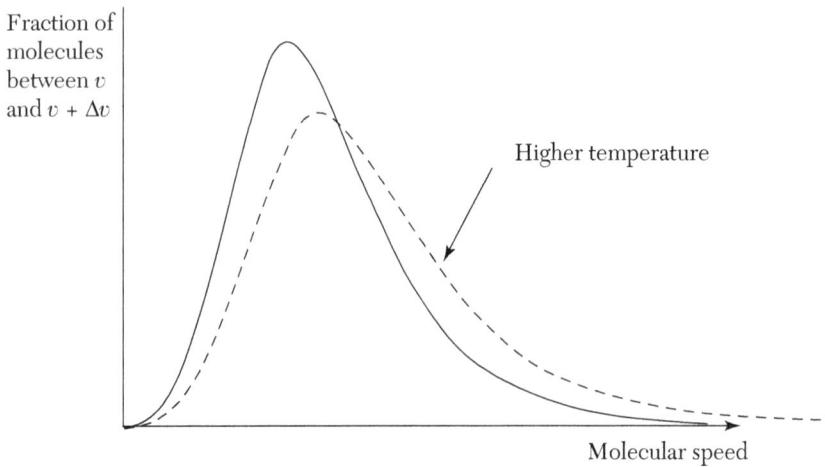

9.5 Boltzmann Factor and Activation Processes

For particles in equilibrium at temperature T with a Maxwell distribution of speeds, the probability of a particular particle gaining an additional energy DE is proportional to the Boltzmann factor:

Boltzmann factor $f = e^{-\frac{\Delta E}{kT}}$

This is equal to the ratio of number of particles in the higher energy state $(E+DE)$ to the number of particles in the lower energy state (E).

$$f = \frac{\text{number of particles with energy } E + \Delta E}{\text{number of particles with energy } E}$$

Many physical and chemical processes depend on a fraction of particles gaining an "**activation energy**" E_A. Evaporation is a good example. A molecule with the mean energy does not have enough energy to escape from the surface, but molecules close to the end of the distribution can. The greater this fraction, the greater the rate of evaporation.

The fraction of molecules with at least the activation energy E_A is given by the Boltzmann factor $f = e^{-\frac{\Delta E}{kT}}$. Since the rate of evaporation is directly proportional to this fraction, we have:

Rate of evaporation $\propto e^{-\frac{\Delta E}{kT}}$

Other activation processes include chemical reactions, conductivity of a semiconductor, viscous flow, and creep. In all of these cases we would expect the rate R to be directly proportional to the Boltzmann factor:

Rate of activation process, $R \propto e^{-\frac{\Delta E}{kT}}$

Fraction of molecules between v and $v + \Delta v$

Molecules in the shaded region can escape from the liquid surface

Molecular speed

Minimum speed needed to escape

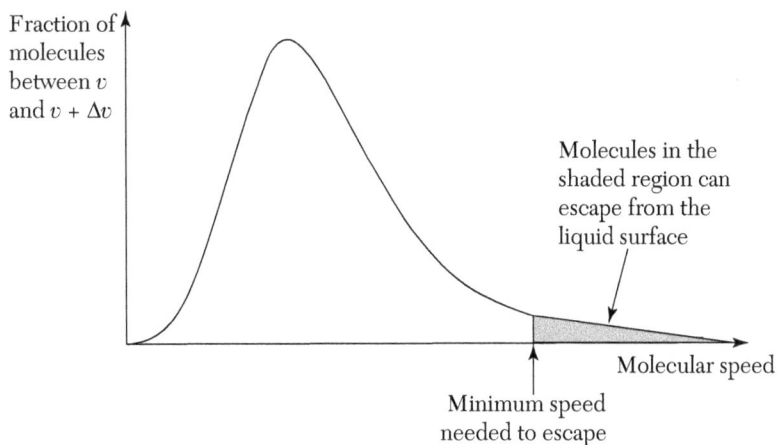

If this is the case then a graph of $\ln (R)$ against $1/T$ should be a straight line with a negative gradient and positive y-axis intercept:

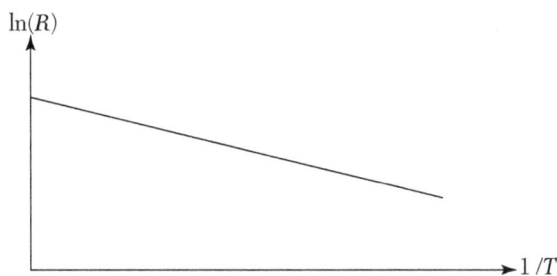

$\ln(R)$

$1/T$

In chemistry, a graph of $\ln$ (reaction rate) against 1/(temperature) can be used to determine rate constants.

9.6 First Law of Thermodynamics

9.6.1 Internal Energy

- The internal energy U of a thermodynamic system is the sum of the random thermal kinetic energies and potential energies of all particles in the system.

For particles in an ideal gas, there are no long-range interactions, so the potential energy term is zero and the internal energy is simply equal to the sum of kinetic energies.

Internal energy of an ideal gas, U = Total KE $= \dfrac{1}{2}nN_{A}mv^{2} = \dfrac{3}{2}nRT$

The fact that, for a fixed amount of gas, this depends only on the absolute temperature of the gas has already been noted. Therefore, the internal energy of the gas is constant for all isothermal changes (changes at constant temperature).

9.6.2 Heating, Working, and the First Law of Thermodynamics

Consider a system with total internal energy U. The internal energy can be changed if the system is heated or if work is done on the system.

▪ Heating: energy transfer as a result of a temperature difference (e.g., placing the system in thermal contact with another system at a different temperature).

▪ Working: energy transfer as a result of movement of an applied force (e.g., compressing the system).

The first law of thermodynamics is a statement of energy conservation for the system and its surroundings.

▪ **First law of thermodynamics:** The change in internal energy of the system DU is the difference of the heat supplied to the system Q and the work done by the system W:

$$\Delta U = Q - W$$

Note that we have defined W as positive when the system does external work. Sometimes W is defined as work done on the system, in which case the sign of W in the equation above changes.

9.6.3 Work Done by an Ideal Gas

Consider an ideal gas expanding against a piston. The gas is at temperature T, has pressure p, and expands by an amount δx. The cross-sectional area of the cylinder is A.

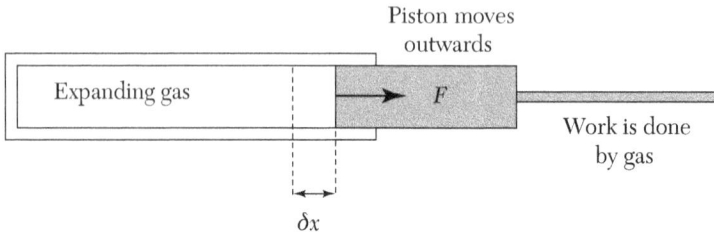

Work done by gas as piston moves a distance δx:

$$\delta W = F\delta x = pA\delta x = p\delta V$$

where δV is the small increase in volume when the piston moves a distance δx.

Work done as piston moves from $x = x_1$ to $x = x_2$:

$$W = \int_{x_1}^{x_2} p dV$$

This is also equal to the area under a graph of pressure against volume.

If the pressure is constant, the work done is simply:

$$W = p\Delta V$$

9.6.4 Thermodynamic Changes

Isothermal changes

An isothermal change is one in which there is no change of temperature. For an ideal gas the internal energy is also unchanged, so $\Delta U = 0$. Therefore $Q - W = 0$ and $Q = W$; the heat supplied to the system must be equal to the work done by the system; or if work is done on the system, an equal amount of heat must flow out of the system. An example of this is a slow compression of a gas that is in thermal contact with its surroundings:

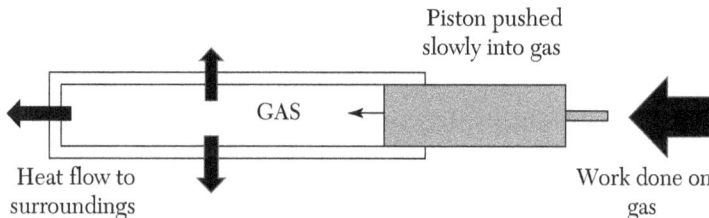

Isothermal compression of an ideal gas obeys Boyle's law:

$$pV = \text{constant}$$

Adiabatic changes

In an adiabatic change no heat flows into or out of the system, so $Q = 0$. This is often because the change takes place very rapidly and there is not enough time for a significant heat transfer. Using the first law of thermodynamics we can see that, when $Q = 0$, $\Delta U = -W$, so the increase of internal energy is equal to the work done *on* the system. An example of this would be a rapid compression of a gas inside an insulated cylinder. Since the internal energy increases, the temperature of the system also increases.

Boyle's law does not apply when an ideal gas is compressed adiabatically (because temperature is not constant). However, adiabatic compressions obey another rule:

$$pV^{\gamma} = \text{constant}$$

where γ is the adiabatic gas constant ($\gamma = 1.4$ for a monatomic gas).

Isochoric changes

An isochoric change is one that takes place with no change in the volume of the system. If the volume is constant, there is no work done on or by the system, so $W = 0$ and $\Delta U = Q$. The change in internal energy is equal to the heat supplied to the system. An example of an (approximately) isochoric change is the explosive combustion of a petrol and air mixture inside the cylinder of a petrol engine.

Isobaric changes

An isobaric change is one that takes place at constant pressure. For a fixed amount of an ideal gas, Charles's law is obeyed: V/T = constant. An example of an (approximately) isobaric change is the expansion of a diesel–air mixture as fuel is injected and ignited inside the cylinder of a diesel engine. Another example is the change of state of a liquid to a vapor at constant temperature and pressure. The vapor occupies a much larger volume than the liquid, and as it expands against external pressure, it does work (e.g., pushing back the atmosphere).

When an ideal gas expands isobarically, work done is given by:

$$W = p\Delta V$$

9.7 Heat Engines and Indicator Diagrams

9.7.1 What is a Heat Engine?

A heat engine uses thermal energy to do useful work, dumping waste heat in the surroundings. Petrol and diesel engines are examples, so are thermal power stations and steam engines. Heat engines work between a high-temperature source and a low-temperature sink, and the work they do is effectively diverted from the flow of heat between these two reservoirs.

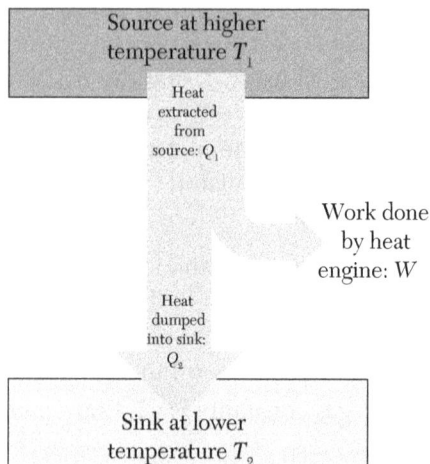

For an heat engine,

$$\text{efficiency} = \frac{\text{useful work output}}{\text{total heat input}} 100\% = \frac{W}{Q_1} = \frac{(Q_1 - Q_2)}{Q_1} = 1 - \frac{Q_2}{Q_1}$$

The second law of thermodynamics states that it is impossible to convert heat to work with 100% efficiency (see Section 10.2.3). The French engineer Sadi Carnot analyzed the cycle of thermodynamic changes (the Carnot cycle) in an ideal heat engine and showed that it cannot exceed a maximum theoretical efficiency determined by the temperatures of the hot and cold reservoirs:

$$\text{efficiency} \leq 1 - \frac{T_2}{T_1}$$

To maximize theoretical efficiency, we need a high source temperature and a low sink temperature.

9.7.2 Indicator Diagrams

Heat engines using a working fluid (e.g., steam or burnt gases) take that fluid around a repeated cycle of changes. These changes might involve heating or cooling, expansion or compression, and work might be done on or by the fluid. Each stage in the cycle can be plotted on an indicator diagram, which is a graph of pressure against volume. This helps us understand the changes of state and energy transfers that take place inside the heat engine.

Here is an example of a simplified indicator diagram. Real heat engines have more complex cycles.

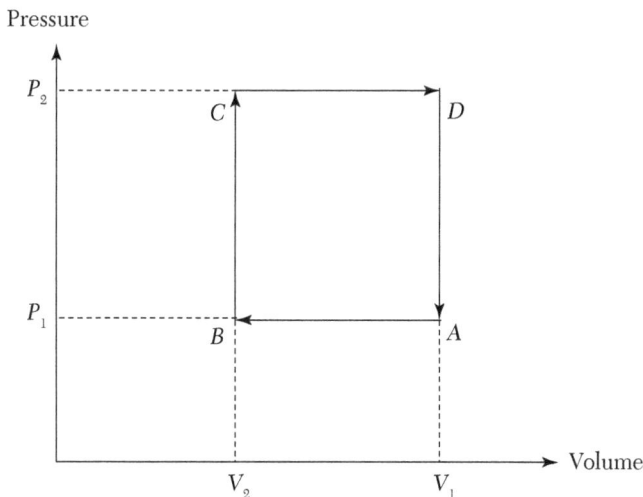

AB: The fluid is compressed at constant pressure. Work must be done by an external agent. The work done on the fluid is equal to the area under the line *AB*. The work done on the fluid is W_1.

BC: The fluid pressure increases rapidly at constant volume. This could be as a result of sudden heating (e.g., ignition of a fuel–air mixture). The fact that the volume remains constant means that there is no work done on or by the fluid. The fluid gains heat Q_1.

CD: The fluid expands at constant pressure and so does work against external forces. The work done is equal to the area under the line *CD*. The work done by the fluid is W_2, so the net work done by the engine in one cycle of changes is $W = W_2 - W_1$. This is equal to the area contained inside the loop on the indicator diagram.

DA: The fluid loses pressure at constant volume. This could be the result of a sudden transfer of heat to the surroundings (e.g., as a result of exhaust). The fluid loses heat Q_2.

If the working fluid is an ideal gas then the gas laws and the first law of thermodynamics can be used to analyze each stage in the cycle in order to calculate the efficiency of the heat engine. While real heat engines do not have simple indicator diagrams, there are good approximations for the way in which certain heat engines work. For example, the Otto cycle is used to model the behavior of spark ignition petrol engines, and the Diesel cycle is used to model the behavior of a compression ignition engine such as a diesel engine.

9.7.3 The Otto Cycle

This cycle models the behavior of the working fluid inside one of the cylinders of a typical four-stroke petrol engine.

- Intake stroke: inlet valve opens and an explosive mixture of petrol and air is drawn into the cylinder.

- Compression stroke: the mixture is compressed.

- Power stroke: following ignition the hot burnt gases push the piston back.

- Exhaust stroke: exhaust valve opens and burnt gases leave the cylinder.

After this the exhaust valve closes and the cycle begins again. In reality the fluid inside the cylinder is not constant; it is drawn in at the start of the cycle and ejected as exhaust at the end of the cycle.

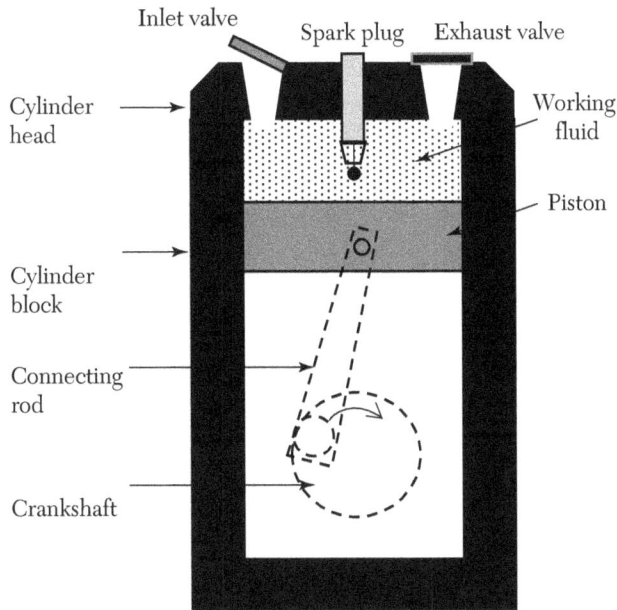

The diagram shows an idealized indicator diagram for the Otto cycle.

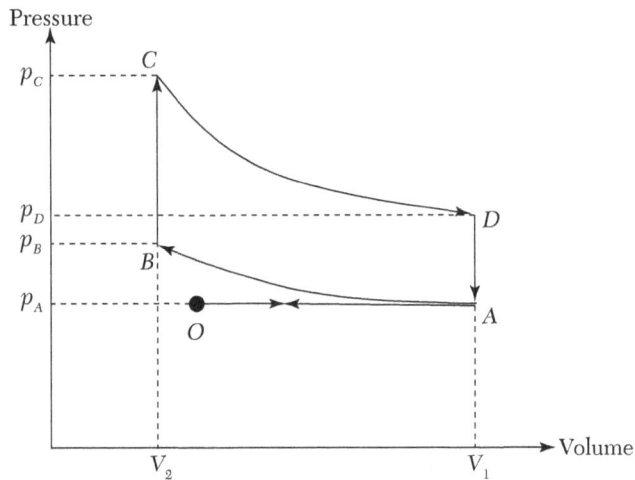

- *AB*: adiabatic compression of gas (work done on gas but no heat transfer)—compression stroke

- *BC*: isochoric heating of gas (heat is supplied as fuel–air mixture explodes but no work is done)

- *CD*: adiabatic expansion of gas (work is done by the gas but no heat is transferred)—power stroke

- *DA*: isochoric cooling of gas (heat is lost from the gas but no work is done by it)

AB corresponds to the compression stroke and *CD* to the power stroke. The closed loop *OAO* represents the intake and exhaust strokes. The area contained by the loop *ABCD* represents the net work done in one cycle.

Analysis of the Otto cycle

For an heat engine,

$$\text{efficiency} \leq 1 - \frac{T_2}{T_1}$$

where T_2 is the sink temperature and T_1 is the source temperature. In the idealized diagram above, $T_1 = T_C$ and $T_2 = T_D$, so the theoretical efficiency of the Otto cycle is:

$$\text{efficiency} \leq 1 - \frac{T_D}{T_C}$$

This can be linked to the **compression ratio,** V_1/V_2, by using the equation of state for an ideal gas and the relation $pV^\gamma = \text{constant}$ for adiabatic changes.

We need to find an expression for T_D/T_C. Consider the power stroke *CD*. This is an adiabatic change, so:

$$p_C V_2{}^\gamma = p_D V_1{}^\gamma \tag{1}$$

The ideal gas equation also applies:

$$\frac{p_C V_2}{T_C} = \frac{p_{D V_1}}{T_D} \tag{2}$$

Combining these equations,

$$\frac{T_D}{T_C} = \left(\frac{p_D}{p_C}\right)\left(\frac{V_1}{V_2}\right) = \left(\frac{V_2}{V_1}\right)^\gamma \left(\frac{V_1}{V_2}\right) = \left(\frac{V_2}{V_1}\right)^{\gamma-1}$$

The theoretical efficiency of the Otto cycle is, therefore,

$$\text{efficiency} \leq 1 - \left(\frac{V_2}{V_1}\right)^{\gamma-1}$$

Using a typical compression ratio $V_1/V_2 = 10$ and taking $\gamma = 1.4$ gives a maximum theoretical efficiency of $(1 - 0.10^{0.4}) = 0.60$ (or 60%).

A real internal combustion petrol engine is only about 25–35% efficient at transferring thermal energy from the fuel into useful mechanical energy. Here are some of the ways in which energy is wasted:

▪ The thermodynamic cycle of a real engine differs from the ideal Otto cycle–valves take time to open and close, and changes in pressure and temperature cannot take place instantaneously.

▪ The intake and exhaust strokes do not form a closed loop and work must be done by the engine during these phases.

▪ There is heat transfer from the engine to the surroundings (in addition to the heat lost during the exhaust stroke).

▪ There is friction between moving parts, which transfers energy to heat.

9.7.4 Diesel Cycle

The most important practical difference between a petrol engine and a diesel engine is that the fuel–air mixture in a petrol engine is ignited by a spark, whereas in a diesel engine it is ignited by compression. Diesel engines have a larger compression ratio than petrol engines (typically 15–20 compared to 7–10), and this contributes to their higher thermodynamic efficiency. The combustion process is also different. Whereas the petrol–air mixture is ignited explosively and the pressure and temperature of the working fluid rise almost instantaneously, in the diesel engine the fuel is sprayed into the cylinder and ignited so that the first part of the power stroke takes place at almost constant pressure.

The idealized cycle is illustrated below.

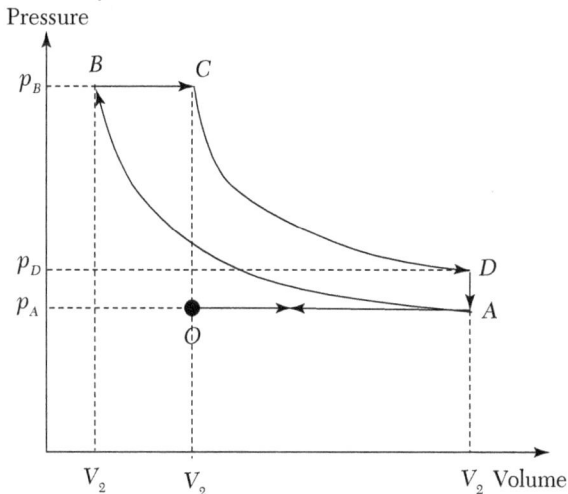

- *AB*: adiabatic compression of air (work done on gas but no heat transfer)—compression stroke. The fuel has not yet been injected, so there is no danger of pre-ignition (which limits the compression ratio for a petrol engine) and the compression ratio V_1/V_2 for a diesel engine can be significantly higher than for a petrol engine.

- *BC*: Fuel is injected and burnt at constant pressure. Work is done by the engine. This is the first part of the power stroke.

- *CD*: adiabatic expansion of gas (work is done by the gas but no heat is transferred). This completes the power stroke.

- *DA*: isochoric cooling of gas (heat is lost from the gas but no work is done by it).

A real diesel engine only approximates the idealized cycle shown above, and additional energy losses occur in much the same way as for a petrol engine.

9.8   Exercises

1. 4.0 m³ of an ideal gas at a temperature of 15 °C and a pressure of 1.10×10^5 Pa is compressed isothermally (i.e., at constant temperature) to one quarter of its original volume.
   (a) What will be its final pressure and temperature?
   (b) What happens to the internal energy of the gas during this compression? Explain by referring to work done and heat transferred.
   (c) Explain how the first law of thermodynamics applies to an isothermal compression (you should say what happens to each term in the equation).
      The gas is then allowed to expand at constant pressure until it again occupies a volume of 4.0 m³.
   (d) What is the final temperature of the gas?
   (e) Explain, in terms of molecular motions, why the final temperature is not 15 °C.

2. A bicycle tire contains 9.2×10^{-4} m³ of air at 20 °C. The pressure in the tire is 480 kPa.
   (a) Calculate the number of moles of air in the tire.
   (b) After the bicycle has been ridden the pressure inside the tire is higher. Explain why.
   (c) Estimate the final temperature of the air if the final pressure is 500 kPa.
   (d) Explain why your answer to (c) can only be approximate.

3. A fixed mass of an ideal gas is trapped in a syringe.

(a) Sketch a graph to show how the pressure inside the syringe varies with volume as the gas is *slowly** compressed (e.g., as if someone put their finger over the end of the syringe and slowly pushed the plunger in).

*Assume this is slow enough for the temperature to remain constant.

(b) On the same graph draw a second line to indicate how you would expect pressure to change with volume if the plunger is pushed in quickly.

(c) Explain the difference between the lines in (a) and (b) by referring to work and internal energy.

4. Air at atmospheric pressure and temperature has a pressure of about 10^5 Pa and a density of about 1.2 kgm^{-3}.

(a) Calculate the rms speed of an air molecule.

(b) Air consists mainly of nitrogen and oxygen and the rms speeds of oxygen and nitrogen molecules in the atmosphere are different. Explain why.

5. Suggest why the ideal gas equation will cease to apply:

(a) at very high pressures

(b) at very high temperatures

(c) at very low temperatures

6. The speed of sound in air is given by:

$$v = \sqrt{\frac{\gamma p}{\rho}}$$

where γ is a dimensionless constant, p is the air pressure, and ρ is the air density. By treating air as an ideal gas, show that the speed of sound in air is independent of pressure but directly proportional to the square root of absolute temperature. (**Hint:** consider one mole of air and use the ideal gas equation to relate density to pressure and temperature by introducing the molar mass M).

7. Eight molecules in a gas are moving parallel to the x-axis and have x-components of velocity equal to 510.0 m/s, −550.0 m/s, −495.0 m/s, 548.0 m/s, −498.0 m/s, 502.0 m/s, 518.0 m/s, −535.0 m/s

Each molecule has a mass of 5.4×10^{-26} kg. Calculate:

(a) the average molecular velocity
(b) the average molecular speed
(c) the rms molecular velocity
(d) the mean molecular kinetic energy.

8. An isothermal change takes place at constant temperature. Explain why isothermal compression results in a flow of heat to the surroundings exactly equal to the work done in compressing the gas.

9. (a) State the first law of thermodynamics and define all the terms in the equation.

(b) A trapped gas inside a cylinder is heated. It expands isothermally (at constant temperature). The heat supplied is H.
  (i) State the change in internal energy of the gas.
  (ii) State the work done by the gas as it expands.

10. The Boltzmann factor is given by $e^{-\frac{\Delta E}{kT}}$.

(a) Define the terms that appear in the equation and explain the meaning of the Boltzmann factor.
(b) Sketch graphs to show how the Boltzmann factor varies with (i) T, (ii) DE, and (iii) DE/kT
(c) Explain why chemical reactions proceed more rapidly at higher temperatures.

11. By what factor, roughly, would you expect the rate of a chemical reaction to change if the temperature increases by 10 K (assume the reaction is taking place at around 350 K and that the activation energy is about 1 eV).

12. In human blood, oxygen molecules bind to hemoglobin molecules. The bond energy is about 0.30 eV. Calculate the ratio of free oxygen molecules to bound oxygen molecules in the blood at body temperature (310 K).

13. (a) Explain why the conductance ($G = 1/R$) of a semiconductor NTC increases with temperature. Your answer should refer to the Boltzmann factor.

(b) A semiconductor NTC thermistor is connected to a source of constant low voltage V and a current I passes through it. I will vary with temperature. Explain why a graph of ln I versus $1/T$ will be a straight line. What is the gradient of this line?

14. Here is the Otto cycle for an ideal four-stroke petrol engine:

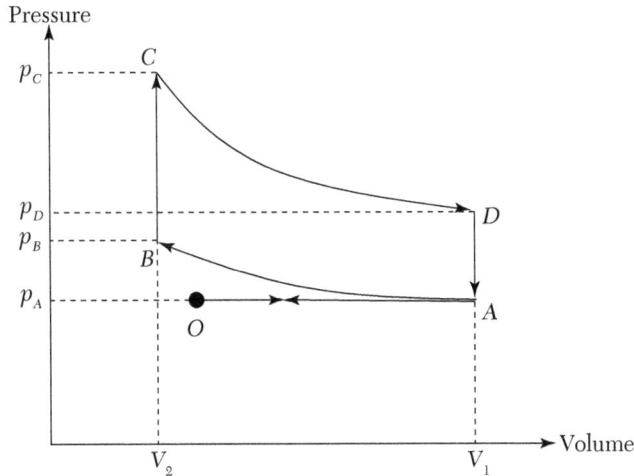

Here is some data for the same engine:

- mean temperature of gases during combustion stroke = 800 °C
- mean temperature of exhaust gases = 80 °C
- area enclosed by indicator diagram loop = 420 J
- calorific value of fuel = 45 MJ kg^{-1}

(a) Describe the thermodynamic changes and energy flows taking place during the stages AB, BC, CD, DA on the diagram.

(b) Calculate the thermodynamic efficiency of the engine.

(c) Calculate the rate of burning fuel (kgs^{-1}) when the engine, which has four cylinders, is rotating at 1500 rpm.

10

STATISTICAL THERMODYNAMICS AND THE SECOND LAW

10.0 Introduction

The gas laws and the ideal gas equation are all derived empirically, that is, from experimental data. They give a good mathematical description of the behavior of an ideal gas, but they offer no explanation of why the gas behaves in this way. The reason for this is that they deal with macroscopic properties such as volume, temperature, and pressure and do not engage with the microscopic behavior of the molecules that constitute the gas. Kinetic theory (see Section 9.3) provides a more fundamental explanation of the behavior of an ideal gas, despite being based on the random motion of particles. Statistical thermodynamics takes this further and provides one of the most important and enigmatic of all physical laws—the second law of thermodynamics.

10.1 Reversible and Irreversible Processes

Consider a dark blue ink droplet dropped into a glass of water. For a while the dark ink forms recognizable and distinct patterns in the water, but as time goes on, these become less and less distinct and after an hour or so the droplet disappears and the water appears uniformly pale blue. The particles of the ink spread more or less uniformly throughout the water. And so it remains.

You could wait a thousand years and the water would continue to appear uniformly pale blue. It is as if, once the ink has spread out, nothing further happens. In one sense this is true. Macroscopically the ink droplet has spread uniformly throughout the water and that is the end of it. However, on the microscopic scale, particles of water and ink are continually moving, colliding, and changing positions. These microscopic collisions are all governed by Newton's laws and are *completely reversible*. This means that if you were to analyze a particular collision and then run it backwards, the reversed collision would be a perfectly acceptable physical process. In fact, if you were presented with videos of a single collision running forwards and backwards, you would be unable to tell which one goes forward in time. Both would obey Newton's laws and conserve energy, momentum, etc.

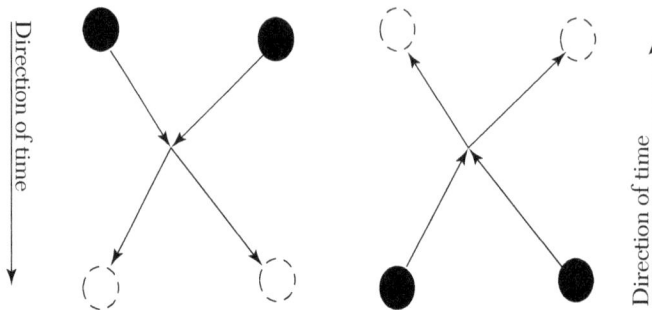

Newton's laws are reversible. So are the laws of electromagnetism and gravity. They do not distinguish between the past and the future. However, macroscopic processes are almost always irreversible. The pale blue water does not resolve itself back into separate clear water and a dark drop, a broken glass remains broken, a scrambled egg remains scrambled and, more personally, we remember the past but not the future and we all eventually grow old and die. This raises two very important questions.

1. If the underlying laws of physics are fundamentally reversible, then how do we explain the irreversibility of the macroscopic world?

2. What distinguishes the past from the future?

Another way to pose this question is in terms of an **arrow of time** that points from the past to the future. Why is there an arrow of time and what determines its direction?

These questions about irreversible processes first emerged in the context of thermodynamics when physicists tried to understand how to maximize the efficiency of heat engines. This led them to the second law of thermodynamics, one of the most important ideas in physics.

10.2 Second Law of Thermodynamics as a Macroscopic Principle

10.2.1 Macroscopic Statements of the Second Law

Heat engines are used to extract heat from work. If it was possible to do this with 100% efficiency then we could build power stations that extract energy directly from the oceans. This would not violate the law of conservation of energy, it would simply lower the temperature of some seawater by a few degrees and the sun would soon re-warm the water. Our energy problems would be solved.

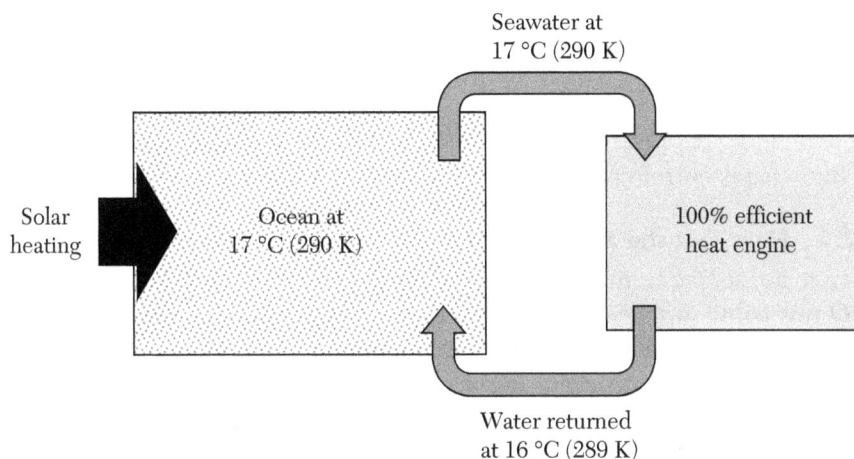

Unfortunately, no such heat engine exists and there is a fundamental limit to the possible efficiency of any heat engine. This is stated qualitatively in a macroscopic version of the second law of thermodynamics:

- It is impossible to construct a heat engine that, operating in a cycle, can extract heat from a source and transfer it completely into work.

This implies that any heat engine MUST dump some waste heat into the environment. It cannot be 100% efficient.

The work generated by a heat engine comes as a by-product of the transfer of heat from the hot source to the cooler sink. If heat could flow back from the sink to the source we could continue to operate the heat engine and increase its efficiency. However, heat only flows spontaneously from systems at higher temperatures to systems at lower temperatures, so this is not possible. This leads to a second, equivalent, statement of the second law of thermodynamics.

▪ It is impossible for heat to flow from a cooler to a hotter body with no other change taking place.

This is a matter of common experience. If you leave a cup of hot coffee on the table it will soon transfer heat to its surroundings until it reaches the same temperature as the room. However long you wait it does not reheat itself. This seems similar to the example of the ink droplet in water, it is another example of an irreversible macroscopic process. Once again the macroscopic one-way process of approaching thermal equilibrium is actually driven by random microscopic collisions that continue to take place after a uniform equilibrium temperature has been reached.

In order to explain these one-way irreversible changes physicists introduced a new quantity, **entropy**. The second law can then be stated in terms of entropy:

▪ The entropy of an isolated system tends to a maximum.

But what is entropy?

10.2.2 Heat Transfer and Entropy

In macroscopic terms the entropy change of a system is related to the heat Q reversibly supplied to it divided by the temperature at which that heat is supplied. This can be written as an integral:

$$\Delta S = \int_{\text{state } A}^{\text{state } B} \frac{dQ}{T}$$

For heat supplied at constant temperature this becomes simply $\Delta S = Q/T$.

The best way to get a feel for this new quantity is to see how it applies to the examples above. We will start with heat flow from hot to cold. Consider two systems A and B at temperatures T_1 and T_2 respectively. If these are placed in thermal contact then they form an isolated system and the total entropy of the system must either increase or stay the same. However, molecular collisions at the boundary between the two systems will result in a flow of heat. How does the transfer of heat δQ from A to B affect the entropy?

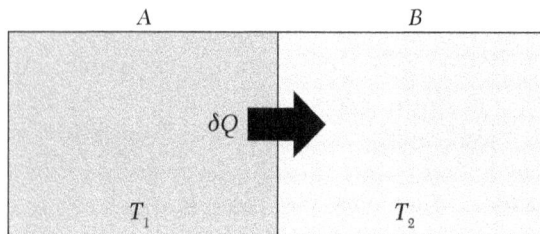

$$\delta S_A = -\frac{\delta Q}{T_1}$$

$$\delta S_B = +\frac{\delta Q}{T_2}$$

$$\delta S_{\text{system}} = -\frac{\delta Q}{T_1} + \frac{\delta Q}{T_2} \geq 0 \text{ (second law of thermodynamics)}$$

Therefore, $\dfrac{\delta Q}{T_2} \geq \dfrac{\delta Q}{T_1}$, so $T_1 \geq T_2$

If entropy cannot decrease then either $T_1 > T_2$ and heat flows from hot to cold or $T_1 = T_2$ and the system is in thermal equilibrium with maximum entropy. This shows that the second law expressed in terms of entropy is in agreement with the statement about direction of heat transfer. Heat cannot, in isolation, flow spontaneously from a colder to a hotter body because this would decrease the entropy of the system.

10.2.3 Entropy and Maximum Efficiency of a Heat Engine

The other macroscopic statement of the second law referred to the efficiency of a heat engine. Let's analyze this using the concept of entropy (and assume that, for the ideal heat engines under consideration, heat is transferred reversibly between the source and the sink).

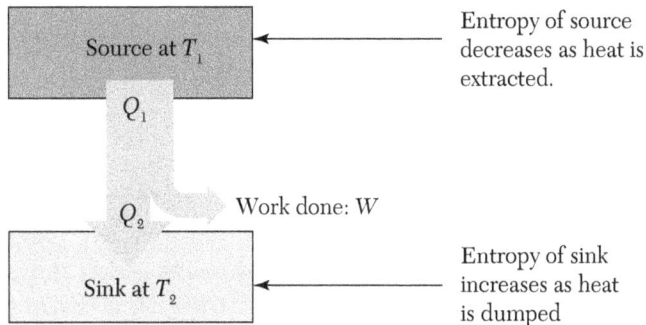

Entropy of source decreases as heat is extracted.

Entropy of sink increases as heat is dumped

Work done has no effect on the entropy of the system, so the entropy changes of the system are:

$$\delta S_{\text{source}} = -\frac{Q_1}{T_1}$$

$$\delta S_{\text{sink}} = +\frac{Q_2}{T_2}$$

$$\delta S_{\text{system}} = -\frac{Q_1}{T_1} + \frac{Q_2}{T_2} \geq 0 \text{ (second law of thermodynamics)}$$

Rearranging this gives:

$$\frac{Q_2}{Q_1} \geq \frac{T_2}{T_1}$$

The efficiency of the engine is:

$$\frac{W}{Q_1} = \frac{Q_1 - Q_2}{Q_1} = 1 - \frac{Q_2}{Q_1}$$

Therefore, efficiency $\leq 1 - \dfrac{T_2}{T_1}$

This result was first derived by a French engineer called Sadi Carnot based on the operation of an ideal reversible heat engine. It sets a limit to the theoretical efficiency of any heat engine and shows that the theoretical efficiency increases as the ratio of T_1 to T_2 increases. This is in agreement with our previous macroscopic statement of the second law and also explains why our seawater heat engine was doomed to failure. Whilst a heat engine working on those principles could be constructed, if it worked between a source temperature of 290 K and a sink temperature of 289 K its maximum theoretical efficiency would be (1 − 289/290) = 0.0034 or 0.34%, and this is before taking into account the unavoidable additional losses caused by friction, etc. For a petrol engine the source temperature (of the hot gases following ignition) is about 1000°C (about 1300 K) and the sink temperature at which the gases are exhausted is about 500°C (about 800 K) giving a maximum theoretical efficiency of (1 − 800/1300) = 0.38 or 38%.

10.3   Entropy and Number of Ways

10.3.1   Macro-State and Micro-States

To understand what entropy actually represents we will need to consider the relationship between macroscopic states and the microscopic configurations that make them up. When we look at a system from the outside (macroscopically) we do not see its microscopic configuration. The uniformly pale blue water left after an ink droplet has spread and diffused throughout the volume might be realized by a very large number of different microscopic arrangements of water and ink particles, all of which appear, from the outside, to be uniformly spread. The key point is this: the same macro-state may correspond to a large number of micro-states. This means that a stable and unchanging macroscopic state does not necessarily imply that the microscopic configurations that make up that state are also stable and unchanging.

Here is a simple model to illustrate what is going on. Imagine a box which contains a large number N of identical gas particles that are initially held in one half of the box by a barrier. Then the barrier is suddenly removed and the particles are free to explore the whole box. For a short while an imbalance will persist but very soon the particles will spread uniformly throughout the box and they will not go back onto one side of it. This is an irreversible process similar to the mixing of ink and water or the transfer of heat from a hotter to a colder body. There is a clear arrow of time from the asymmetric state (all particles on one side of the barrier) to the symmetric state (particles spread evenly throughout the container).

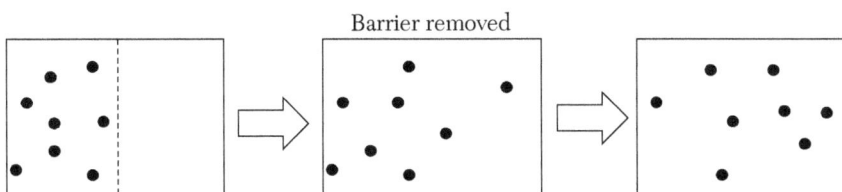

Barrier removed

To understand why this is irreversible we need to consider the microstates of the system. Assume that, after the barrier has been removed, each particle has a 50% chance of being on either side of the barrier. This seems reasonable since the particles move randomly. What is the probability that all N particles are found on the left hand side of the barrier? It is ½ N, which is vanishingly small for large N. This is because there is just one way in which this can occur—just one microstate of the system which has all particles on the left (ignoring rearrangements within the N particles themselves!), in the same way that if you toss a coin N times there is just one way that all N coins can land heads. Now consider the probability that $(N - 1)$ particles are on the left of the barrier and just 1 is on the right. There are N possible ways in which this can happen (each of the particles could be the one that is on the right) so the probability is N times larger. For a distribution with $(N - 2)$:2 the probability is much larger again (by a factor of ½($N - 1$) times. This is because there are $\dfrac{(N(N-1))}{2!}$ ways in which this can occur. In fact the number of ways continues to increase up to a distribution with $N/2$ particles on each side and then falls back again until there is just one way in which all the particles can be found on the right-hand side. The "number of ways" we have referred to here is the number of microstates that represents each macro-state. The final macro-state (a 50:50 distribution) is also the macro-state that can be realized in the largest number of ways, i.e., the macro-state that can be realized in the largest number of distinct micro-states. The graph below shows how the "number of ways," W, changes with the distribution of particles:

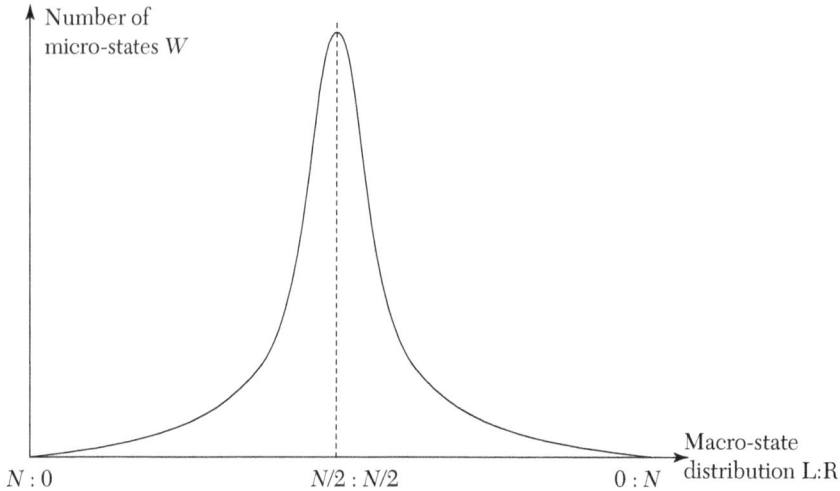

As N increases, the peak becomes incredibly sharp (tall and narrow) so that virtually all of the micro-states cluster close to the $N/2:N/2$ equilibrium distribution. Most of the micro-states correspond to macro-states that are indistinguishable from this 50:50 distribution. Since the particles move randomly we can assume that, left alone, the system will explore all of the microstates available to it. The system we considered started off in a macro-state which corresponds to only 1 micro-state. As time goes on it explores other adjacent micro-states, most of which lie closer to the peak because the number of micro-states increases that way. This means the system tends to evolve toward the peak. Given the fact that the number of micro-states close to the peak vastly outweighs all of those even a short distance away from it then, if we leave the system for any length of time, the overwhelming probability will be that we will find it in a macro-state close to the peak of the distribution—i.e., close to equilibrium. And once in equilibrium it is highly unlikely to fluctuate far from it because the number of micro-states drops rapidly away from equilibrium. For the systems we usually interact with, N is enormous (e.g.,$>10^{20}$), so the probability of a large fluctuation from equilibrium is so small that it can be assumed to be zero.

We can now give a more fundamental, microscopic, explanation of the irreversible evolution toward equilibrium. It is based on three assumptions.

- Macroscopic systems can be composed of a number (usually a very large number) W of microscopic configurations or micro-states (sometimes called "number of ways").

- All micro-states of a macroscopic system are equally probable.

- An isolated system will explore all accessible micro-states because of the random thermal motions of its particles.

The consequence is that we expect a system that is prepared in a non-equilibrium state to evolve toward equilibrium and then stay there. Equilibrium is the macro-state that corresponds to the largest number of accessible micro-states, i.e., the macro-state for which W has a maximum value.

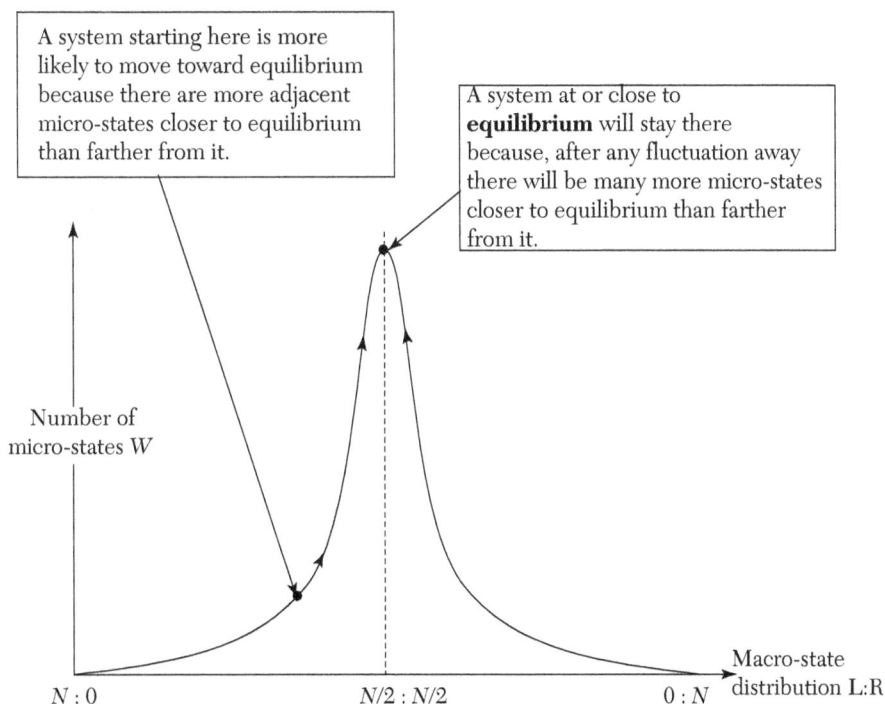

A system starting here is more likely to move toward equilibrium because there are more adjacent micro-states closer to equilibrium than farther from it.

A system at or close to **equilibrium** will stay there because, after any fluctuation away there will be many more micro-states closer to equilibrium than farther from it.

Number of micro-states W

Macro-state distribution L:R

$N : 0$ $N/2 : N/2$ $0 : N$

10.3.2   Entropy and Number of Ways

The idea that systems evolve toward a macroscopic state that can be realized in the maximum number of microscopic configurations or ways corresponds to the evolution of a system toward maximum entropy. If we are to explain irreversibility we will need to link entropy, S, and number of ways, W. The first person to do this was Ludwig Boltzmann and the equation linking the two quantities is:

$$S = k \ln W$$

where k is the Boltzmann constant and W is the number of micro-states.

In this context a micro-state is a distinct way of arranging energy and particles in the system.

The S.I. unit of entropy is JK^{-1}.

It might seem surprising to see the natural logarithm in this equation. However, consider joining two systems each having W configurations, together. The number of possible configurations of the combined system is not $2W$ but W^2. However, if the original entropy of each system is S then the entropy of the two systems together must be $2S$, so the use of a logarithm reduces this multiplicative property to one of addition: $\ln(W^2) = 2\ln(W)$. The logarithmic function makes entropy an extensive quantity like mass so that the total entropy of a combination of different systems is equal to the sum of their individual entropies.

If entropy is defined in this way then the second law, that the entropy of an isolated system tends to a maximum value, follows simply from probability. The equilibrium macro-sate is the most probable macro-state, i.e., the one that can be realized in the largest number of micro-states This is the maximum value of W and therefore maximum entropy. The arrow of time is the direction from low entropy to high entropy and from small W to large W.

10.3.3 Poincaré Recurrence

The assumption that a system will explore all possible microstates implies that, if left for a long enough time, it will eventually return to its initial state. In fact, in an infinite time we would expect it to return to all configurations an infinite number of times. This means that whilst the statistical interpretation of entropy does explain why it is *overwhelmingly likely* that a system will evolve toward maximum entropy it is not impossible for entropy to decrease. Poincaré pointed out that any particular system will have a characteristic average time between returning to its initial state. This is called the Poincaré recurrence time. For the type of macroscopic system that is usually encountered (e.g., a flask of gas) this is enormous even compared to the age of the universe, so it is safe to assume that the second law will hold and these systems will evolve toward states of maximum entropy.

10.4 What is Temperature?

So far we have not given a fundamental explanation of temperature. However we have identified heat transfer from hot to cold with an increase in entropy. If we take a closer look at this we will be able to gain a more fundamental understanding of the concept of temperature.

When two systems at temperatures T_{hot} and T_{cold} are placed in thermal contact heat flows from the system at higher temperature to the system at lower temperature. We showed (see Section 10.2.2) that this is a consequence of increasing entropy. Removing heat δQ from the hotter system reduces its entropy by an amount $\delta Q/T_{hot}$ which is smaller than the increase in entropy of the cold system, $\delta Q/T_{cold}$ resulting in a net increase in entropy of the combined system and therefore satisfying the second law. What this must mean is that transfer of heat δQ to or from a hotter system has a smaller effect on its entropy than transfer of the same amount of heat to or from a colder system. When two systems are in thermal equilibrium the entropy change when heat δQ is transferred between them is zero. Richard Feynman likened it to drying yourself with a small towel. At first you are wetter than the towel and water is transferred from you to the towel. However, once you have used the towel for a while it becomes so wet that it no longer dries you—as you rub yourself with the towel as much water transfers from you to the towel as transfers from the towel to you—both the towel and your skin have reached the same degree of "wetness." Continuing to use the towel results in no further change in wetness. This is analogous to the approach to thermal equilibrium.

Temperature is therefore related to the rate at which entropy changes when heat is transferred to or from a system, and we can define the thermodynamic temperature using this equation:

$$\frac{1}{T} = \frac{dS}{dQ}$$

How does this relate to the statistical description of entropy in terms of microscopic configurations? This can be understood by considering the effect of adding 1 quantum of energy to a system that already contains N quanta. Whilst this will always increase the number of configurations of the system it has a greater effect on the entropy of a system with an initially smaller number of configurations than on one with an initially larger number. Mathematically this is a consequence of the natural logarithm in Boltzmann's formula.

10.5   Absolute Zero and Absolute Entropy

10.5.1   Entropy at Absolute Zero

As the temperature of a system is reduced it will approach a state of minimum energy. In principle, for an ideal crystalline material, this state exists in a single configuration.

Therefore, $W = 1$ and $S = k \ln (W) = 0$

This is a statement of the **third law of thermodynamics**:

▪ The entropy of a perfect crystal is zero at absolute zero.

In practice it is impossible to cool anything to absolute zero so the third law serves as a theoretical starting point for the calculation of absolute entropies.

10.5.2 Calculating Absolute Entropy

Whilst Boltzmann's formula provides a clear idea of what entropy represents it is not always straightforward to calculate entropies based on number of configurations. However, since entropy determines how a system will evolve, e.g., whether a chemical reaction will proceed or not, we need to know the absolute entropy values for different substances. In order to do this we must return to the macroscopic definition of entropy change:

$$\delta S = Q \delta T \text{ (for a reversible heat transfer)}$$

The absolute entropy at temperature T of a substance is then defined as:

$$S = \int_0^T \frac{dQ}{T} = \int_0^T \frac{c(T)dT}{T}$$

where $c(T)$ is the temperature-dependent heat capacity of the substance.

10.5.3 Entropy Changes for an Ideal Gas

When a gas is heated, compressed or allowed to expand, its entropy changes. This is because the number of ways that energy can be distributed amongst the gas particles, and the number of ways the particles themselves can be distributed in the volume available to them, changes. Formulae for entropy changes of an ideal gas under different types of change are derived below.

Isochoric changes (constant volume)

$$\delta U = Q - W \quad \text{but} \quad (W = 0), \quad \text{therefore} \quad \delta U = Q = c_V \delta T$$

$$\Delta S = \int_{T_1}^{T_2} \frac{dQ}{T} = \int_{T_1}^{T_2} \frac{c_V dT}{T} = \int_{T_1}^{T_2} \frac{3R dT}{2T} = \frac{3}{2} R \ln\left(\frac{T_2}{T_1}\right)$$

Isobaric changes (constant pressure)

$$\delta U = Q - W = Q - p \delta V$$

$$Q = \frac{3}{2} R \delta T + p \delta V = \frac{3}{2} R \delta T + \frac{RT \delta V}{V} \text{ (using the ideal gas equation)}$$

$$\Delta S = \int\limits_{T_1}^{T_2} \frac{3R dT}{2T} + \int\limits_{V_1}^{V_2} \frac{R \delta V}{V} = \frac{3}{2} R \ln\left(\frac{T_2}{T_1}\right) + R \ln\left(\frac{V_2}{V_1}\right)$$

Isothermal changes (constant temperature)

For a constant temperature change $\delta U = 0$ and $Q = W = p \delta V$.

$$Q = p \delta V = \frac{R T \delta V}{V} \qquad \text{(using the ideal gas equation)}$$

$$\Delta S = \int\limits_{V_1}^{V_2} \frac{R \delta V}{V} = R \ln\left(\frac{V_2}{V_1}\right)$$

10.6 Refrigerators and Heat Pumps

10.6.1 Refrigerators

The purpose of a refrigerator is to cool things down. To do this heat must flow out of a cooler body (the food you place inside the refrigerator) and into a warmer one (the surrounding air in the room outside the refrigerator). The second law of thermodynamics states that this cannot occur in isolation because it would lower the entropy of the universe. However, it can occur if work is done.

A refrigerator is actually a reversed heat engine:

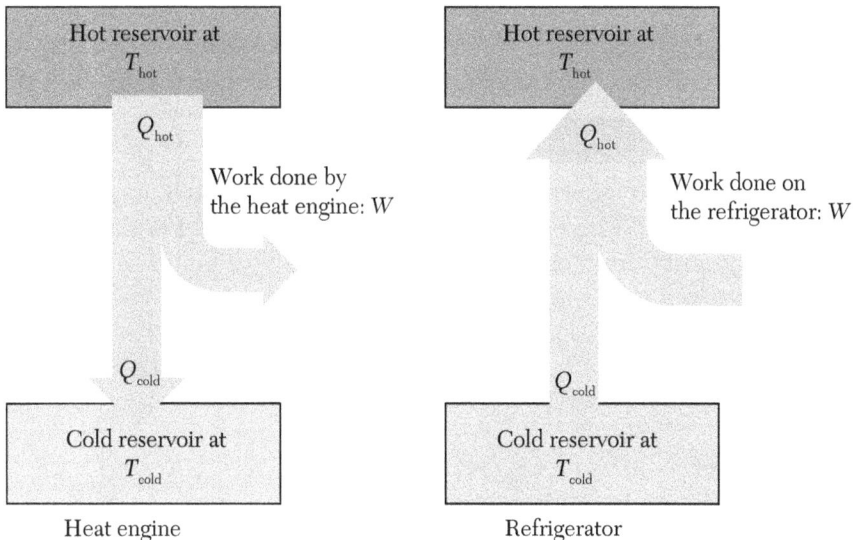

Heat engine   Refrigerator

For a heat engine, the heat reservoirs are burnt fuel (hot) and the environment (cold). For a refrigerator the heat reservoirs are the foodstuff that must be cooled down (cold) and the environment (hot). Typically, a domestic refrigerator removes heat from food at about 4°C and dumps heat into a room at about 20 °C although different types of refrigerator can work in different temperature regimes. The reason you need to plug an electric refrigerator into a mains supply is so that electricity can supply the work that will ultimately be dumped as additional heat in the environment.

For a refrigerator to operate it must obey the second law of thermodynamics so the entropy increase caused by the heat dumped in the environment must be equal to or greater than the entropy decrease caused by extracting heat from the stuff inside the refrigerator:

$$\Delta S_{environment} = \frac{Q_{hot}}{T_{hot}}$$

$$\Delta S_{food} = -\frac{Q_{cold}}{T_{cold}}$$

$$\Delta S_{universe} = \frac{Q_{hot}}{T_{hot}} - \frac{Q_{cold}}{T_{cold}} \geq 0$$

$$\frac{Q_{hot}}{T_{hot}} \geq \frac{Q_{cold}}{T_{cold}}$$

which implies that $\frac{Q_{hot}}{Q_{cold}} \geq \frac{T_{hot}}{T_{cold}}$, so Q_{hot} (the heat dumped in the environment) must be greater than Q_{cold} (the heat extracted from the food). The refrigerator can only operate if we supply additional energy. This is the work $W = Q_{hot} - Q_{cold}$. As far as the universe is concerned a refrigerator is a heater!

The coefficient of performance (CoP) is a measure of how effective the refrigerator is and is the ratio Q_{cold}/W. The higher this is, the more joules of heat are removed from objects inside the refrigerator per joule of electrical work done, so it is like the "efficiency" of the refrigerator. The second law sets a limit to the theoretical thermodynamic efficiency of a heat engine and it also sets a limit to the theoretical coefficient of performance for a refrigerator.

$$CoP_{refrigerator} = \frac{Q_{cold}}{W} = \frac{Q_{cold}}{Q_{hot} - Q_{cold}} = \frac{1}{\frac{Q_{hot}}{Q_{cold}} - 1} \leq \frac{1}{\frac{T_{hot}}{T_{cold}} - 1}$$

10.6.2 Heat Pumps

Heat pumps are used to extract heat from a cooler environment and dump it into a warmer environment and so, like a refrigerator, they are really a reversed heat pump. However, the point of the heat pump is not the amount of heat extracted from the cold reservoir but the amount of heat delivered to the hot reservoir. The coefficient of performance for a heat pump is therefore defined as the ratio Q_{hot}/W and the second law once again sets a limit on this:

$$\text{CoP}_{\text{heat pump}} = \frac{Q_{\text{hot}}}{W} = \frac{Q_{\text{hot}}}{Q_{\text{hot}} \; Q_{\text{cold}}} = \frac{1}{\frac{Q_{\text{cold}}}{\text{hot}}} \leq \frac{1}{\frac{T_{\text{cold}}}{\text{hot}}}$$

10.7 Implications of the Second Law

10.7.1 The Second Law, the Arrow of Time, and the Universe

When the second law of thermodynamics is applied to the universe it states:

▪ The entropy of the universe tends to a maximum value.

This implies that the arrow of time points in the direction of increasing entropy. The past is a low entropy state and the future is a high entropy state.

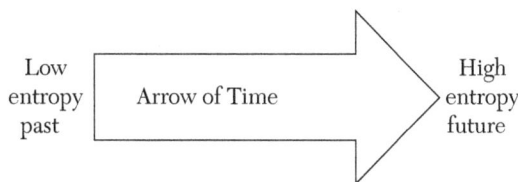

On a microscopic scale the universe is evolving from a macro-state that exists in a low number of ways (small number of configurations) to a macro-state that exists in a much larger number of ways and ultimately to a state of maximum entropy in which the number of possible microscopic configurations reaches its maximum value. Whilst this final state of the universe is very far in the future, once reached heat engines would no longer be able to do useful work because everything would be at the same temperature. Energy would become **unavailable** (for work) because heat engines operate between heat reservoirs at different temperatures and increase the entropy of the universe. When the universe reaches thermal equilibrium entropy is at a maximum. This final state is sometimes called the **heat death** of the universe.

So, have we explained the arrow of time? Not entirely. The microscopic description of the second law shows that, *if* a macroscopic system starts off in a low entropy state it is overwhelmingly likely to move to the high entropy states that can be realized in a larger number of ways. However, the arrow of time ceases to exist once the system has reached equilibrium, so one large question remains: why did the universe start in a low entropy state? According to our analysis the initial configuration of the universe must have been one that can be realized in only a relatively small number of ways (low W). This was a state of low probability in the sense that if we considered the totality of macro-states the universe can have then the actual state in which it began belongs to a tiny subset of these. This is rather like the first example we considered, the ink droplet in water—the system starts in a very special low probability, low entropy state and evolves toward a higher probability, high entropy equilibrium. Our universe is also evolving toward a high entropy equilibrium state. The assumption that the universe began in a low entropy state is sometimes called the "**past hypothesis**."

10.7.2   Second Law and Living Things

On the face of it, living things do not seem to obey the second law of thermodynamics. They gather resources and energy and grow from simple beginnings into complex functioning beings. They seem to create order from disorder and lower their own entropy. However, living things are not isolated systems. They exchange energy and matter with their environments. While it is true that constructive processes within our cells might lower the entropy of certain internal structures, it is also true that the metabolism that drives this generates heat and dumps it into the environment. This increases the entropy of our surroundings by an amount that is much greater than any local reduction of entropy inside our bodies so the net effect of a living thing is to increase the entropy of the universe. When living things are considered along with their surroundings they do obey the second law of thermodynamics just like anything else!

10.7.3   Entropy and Energy Availability

There is a continual search for new sources of energy to satisfy our energy-hungry civilization. However, it is often said that we are fast approaching an energy crisis. Given that energy is never created or destroyed this seems almost contradictory. How can our energy run out? The answer is actually quite simple. Every time we harness a non-renewable source such as oil, natural gas, or uranium and use it in a heat engine (e.g., a thermal power station) to do work there is an increase in the entropy of the universe. The primary energy source we used has transferred its energy to a large number of highly dispersed particles (e.g., in heating the atmosphere) at a much lower temperature. This makes the original energy much harder to harness and use in another heat engine to do more work. The original energy has not disappeared but has become unavailable. Increasing entropy can be thought of as the increasing unavailability of energy to do work.

10.8 Exercises

1. (a) Explain the following terms:
- Macroscopic state of a system
- Microscopic state of a system
- Number of ways

(b) Explain the link between number of ways and entropy.

(c) Use the terms above to explain why, when a small crystal of potassium permanganate is placed into a large beaker of water, the system become mixed over time but is never observed to un-mix.

(d) Discuss whether, if left long enough, the system could un-mix.

2. (a) What is meant by "The Arrow of Time"?

(b) Explain why Newton's laws do not provide such an arrow.

(c) Explain how thermodynamics does provide such an arrow.

(d) Discuss whether cosmology provides another arrow of time.

3. (a) What does the second law of thermodynamics imply about the entropy of the early universe? Explain.

(b) What does the second law of thermodynamics imply about the macro-state of the early universe (in terms of probability)?

4. (a) Explain why living things might *seem* to violate the second law of thermodynamics.

(b) Explain how in fact living things do not violate it and are in fact governed by it.

5. (a) Explain how each of the following statements is consistent with the fact that the entropy of an isolated system cannot decrease:

(i) It is impossible to transfer thermal energy to mechanical work with 100% efficiency.

(ii) It is impossible to transfer heat from a colder to a hotter body with no other changes taking place.

(b) Explain how a refrigerator *can* transfer heat from a colder to a hotter body without violating the second law of thermodynamics.

6. Show that the efficiency of an ideal heat engine is limited by the equation, efficiency $\leq 1 - \dfrac{T_2}{T_1}$.

CHAPTER 11

OSCILLATIONS

11.0   Introduction

An oscillator undergoes regular periodic motion about a fixed equilibrium position (or value). The simple model of a mass oscillating on a spring can be adapted to explain a wide variety of physical phenomena, from lattice vibrations in a crystalline solid to the effects of seismic waves. The mathematical analysis developed to describe mechanical oscillators can also be used for electrical and electromagnetic oscillations and is the starting point for understanding all kinds of wave motion.

11.1   Capturing and Displaying Oscillatory Motion

A motion sensor can be used to record the position and time of a simple oscillator consisting of a mass suspended from a light spring that obeys Hooke's law.

The diagram shows a graph of position against time. This is sinusoidal with an amplitude A and time period T. An oscillation that is purely sinusoidal is called **simple harmonic**.

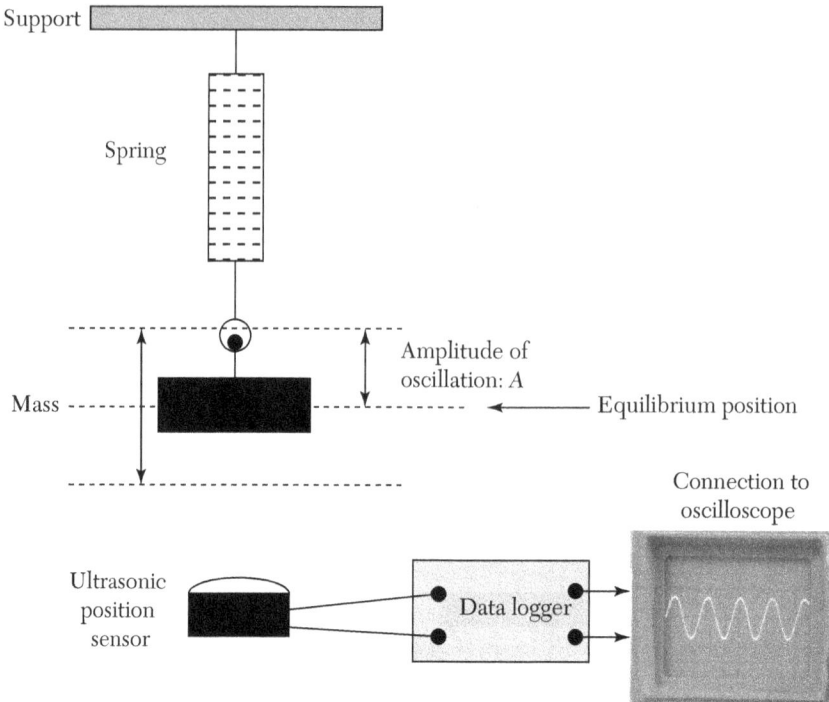

- Amplitude A: maximum displacement from equilibrium

- Time period T: time for one complete cycle of oscillation

- Frequency f: number of oscillations per second, $f = \dfrac{1}{T}$

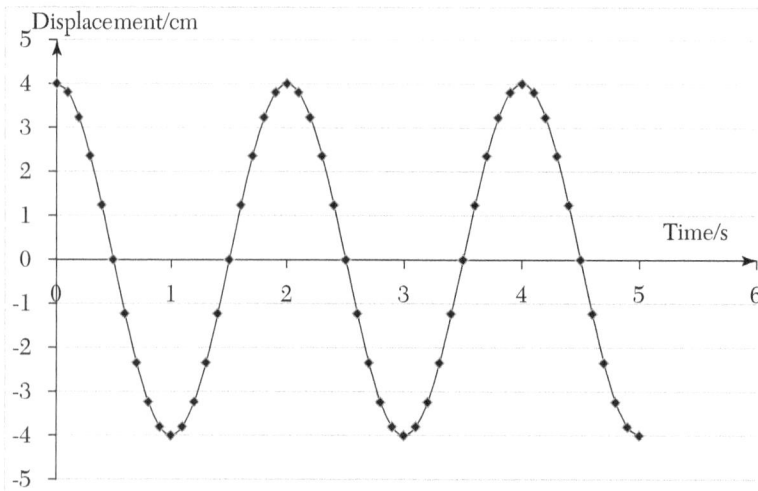

The graph above shows a simple harmonic oscillation with amplitude 4.0 cm, time period 2.0 s, and frequency 0.50 Hz. At $t = 0$ the oscillator is at its maximum positive amplitude, so the displacement x of the oscillation can be represented by cosine function:

$$x = 4.0 \text{ cosine } (\pi t)$$

(πt is in radians, so when $t = 2s$ the oscillator completes one cycle of oscillation.)

In general the displacement x of a simple harmonic oscillator can be represented by an equation of the form:

$$x = A\cos(\zeta t + \delta)$$

The term in the parentheses is called the phase of oscillation. It is an angle in radians.

The oscillator must complete one cycle in a time T, so $\omega T = 2\pi$ and $\omega = 2\pi/T = 2\pi f$. This is called the *angular frequency* of the oscillation.

$$x = A \cos(2\pi f t + \delta)$$

δ is a variable phase angle that affects the displacement at $t = 0$. If $\delta = 0$ the graph will appear like the one above. If $\delta = -\pi/2$ then the graph will be a sine curve, starting at $x = 0$. Other values of d simply change the value of displacement at $x = 0$ without changing the shape of the curve.

11.1.1   Graphs and Equations of Displacement, Velocity, and Acceleration

Most motion data loggers connected to a position sensor can also be used to display velocity and acceleration. Mathematically these are related to displacement by differentiation:

$$x = A \cos \omega t$$

$$v = \frac{dx}{dt} = -\omega A \sin \omega t$$

$$a = \frac{dv}{dt} = -\omega^2 A \cos \omega t = -\omega^2 x$$

There is a $\pi/2$ phase difference between v and x and between a and v. There is a π phase difference between a and x.

The three graphs that follow show how these graphs are related to one another. Note the maximum values of velocity and acceleration, and when these occur,

- $v_{max} = \omega A$ when displacement is zero

- $a_{max} = \omega^2 A$ when displacement is maximum and velocity is zero

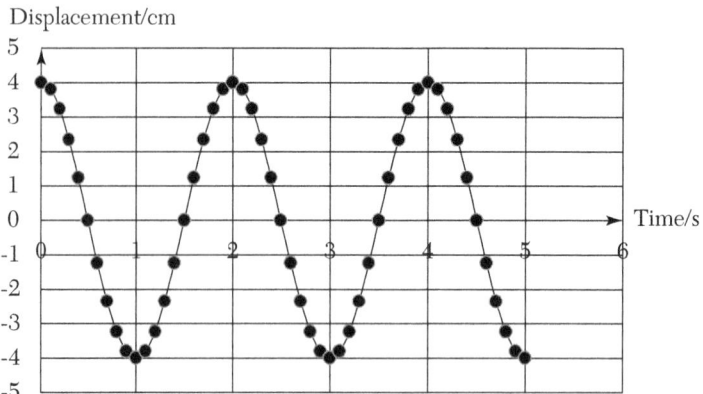

$v = dv/dt$

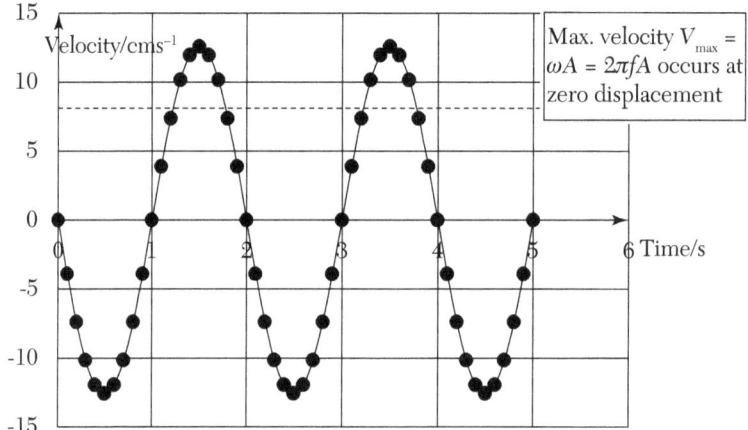

Max. velocity V_{max} = ωA = $2\pi fA$ occurs at zero displacement

$a = dv/dt$

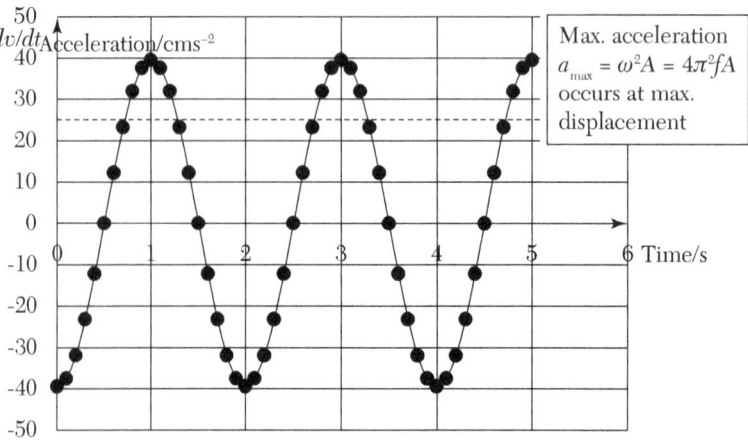

Max. acceleration a_{max} = $\omega^2 A$ = $4\pi^2 fA$ occurs at max. displacement

11.1.2 Phase and Phase Difference

Oscillations are cyclic repeated motions, and the position of an oscillator within its cycle of oscillations is called its phase. Once complete, a cycle corresponds to a phase change of 360° or 2π radians. In graphical terms, a phase shift is equivalent to moving the graph along the time axis by a certain amount. A shift of time period T corresponds to a phase change of 2π. The graph below shows two oscillations with a phase difference of $\pi/3$.

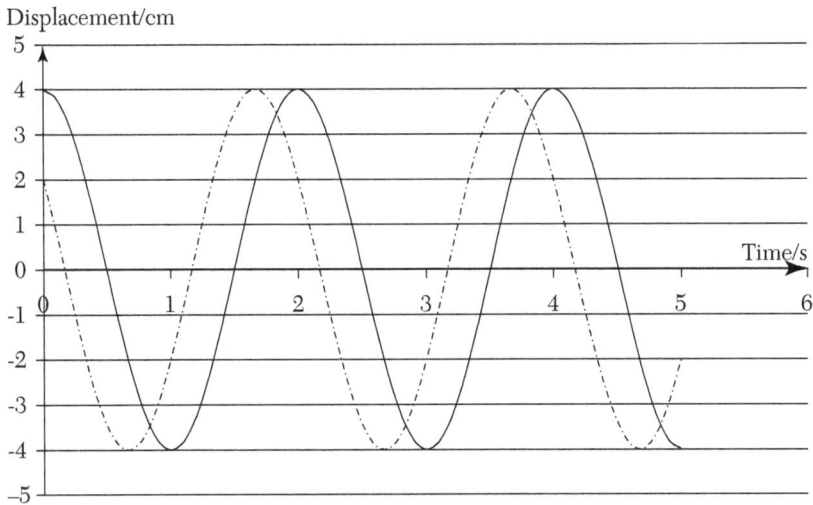

11.2 Simple Harmonic Motion

11.2.1 Equation of Motion for Simple Harmonic Motion

A simple harmonic motion is defined by two characteristics:

- Acceleration is directed toward a fixed equilibrium position.

- Acceleration is directly proportional to displacement from that position.

The equation of motion for simple harmonic motion is:

Acceleration = − (constant) × (displacement from equilibrium)
$$a = -\omega^2 x$$

This can be written as a second-order differential equation:

$$\frac{d^2x}{dt^2} = -\omega^2 x$$

The constant is written as ω^2 because this turns out to be physically significant (it corresponds to the angular frequency of the oscillation). A general solution to this equation is:

$$x = A\cos(\zeta t + \delta)$$

as can be easily shown by differentiating this twice. It then follows that:

$$\omega = 2\pi f = 2\pi/T$$

Notice that if we did not use a squared term for the constant in the equation of motion, we might have had a square-root in this solution.

11.2.2 Physical Conditions for Simple Harmonic Motion

To show that a mechanical oscillator will exhibit simple harmonic motion, we must demonstrate that the forces acting on the oscillator will result in an equation of motion of the form $a = -\omega^2 x$. For an oscillator of constant mass m, this will be true if $F = ma = -m\omega^2 x$. The conditions are:

- Resultant force is directed toward a fixed equilibrium position.
- Resultant force is directly proportional to displacement from that position.

Once we have established that this is the case, we can find ω and hence f from the constant of proportionality between force and displacement:

If $\quad F \propto -x$, then $F = -kx = -m\omega^2 x$

So $$\omega = \sqrt{\frac{k}{m}} \quad \text{and} \quad f = \frac{1}{2\pi}\sqrt{\frac{k}{m}}$$

11.3 Mass-Spring Oscillator

An ideal mass-spring oscillator has a mass m attached to a massless spring with spring constant k. In equilibrium the spring has an extension x_0. At this point the weight of the mass is supported by the tension in the spring.

$$kx_0 = mg$$

When the spring is displaced from equilibrium by an additional amount x, the magnitude of the resultant force on the mass becomes:

$$k(x + x_0) - mg = kx$$

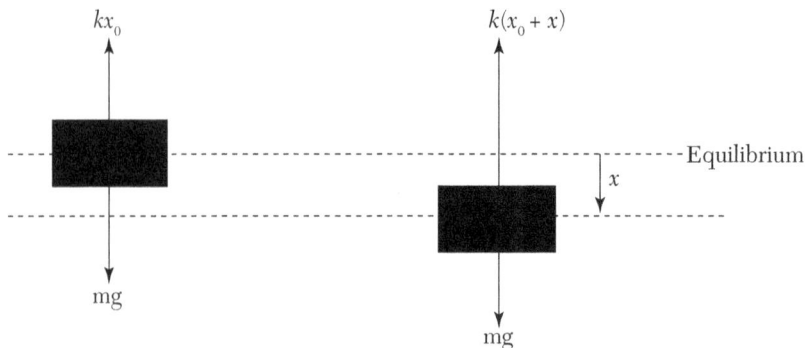

Resultant force is in the opposite direction to the displacement from equilibrium, so:

$$F = -kx$$

This satisfies our conditions for simple harmonic motion, so an ideal mass-spring oscillator *is* a simple harmonic oscillator. Now we can use angular frequency to find the frequency and time period:

For simple harmonic motion, $F = -m\omega^2 x$

For the mass-spring oscillator, $F = -kx$

Therefore, $\omega = \sqrt{\dfrac{k}{m}}$

which gives $f = \dfrac{1}{2\pi}\sqrt{\dfrac{k}{m}}$ and $T = 2\pi\sqrt{\dfrac{m}{k}}$

11.4   Simple Pendulum

A simple pendulum consists of a point mass suspended from a light in extensible string of length l. When it is displaced by a small angle from its equilibrium position, it undergoes regular oscillations that are approximately simple harmonic.

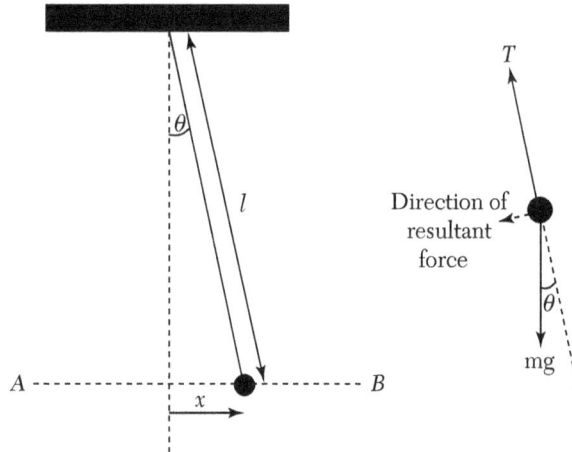

The free body diagram on the right shows that the resultant of the two forces (tension and weight) acting on the bob will be $mg \sin\theta$ along the line shown. If we consider the horizontal motion (along AB), then the resultant force acting horizontally is:

$$F = -mg \sin\theta \cos\theta$$

It is clear from the diagram on the left that $\sin\theta = x/l$, so

$$F = -\frac{mgx}{l}\cos\theta$$

This does *not* satisfy our condition for simple harmonic motion because the resultant force is proportional to $x\cos\theta$ and not just to x ($\cos\theta$ is itself a function of x). However, for small angles, $\cos\theta$ is approximately equal to 1; so the smaller the angular amplitude of the oscillation, the closer it will be to an ideal simple harmonic oscillator:

$$\text{As } \theta \to 0, \quad \cos\theta \to 1 \quad \text{and} \quad F \to -\left(\frac{mg}{l}\right)x$$

Under these conditions the constant of proportionality between F and x is (mg/l).

Therefore, $\omega = \sqrt{\dfrac{mg}{l}}$

which gives $f = \dfrac{1}{2\pi}\sqrt{\dfrac{g}{l}}$ and $T = 2\pi\sqrt{\dfrac{l}{g}}$

Notice that this is independent of the mass of the bob. This is because the resultant force is directly proportional to the mass, so mass will cancel when calculating accelerations:

$$a = F/m = (\text{constant} \times m)/m.$$

This is for the same reason that objects of different mass fall with the same acceleration in the same gravitational field.

11.5   Energy in Simple Harmonic Motion

11.5.1   Variation of Energy with Time

A simple harmonic oscillator continually transfers energy between kinetic energy and potential energy. For the simple pendulum the potential energy is purely gravitational; for the mass-spring oscillator it is a combination of gravitational and potential. If frictional forces (damping) can be neglected, the total energy of the oscillator, which is the sum of its kinetic and potential energies, remains constant. Under these conditions the oscillator is described as a **free oscillator**.

It is easy to analyze the energy of the oscillator by starting with its kinetic energy.

$$KE = \frac{1}{2}mv^2 = \frac{1}{2}m(-\omega A\sin(\omega t))^2 = \frac{1}{2}m\omega^2 A^2\sin^2(\omega t)$$

If we take potential energy to be zero at the equilibrium position, then the maximum kinetic energy is equal to the total energy TE:

$$TE = \frac{1}{2}m\omega^2 A^2$$

$$TE = KE + PE,$$

So $PE = TE - KE = \frac{1}{2}m\omega^2 A^2\left(1 - \sin^2(\omega t)\right) = \frac{1}{2}m\omega^2 A^2 \cos^2(\omega t)$

The graphs below shows how each type of energy varies during one cycle of oscillation.

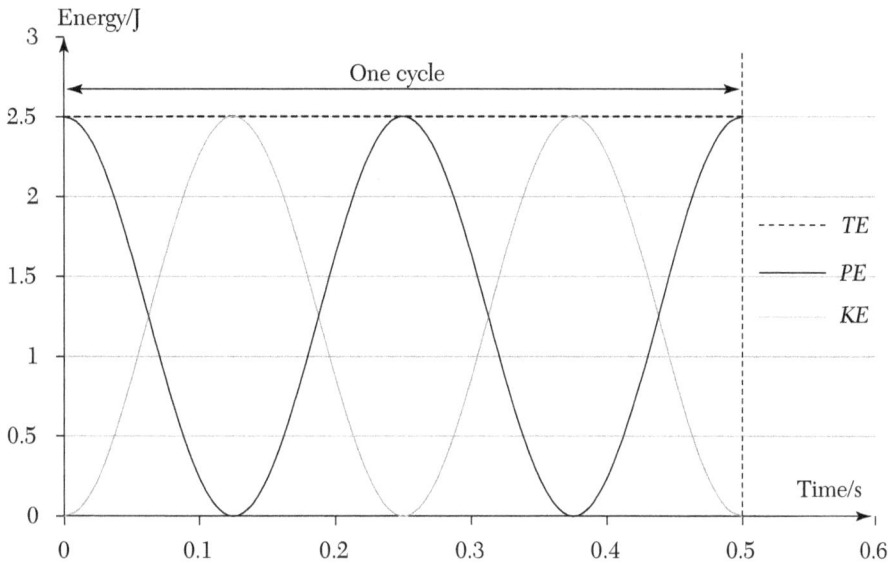

Notice that both KE and PE reach their maximum values twice per oscillation; this is because they are related to the square of the sine and cosine.

11.5.2   Variation of Energy with Position

The graphs above show how energy varies *with time* during one oscillation. It is also interesting to see how the different types of energy vary *with position*. To do this we will look at the work done by the resultant force F.

Since $F = -kx$ for simple harmonic motion, we can find the total energy of the oscillator by integrating the work done by the oscillator as it moves from equilibrium to one amplitude.

$$W = \int_{x=0}^{x=A} F dx = \int_{x=0}^{x=A} -kx dx = -\frac{1}{2}kA^2$$

The oscillator had maximum *KE* at equilibrium and has zero *KE* at the amplitude, so the maximum *KE*, and therefore total energy of the oscillator, is given by:

$$TE = \frac{1}{2}kA^2$$

By the same analysis, the work done in moving to a displacement *x* is ½ *kx²* therefore:

$$KE = \frac{1}{2}k(A^2 - x^2)$$

$$PE = \frac{1}{2}kx^2$$

The variation of each type of energy with position is shown below:

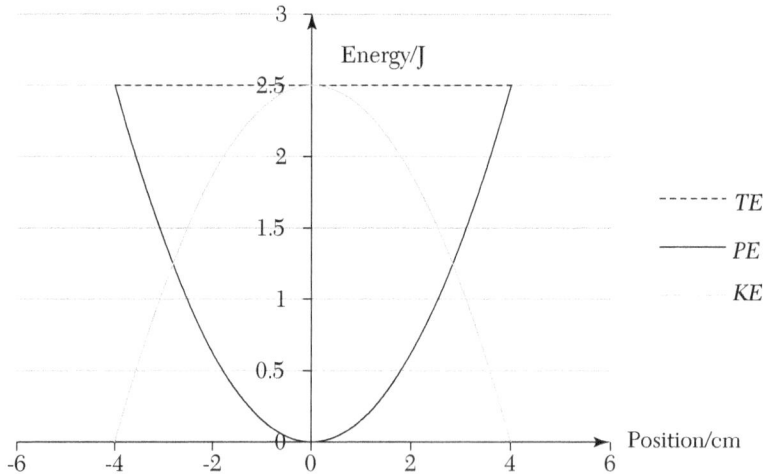

11.5.3 Damping

Real oscillators are subject to frictional forces that oppose their motion. The oscillator must do mechanical work against these forces, so (unless it is driven by an external energy source) its total energy decreases with time and so does its amplitude. The oscillator is **damped**. The heavier the damping, the greater the rate at which the oscillator loses energy and the greater the rate of decay of its amplitude. In many cases the amplitude decays approximately exponentially. This occurs when the oscillator loses the same fraction of its total energy on each oscillation.

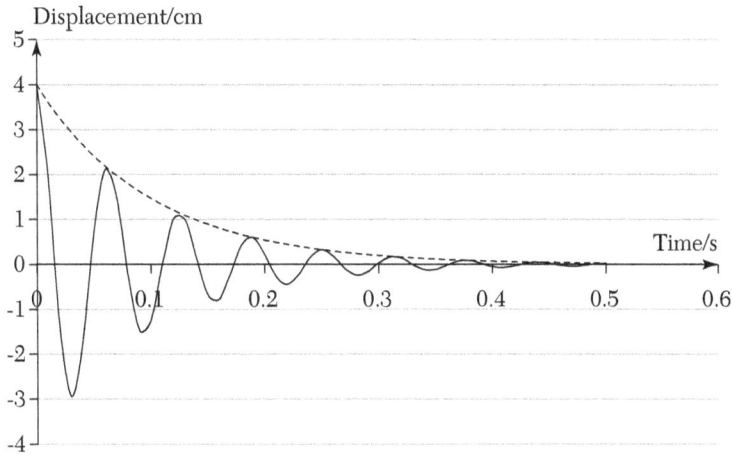

If the oscillation does decay exponentially, its displacement varies according to an equation of the form:

$$x = Ae^{-\gamma t} \cos(\omega t)$$

where γ is a damping coefficient.

Light damping has very little effect on the natural frequency of the oscillator, but heavy damping reduces this frequency.

11.6   Forced Oscillations and Resonance

If an oscillator is driven by an external periodic force, it is called a forced (or driven) oscillator. Its response to the driver depends on the relationship between the frequency of the driver f_d and its natural frequency f_0 for free oscillations. It will also depend on the strength of coupling to the driver and on the amount of damping in the system.

The qualitative effects on a forced oscillator can be demonstrated very simply. Take a simple pendulum of about 1 m length and hold the end of the string in your hand. Move your hand back and forth along a line at a very low frequency with an amplitude of about 2.0 cm. The string remains more or less vertical, and the bob of the pendulum moves with the same amplitude phase and frequency as your hand. Gradually increase the frequency of hand movement while keeping the same amplitude of motion. The bob continues to move with the driving frequency but with an increasing amplitude; and at a certain frequency, the amplitude of the bob's motion increases dramatically while lagging about $\pi/2$ behind in phase. If you continue to increase the frequency beyond this point, the amplitude of response gets smaller; and when you shake the string at a high frequency, the bob stays almost still at the center of its motion. Closer inspection shows that it does oscillate with the driver frequency but with a tiny amplitude and a phase lag of almost π. The strong response occurs when the driving frequency is equal to the natural frequency of the oscillator and is called **resonance**. This pattern of response is shown in the graph below.

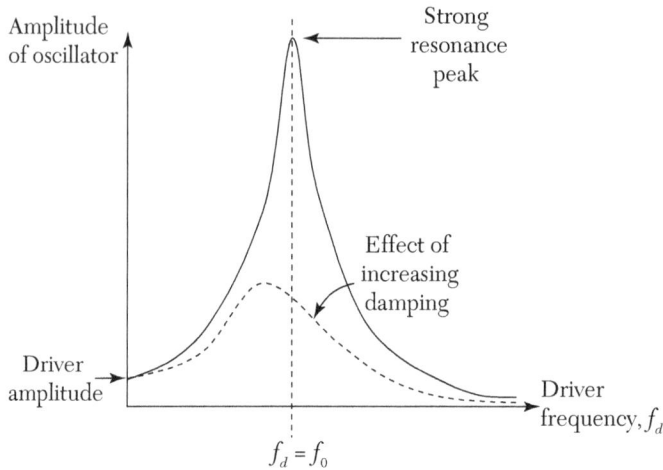

Increasing damping has two effects:

◾ the amplitude at resonance is lower

◾ the frequency of the resonance is slightly below the natural frequency, f_0

At resonance the oscillator absorbs energy from the driver and the amplitude grows until energy losses due to damping occur at the same rate as energy is supplied from the driver. As damping is reduced, this balance occurs at ever-increasing amplitudes, and if there was no damping, it would grow without limit. This can be destructive, and this kind of destructive resonance is applied in ultrasonic devices used for cleaning jewelry or breaking up kidney stones. In other situations damping is deliberately increased in order to reduce the amplitude at resonance and protect the structure, for example, in the design of earthquake-resistant buildings.

The phase relation between the driver and the driven oscillator varies from 0 at very low frequencies to π at high frequencies. At resonance the driven oscillator lags the driver by $\pi/2$.

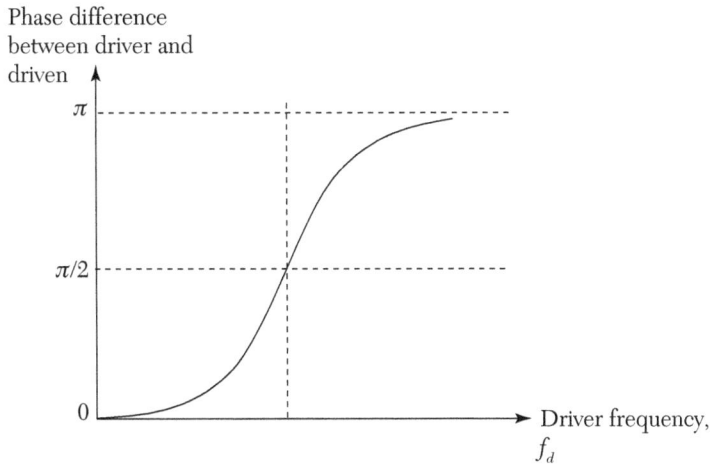

Phase difference between driver and driven (vertical axis, with values π, $\pi/2$, 0) versus Driver frequency, f_d (horizontal axis).

11.7   Exercises

1. A particle moves with simple harmonic motion between points A and C in the diagram below. Where is it at the instance when:

(a) it is stationary?
(b) it has maximum velocity to the right?
(c) it has zero acceleration?
(d) it has maximum acceleration to the left?
(e) it has maximum kinetic energy?
(f) it has maximum potential energy?

2. The diagram below shows a vehicle tethered between two similar springs. When it is displaced from equilibrium and released, it undergoes periodic oscillations about an equilibrium position. Neither spring goes slack during these oscillations and both springs obey Hooke's law in compression and extension. The vehicle has a mass of 0.80 kg.

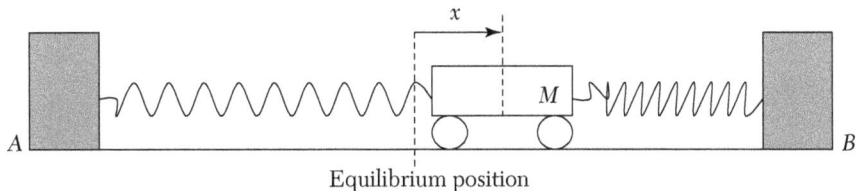

Equilibrium position

(a) Explain what is meant by the "equilibrium position."
(b) Explain why the oscillations will be simple harmonic.

   The trolley is displaced 6.0 cm to the right and released. It oscillates with a period of 2.0 s.

(c) Calculate the frequency of the oscillations.
(d) State the initial amplitude of the oscillations.
(e) Write down an equation for the displacement of the trolley as a function of time. Take $t = 0$ to be the moment of release.
(f) Sketch graphs of (i) displacement versus time, (ii) velocity versus time, and (iii) acceleration versus time for the motion of the trolley when it is released from an initial displacement of 6.0 cm and then oscillates with a period of 2.0 s. Draw the graphs one above the other and label them as completely as you can.
(g) Calculate the maximum velocity and total energy of the oscillations.
(h) What would happen to the period of oscillation if the trolley is stopped and then released from a displacement of 3.0 cm?
(i) The period of oscillations will change if the mass is changed or the stiffness of the springs is changed. Why?
(j) After the trolley is released, its oscillations gradually die away. Explain why this happens.
(k) As the oscillations decay, successive amplitudes are:6.0, 5.4, 4.9, 4.4, 3.9, 3.5, 3.2 cm

   A student suggests that the amplitude is decaying exponentially. Devise and carry out a test on this data to check this suggestion.
(l) Predict the amplitude of the oscillator after 10 oscillations.

3. (a) Show that $x = A \cos \omega t$ is a solution to the simple harmonic equation of motion $\dfrac{d^2x}{dt^2} = -\omega^2 x$

(b) A particular oscillation can be represented by the equation: $x = 2.5 \cos 10\, \pi t$.
   (i) State the amplitude of this oscillation.

   (ii) Calculate the time period and frequency of the oscillation.

   (iii) Calculate the maximum velocity of the oscillator.

   (iv) Calculate the maximum acceleration of the oscillator.

(c) (i) Derive an equation for the variation of kinetic energy with time for this oscillator given that the mass of the oscillator is 0.50 kg.
   (ii) Sketch a graph of kinetic energy with time for this oscillator.

(iii) Add a second line to your graph to show the variation of potential energy with time.

4. During an earthquake the floor of a building oscillates vertically with an amplitude A and frequency f. Derive a formula for the frequency at which objects in contact with the floor will just lose contact during the earthquake.

5. An atom of mass 5×10^{-26} kg is in a cubic lattice with all bonds between adjacent pairs of atoms having a spring constant about $100 \, \text{Nm}^{-1}$.

  (a) Estimate the natural frequency of oscillation of the atom.

  (b) If it is able to absorb radiation at this frequency, what part of the electromagnetic spectrum does it absorb?

6. A simple pendulum has a period of 1.6 seconds on Earth.

  (a) Calculate its length.
  (b) As the temperature rises, its length increases by 1%. Calculate the percentage change in its time period and state whether this is an increase or a decrease.
  (c) Explain why the mass of the pendulum bob does not affect its time period.
  (d) A mass-spring oscillator is set up so that it has the same time period as the pendulum. Both are then transported to the Moon. How do their time periods compare on the moon? Explain.

CHAPTER 12

ROTATIONAL DYNAMICS

12.0 Introduction

The circle is a simple geometric shape that is used throughout science to model cyclic processes and is often the starting point for more complex theories (e.g., of planetary motion or electron orbits). Simple harmonic motion can be regarded as a projection of circular motion onto a diameter, making the concept of phase very clear. Rotating vectors, or phasors, are powerful ways to model oscillations and waves.

12.1 Angles

12.1.1 Measuring Angles in Radians

The degree is an arbitrary division of the circle into 360 equal parts. This is a useful measure of angle, but in physics it is often simpler to work in a different unit, the radian. The reason for this is that the radian is defined directly from the geometry of the circle.

- **Definition:** The angle θ at the center of a circle subtended by an arc of length l is equal to the ratio l/r where r is the radius of the circle.

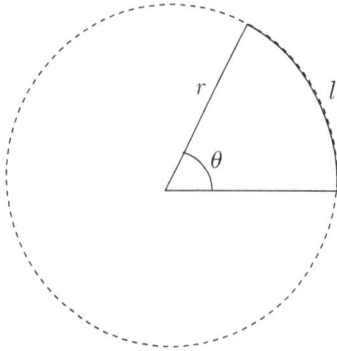

$\theta = l/r$
For a complete circle, $l = 2\pi r$
Therefore, $\theta_{circle} = 2\pi$ radians
To convert between degrees and radians,
2π radians = 360°; π radians = 180°;
$\pi/2$ radians = 90°

12.1.2 Small-Angle Approximations

For small angles there are some useful approximations that can simplify calculations involving trigonometric functions such as sine and cosine. Consider a right-angled triangle used to define the trigonometric functions (a = adjacent side, o = opposite side, h = hypotenuse).

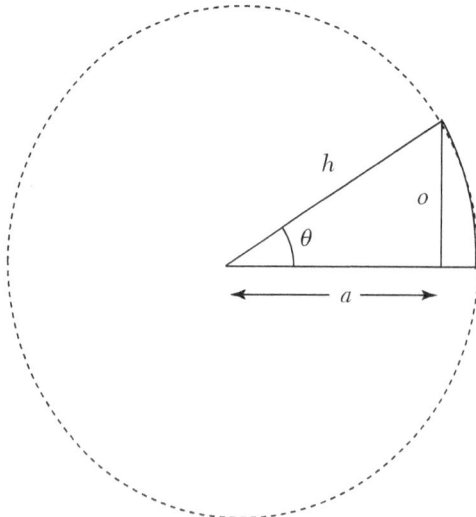

$\theta = \dfrac{l}{h}$ (h = radius of circle)

$\sin\theta = \dfrac{o}{h}$

$\cos\theta = \dfrac{a}{h}$

As $\theta \to 0$ $o \to l$ and $a \to h$ (radius)

therefore:
$\sin\theta \to \theta$ (in radians)
$\cos\theta \to 1$
$\tan = \dfrac{\sin\theta}{\cos\theta} \to \theta$

The significance of this is that, for small angles, we can replace the sine or tangent of the angle with the angle itself (in radians). But how small is small? Like all approximations this depends on how precise a value is needed. The table below shows that the approximations work to better than 1% when the angle is 0.1 radian and better than 2% for 0.2 radian.

Angle in radians	sine of angle	cosine of angle	tangent of angle	Angle in degrees
1.00	0.842	0.540	1.56	57.3
0.50	0.479	0.878	0.546	28.6
0.30	0.296	0.955	0.309	17.2
0.20	0.199	0.980	0.203	11.5
0.10	0.0998	0.995	0.100	5.73

12.2 Describing Uniform Circular Motion

An object moving in uniform circular motion has constant speed and turns through the same angle every second.

12.2.1 Angular Displacement, Angular Velocity, and Angular Acceleration

Consider an object of mass m moving in uniform circular motion of radius r at a constant speed v. There is a linear displacement δs and angular displacement $\delta\theta$ in time δt.

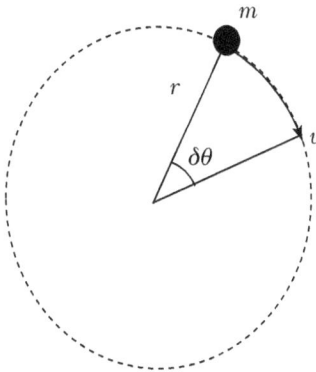

Angular velocity is defined as the rate of change of angle, $\omega = \dfrac{\delta\theta}{\delta t}$ (rads^{-1})

From the diagram, $\gamma\theta = \dfrac{\delta s}{r} = \dfrac{v\delta t}{r}$

Therefore, $\dfrac{\delta\theta}{\delta t} = \dfrac{v}{r}$

Taking $\delta t \to 0$ gives:

$\omega = \dfrac{v}{r}$ or $v = r\omega$

This is a particularly useful relation.

For uniform circular motion, ω = constant, so the angular displacement $\Delta\theta = \omega\Delta t$. If the period of rotation is T, then:

$$2\pi = \omega T \quad \text{and} \quad \omega = \frac{2\pi}{T} = 2\pi f$$

which relates the angular frequency ω(rads^{-1}) to the rotation frequency f (Hz) or the time period of rotation T(s).

The angular acceleration a is defined as the rate of change of angular velocity:

$$\alpha = \frac{d\omega}{dt} = \frac{d^2\theta}{dt^2}$$

For circular motion, the radius is constant, so:

$$\alpha = \frac{d\omega}{dt} = \frac{d}{dt}\left(\frac{v}{r}\right) = \frac{1}{r}\frac{dv}{dt} = \frac{a}{r}$$

where a is the tangential acceleration (not to be confused with centripetal force).

12.3 Centripetal Acceleration and Centripetal Force

In uniform circular motion, angular velocity and speed are both constant. Velocity, however, is continually changing. This is because velocity is a vector quantity and its direction is changing. Acceleration is defined as the rate of change of velocity, so even though speed is unchanging, the object is accelerating. The direction of this acceleration (as we shall see later) is toward the center of the circle: this is called a **centripetal acceleration**. From Newton's second law there must be a resultant force acting toward the center of the circle. This is called a centripetal force, and it is this force that is responsible for circular motion.

12.3.1 Centripetal Acceleration

Consider a mass m moving in uniform circular motion at constant angular velocity ω. The vector diagram on the right shows how the velocity changes during a short time δt.

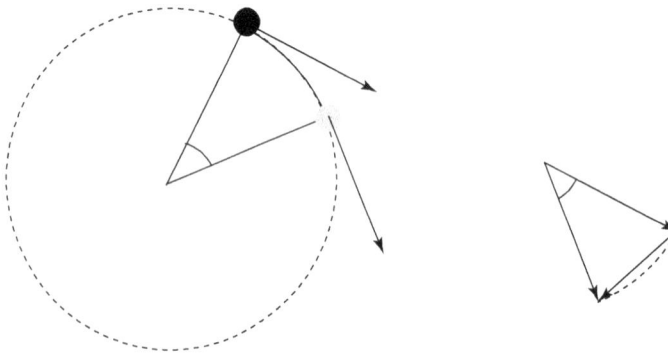

For small angles, δv in the vector triangle becomes equivalent to an arc of a circle (shown by the dotted line), so:

As $\delta\theta \to 0$, $\qquad \dfrac{\delta v}{v} \to \delta\theta$

We also have $\qquad \delta\theta = \dfrac{v\delta t}{r}$

Equating the two expressions for $\delta\theta$:

$$\frac{v\delta t}{r} \approx \frac{\delta v}{v} \quad \text{and} \quad a = \frac{\delta v}{\delta t} \approx \frac{v^2}{r}$$

In the limit of $d\theta \rightarrow 0$,

$$a = \frac{dv}{dt} = \frac{v^2}{r}$$

This is an expression for the acceleration. This can also be written in terms of the angular velocity, $a = r\omega^2$.

It is also clear from the vector diagram that in the limit of $\delta\theta \rightarrow 0$ the vector change in velocity δv becomes perpendicular to both v_1 and v_2 and is directed inwards of the center of the circle. It is a **centripetal** (center-seeking) acceleration.

Centripetal acceleration, $a = \dfrac{v^2}{r} = r\omega^2$, is toward the center of the circle.

12.3.2 Centripetal Force

Newton's first law of motion states that an object will continue to move in a straight line at constant velocity unless acted upon by a resultant external force. This implies that, when an object changes direction of motion, even if its speed does not change, there must be a resultant external force. Newton's second law relates the resultant force to acceleration by the equation $F = ma$, so the magnitude of the force acting in uniform circular motion is given by:

Centripetal force, $F = ma = \dfrac{mv^2}{r} = mr^2$, is toward the center of the circle.

12.3.3 Centripetal Not Centrifugal!

It is a matter of common experience that one feels forced to the outside of the curve when traveling inside a turning vehicle. This is often explained by saying that a centrifugal ("center fleeing") force acts outward from the center of the turning circle. This force does not exist. It is an apparent or "inertial" force that *seems* to be needed to make sense of motion in a non-inertial (accelerating) reference frame. To understand this more clearly, consider how a loose object would behave inside a moving vehicle that is initially moving in a straight line and then begins to turn in an arc of a circle. The object is free to slide inside the vehicle and is initially traveling at the same speed as the vehicle. The upper diagrams show the view from outside the vehicle, and the lower diagrams show the view from inside the vehicle.

Vehicle and object moving at the
same constant velocity

Vehicle begins to turn, but there is no resultant force
on object, so it continues to move in a straight line

Vehicle collides with an object providing inward force:
object begins to move in circular motion

Object remains at rest
inside vehicle

Object collides with wall and
exerts an outward force on it

Relative to vehicle the object begins to
accelerate toward side. This is explained by
assuming there is an outward centrifugal force

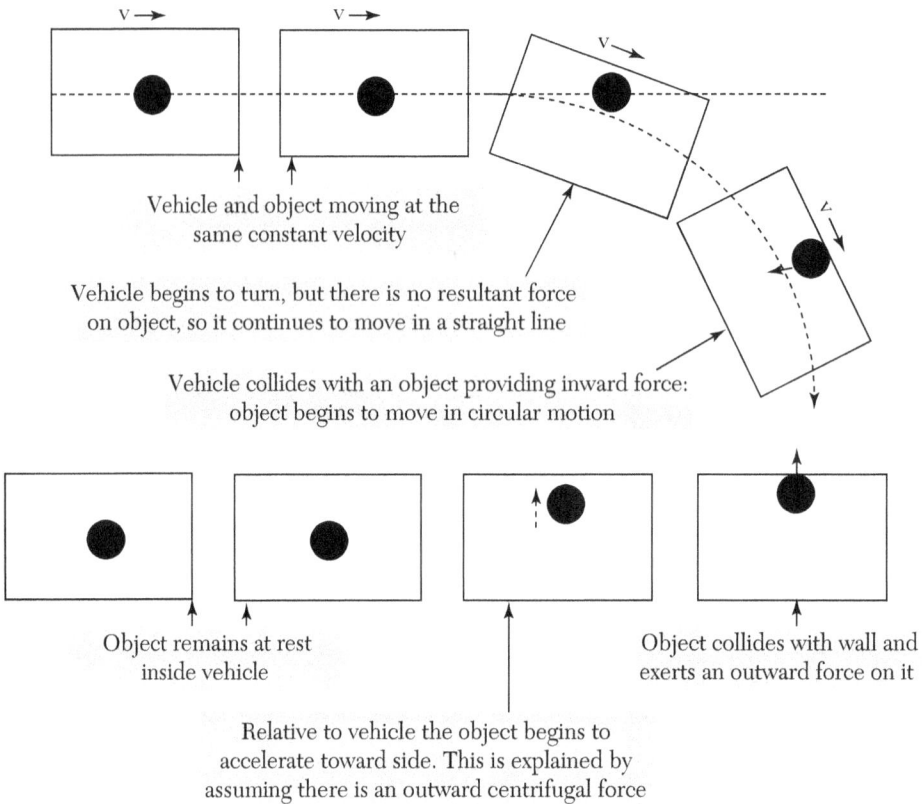

In the top case we are describing the physics from an external non-accelerating (inertial) frame of reference. Newton's laws of motion apply, so the object continues to move in a straight line in the absence of a resultant force. There is no acceleration or resultant force until the side of the vehicle has moved inwards and collided with the sliding object.

In the bottom case the apparent acceleration is because the observer is inside the vehicle and does not take into account his own acceleration. In order to explain the apparent acceleration of the object, he introduces an imaginary outward force, the centrifugal force.

Centrifugal forces are examples of **inertial forces**, introduced in order to explain observed physics from a non-inertial (accelerating) frame of reference. Unlike the centripetal forces in an inertial reference frame, inertial forces have no physical origin and that is why they are referred to as imaginary. They are helpful if we need to solve physical problems inside a rotating reference frame, but we must always bear in mind that they are an artifact of our reference frame and do not arise from physical causes.

12.3.4   Moving in Uniform Circular Motion

When an object moves in uniform circular motion, velocity and acceleration are always perpendicular. This means that force is also perpendicular to velocity, so there is never a displacement in the direction of the resultant force and the resultant force does no work on the object. The diagram shows velocity and acceleration vectors at different positions in the circle.

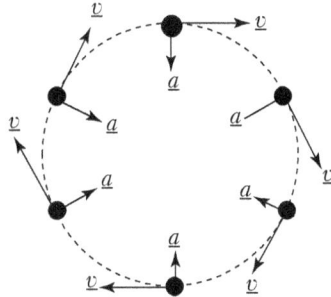

When you swing a stone in a circle on the end of a string, the forces acting on the stone are directed toward the center of the circle. If the string suddenly breaks, the stone has no resultant force acting on it and flies off along a tangent; it does not accelerate outwards because there is no centrifugal force. It moves in a straight line at constant velocity. (Here we ignore the effects of other external forces such as gravity.)

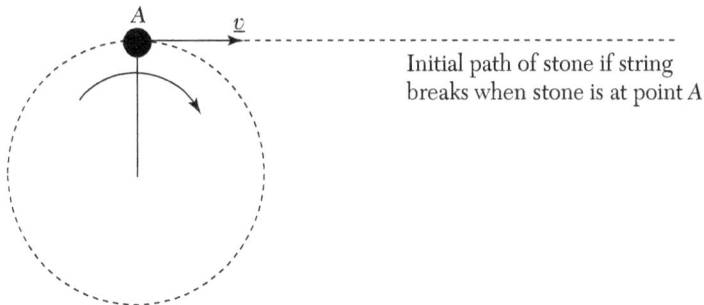

Initial path of stone if string breaks when stone is at point A

Note that centripetal force is not a new kind of force. It is the magnitude of the resultant force on a body moving with constant uniform circular motion. It arises as a result of the real physical forces that act on it. In some cases this is simple. For example, the Moon's orbit around the Earth is approximately circular, and the centripetal force is provided by the gravitational attraction toward the Earth.

$$F_{\text{grav}} = mr\omega^2 = \frac{4\pi^2 mr}{T^2}$$

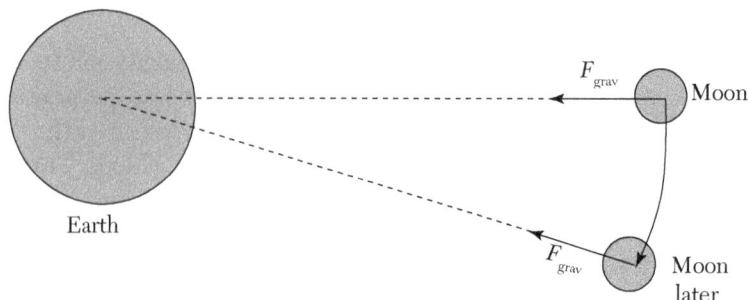

T is the Moon's orbital period, *r* is its orbital radius, and *m* is its mass.

A more complex situation involves an object moving in a vertical circle in a uniform vertical gravitational field. Examples might be a person on a fun fair ride, a plane looping the loop, or just a stone on a string. If the motion is uniform (constant angular velocity) the resultant force stays constant in magnitude, but the forces that contribute to it change. The example below shows a person standing in a capsule on a fairground ride and indicates the forces in four different positions. The capsule is rotating at constant angular velocity.

Only two forces act on the man. His weight *mg* is the same in all positions, but the contact force *R* from the floor of the capsule changes with position. The resultant force at all points is the vector sum of weight and reaction force and must be $F = mr\omega^2$ toward the center of the circle.

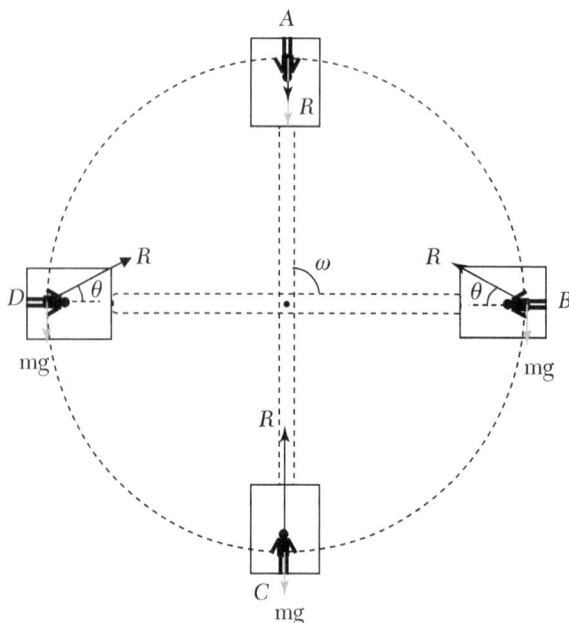

Position *A*

Weight and contact force act in the same direction, so $mr\omega^2 = mg + R_A$

$$R_A = mr\omega^2 - mg$$

For one particular angular velocity ω_0, the contact force falls to zero, $R_A = 0$. This is when $mr\omega_0^2 = mg$, so $\omega_0 = \sqrt{\dfrac{g}{r}}$.

Under these conditions the contact force between the man and the floor is zero and he feels "weightless." This is only an *apparent* weightlessness because he is in fact free-falling at this moment and the capsule is accelerating downwards at g. For higher values of angular velocity, there must be a contact force from the floor. For lower values of angular velocity he will lose contact with the floor and begin to fall downwards toward the roof of the capsule. In practice this sets a minimum value for the practical angular velocity in such a fairground ride.

Position *C*

Weight and contact force act in opposite directions, so $mr\omega^2 = R_A - mg$

$$R_A = mr\omega^2 + mg$$

The contact force is greater than his weight. He experiences this through the reaction force from the floor pushing up on his feet. This makes him feel heavy at the bottom, as if his weight has increased. Once again this is only an apparent increase in weight since the gravitational forces have not changed.

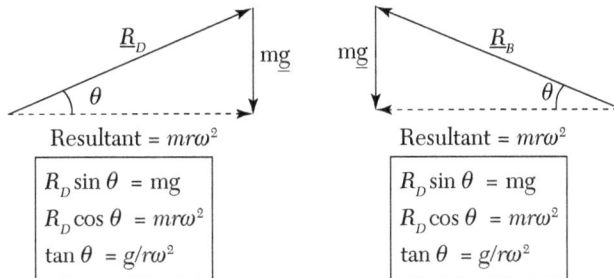

$R_D \sin \theta = mg$	$R_D \sin \theta = mg$
$R_D \cos \theta = mr\omega^2$	$R_D \cos \theta = mr\omega^2$
$\tan \theta = g/r\omega^2$	$\tan \theta = g/r\omega^2$

Positions *B* and *D*

In both positions the contact force has a vertical component that balances the weight and a horizontal component that provides centripetal force.

12.4   Circular Motion, Simple Harmonic Motion, and Phasors

Simple harmonic motion is equivalent to the projection of circular motion onto a diameter. This can be shown using a phasor, a rotating vector. The length of the phasor, A, defines the radius of the circle and is equal to the amplitude of the simple harmonic motion. The constant angular velocity ω of rotation of the phasor is equal to the angular frequency of the simple harmonic motion. The angle the phasor turns through is the phase of the simple harmonic motion.

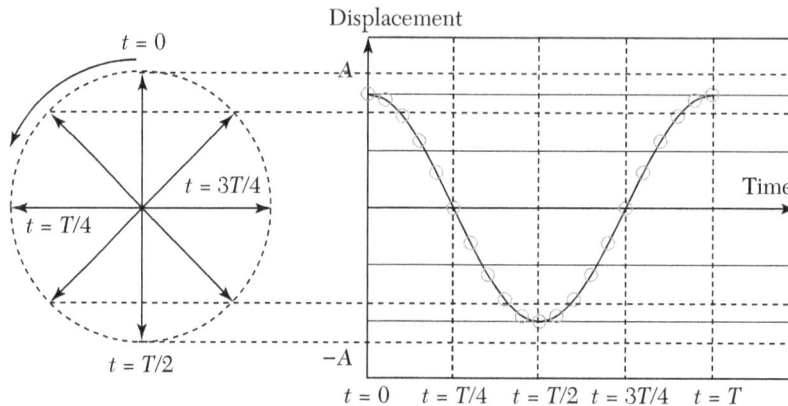

Phasors are a useful mathematical tool that is particularly useful when analyzing wave superposition.

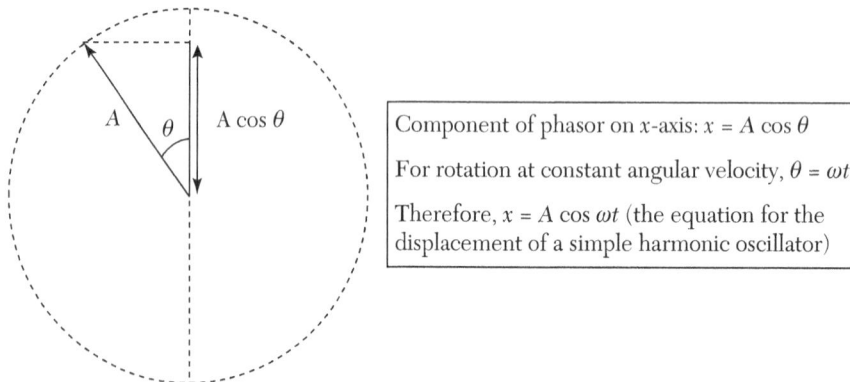

Component of phasor on x-axis: $x = A \cos \theta$

For rotation at constant angular velocity, $\theta = \omega t$

Therefore, $x = A \cos \omega t$ (the equation for the displacement of a simple harmonic oscillator)

12.5   Rotational Kinematics

12.5.1   Equations of Uniform Angular Acceleration

The *suvat* equations are invaluable when dealing with problems of constant acceleration. There is a corresponding set of equations for rotation under constant angular acceleration. Since the underlying definitions of linear and angular motion are mathematically equivalent, all the equations have the same form; we simply replace linear variables with rotational variables.

Linear definitions		Rotational definitions	
Displacement	s	Angular displacement	θ
Velocity	$v = \dfrac{ds}{dt}$	Angular velocity	$\omega = \dfrac{d\theta}{dt}$
Acceleration	$a = \dfrac{dv}{dt} = \dfrac{d^2 s}{dt^2}$	Angular acceleration	$\alpha = \dfrac{d\omega}{dt} = \dfrac{d^2\theta}{dt^2}$

This allows us to use the following analogy:

$$s \to \theta$$
$$u \to \omega_i$$
$$v \to \omega_f$$
$$a \to \alpha$$
$$t \to t$$

to generate the corresponding rotational equations for uniform angular acceleration:

$$v = u + at \qquad \to \qquad \omega_f = \omega_i + \alpha t$$

$$s = \frac{(u+v)t}{2} \qquad \to \qquad \theta = \frac{(\omega_i + \omega_f)t}{2}$$

$$s = ut + \frac{1}{2}at^2 \qquad \to \qquad \theta = \omega_i t + \frac{1}{2}\alpha t^2$$

$$s = vt - \frac{1}{2}at^2 \qquad \to \qquad \theta = \omega_f t - \frac{1}{2}\alpha t^2$$

$$v^2 = u^2 + 2as \qquad \to \qquad \omega_f^2 = \omega_i^2 + 2\alpha\theta$$

These are used in exactly the same way as the original *suvat* equations.

12.5.2 Rotational Kinetic Energy

When a rigid body rotates about a fixed axis through its center of mass, it has no translational kinetic energy. However, every point mass inside the body has a velocity and, therefore, has its own kinetic energy. The sum of all of these kinetic energies is equal to the rotational kinetic energy of the body. When a body is rolling, it has both translational kinetic energy (because its center of mass is moving) and rotational kinetic energy (because it is rotating about its center of mass). The total kinetic energy is equal to the sum of translational and rotational kinetic energies.

The diagram below shows a rigid body rotating about its center of mass (CM). It consists of N particles and the ith particle has mass m_i and is at a distance r_i from the axis.

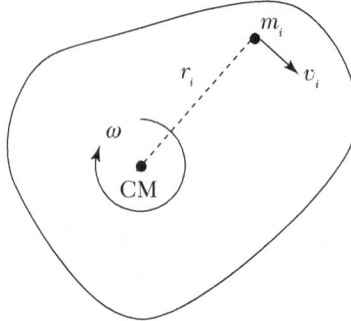

The kinetic energy of the ith particle is:

$$ke_i = \tfrac{1}{2}\,m_iv_i^2 = \tfrac{1}{2}\,m_ir_i^2\omega^2$$

The rotational kinetic energy of the body is:

$$\text{RKE} = \sum_{i=1}^{i=N} ke_i = \sum_{i=1}^{i=N} \frac{1}{2}m_ir_i^2\omega^2$$

The angular velocity is the same for all particles, so:

$$\text{RKE} = \frac{1}{2}\left(\sum_{i=1}^{i=N} m_ir_i^2\right)\omega^2$$

The term in parentheses is called the **moment of inertia I** of the body:

$$I = \left(\sum_{i=1}^{i=N} m_ir_i^2\right)$$

The S.I. unit of moment of inertia is kgm^2.

Moment of inertia in rotational dynamics plays an analogous role to mass (inertia) in linear mechanics. However, unlike mass, the moment of inertia is not a fixed quantity for a body because it depends on both the mass and its distribution about the axis of rotation.

Using moment of inertia we can extend the analogy between linear and rotational motion:

Linear kinetic energy = $\tfrac{1}{2}\,mv^2$ $\rightarrow$ Rotational kinetic energy = $\tfrac{1}{2}\,I\omega^2$

Here is an example that involves both translational and rotational kinetic energy: a cylinder of radius r, mass m, and moment of inertia I rolling without slipping down an inclined plane so that its center of mass drops through a vertical height h.

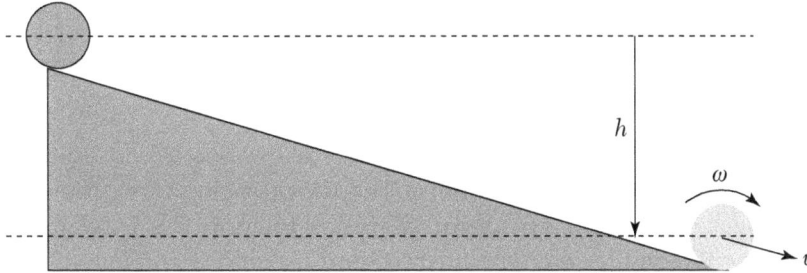

Neglecting energy losses due to friction, the gain in total kinetic energy must equal the loss of gravitational potential energy. If the cylinder starts from rest, then its final kinetic energy will be (using $\omega = v/r$):

$$mgh \quad = \text{translational KE} + \text{rotational KE}$$

$$= \tfrac{1}{2}\, mv^2 + \tfrac{1}{2}\, I\omega^2 = \tfrac{1}{2}\, mv^2 + \tfrac{1}{2}\, Iv^2/r^2$$

This can be rearranged to give the linear velocity at the bottom:

$$v = \sqrt{\frac{2mgh}{\left(m + \dfrac{I}{r^2}\right)}}$$

The larger the moment of inertia, the lower the final linear velocity. This is because a larger fraction of the energy goes into rotation. A cylinder has a larger moment of inertia than a sphere because more of the mass is farther from the axis. If a cylinder and ball are rolled down the same slope side by side, then the ball will reach the bottom first.

If the surface is completely smooth and the object slides without rolling, then there is no rotational kinetic energy and the expression above reduces to:

$$v = \sqrt{2gh}$$

This shows that a block sliding down a smooth slope will always get to the bottom faster than any round rolling object on the same slope.

12.5.3 Angular Momentum

The angular momentum of a point mass is defined as the moment of its momentum about a point. That is the linear momentum multiplied by its perpendicular distance from the point considered.

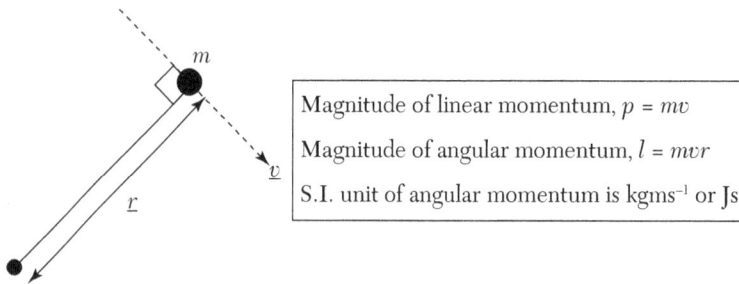

Magnitude of linear momentum, $p = mv$

Magnitude of angular momentum, $l = mvr$

S.I. unit of angular momentum is $kgms^{-1}$ or Js

Angular momentum is actually the vector cross-product of the displacement vector $\underline{r}$ and the linear momentum $\underline{p}$. This is written as $\underline{l} = \underline{r} \wedge \underline{p}$. The direction of angular momentum is related to the axis of rotation. For 2D problems we need to only consider whether the direction is clockwise or counterclockwise.

The angular momentum of a rigid body can be found by summing the individual contributions from each point particle inside the body (in the same way that we derived a formula for rotational kinetic energy).

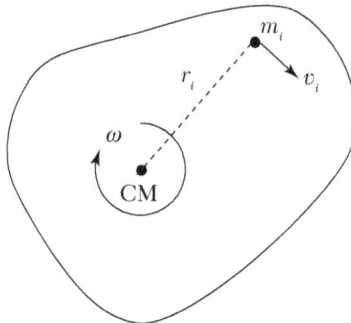

The angular momentum of the ith particle is:

$$l_i = m_i v_i r_i = m_i r_i^2 \omega$$

The angular momentum of the whole body is, therefore,

$$L = \sum_{i=1}^{i=N} l_i = \sum_{i=1}^{i=N} m_i r_i^2 \omega$$

The angular velocity is the same for all particles, so:

$$L = \left(\sum_{i=1}^{i=N} m_i r_i^2 \right) \omega = I\omega$$

Notice that the **moment of inertia** I of the body, $\sum_{i=1}^{i=N} m_i r_i^2$, once again plays a role analogous to mass in linear mechanics.

Linear momentum $= mv \quad \rightarrow \quad$ Angular momentum $= I\omega$

12.5.4   Second Law of Motion for Rotation

In linear mechanics, resultant force is equal to the rate of change of linear momentum. An analogous relation exists in rotational motion and depends on the torque or moment of a force.

Torque (moment of a force)

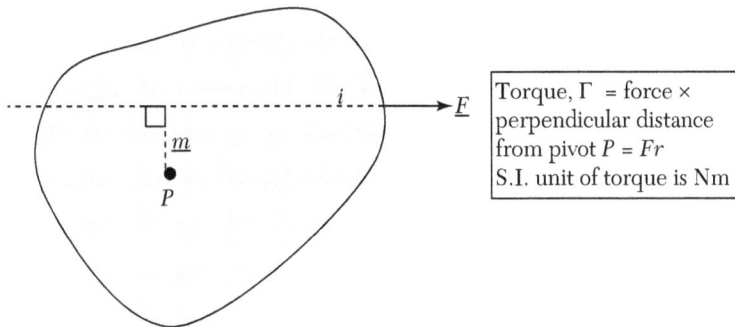

(Mathematically torque is the vector cross-product of force and displacement from the pivot to the point of application of the force.)

When a resultant torque is applied to a rigid body, each point particle inside the body experiences a resultant force that causes a linear acceleration. The sum of torques from these individual forces must equal the total resultant force on the body.

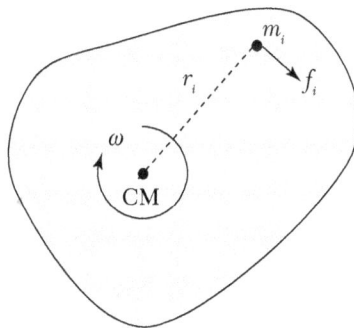

For the ith particle, $f_i = m_i a_i = m_i r_i \alpha$ (using $a = r\alpha$)

$$\Gamma = \sum_{i=1}^{i=N} f_i r_i = \sum_{i=1}^{i=N} m_i a_i r_i = \sum_{i=1}^{i=N} m_i r_i^2 \alpha = I\alpha$$

This has the same form as Newton's second law in linear mechanics:

Linear case, $F = ma$ $\rightarrow$ Rotational case, $\Gamma = I\alpha$

We can use $\alpha = \dfrac{d\omega}{dt}$ to relate this to angular momentum.

Linear case, $\qquad F = \dfrac{d(mv)}{dt} \rightarrow \Gamma = \dfrac{d(I\omega)}{dt}$

We can now state Newton's laws of motion in a form that applies to rotational motion.

◾ **Newton's first law for rotational motion**

An object continues to rotate with constant angular momentum until acted upon by a resultant external torque.

◾ **Newton's second law for rotational motion**

The resultant external torque on a system is equal to the rate of change of angular momentum of that system.

◾ **Newton's third law for rotation**

When object A exerts a torque on object B, B exerts an equal but opposite torque on A.

12.5.5 Conservation of Angular Momentum

Newton's laws of motion for rotation show that angular momentum can only change as a result of an external resultant torque.

$$\Gamma = \dfrac{d(I\omega)}{dt}, \text{ so } \Delta(I\omega) = \int \Gamma dt$$

That is, change of angular momentum = angular impulse

They also show that when systems interact, the torques experienced by each system are equal and opposite, so while they interact they exert equal and opposite angular impulses on one another. Consequently any change of momentum of system A is equal and opposite to the change of momentum of system B, and the total angular momentum of the combined system does not change. This can be stated as the law of conservation of angular momentum:

The angular momentum of an isolated system (no external resultant torque) is constant.

An interesting example of this is when an ice-skater pirouettes. She balances on the point of one skate while rotating relatively slowly with arms and one leg outstretched. Gradually she draws her arms and leg closer to the axis of rotation, thus reducing her moment of inertia. There is no external torque, so her angular momentum ($L = I\omega$) cannot change. If I falls, ω must increase, so she spins faster. Another similar example is the collapse of the core of a massive star at the end of its life. Its radius reduces by many orders of magnitude, so the rotation rate can be very high—some neutron stars have rotation periods of less than one millisecond.

12.6 Deriving Expressions for Moments of Inertia

Moment of inertia is defined by the equation $I = \left(\sum_{i=1}^{i=N} m_i r_i^2 \right)$, and we can use this to derive useful expressions for the moments of inertia of a range of standard mass distributions.

12.6.1 Moment of Inertia of One or More Point Masses

For a single-point mass distance r from the center of rotation,

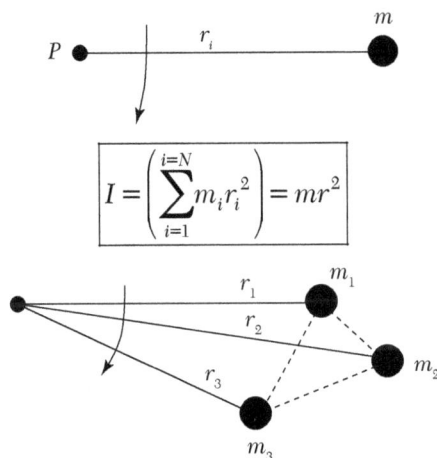

$$I = \left(\sum_{i=1}^{i=N} m_i r_i^2 \right) = mr^2$$

For several point masses at various distances from the center of rotation,

$$I = \left(\sum_{i=1}^{i=N} m_i r_i^2 \right) = m_1 r_1^2 + m_2 r_2^2 + m_3 r_3^2$$

12.6.2 Moment of Inertia of a Rod

For a continuous mass distribution we need to convert the sum into an integral. For a thin rod lying along the x-axis this can be done by considering a short section of length δx.

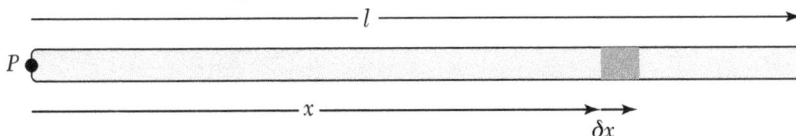

Let the total mass of the rod be m. The mass of the short section of length δx at distance x from the center of rotation is $\delta m = \dfrac{m\delta x}{l}$, and the moment of inertia of this section (treated as a point mass) is $\delta I = \dfrac{mx^2\delta x}{l}$.

The moment of inertia of the rod about one end is, therefore,

$$I_{\text{rod end}} = \left(\sum_{i=1}^{i=N} m_i r_i^2\right) = \sum_{x=0}^{x=l} \delta I = \int_{x=0}^{x=l} \frac{mx^2 dx}{l} = \frac{1}{3} ml^2$$

A similar approach can be used to determine the moment of inertia about any point in the rod. For example, the moment of inertia about the center of the rod would be found by taking $x = 0$ at the center and then integrating from $-l/2$ to $+l/2$:

$$I_{\text{rod CM}} = \left(\sum_{i=1}^{i=N} m_i r_i^2\right) = \sum_{x=0}^{x=l} \delta I = \int_{x=-l/2}^{x=+l/2} \frac{mx^2 dx}{l} = \frac{1}{12} ml^2$$

This moment of inertia is smaller than the moment of inertia about one end because more of the mass is now closer to the axis of rotation.

You can get a qualitative "feel" for moment of inertia by taking a rod (e.g., ruler) holding it in the center and trying to rotate it back and forth. You will feel a resistance to rotation. Now do the same thing while holding it near one end. The resistance to rotation has increased significantly—the moment of inertia is larger.

12.6.3 Moment of Inertia of a Cylindrical Shell and a Uniform Cylinder

A cylindrical shell or ring with thickness much less than its radius rotating about its center has all of its mass at a constant distance from the center of rotation. Its moment of inertia is therefore simply $I = \delta m r^2$ where δm is the mass of the shell. We can find the moment of inertia of a disc or cylinder by integrating over all of the thin cylindrical shells that make it up.

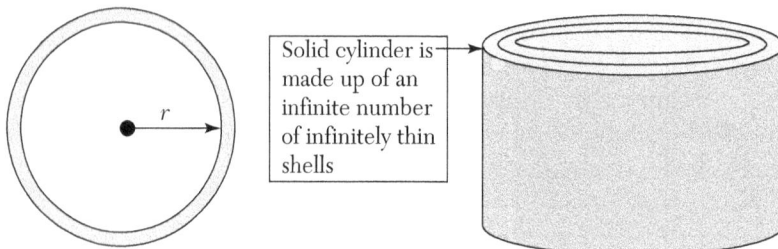

Solid cylinder is made up of an infinite number of infinitely thin shells

The diagram that follows shows an end view of a solid cylinder of total mass m and radius r. A thin shell at radius x is shown.

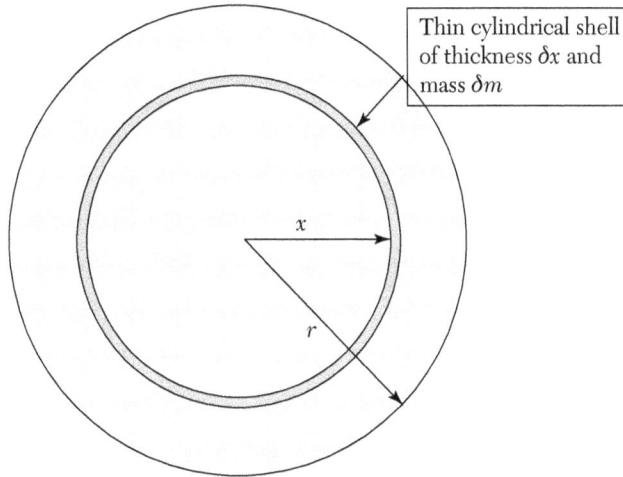

Thin cylindrical shell of thickness δx and mass δm

The moment of inertia of a cylindrical shell is $\delta I = \delta m x^2$, and the mass of the shell is a fraction of the mass of the cylinder given by $\delta m = \dfrac{2\pi x m \delta x}{\pi r^2} = \dfrac{2 x m \delta x}{r^2}$.

This is because the area of the shaded strip above is effectively $2\pi x \delta x$ if treated like a long thin rectangle.

The moment of inertia of the entire cylinder is then:

$$I_{cylinder} = \sum_{x=0}^{x=r} \delta I = \int_{x=0}^{x=r} \frac{2mx^3 dx}{r^2} = \frac{1}{2}mr^2$$

12.6.4 Moment of Inertia of a Uniform Sphere

A uniform sphere of mass m and radius r can be considered to be made up of an infinite number of thin discs of varying radius. The total moment of inertia is the sum of moments of inertia of all such discs.

The moment of inertia of the sphere is:

$$I_{sphere} = \int_{\theta=0}^{\theta=\pi} \frac{1}{2}x^2 dm$$

In order to integrate this we must express δm in terms of $d\theta$. It simplifies things if we introduce the density of the material of the sphere ρ at this stage.

$$\delta m = \rho \pi r^2 \sin^2 \tau \delta x$$

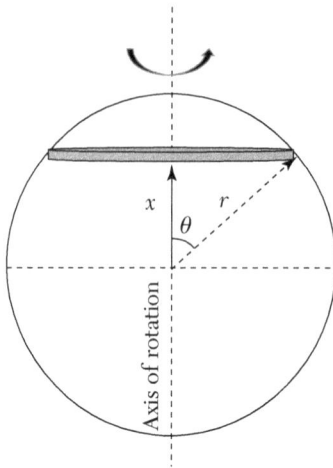

Thin disc of thickness δx a distance x above center of sphere has mass δm and radius $r \cos \theta$.
Its moment of inertia about the rotation axis is $\delta l = 1/2\, x^2\, \delta m$.

However, x and θ are related by:

$$x = r \cos \theta,$$

so

$$\frac{dx}{d\theta} = -r \sin \theta$$

So we can replace dx by $-r \sin \theta\, d\theta$

$$I_{\text{sphere}} = \int_{\theta=0}^{\theta=\pi} -\frac{1}{2} \rho \pi r^5 \sin^5 \theta d\theta$$

This can be integrated using standard techniques to give:

$$I_{\text{sphere}} = \frac{8 \rho \pi r^5}{15}$$

Now replace the density with $\rho = \dfrac{3m}{4\pi r^3}$ to give:

$$I_{\text{sphere}} = \frac{2}{5} mr^2$$

12.7   Torque Work and Power

It should come as no surprise that we can extend the analogy between linear and rotational motion to include work and power. This can be shown using the same summation techniques as we used in Sections 12.5.2 and 12.5.3. The important results are:

Work: $W = Fs$ (constant force) $\rightarrow W = \Gamma\theta$ (constant torque)

$$W = \int F ds \rightarrow W = \int \vartheta dt$$

Power: $P = Fv \rightarrow P = \Gamma\omega$

12.8   Rotational Oscillations—The Compound Pendulum

A compound pendulum consists of an extended rigid body pivoted at one point and undergoing regular periodic oscillations. In order to analyze these oscillations, we need to consider the resultant torque when the pendulum is displaced through a small angle from its equilibrium position.

The center of mass (CM) is a distance h below the pivot P.

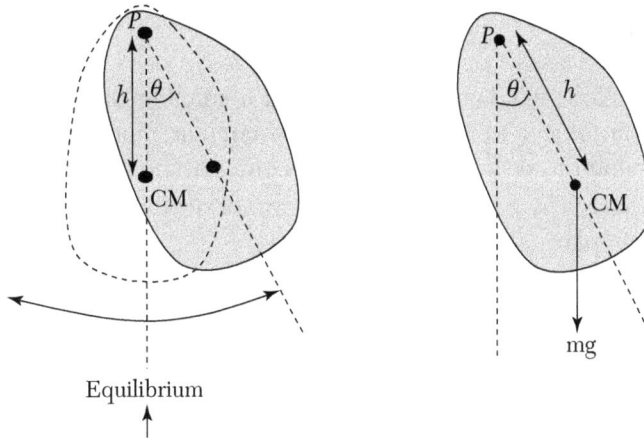

Equilibrium

The right-hand diagram shows how the weight produces a restoring torque $\Gamma = -mgh \sin \theta$ (negative sign indicates that this is directed toward equilibrium) when the body is displaced through angle θ.

From Newton's second law,

$$\vartheta = I\frac{d\omega}{dt} = I\frac{d^2\theta}{dt^2}$$

The equation of motion for this compound pendulum is, therefore,

$$\frac{d^2\theta}{dt^2} = -\frac{mgh\sin\theta}{I}$$

For small angles, $\sin \theta \to \theta$ (in radians), so for small angles we can write (approximately):

$$\frac{d^2\theta}{dt^2} = -\left(\frac{mgh}{I}\right)\theta$$

This has the same mathematical form as the equation of motion for simple harmonic motion:

$$\frac{d^2x}{dt^2} = -\omega^2 x$$

So it will have sinusoidal solutions of the same kind.

The compound pendulum undergoes simple harmonic angular oscillations (for small angles). Its period and frequency can be found using:

$$\omega^2 = \left(\frac{mgh}{I}\right)$$

giving $T = 2\pi\sqrt{\dfrac{I}{mgh}}$ and $f = \dfrac{1}{2\pi}\sqrt{\dfrac{mgh}{I}}$

12.9 Exercises

1. (a) Use Newton's laws of motion and a suitable diagram to show that for an object to move in uniform circular motion there must be a resultant force acting toward the center of the circle.
   (b) Explain why a centripetal force cannot do work on an object moving in circular motion.

2. A small ball of mass m is released from a height h on a track that leads to a looping section as shown below.

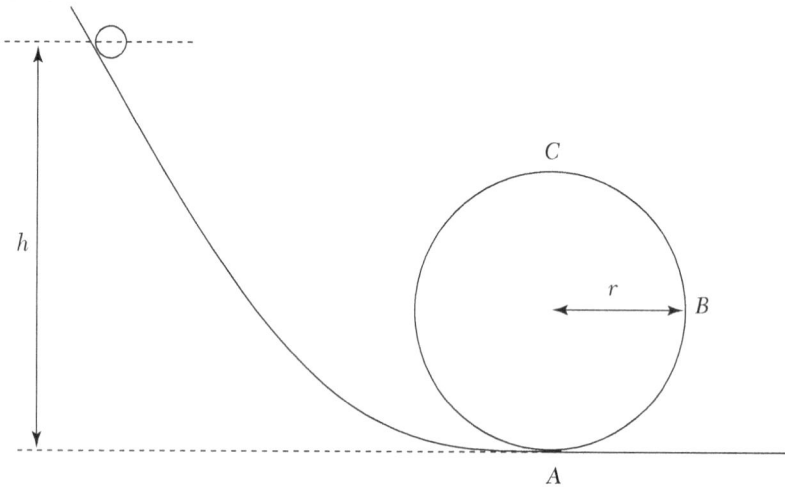

(a) Derive an expression in terms of r for the minimum height h from which the ball can be released if it is to complete the loop without losing contact with the track. Ignore the rotational motion of the ball and assume friction is negligible.
(b) Draw free body diagrams to show the forces acting on the ball at points A, B, and C.
(c) Discuss whether the resultant force on the ball is toward the center of the circle at all points as it completes the loop.
(d) Derive an expression for the velocity of the ball at point C.

3. A car tire of radius r is rolling along a flat horizontal surface at constant speed v. Copy the diagram below and add labeled arrows to show the velocity and acceleration of a particle fixed to the tire at each of the points A to D.

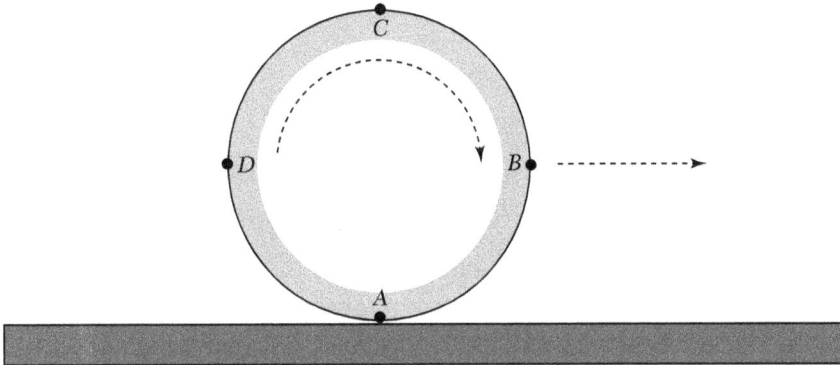

4. A vinyl record rotates at 33 revolutions per minute. Its diameter is 30 cm and its mass 200 g.
(a) Calculate the time period of the rotation.
(b) Calculate the frequency of the rotation.
(c) Calculate the angular velocity of the record.
(d) Calculate the angular momentum of the record.
(e) Calculate the average torque needed to accelerate the record to its playing speed in 0.50 s.
(f) Calculate the rotational kinetic energy of the record.

5. Derive an expression for the moment of inertia of a uniform rod of length l rotating about an axis perpendicular to the rod and through a point one third of the way along the rod.

6. A space station consists of two spherical accommodation pods of radius 20 m separated by a connecting tunnel of length 400 m. The mass of each pod is 50,000 kg and the mass of the connecting tunnel is 80,000 kg. The space station is rotating about its center of mass.

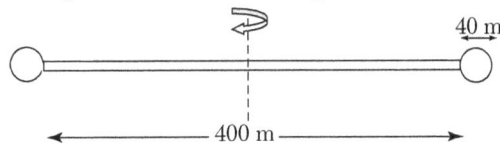

(a) Estimate the moment of inertia of the space station about its rotation axis by treating the pods as point masses concentrated at their centers of mass and by treating the connecting tunnel as a rod. The moment of inertia of a rod of mass m and length l about its CM is given by $I = \dfrac{1}{12}ml^2$.

(b) Explain why an astronaut standing on the outer edge of a pod would experience artificial gravity.

(c) Calculate the angular velocity of rotation that will create an effect of artificial gravity of strength 9.8 Nkg^{-1} at the outer edge of the pods.

(d) Calculate the rotational kinetic energy of the space station when its angular velocity is 0.20 rads^{-1}.

7. A small flywheel has a moment of inertia of 2.0×10^{-3} kgm^2 and it is rotating at 50 revolutions per second. The frictional torque working against rotation is 2.4×10^{-2} Nm.

(a) Calculate the angular momentum of the flywheel including appropriate units.

(b) Calculate the time taken for the flywheel to come to rest.

8. A group of children are sitting on the outside of a roundabout that is rotating at a constant rate. They all move toward the center of the roundabout and it speeds up. Explain why this occurs and discuss the angular momentum and energy changes that take place in this system. Assume that external torques can be ignored.

9. The Earth's rotation is slowing because of its tidal interaction with the Moon. This causes the day length to gradually increase. Day length increases by 1 second roughly every 18 months.

(a) Use this information and the data below to calculate the torque acting on the Earth to slow down its rotation. Assume the Earth is of uniform density.

Mass of Earth $= 6.0 \times 10^{24}$ kg

Radius of Earth $= 6400$ km

Number of seconds in a day $= 86400$

Moment of inertia of a solid sphere of mass m and radius a is $I = 2/5\ ma^2$

(b) The Earth is not uniform; its density increases toward its center. State and explain how this would affect your answer to (a).

CHAPTER 13

WAVES

13.0 Introduction

There are many different types of waves, but the underlying physics is common to all of them, whether they are mechanical vibrations in a medium, like sound, or vibrations of an electromagnetic field, like light. They all transfer energy, reflect, refract, diffract, and interfere. The wave model is one of the most important models in physics, and in the next four chapters, we will investigate waves in considerable detail.

13.1 Describing and Representing Waves

13.1.1 Basic Wave Terminology

Imagine starting circular water waves using a dipper oscillating up and down at the center of a pond. A series of circular, equally spaced crests moves outwards. The source of these waves is the oscillation of water at the center, and they are formed because, as those water molecules move up and down, they exert forces on the ones adjacent to them pulling them up and down slightly later so that these molecules oscillate with the same frequency and amplitude but with a small phase delay relative to the source. The phase delay increases with distance from the source, and when this delay is 2π radians, the molecules are once again oscillating in phase with the source. Around any circle centered on the source, all of the molecules are oscillating in phase with one another because they are all equidistant from the source. The visible circles are formed by crests, where all of the oscillations are simultaneously at a positive amplitude. These positions move outwards at constant velocity. This is called the **phase velocity** of the wave. It is the movement of the wave disturbance but not the outward movement of matter. The diagram below shows how this appears from above. The circular lines of constant phase are called **wave fronts** and the arrows are called **rays**. Rays and wave fronts are perpendicular to one another and are alternative ways to represent the wave pattern. The separation between two adjacent wave fronts is called the wavelength λ of the wave.

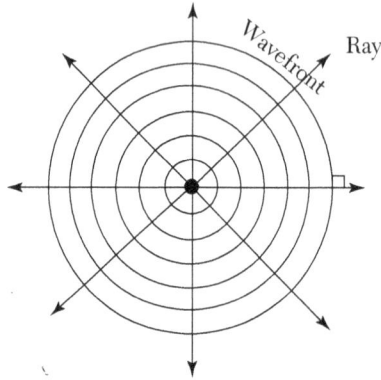

The time period T for the formation of one wave is equal to the time period of the source oscillations, so all particles in the wave oscillate with the same frequency f as the source. During one oscillation of the source the wave moves forward a distance equal to its wavelength, so the phase velocity v of the wave is given by:

$$v = \frac{\lambda}{T} = f\lambda$$

We can now define some terms:

▣ Traveling or progressive wave

A wave where all the particles oscillate with the same (or a decaying) amplitude but with a progressive phase delay in direct proportion to their distance from the source.

▣ Wavelength λ

Shortest distance between two particles oscillating in phase in the wave.

▣ Time period T

Time for one complete wave to leave the source or time for a particle to complete one cycle of oscillation.

▣ Frequency f

Number of waves leaving the source in 1 second, $f = \frac{1}{T}$.

■ **Phase velocity v**

Velocity at which a point of constant phase (e.g., a wave crest) travels away from the source, $v = f\lambda$

■ **Wavefront**

Line of constant phase in the wave pattern; perpendicular to rays.

■ **Ray**

Arrow in the direction of energy transfer; perpendicular to wave fronts.

13.1.2   Transverse and Longitudinal Waves

The source of a wave is a vibration or oscillation. There are two distinct ways in which the oscillations can be related to the direction of energy transfer (direction of travel of the wave):

Transverse waves

Vibration directions are perpendicular to the direction of energy transfer.

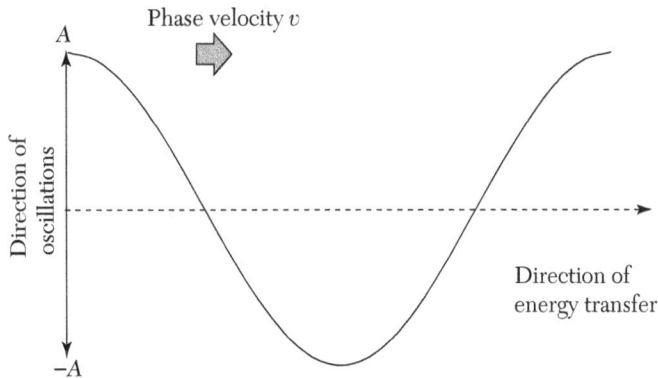

As the wave moves to the right, the disturbance at each position varies vertically with the same amplitude as the source (assuming no energy dissipation). All electromagnetic waves are transverse, as are seismic S-waves (secondary or shear waves).

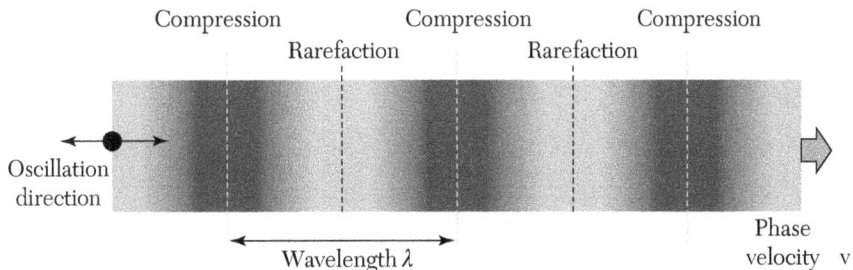

Longitudinal waves

Vibration directions are parallel to the direction of energy transfer. This results in regions of compression (shown as dark areas below) and rarefaction (low density, shown as lighter areas below) in the medium through which the wave passes.

As the wave moves to the right, the disturbance at each position also varies horizontally, with the same amplitude as the source (assuming no energy dissipation). Sound and ultrasound are longitudinal waves, so are seismic P-waves (primary or pressure waves).

Some waves are a combination of longitudinal and transverse waves. This results in particles undergoing elliptical motions as the wave passes. Surface water waves are like this.

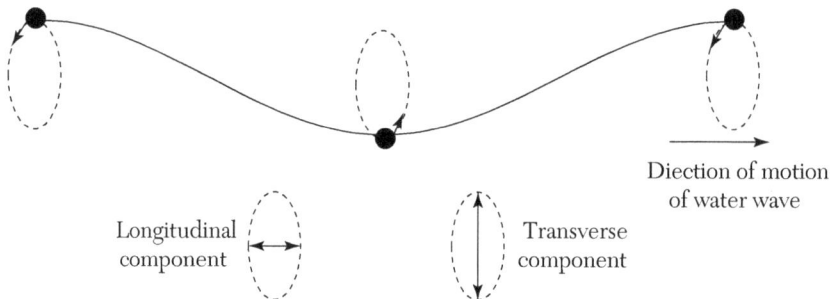

Diection of motion
of water wave

Longitudinal
component

Transverse
component

13.1.3   Graphs of Wave Motion

For a one-dimensional wave, e.g., a wave traveling along the x-axis, we can plot a graph of how the wave disturbance (y) varies with position (x) at any fixed moment of time or of how the disturbance varies with time at any particular position.

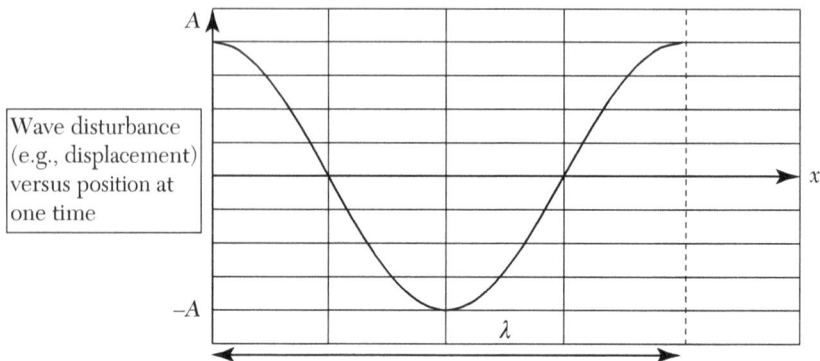

Wave disturbance (e.g., displacement) versus position at one time

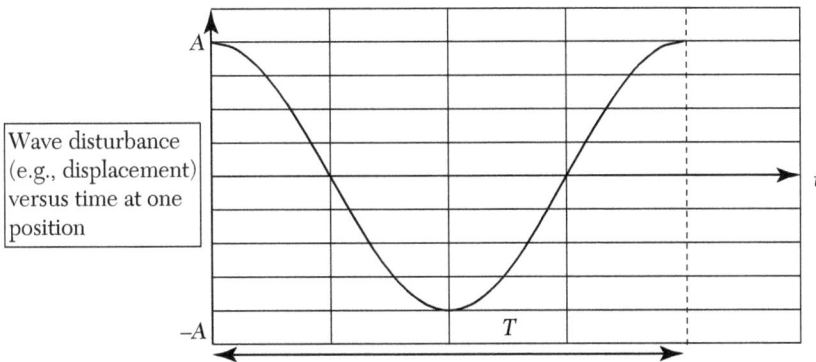

Wave disturbance (e.g., displacement) versus time at one position

These graphs are similar to each other but represent different things. The repeat distance in space is the wavelength, and the repeat distance in time is the time period of the wave.

13.1.4 Equation for a One-Dimensional Traveling Wave

The displacement of a one-dimensional wave varies with both time and position, so the equation to represent the whole wave will be a function of these two variables. For a traveling wave, all points oscillate with the same amplitude but with a progressive phase delay with respect to the source. If the source oscillates with simple harmonic motion, then the displacement at the source $(x = 0)$ can be written as:

$$y = A \cos (\omega t)$$

where $\omega = 2\pi f$.

The oscillation at any other point on the positive x-axis will be given by an equation of the form:

$$y = A \cos (\omega t - \delta)$$

where δ is a phase delay that depends on x (distance from the source). When $x = \lambda$ the phase delay is 2π (two particles separated by one wavelength oscillating in phase). Therefore, $\delta = \dfrac{2\pi x}{\lambda}$ and so:

$$y = A \cos\left(\omega t - \frac{2\pi x}{\lambda}\right)$$

The term $\dfrac{2\pi}{\lambda}$ is called the wave number k, so we can write the equation for a one-dimensional traveling wave of amplitude A moving in the positive x-direction thus:

$$y = A \cos(\omega t - kx)$$

If the wave is moving in the negative x-direction, the sign in the equation changes:

$$y = A \cos(\omega t + kx)$$

Derivation of phase velocity

The phase velocity is the velocity at which a point of constant phase, e.g., a wave crest, moves along the axis. The phase of the wave is given by:

$$\phi = (\omega t - kx)$$

$$\frac{d\phi}{dt} = \omega - k\frac{dx}{dt} = \omega - kv = 0 \qquad \text{(for position of constant phase)}$$

$$v = \frac{\omega}{k} = \frac{2\pi f \lambda}{2\pi} = f\lambda$$

$$v = f\lambda$$

13.1.5 Amplitude and Intensity

The intensity of a wave is the power delivered per unit area $(\mathrm{Wm^{-1}})$. This will be proportional to the energy of each oscillator in the wave. For a harmonic wave the disturbance at each point is simple harmonic, so the total energy of each oscillator is given by an expressions of the form:

$$E = \frac{1}{2} m\omega^2 A^2$$

It follows that the intensity of a harmonic wave is directly proportional to the square of its amplitude. This is a general and very useful relation:

$$I \propto A^2$$

Doubling amplitude increases intensity by a factor of four, etc.

13.2 Reflection

When a wave strikes a boundary it can be wholly or partially reflected. The law of reflection states that:

- The incident and reflected rays make equal angles to the normal to the surface at the point of incidence, and both rays and the normal lie in the same plane.

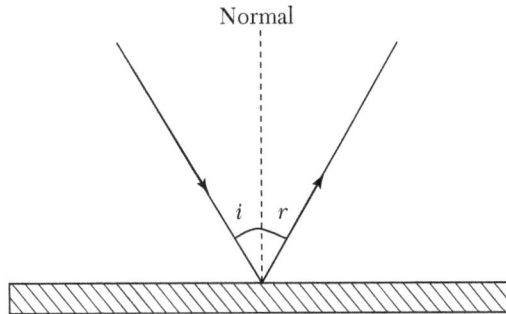

If two plane reflectors are placed at 90° to each other, then a ray striking either one of them will return parallel to its original path. This is used in car reflectors and was used by the Apollo astronauts who left an array of corner reflectors on the surface so that lasers sent from Earth would reflect back and allow the distance between the Earth and Moon to be measured precisely.

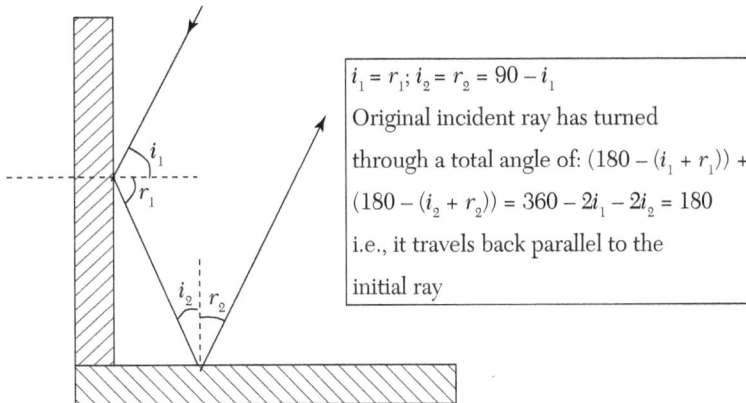

$i_1 = r_1; i_2 = r_2 = 90 - i_1$

Original incident ray has turned

through a total angle of: $(180 - (i_1 + r_1)) +$

$(180 - (i_2 + r_2)) = 360 - 2i_1 - 2i_2 = 180$

i.e., it travels back parallel to the

initial ray

13.3   Refraction

13.3.1   Refraction at a Boundary Between Two Different Media

Wave velocity depends on the medium through which the wave is traveling. For example, light slows down when it travels from air into glass, and surface water waves slow down when they travel from deeper to shallower water. If the wave strikes the boundary between two different media at an angle to the normal, then the wave direction changes. This is called refraction.

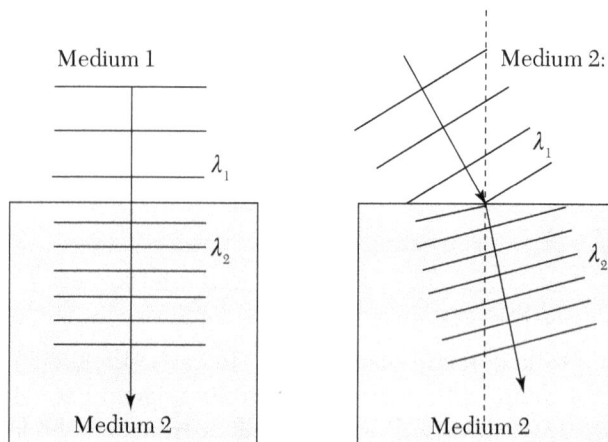

No waves are created or destroyed, so the frequency must be the same in both media. Therefore,

$$v_1 = f\lambda_1 \quad \text{and} \quad v_2 = f\lambda_2$$

$$\frac{v_2}{v_1} = \frac{\lambda_2}{\lambda_1}$$

13.3.2 Snell's Law of Refraction

A simple experiment can be used to find the relationship between the incident and refracted angles at a boundary between two media, in this case air and glass.

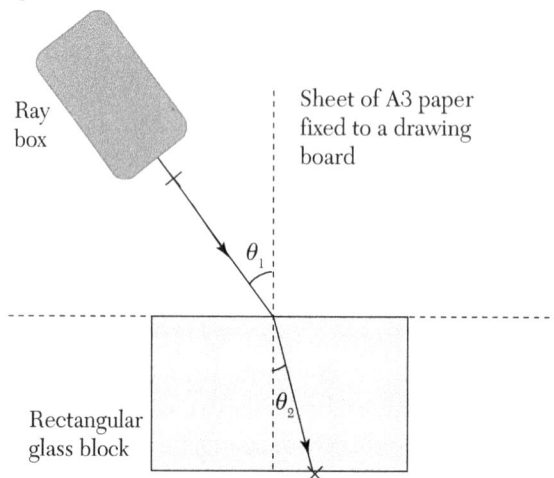

The position of the edge of the block and the normal to this line are marked onto the white paper, and the ray box is used to direct a single fine ray at the point where these two lines intersect. The ray is traced using optical pins (shown by X on the diagram) pushed into the board. The block can then be removed and the path of the ray outside and inside the block can be drawn using a pencil and ruler. The incident angle θ_1 and the refracted angle θ_2 are then measured using a protractor. If this is done for incident angles in the range 0–90° then a graph of $\sin \theta_1$ against $\sin \theta_2$ is a straight line.

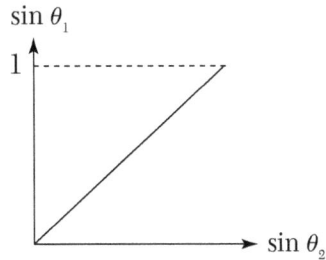

This shows that:

$$\frac{\sin \theta_1}{\sin \theta_2} = \frac{\sin \text{ (incident angle)}}{\sin(\text{refracted angle})} = \text{constant}$$

This is called **Snell's Law**, and the constant is called the **relative refractive index** $_1n_2$ for a ray passing from medium 1 into medium 2.

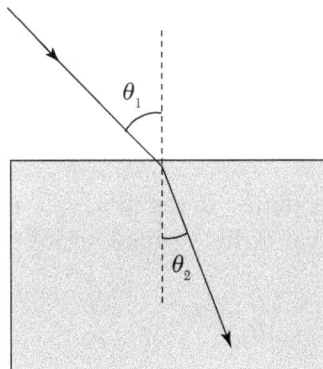

Snell's law:

$$\frac{\sin \theta_1}{\sin \theta_2} = {}_1n_2$$

If medium 1 is the vacuum, then this constant is the absolute refractive index for medium 2.

In practice, since the speed of light in air is almost the same as the speed of light in a vacuum, the absolute refractive index is usually used when medium 1 is air.

Refraction occurs because the wave velocity changes at the boundary. This implies that refractive index must be related to this change of velocity.

The diagram below can be used to derive this relationship.

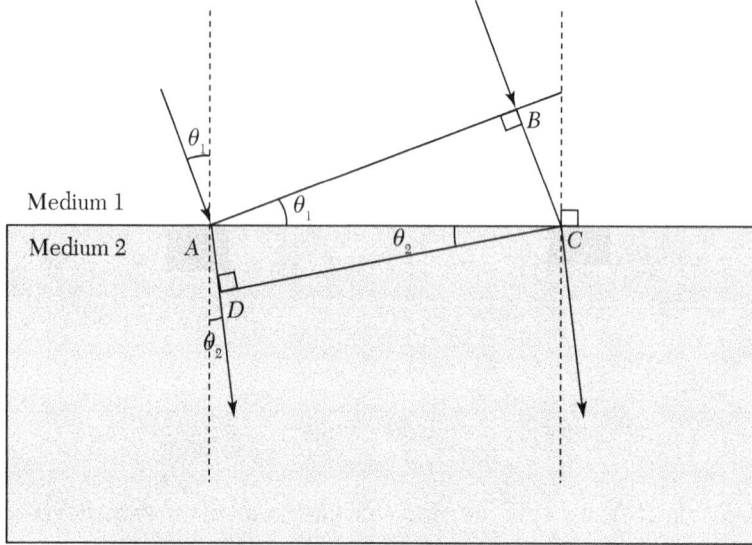

AB and *DC* are adjacent wave fronts so they are separated by one wavelength:

$$BC = \lambda_1 = \frac{v_1}{f}\frac{\partial^2 \Omega}{\partial u^2}$$

$$AD = \lambda_2 = \frac{v_2}{f}$$

AC is the common hypotenuse to the two right-angled triangles containing angles θ_1 and θ_2, so we can write expressions for the sines of each angle:

$$\sin\theta_1 = \frac{BC}{AC} = \frac{v_1/f}{AC}$$

$$\sin\theta_2 = \frac{AD}{AC} = \frac{v_2/f}{AC}$$

Dividing the first equation by the second gives us Snell's law in terms of the wave velocities in each medium:

$$\frac{\sin\theta_1}{\sin\theta_2} = \frac{v_1}{v_2} = 1n_2$$

The index of refraction is the ratio of wave speeds in the two media on either side of the boundary. It is obvious from this equation that if the ray direction is reversed, it will retrace its path. In other words, the refractive index when going from medium 2 to medium 1 is the reciprocal of the refractive index when going from medium 1 into medium 2:

$$_2n_1 = \frac{v_2}{v_1} = \frac{1}{_1n_2}$$

It is also clear that with a greater ratio of velocities, more refraction will occur and that rays of light will bend toward the normal as they enter a medium with a lower speed of light and away from the normal when they enter a medium with a higher speed of light.

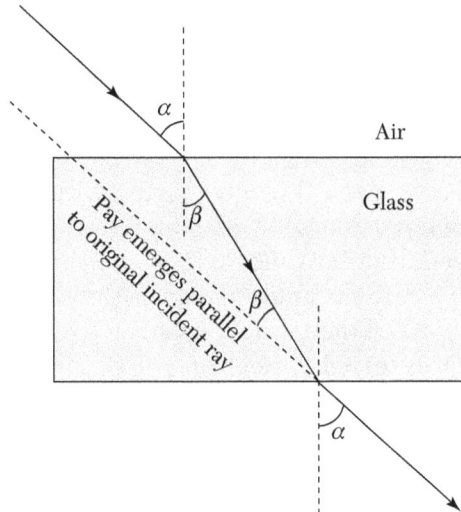

13.3.3   Absolute and Relative Refractive Indices

The absolute refractive index of a medium is the value of $_1n_2$ when medium 1 is the vacuum. The speed of light in a vacuum is c and in a medium is v ($< c$).

Absolute refractive index, $n = \dfrac{c}{v}$

Therefore, $v = \dfrac{c}{n}$

The speed of light v in a medium of absolute refractive index n is equal to the speed of light in a vacuum c divided by the absolute refractive index of the medium. For example, the absolute refractive index of glass is about 1.5 and the speed of light in a vacuum is 3.0×10^8 ms^{-1}, so the speed of light in glass is about 2.0×10^8 ms^{-1}. The greater the refractive index, the slower the speed of light in that medium. The speed of light depends on the electron density inside the material; so the denser the medium, the slower the speed of light and the larger the refractive index.

We can also express the relative refractive index at a boundary in terms of the two absolute refractive indices of the media on either side of that boundary. Let n_1 and n_2 be the absolute refractive indices and let $_1n_2$ be the relative refractive index for a ray of light passing from medium 1 into medium 2.

$$_1n_2 = \frac{v_1}{v_2} = \frac{c/n_1}{c/n_2} = \frac{n_2}{n_1}$$

This gives us the most useful form of Snell's law:

$$\frac{n_2}{n_1} = \frac{\sin\theta_1}{\sin\theta_2}$$

$$n_1\sin\theta_1 = n_2\sin\theta_2$$

13.3.4 Total Internal Reflection

When a ray of light moves into a medium with lower refractive index and therefore higher wave speed, it bends away from the normal. However, the maximum angle of refraction is 90°, and this occurs when the incident angle inside the first medium is less than 90°. The incident angle at which this occurs is called the **critical angle** c, and for larger incident angles there is no refracted ray. All of the incident light is reflected from the boundary. This is called **total internal reflection** (TIR). This can be demonstrated by directing a ray of light at a semicircular glass prism. If the ray is aimed along a radius, there is no change of deflection on entering the prism.

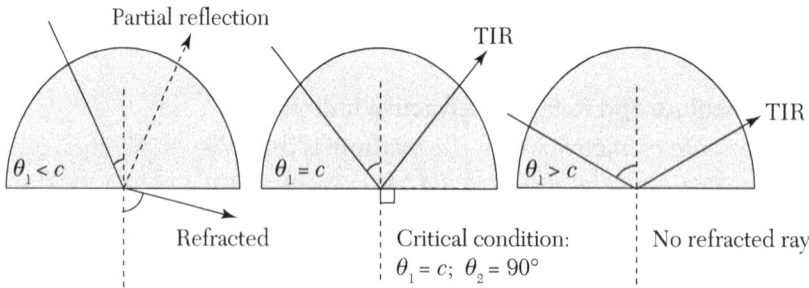

Partial reflection $\quad$ TIR $\quad$ TIR

$\theta_1 < c \qquad \theta_1 = c \qquad \theta_1 > c$

Refracted $\qquad$ Critical condition: $\qquad$ No refracted ray
$\theta_1 = c$; $\theta_2 = 90°$

The condition for the critical angle is that the refracted angle must be equal to 90°.

$$n_1\sin\theta_1 = n_2\sin\theta_2$$

$$n_1\sin c = n_2\sin 90 = n_2$$

$$\sin c = \frac{n_2}{n_1}$$

If medium 2 is the vacuum (or air), then:

$$\sin c = \frac{1}{n}$$

where n is the absolute refractive index of the transparent block.

Total internal reflection can only occur when light travels from a material of higher refractive index to one of lower refractive index.

13.3.5   Optical Fibers

Total internal reflection is a particularly efficient form of reflection and is used inside optical fibers for data transmission. The basic principle of an optical fiber is to have a transparent core surrounded by a transparent cladding material of lower refractive index so that total internal reflection can occur at the core–cladding boundary.

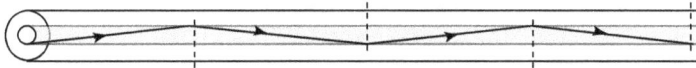

Each time the ray reaches the boundary, its incident angle is greater than the critical angle, so it is repeatedly totally internally reflected. Light has a very high frequency, so it can be modulated to carry a great deal of information.

There are two main types of optical fiber: mono-mode fibers, which effectively only allow a single path for the light (by having an extremely narrow core of about 10 mm), and multi-mode fibers, which are much thicker and allow multiple light paths. The disadvantage of multi-mode fibers is that, over a long distance, different parts of the signal have traveled significantly different distances and developed time delays. If these become comparable to the time between ones and zeroes in the digital signal, then the information is lost. The longer the fiber, the lower the maximum data transfer rate, so they tend to be used over shorter distances, e.g., within a single building. The advantage of multi-mode fibers is that they are cheaper and can carry light of multiple wavelengths, so they are often used with LEDs rather than lasers. Mono-mode fibers use laser sources working at a single wavelength and can transmit high data rates over great distances (up to thousands of kilometers).

13.3.6   Dispersion

The amount of refraction at a boundary depends on the refractive index at that boundary. However, absolute refractive indices depend on the wavelength of light (because the speed of light in a medium depends on wavelength). This means that if a polychromatic ray (i.e., one with a range of wavelengths present, such as white light) refracts at a boundary, different wavelengths will refract different amounts. This is called **dispersion** and is familiar from the way a triangular glass prism can disperse white light into a **spectrum** of colors.

For glass, the (higher frequency)shorter wavelength (blue/violet) end of the spectrum travels more slowly than the (lower frequency)longer wavelength end (red/orange) of the spectrum, and so has a higher refractive index. For a certain type of crown glass, $n_{red} = 1.509$, while $n_{violet} = 1.521$, a small difference but one that is clearly demonstrable.

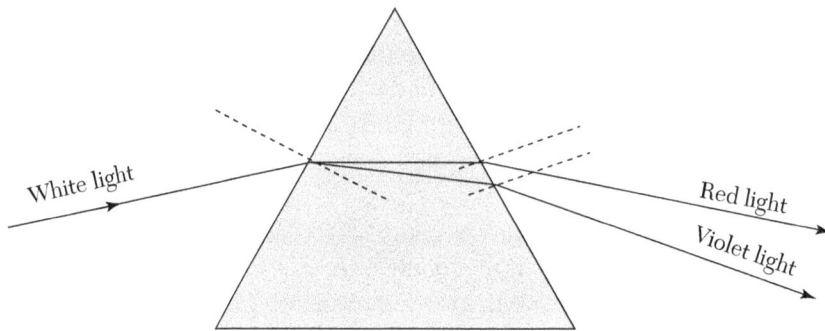

13.4   Polarization

13.4.1   What is Polarization?

Longitudinal waves vibrate parallel to the direction in which they transfer energy, so there is a unique vibration direction. Transverse waves, however, vibrate at 90° to the direction in which they transfer energy, so they can vibrate in any direction perpendicular to the direction in which the wave is traveling. If a transverse wave is confined to oscillate only in one plane, it is said to be plane-polarized. Transverse waves can be polarized, but longitudinal waves cannot.

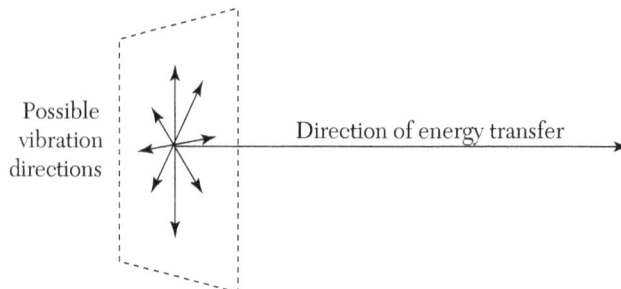

For example, if the only vibration direction is vertical, then the wave is vertically plane-polarized.

13.4.2 Polarizing Filters

An ideal polarizing filter will transmit one direction of polarization and absorb the perpendicular direction of polarization. If the direction of polarization of the incident wave is at an angle to the polarizing direction of the filter, then only a component of the wave's amplitude is transmitted.

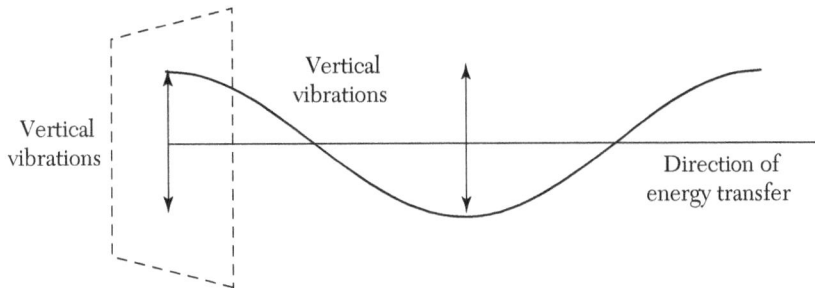

In the diagram below, we are looking in the direction of wave travel with the wave moving away from us. The diagram shows a polarizing filter that will transmit vertically plane polarized light. The light incident on the filter is plane-polarized at an angle θ to the vertical.

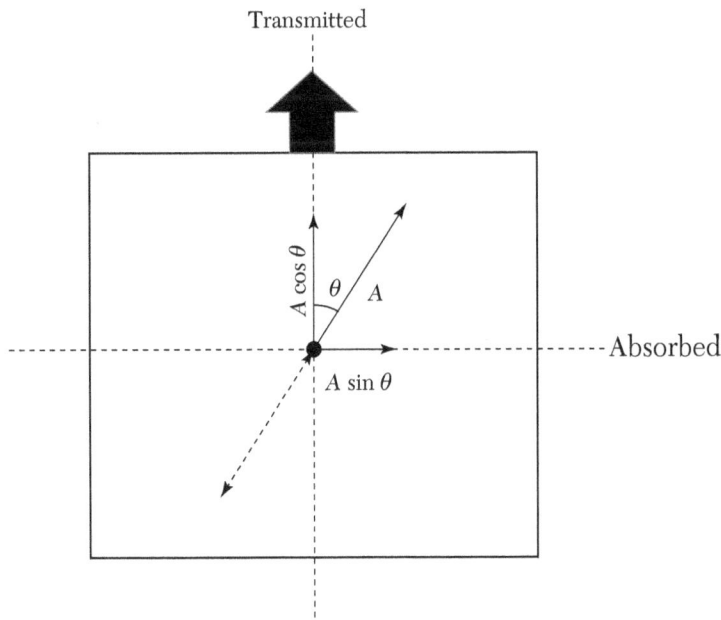

- The transmitted amplitude is: $A_{trans} = A \cos \theta$
- The transmitted intensity is: $I_{trans} = I_0 \cos^2\theta$ (using $I \propto A^2$)
- The transmitted polarization direction is vertical (i.e., the same as the filter).

Rotation of Polarizing Filter

If vertically plane polarized light in incident on a vertical polarizing filter that is slowly rotated around the axis of the beam, the intensity variation maps out the $\cos^2\theta$ function.

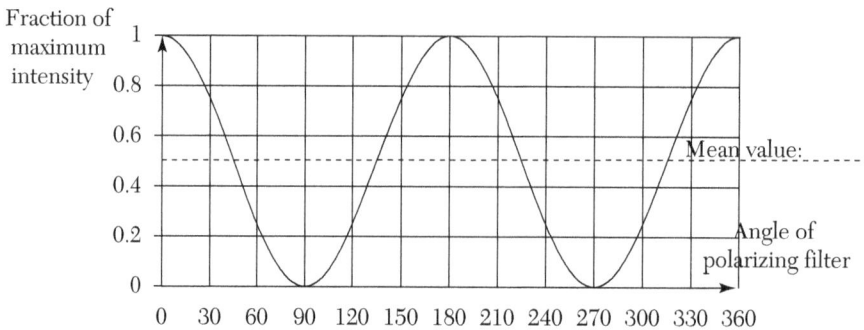

Unpolarized Light Incident on a Polarizing Filter

Unpolarized light contains all possible vibration directions perpendicular to the direction of travel of the wave. When this falls onto an ideal vertical polarizing filter, the intensity is reduced to 50% of its original value. This follows from the equation for intensity above. The intensity at any angle is proportional to $\cos^2\theta$, so the intensity of the transmitted light will be the incident intensity multiplied by the average value of $\cos^2\theta$ over one rotation (θ from 0 to 2π). If we use the intensity $\cos^2\theta = (\frac{1}{2} \cos 2\theta + \frac{1}{2})$ we can see that the average value of the first term is zero (since cosine is equally positive and negative over one cycle), so the average of $\cos^2\theta$ must be $\frac{1}{2}$. Therefore, the transmitted intensity is $I_{trans} = 0.5 \, I_0$. This remains the case if the polarizing filter is rotated.

Crossed polarizing filters

If light falls on two polarizing filters, one placed behind the other and the second filter is rotated with respect to the first, there will be maxima of intensity when the angles between their polarizing filters are 0°, 180°, and 360° and minima at 90° and 270° (as in the graph above).

13.4.3 Rotation of the Plane of Polarization

Some transparent media can rotate the plane of polarization around the direction of travel of the wave. Sugar solutions do this because the molecules are themselves asymmetric and the amount of rotation can be used to measure the concentration of the solution (higher concentrations produce a larger rotation angle per unit distance). A **polarimeter** consists of two polarizing filters mounted either side of the sample to be tested. One of the filters can be rotated and the angle between the two filters can then be read off from a fixed scale. With no sample, maximum intensity will be when the filters are aligned. If the sample rotates the plane of polarization, the second filter can be rotated until the intensity is once again a maximum and the angle of rotation can be measured.

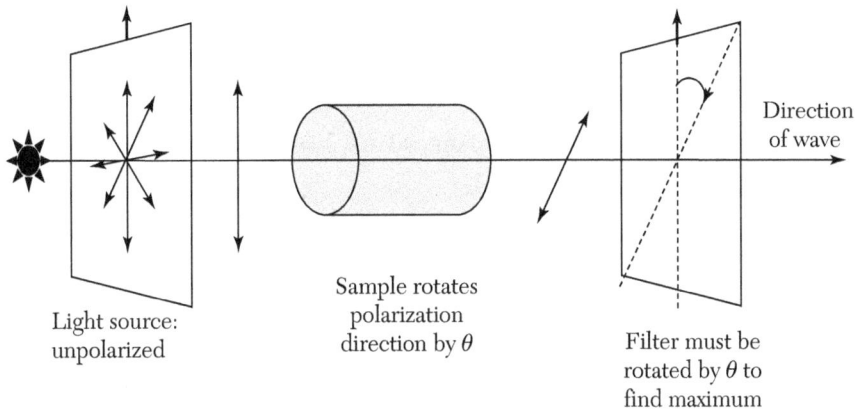

Light source: unpolarized

Sample rotates polarization direction by θ

Filter must be rotated by θ to find maximum

Direction of wave

13.4.4 Polarization by Reflection and Scattering

When unpolarized light is incident on a non-metallic transparent medium, some of the light is refracted into the medium and some is reflected from the boundary. The interaction between light and matter causes dipoles in the material surface to absorb energy, vibrate, and then re-radiate the light. However, light is a transverse wave, so these vibrations determine the possible polarization of the refracted and reflected light rays. At a particular angle of incidence, called the Brewster angle, the reflected and refracted angles will be at 90° and the reflected ray is plane-polarized parallel to the surface of the material. The refracted ray is also partially plane-polarized along a line parallel to the direction of the reflected ray.

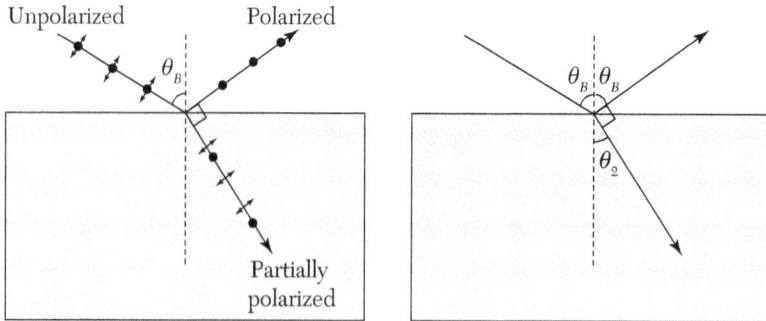

We can derive an expression for the Brewster angle θ_B by using Snell's law and the law of reflection. Let the refractive index of the first and second media be n_1 and n_2, respectively.

$$n_1 \sin \theta_B = n_2 \sin \theta_2$$

But
$$\theta_2 = 180 - (90 + \theta_B) = 90 - \theta_B$$

$$n_1 \sin \theta_B = n_2 \sin (90 - \theta_B) = n_2 \cos \theta_B$$

$$\tan \theta_B = \frac{n_2}{n_1}$$

This is **Brewster's law**.

Photographers use polarizing filters to enhance contrast, e.g., to darken the sky compared to the clouds, or to reduce glare from reflective surfaces (e.g., water or windows). Polarizing sunglasses are also used to reduce reflected glare by absorbing one component of polarization.

13.5 Exercises

1. A traveling wave is described by the equation below (distances measured in meters):

$$y = 0.25 \cos(30t + 2x)$$

(a) State the amplitude, frequency, and wavelength of this wave.
(b) Calculate the speed of the wave.
(c) State the direction in which the wave is traveling.

2. A swimmer is treading water in the sea. Waves traveling at 1.2 ms^{-1} cause her to bob up and down six times a minute. The distance between her highest and lowest positions as the wave passes is 2.4 m.
(a) Sketch a graph to show how her displacement varies with time over a period of 20 s.
(b) Calculate the frequency and wavelength of the waves.

(c) Her friend is also treading water but is 15 m further out to sea. What is the phase difference between the oscillations of the two swimmers? Assume that the waves are traveling directly toward the shore.

3. The diagram below shows a pin placed symmetrically between two plane mirrors that are perpendicular to each other.

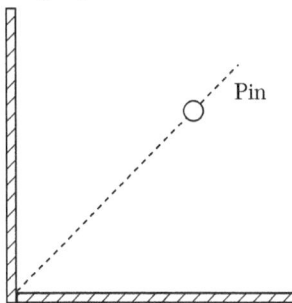

An observer looking into the mirrors can see three images of the pin. Draw a careful ray diagram to locate the positions of these images.

4. Here are some absolute refractive indices for different media.

Vacuum $n = 1$ (exactly)
Air $n = 1.000293$ (at s.t.p.) .... usually taken to be 1.00
Diamond $n = 2.42$
Glass $n = 1.50$
Water $n = 1.33$
Speed of light in a vacuum $c = 3.0 \times 10^8$ ms⁻¹

The table below refers to a ray of light traveling from medium 1 to medium 2 as shown in the diagram.

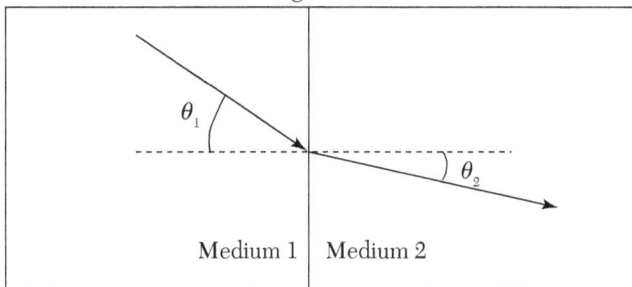

(a) Complete the table using values from the list above:

Medium 1	Medium 2	v_1	v_2	θ_1	θ_2
	Glass	3.0×10^8 ms⁻¹		40°	
Glass	Air			40°	
Water	Diamond				20°
			2.0×10^8 ms⁻¹	55°	70°

(b) Explain what is meant by total internal reflection and state the conditions under which it occurs.
(c) Calculate the critical angle for an interface between:
  (i) water and air
  (ii) glass and air
  (iii) diamond and air
(d) Suggest why a real diamond sparkles more than a fake glass "diamond."

5. The refractive index n of glass can be determined by measuring the minimum angle of deviation D when light passes through a triangular prism of apex angle A:

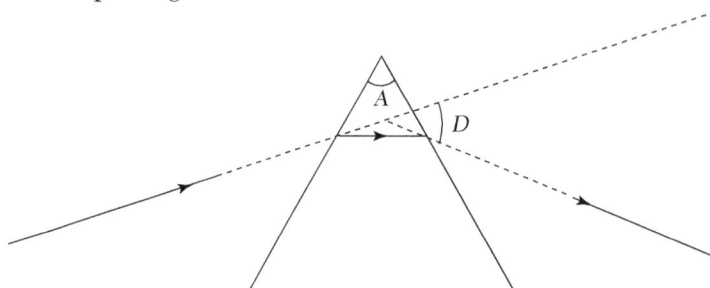

The minimum deviation occurs when the light passes symmetrically through the prism (the ray inside the prism is parallel to the base). The formula used to find the refractive index is:

$$n = \frac{\sin\left(\dfrac{A+D}{2}\right)}{\sin\left(\dfrac{A}{2}\right)}$$

Derive this formula.

6. Two polarizing filters are placed at 90° to each other. A third polarizing filter is placed between them and slowly rotated. Unpolarized light is directed into the system of three filters.
(a) Explain why the intensity of light passing through the system of three filters reaches a maximum when the third filter is at 45° to the direction of the first polarizing filter.
(b) Calculate the fraction of the incident intensity that passes through the system when the middle polarizing filter is at 30° to the first polarizing filter.

CHAPTER 14

LIGHT

14.1 Light as an Electromagnetic Wave

14.1.1 Waves or Particles?

Properties such as reflection and refraction of light have been known since antiquity, but the nature of light remained a mystery. Newton thought that light must consist of streams of tiny corpuscles (particles) emitted from bright objects, but Huygens (a Dutch physicist) thought that light is a form of wave motion like ripples on the surface of a pond. It turns out that neither the wave model nor the particle model is a complete model of the nature of light (or any other electromagnetic wave). However, aspects of both the wave model and the particle model can be used to explain how light behaves, but we have to be careful where we use each model. We will look at wave–particle duality in more detail later (see Section 27.3), but in this chapter we concentrate on the wave model.

14.1.2 Electromagnetism

In the 19th century, Michael Faraday and James Clerk Maxwell tried to make sense of all the different electromagnetic phenomena that had been discovered up that time. This included:

- how charges exert forces on one another (see Section 17.3.1)

- how electric currents create magnetic fields (see Section 20.3.1)

- how magnetic fields exert forces on moving charges and electric currents (see Section 20.2.2)

■ how changing magnetic fields can induce voltages (see Section 21.1.1)

In order to do this Faraday introduced the idea of an **electromagnetic field**. Electromagnetic fields are created by electric charges and exert forces on electric charges. For example, a positive charge at one point in space creates an electric field through all of space, and another charge some distance from the first experiences a force *from the field*. The beauty of this model is that the field itself has properties, and charges do not interact by "instantaneous action-at-a-distance" but by interacting with the *local* electromagnetic field.

Maxwell discovered a set of four equations—the Maxwell equations that describe how fields are created by, and interact with, charges and how the field in one place affects the field nearby. The equations showed that if the field is disturbed in one place, e.g., because a charge is vibrating, that disturbance spreads outwards from the source at a constant speed. The speed could be calculated from the equations and is equal to the measured speed of light. This suggested very strongly that light must be an electromagnetic disturbance. It also showed that visible light is just part of a much wider spectrum of electromagnetic waves, all of which travel at the same constant speed in a vacuum.

14.1.3   Electromagnetic Waves

Electromagnetic waves travel through a vacuum, so they cannot be mechanical waves—there are no material particles to vibrate. So, what does vibrate? The electric and magnetic fields at each point in space. An electromagnetic wave is a transverse wave (we know this because light can be polarized) and consists of vibrations of the electric and magnetic field at each point in space. The direction of the electric field vibration is at 90° to the direction of vibration of the magnetic field. The diagram below shows the relationship between the electric and magnetic fields in a plane-polarized electromagnetic wave.

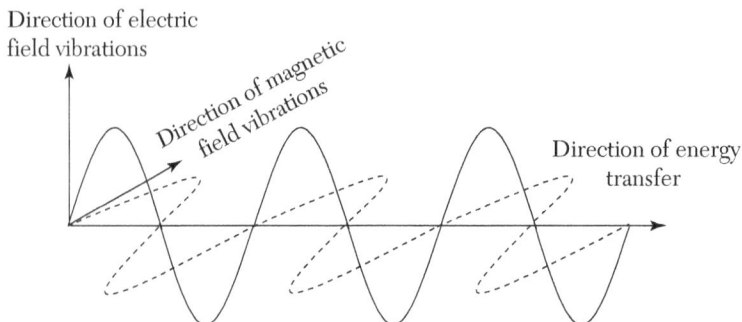

The polarization direction is, by convention, the direction in which the electric field vibrates. In the example above, the polarization is vertical.

It is important to realize that this diagram represents the variation of the two fields at a point ON the red line. They do not represent physical motions of anything off the line. As the wave passes a point, the electric and magnetic fields at that point increase and decrease in strength periodically. If a wave like the one above strikes a material medium, the two fields cause charges in the surface to vibrate at the frequency of the wave. This is how a radio antenna detects radio waves—the vibrating charges in the antenna set up a weak A.C. signal that can be amplified.

Maxwell's equations show that electromagnetic waves are emitted when charges accelerate (e.g., while oscillating) and cause charges to accelerate when they are absorbed.

Visible light consists of part of the much larger electromagnetic spectrum. Our eyes respond to wavelengths from about 390 nm (violet) to 700 nm (red), which corresponds to a frequency range of 4.3×10^{14} to 7.7×10^{14} Hz. The main regions of the electromagnetic spectrum are shown below.

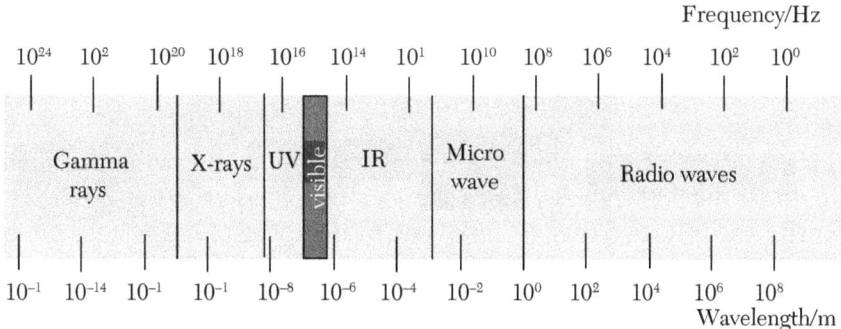

Frequency/Hz

| 10^{24} | 10^{2} | 10^{20} | 10^{18} | 10^{16} | 10^{14} | 10^{1} | 10^{10} | 10^{8} | 10^{6} | 10^{4} | 10^{2} | 10^{0} |

| Gamma rays | | X-rays | UV | visible | IR | | Micro wave | | Radio waves | | | |

| 10^{-1} | 10^{-14} | 10^{-1} | 10^{-1} | 10^{-8} | 10^{-6} | 10^{-4} | 10^{-2} | 10^{0} | 10^{2} | 10^{4} | 10^{6} | 10^{8} |

Wavelength/m

You might still be wondering what it is that actually vibrates when an electromagnetic wave passes through a vacuum—calling it the electric and magnetic field doesn't really answer the question. Nineteenth-century physicists also puzzled about this and assumed that there must be some invisible medium filling all of space so that electric and magnetic fields are distortions of this medium and electromagnetic waves involve vibrations in the medium. They called the medium the **luminiferous ether**, but no attempt to demonstrate its existence was ever successful and Einstein's special theory of relativity abandoned it (see Section 24.1.2).

14.1.4 Measuring the Speed of Light

The first successful measurement of the speed of light was an astronomical one. Jupiter has several moons and each of these completes one orbit in a constant time. However, when viewed from Earth the time of orbit seems to vary by about ±8 minutes over the course of a year. The astronomer, Olaf Römer, realized that as the Earth and Jupiter orbit at different rates the distance between the two varies by a distance equal to the diameter of the Earth's orbit. This means that light takes different times to reach the Earth when the relative positions of Earth and Jupiter change, and it was concluded that the speed of light must be equal to the diameter of the Earth's orbit $(1.5 \times 10^{11}$ m) divided by 8 minutes (480 seconds). This is about $3 \times 10^8 \, \text{ms}^{-1}$.

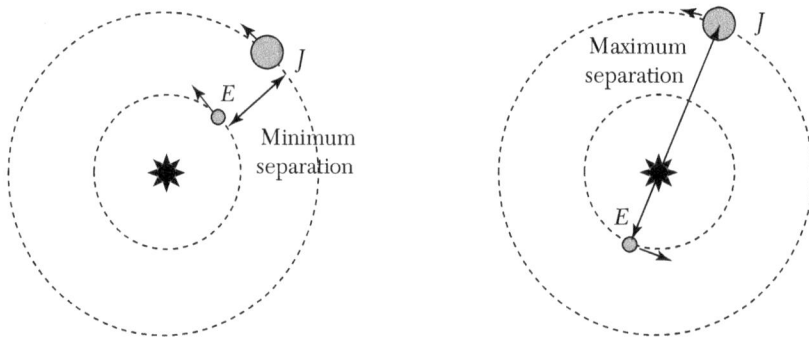

The first terrestrial measurement of the speed of light was made by Armand Fizeau in 1849. He employed an ingenious method using a rapidly spinning toothed wheel to chop light into short pulses. The pulses then hit a distant mirror and reflected back to the wheel. As the speed of the wheel increased, the returning light hit the next tooth and was blocked. This meant that the time taken for the light to travel to the mirror and back was equal to the time taken for the wheel to rotate by the angle between a tooth and the next gap. He measured the rate of rotation at which this occurred and used this to find the time of flight t of the light pulse. The speed of light was then calculated from the distance to the mirror, d, and back divided by the time of flight:

$$c = 2d/t$$

This method was modified and improved by Jean Leon Foucault and then by Albert Michelson. They replaced the toothed wheel with a rotating mirror, but the principle of the method was similar. Foucault even measured the speed of light in water by placing a tube of water in the light path between

To oscilloscope to
record emitted pulse

LED base producing short
pulses at high frequency

Mirror

Detector

d

To oscilloscope to
record detected pulse

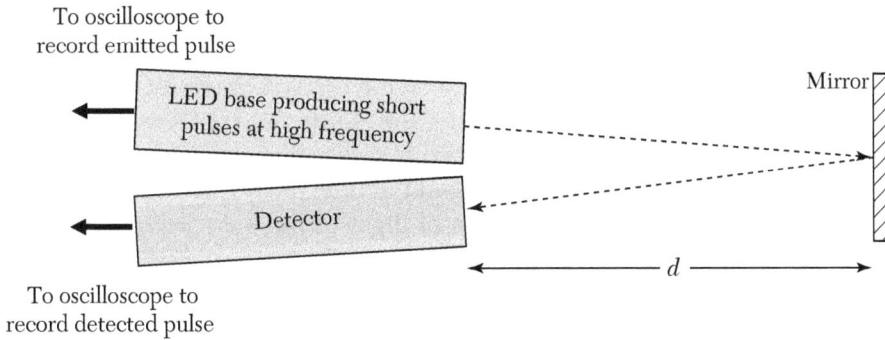

the rotating mirror and the fixed distant mirror. His results showed that light travels more slowly in water than in air, a result that reinforced the wave model of light because it was consistent with the wave explanation of refraction at an air–water boundary (the particle model predicted that the particles of light would speed up as they entered the denser medium). Michelson spent most of his life-refining methods for measuring the speed of light. He also carried out the famous Michelson-Morley experiment, which seemed to show that the speed of light is independent of the motion of the observer, an effect that was eventually explained by Einstein's special theory of relativity (see Chapter 24).

What makes it difficult to measure the speed of light in a laboratory is the fact that it is so fast. This means that we have to measure extremely short time intervals. Nowadays high-speed electronic devices can do this and the measurement can be carried out successfully over a distance of just a few meters. The principle is very similar to that used by Fresnel, Foucault, and Michelson, wherein short pulses of light are generated by an LED laser and a fast oscilloscope is used to detect the emission of the pulse and the detection of its reflection from a mirror placed a few meters away.

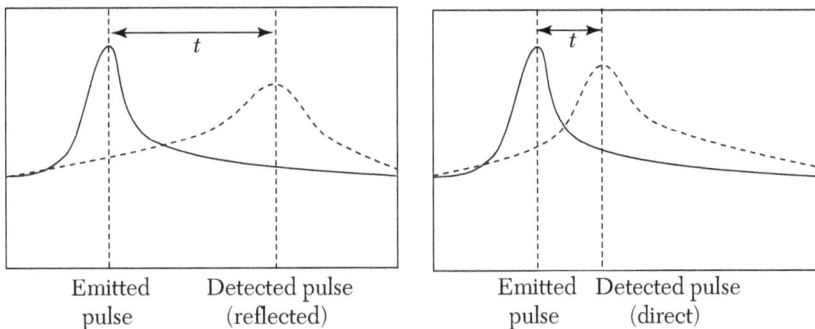

t

Emitted
pulse

Detected pulse
(reflected)

t

Emitted
pulse

Detected pulse
(direct)

One slight complication with this method is that the time it takes for the electronic processing of the signals is comparable to the time of flight of the light pulses, so the time t_1 measured on the oscilloscope must be corrected for this. One way to do this is to place the emitter and detector so that they are facing each other and the light path is virtually zero. There will still be two separate peaks on the screen, and the time delay t_2 is now entirely due to the equipment. The actual time of flight of the light pulse is therefore $(t_1 - t_2)$ and the speed of light is $2d/(t_1 - t_2)$. Typical oscilloscope traces are shown in the diagram.

14.1.5 Maxwell's Equations and the Speed of Light

Maxwell's equations themselves are beyond the scope of this book. However, the equations lead to an expression for the speed of all electromagnetic waves:

$$c = \sqrt{\frac{1}{\eta_0 \mu_0}}$$

where ε_0 is the permittivity of free space (effectively a measure of the ability of a vacuum to support an electric field) and μ_0 is the permeability of free space (effectively a measure of the ability of a vacuum to support a magnetic field).

When electromagnetic waves penetrate matter (e.g., light going through glass), the equation becomes:

$$v = \sqrt{\frac{1}{\eta_0 \varepsilon_r \mu_0 \mu_r}}$$

where ε_r is the relative permittivity of the medium and μ_r is the relative permeability of the medium. Whereas the speed of all electromagnetic waves is the same in a vacuum, the values for the relative permittivity and permeability in a medium vary with frequency, so different parts of the electromagnetic spectrum travel at different speeds inside media, causing **dispersion**. This is responsible for the way white light can be spread into a spectrum by a triangular prism.

We have already met the equation $v = \dfrac{c}{n}$ where n is the absolute refractive index of a medium. It follows that the absolute refractive index is related to permittivity and permeability by:

$$n = \sqrt{\varepsilon_r \mu_r}$$

14.1.6 Defining Speed, Time, and Distance

The speed of light in a vacuum is one of the fundamental physical constants, and it has been measured ever more precisely. Its value is now fixed to be 299,752,498 ms^{-1} and this will not change even if the speed of light is measured even more precisely in future. The second is also a defined quantity, equal to 9,192,631,770 periods of vibration of the radiation emitted between two energy levels in the cesium-133 atom. This implies that the length of the meter cannot be a defined quantity since distance, speed, and time are all related. The meter is a derived quantity equal to the distance traveled by light in a vacuum in a time of 1/(299,752,498) of a second.

The fact that the speed of light can be derived from Maxwell's equations raises an interesting question about its meaning. What is the speed of light measured against, or more specifically, in what reference frame does light travel at c? In the nineteenth century most physicists assumed that light is a vibration in some as-yet-undiscovered and all-pervading medium, which they called the luminiferous ether so that light traveled at speed c relative to this ether. The implication was that the speed measured by an observer who was also moving relative to the ether at speed v would be a relative speed. For example, if both the light and the observer moved in the same direction through the ether at speeds c and v, then the speed of light relative to the observer should be c-v. But this is not the case. *The speed of light is the same for all uniformly moving observers* and is independent of the speed of the observer or the source. This is explained by Einstein's special theory of relativity, which also led to the abandonment of the idea of the luminiferous ether.

14.2 Ray Optics

The behavior of optical instruments can be explained by following the paths of rays. Rays are assumed to travel in straight lines (rectilinear propagation) from a source, only changing direction when they pass through a lens or prism, or when they are reflected.

14.2.1 Thin Lenses

Convex or converging lenses are shaped in such a way that light traveling parallel to their principal axis is refracted to a real focal point beyond the lens. The distance from the optical center (of the lens) to the focal point F is called the focal length f of the lens. The reciprocal of this value is called the power of the lens P and is measured in diopters (1 diopter = 1 m^{-1}).

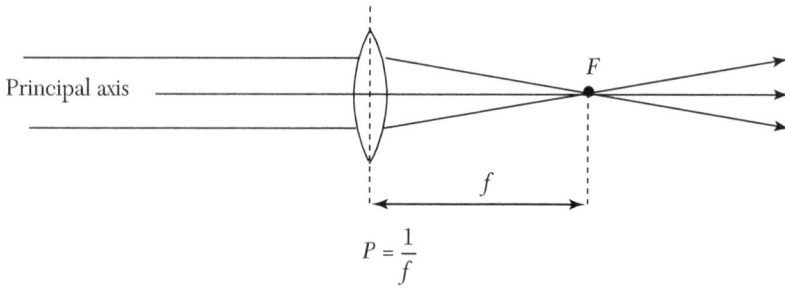

$$P = \frac{1}{f}$$

$$P = \frac{1}{f}$$

Concave, or diverging, lenses are shaped in such a way that light traveling parallel to their principal axis diverges from a virtual focal point before the lens.

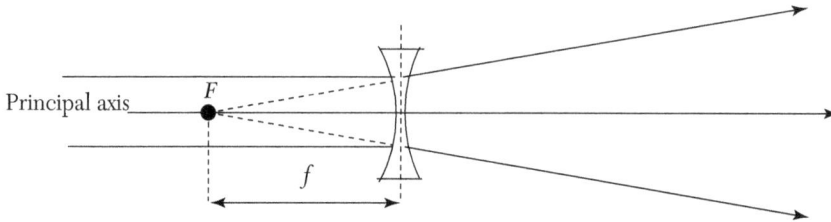

In a real lens, light will refract as it enters the lens and refract again as it leaves on the opposite side. A thin lens is one where both refractions can be assumed to take place at the same point, on a line perpendicular to the principal axis and passing through the optical center of the lens. Ray diagrams involving thin lenses are often drawn with the lens itself shown as a single vertical line passing through the optical center and a small symbol showing the type of lens.

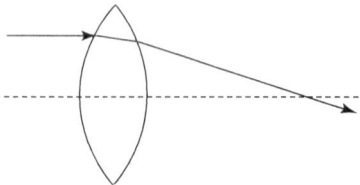

Thick

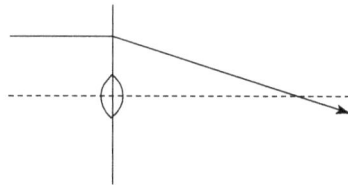

Thin lens

14.2.2 Predictable Rays for Thin Lenses

Ray diagrams can be used to find the positions of objects and images in optical instruments. To draw them we make use of predictable rays.

Convex lens—predictable ray 1: A ray parallel to the principal axis will pass through the focal point on the far side of the lens.

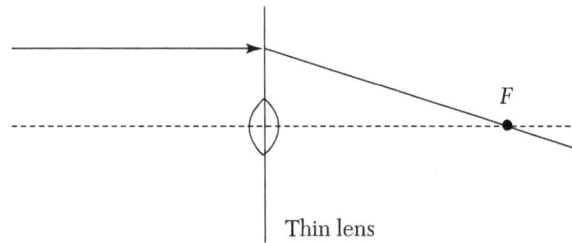

In ray optics the direction of the ray is reversible, so a ray passing through the focal point before it reaches the lens will then travel parallel to the principal axis on the far side of the lens.

Convex lens—predictable ray 2: A ray passing through the optical center of the lens does not change direction.

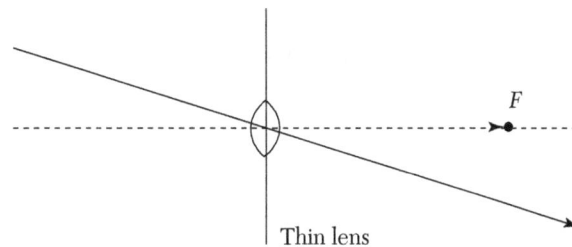

Concave lens—predictable ray 1: A ray parallel to the principal axis diverges from the principal axis on the other side of the lens along a line that traces back to the focal point.

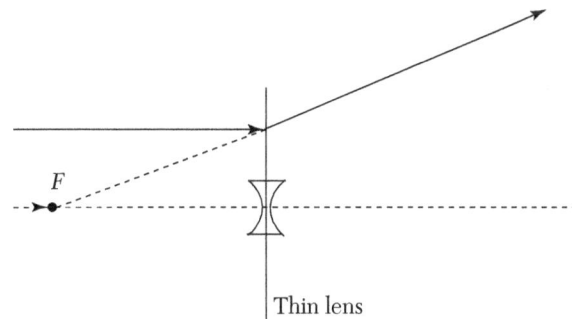

The direction of the ray is reversible, so a ray heading toward the focal point on the far side of the lens will then travel parallel to the principal axis after it passes through the lens.

Concave lens—predictable ray 2: A ray passing through the optical center of the lens does not change direction.

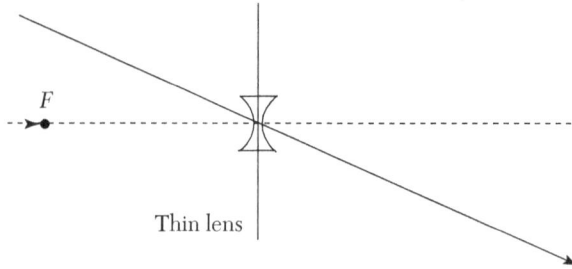

Thin lens

14.2.3   Images

An image is a point-by-point representation of an object. A **real image** is formed when the rays creating the image pass through the image points, for example, the image formed on the retina of the eye, or the image projected onto a cinema screen. A virtual image is formed when the rays creating the image diverge from points on the image but do not pass through those points, for example, the image in a plane mirror, or the image of the bottom of a swimming pool seen through the surface. Since the rays responsible for the image do not pass through it, a **virtual image** cannot be formed onto a screen.

The size and position of an image can be determined by drawing two predictable rays from a single point on an object and tracing their paths. The place where they intersect is the position where the image point is formed. Objects are often shown as vertical arrows and the two rays to be traced are from the tip of the arrow.

14.2.4   Image Formation with a Convex Lens

The distance from the object to the optical center of the lens is called the object distance, u. The distance from the optical center of the lens to the image is called the image distance, v.

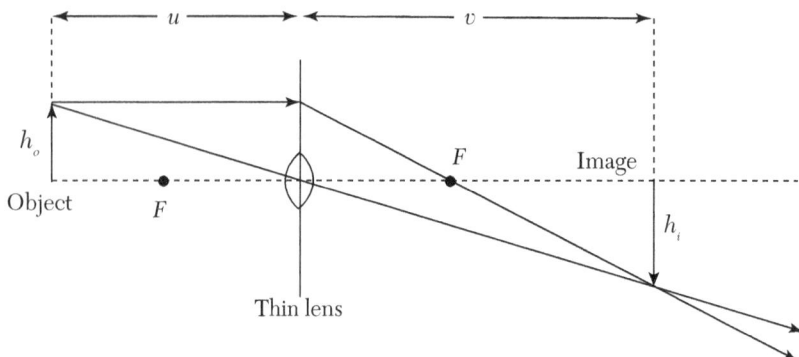

Thin lens

A real, inverted, magnified image has been formed. The linear magnification m is given by the equation:

$$m = \frac{h_i}{h_o}$$

Real images are formed if the object distance is greater than the focal length of the lens $(u > f)$. However, if the object is placed in the focal plane (so that $u = f$), then the rays are parallel after passing through the lens. They are said to form an image at infinity $(v = \infty)$.

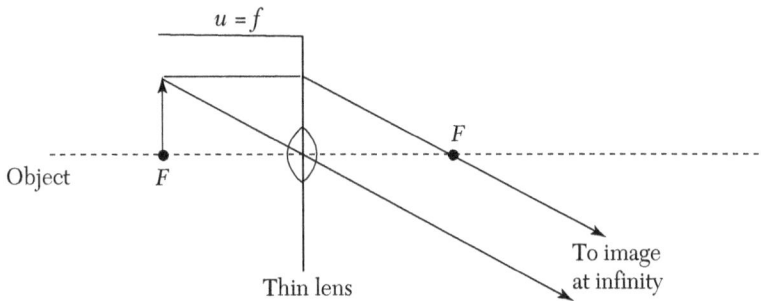

Large magnification can be achieved by placing the object so that u is just greater than f. Magnification of 1 is when $u = 2f$; and for larger object distances, the image is diminished (smaller than the object, $m < 1$).

For object distances less than the focal length, the emerging rays diverge. However, these can be traced backwards to a virtual image on the same side of the lens as the object.

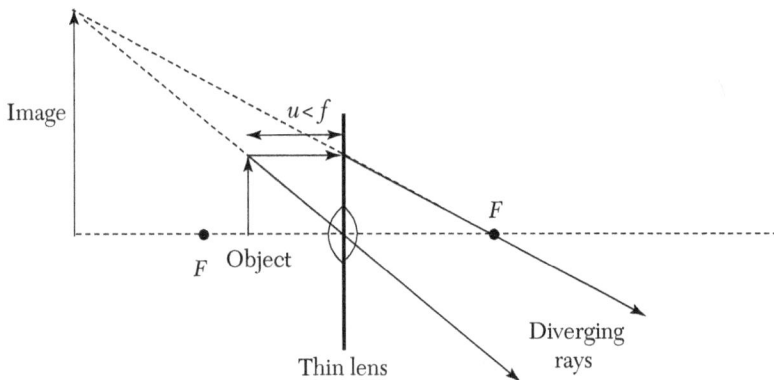

This is an erect (same way up as the object), magnified, virtual image.

This ray diagram can be used to explain how a magnifying glass works. Your eye would be positioned on the optical axis to the right of the lens and would look through the lens to see the magnified image beyond the object.

Summary of the behavior of a convex lens			
Object distance	**Image position**	**Image nature**	**Magnification**
>2f	Between f and 2f	Real	<1
2f	2f	Real	1
Between f and 2f	>2f	Real	<1
f	Infinity	Undetermined	Infinite
<f	>f	Virtual	>1

14.2.5   Image Formation with a Concave Lens

A similar approach can be used to find the image position when a concave lens is used.

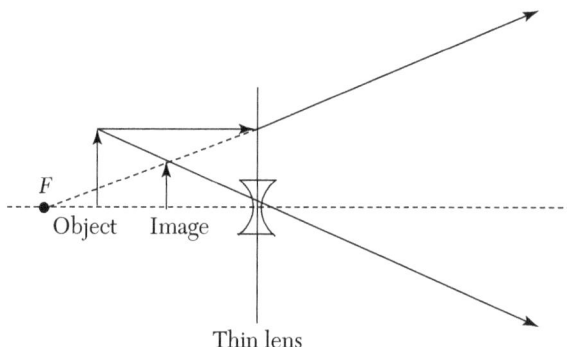

Thin lens

The image is erect (same way up as the object), diminished, and virtual.

14.2.6   Object at Infinity

If an object is at great distance from a convex lens ($u \gg f$), rays diverging from a point on the object will be almost parallel when they reach the lens. The object is at **optical infinity;** and when the rays pass through the lens, they converge to form an image point in the focal plane of the lens. This is the situation when we look at a distant object, and an inverted image of that object is formed on the retina of our eye.

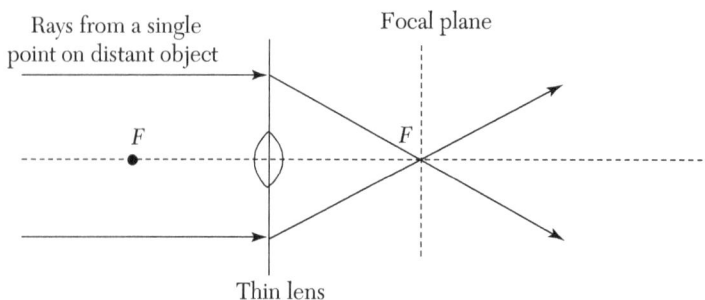

Thin lens

For an extended object at great distance, e.g., the Sun, there will be an angle between the rays from a point on one side of the Sun and the rays from a point on the opposite side of the Sun. They will be focused in the focal plane above and below the focal point of the lens, and an image of the Sun will be formed between them.

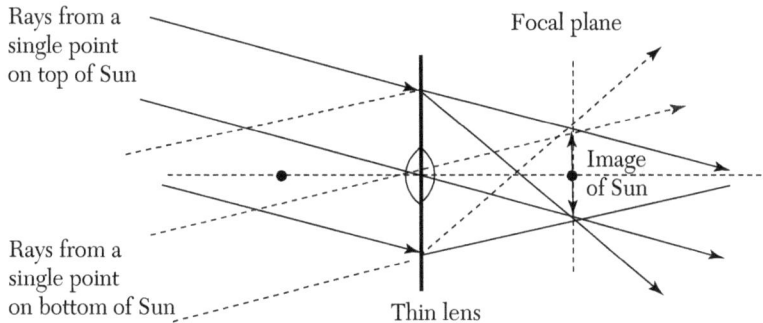

This creates an image of the Sun that can be projected onto a screen—e.g., to look for sunspots.

The power of a lens depends on its shape and on the material from which it is made. The higher the refractive index, the more the light is refracted and the shorter the focal length of the lens. However, the refractive index depends on wavelength, so objects that have a range of colors are not focused sharply at any one point. This is called **chromatic aberration**. Achromatic lenses are composite lenses that use a range of different lenses to compensate for chromatic aberration.

14.2.7 Lens Equation

The equation relating u, v, and f is called the lens equation. It can be derived from the ray diagram for a convex lens, but, with a suitable sign convention, it also applies to concave lenses. Look at the two shaded triangles in the ray diagram below.

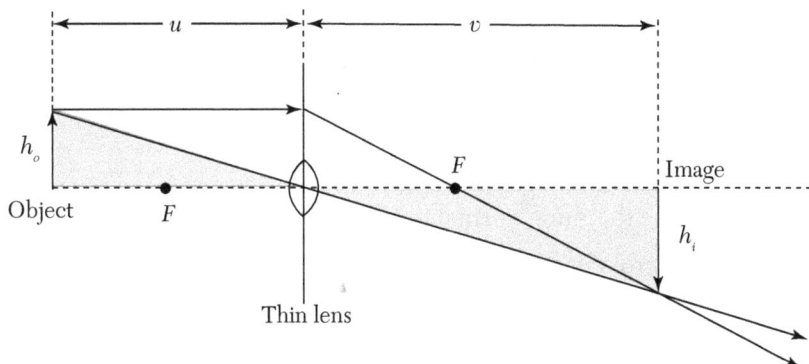

These are similar triangles, so:

$$m = \frac{h_i}{h_o} = \frac{v}{u}$$

Now look at these two shaded triangles that are also similar:

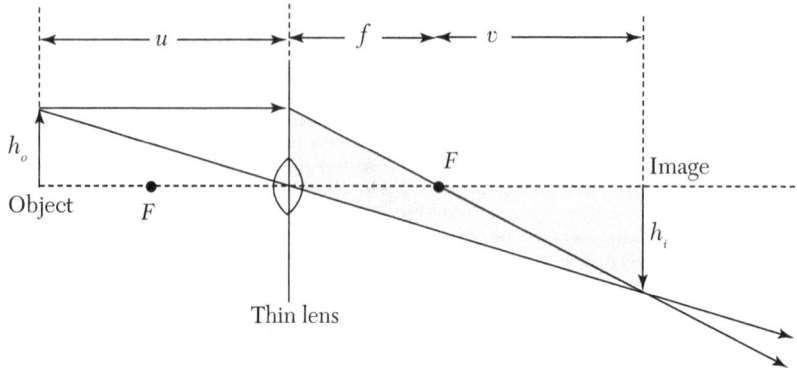

Therefore,

$$\frac{h_i}{h_o} = \frac{(v-f)}{f} = \frac{v}{u}$$

which can be rearranged to give the **lens equation**:

$$\frac{1}{f} = \frac{1}{u} + \frac{1}{v}$$

This can be used for convex and concave lenses using the **REAL IS POSITIVE** sign convention:

- f is positive for a convex lens (real focus) and negative for a concave lens (virtual focus).

- u is positive for a real object and negative for a virtual object.

- v is positive for a real image and negative for a virtual image.

14.2.8 Virtual Image Formed by a Plane Mirror

When an object is placed in front of a plane mirror, the reflected rays diverge from an image point behind the mirror.

The two shaded triangles are similar, so $u = v$ and $m = 1$. The object and the image lie on the same normal to the mirror.

Curved mirrors can be used like lenses to focus and project images. A suitably shaped concave mirror has a real focus, and a convex mirror has a virtual focus.

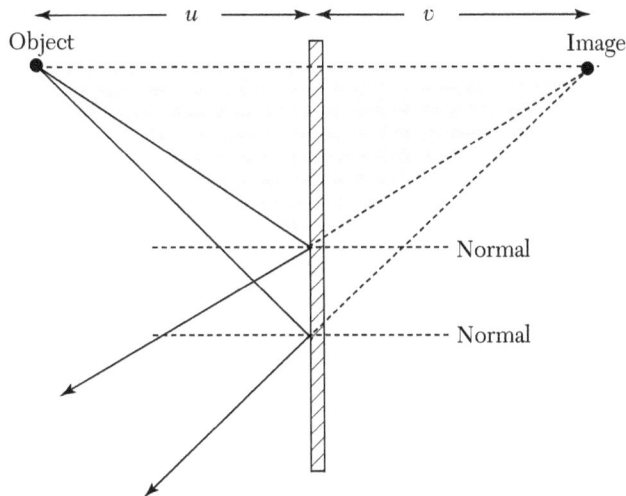

Concave mirrors are used to focus microwaves and radio waves so that a detector placed at or near the focal point receives a strong signal (e.g., for a radio telescope or a satellite TV dish). The way to think about curved mirrors is to realize that the ray diagrams for a concave mirror are like those for a convex lens, but with a reflection, and the ray diagrams for a convex mirror are like those for a concave lens with a reflection.

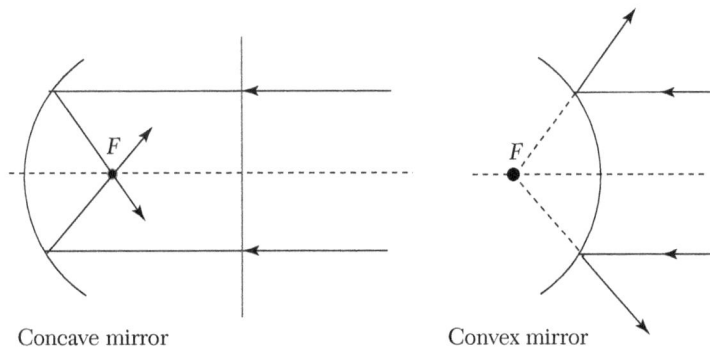

Concave mirror   Convex mirror

If a convex mirror is actually part of a sphere, rays nearer the edge of the mirror focus to a different position than rays near its center. This results in an unfocused image and is an effect called **spherical aberration**. To avoid this the correct shape for a concave mirror is a paraboloid.

14.2.9   Real and Apparent Depth

When you look down into a swimming pool, the floor of the pool appears closer to the surface than it actually is. This reduces the apparent depth of the water. The reason for this is that when light refracts at the water surface it creates a virtual image of the floor of the pool that is closer to the surface. If you look directly down into the pool, the ratio of the real depth to the apparent depth is equal to the refractive index of water.

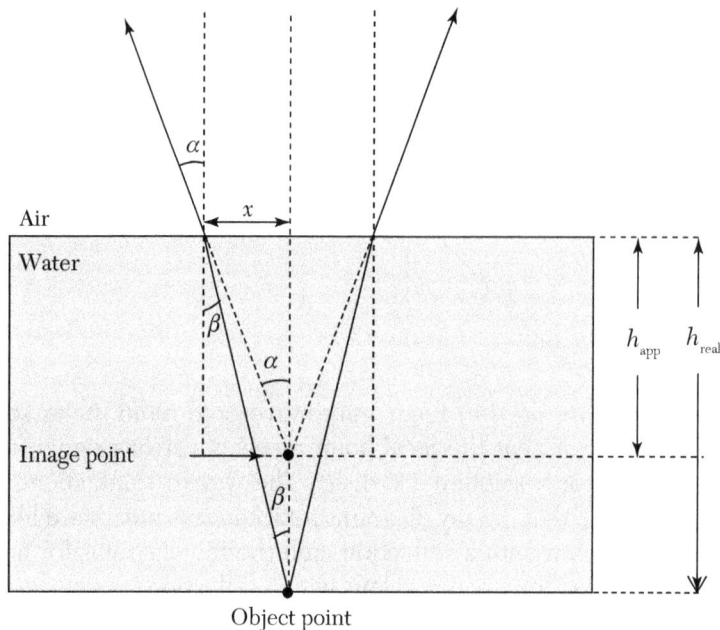

$$\frac{\sin \alpha}{\sin \beta} = n_{\text{water}}$$

From the diagram,

$$\sin \alpha = \frac{x}{h_{appt}} \quad \text{and} \quad \sin \beta = \frac{x}{h_{real}}$$

Therefore,

$$n_{\text{water}} = \frac{h_{\text{real}}}{h_{\text{appt}}}$$

14.3   Optical Instruments

The behavior of an optical instrument (e.g., its magnifying power and the nature and position of the image it produces) can be determined by ray tracing.

14.3.1   Astronomical Refracting Telescope

Telescopes are used to observe distant objects, so the rays of light arriving at the objective of the telescope from a particular point on the object are parallel rays. The telescope is used to increase the visual angle, that is, the angle subtended by the image at the eye. It does this using two convex lenses. The first, the objective lens, is a low-power convex lens that creates a real, inverted image of the distant object on the objective focal plane inside the barrel of the telescope. The second, the eyepiece lens, is a higher-power lens used as a magnifying glass to magnify the intermediate image. In normal adjustment, the telescope accepts parallel rays and parallel rays leave the eyepiece; therefore, the distance from the first image to the eyepiece lens must equal the eyepiece focal length. This makes the total distance between the two lenses equal to the sum of focal lengths of the lenses.

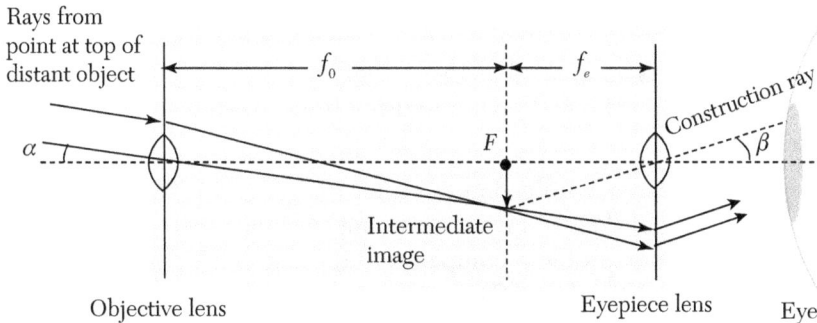

In the ray diagram above, the rays from the top of the object enter the objective at an angle α to the principal axis. Rays from the bottom of the object are assumed to enter along the principal axis but are not shown in the diagram. The angle α is therefore equal to the angle subtended by the object at the objective, and β is the angle subtended by the image at the eye.

- α is the angle subtended by the distant object at the objective lens.

- β is the visual angle—the angle subtended at the eye by the rays forming image.

   The angular magnification M of the telescope is the ratio α/β.

   Using similar triangles and alternate angles, it is easy to see that

$$\frac{\tan \beta}{\tan \alpha} = \frac{f_o}{f_e}.$$

   However, the angles involved are small, so we can use the approximation $\tan \theta \approx \theta$ (in radians) to obtain an expression for angular magnification:

$$M = \frac{\beta}{\alpha} = \frac{f_o}{f_e}$$

The image is, as can be seen in the diagram, inverted. For astronomical work, this is not an issue. However, a terrestrial refracting telescope has a third convex lens in between the intermediate image and the eyepiece, and this lens simply inverts the intermediate image so that the final image is erect.

14.3.2 Astronomical Reflecting Telescope (Newtonian Telescope)

The amount of light captured by the objective of a telescope is directly proportional to its area; so the larger the diameter of an objective, the greater the sensitivity of the instrument. Astronomers need to observe very faint objects, so astronomical telescopes need large-diameter objectives. Very large-diameter lenses are heavy, difficult to manufacture, and have the additional problem of chromatic aberration, so large optical telescopes use concave mirrors as their objective rather than convex lenses. This design was first used by Isaac Newton and is often referred to as a Newtonian telescope.

The angular magnification is equal to f_o/f_e as for the refractor. The secondary mirror does reduce the amount of light received by the primary, but this is a small problem compared to the difficulties of building a large achromatic lens. The largest ground-based optical telescope is in La Palma on the Canary Islands. It is the Gran Telescopio Canaris and the diameter of its objective mirror is 10.4 m. However, the Great Magellan telescope is currently under construction in Chile, and when completed this will have a diameter of 24.5 m. The reflecting surface of these large telescopes is made up of several individual reflectors rather than a single dish.

Objective: concave mirror of focal length f_o

Secondary mirror

Eyepiece: convex lens of focal length

Radio telescopes are also reflecting telescopes, using a large concave dish to reflect radio waves to a detector. Radio waves have much longer wavelengths than light, so radio telescopes must have much larger diameters than optical telescopes (the resolution of a telescope depends on its diameter—see Section 15.3.4). The world's largest radio telescope is the Tianyan (Heavenly Eye) telescope in Guizhou Province, China. Its objective reflector has a diameter of 500 m and is built into a natural depression in the landscape. The signal from a radio telescope is detected by placing a receiver close to the focal point of the objective mirror.

14.3.3 Compound Microscope

A compound microscope uses two convex lenses. The objective lens creates a real magnified inverted intermediate image, and the eyepiece lens is used like a magnifying glass to make this even larger. It creates a final magnified virtual image.

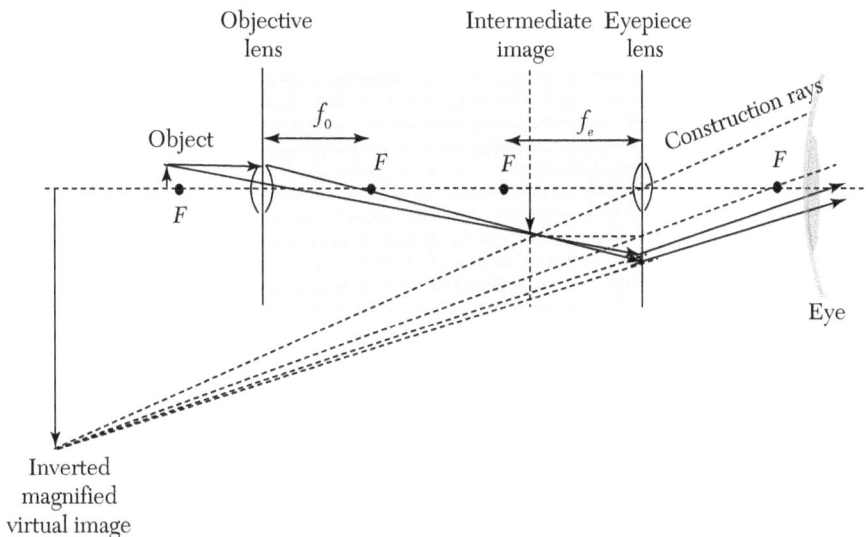

14.4 Doppler Effect

When a source moves relative to an observer or vice versa, the wavelength and frequency of the waves received by the observer changes. This is called the Doppler Effect, and it occurs for any type of wave. An everyday example of this is the change in frequency of a siren as an emergency vehicle approaches and then moves past us. As it approaches we hear a higher frequency, and when it moves away we hear a lower frequency than the frequency we would hear if the vehicle was at rest.

Frequency heard by
stationary observer

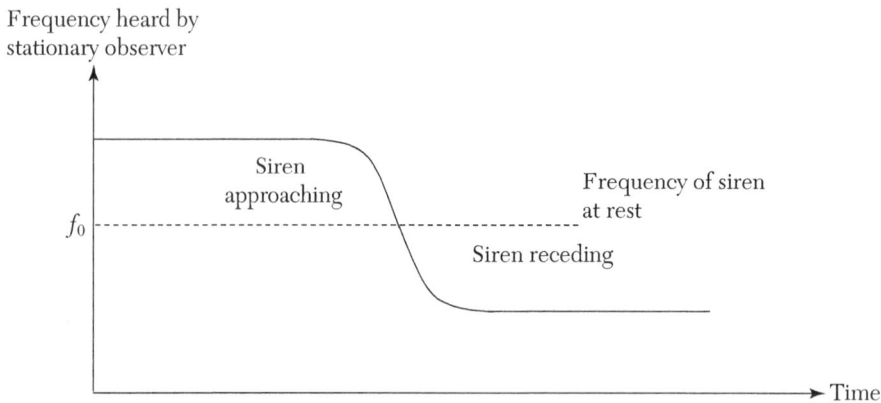

Here we restrict our discussion to electromagnetic waves. For relative velocities significantly lower than the speed of light, the Doppler Effect for electromagnetic waves can be analyzed very simply. This is because while relative velocity affects wavelength and frequency, it has no effect on the velocity of the waves, since this is independent of the velocity of the source or observer. When relative speeds are significant comparable to the speed of light, there is an additional relativistic effect to take into account—the frequency of the source, as measured by the observer, changes as a result of time dilation.

14.4.1 Doppler Effect for Electromagnetic Waves

Consider a source that emits waves of frequency f_0 and wavelength λ_0 as measured by an observer at rest with respect to the source. If the source approaches an observer with velocity v, it will move a distance v/f_0 toward the observer in the time taken to emit one complete wave. This means that the next wave front is emitted from a point $\lambda_0 - v/f_0$ behind the first, so the wavelength has been reduced by v/f_0. However, $f_0 = (v/c)\lambda_0$, so the new wavelength is $\lambda' = (1 - v/c)\lambda_0$ and the change in wavelength is $\Delta\lambda = -(v/c)\lambda_0$. It is clear from this that when the source recedes at velocity v the change in wavelength is of the same magnitude but opposite sign, i.e., an increase in wavelength. This change in wavelength is often referred to as a "Doppler shift."

Doppler shifts

- Source and observer approaching with relative velocity v: $\Delta O = -\dfrac{v}{c} O_0$

- Source and observer receding with relative velocity v: $\Delta O = +\dfrac{v}{c} O_0$

These shifts in wavelength are accompanied by changes of frequency with the frequency increasing for approaching and decreasing for recession. The new frequency f' can be calculated using the fact that the velocity of light is a constant so that $f' = c/\lambda'$.

Doppler shift on reflection

When waves reflect from a moving object, the Doppler shift is doubled. The arriving waves are Doppler-shifted and then these Doppler-shifted waves are effectively emitted from a moving source and so are Doppler-shifted for a second time:

$$\Delta\lambda_{\text{reflection}} = -2\frac{v}{c}\lambda_0$$

The Doppler Effect is often used to measure the velocity of a remote object. Doppler radar can be used to work out the velocity and rotation rate of an asteroid. The Doppler Effect is used by the police in radar speed guns to detect speeding motorists. Astronomers can use the Doppler shifts of radio waves in the spiral arms of the Milky Way to work out the rotation rate of our own galaxy. Doctors use Doppler ultrasound to measure the rate of flow of blood inside capillaries.

14.4.2 "Red Shifts" and "Blue Shifts"

When visible light is Doppler-shifted, the change in wavelength causes a change in color. If the source is moving away from us then the received wavelength increases and the observed color moves toward the red (longer wavelength) end of the spectrum. This is called a "red shift." However, we must be careful—if the shift takes the wavelength beyond the red end of the spectrum, then it actually moves away from red into the infrared. A "red shift" is always a move toward longer wavelengths, but is not always a move toward the red end of the spectrum, nor does it have to involve red light!

The redshift z is defined as:

$$z = \frac{\lambda' - \lambda_0}{\lambda_0} = \frac{v}{c}$$

where λ' is the longer wavelength received by the observer, λ_0 is the wavelength of light that would be observed by an observer at rest with respect to the source, and v is the speed of recession.

In a similar way, when a source moves away from us, the light we receive is said to have been "blue-shifted" (moved to shorter wavelengths).

In the 1920s, the astronomers Edwin Hubble and Vesto Slipher discovered that the spectral lines in light received from distant galaxies are all red-shifted. The implication is that distant galaxies are all moving away from us and from each other. In addition to this, Hubble showed that the red-shift is directly proportional to the distance of the galaxy, so that very distant galaxies are moving away from us more rapidly than closer galaxies. Since redshift is directly proportional to recession velocity, this discovery showed that recession velocity v is directly proportional to distance d. This is called Hubble's law:

$$v = H_0 d$$

where H_0 is the Hubble constant.

This discovery led to the idea of the expanding universe (see Section 28.3). This was eventually explained by Einstein's general theory of relativity, which reinterpreted the galactic redshifts. According to Einstein's model, the reason for the red shifts is not that galaxies are flying apart in pre-existing space but that space itself is expanding so that the separation between the galaxies increases. Either way we can use the redshift to calculate the speed at which any particular galaxy is receding from us.

In 1964, the American radio astronomers Arno Penzias and Robert Wilson discovered that the universe is bathed in low-intensity microwaves with an almost perfect black body radiation spectrum. These are believed to be an "echo" of the Big Bang. According to the Big Bang theory the early universe was filled with high-energy short-wavelength gamma radiation, but in the past 13.7 billion years of expansion, this has been hugely red-shifted and is now present as the cosmic microwave background radiation.

14.5 Exercises

1. The distance to the Moon has been measured by reflecting light from arrays of mirrors left on the Moon's surface by Apollo astronauts. The mirrors are arranged as corner reflectors that reflect incoming beams back on themselves. However, the spreading of the beam means that only about 1 in 10^{17} photons leaving Earth are detected in the reflection! The uncertainty in the measurements of distance are of the order of 1 cm.

The speed of light is 299,792,458 ms^{-1}
The distance to the Moon is (on average) 385,000.6 km.
(a) Calculate the time taken for light to make the return trip to the moon.
(b) How precisely must this time be measured in order to achieve a distance accuracy of ±1 cm?

(c) Why is it impossible to prevent the beam from spreading?

2. A magnifying glass of focal length 4.0 cm is used to form a magnified image of a small insect placed 3.0 cm from the lens.

(a) Draw a ray diagram to show how the magnified image is formed.
(b) Calculate the linear magnification of the image.
(c) Describe and explain what happens to the image as the magnifying glass is slowly moved away from the insect.

3. Complete the table below, which refers to a convex lens of focal length 20 cm.

Object distance	Image distance	Linear magnification	Nature of image (real/virtual)
10 cm			
20 cm			
30 cm			
40 cm			
Infinity			

4. Complete the table below, which refers to a concave lens of focal length 20 cm.

Object distance	Image distance	Linear magnification	Nature of image (real/virtual)
10 cm			
20 cm			
30 cm			
Infinity			

5. An astronomical refracting telescope has an objective lens of diameter 5.0 cm and focal length 120 cm. It is in normal adjustment with an eyepiece lens of focal length 4.0 cm.

(a) What is the magnifying power of the telescope.
(b) The eyepiece is changed for one of focal length 6.0 cm and readjusted so that it is in normal adjustment with the new lens. How does this affect the magnifying power and the length of the telescope?

6. (a) A policeman with a radar gun measures a Doppler shift of 83 parts in a billion for radio waves reflected from an approaching car. How fast is the car moving?

(b) An astronomer measures a redshift for a distant galaxy of 0.025. How far away is the galaxy? The Hubble constant $H_0 = 2.2 \times 10^{-18}$ s^{-1}.

7. Suggest an experimental method, based on the Doppler Effect, that could be used to measure the rotation period of the Sun.

CHAPTER 15

SUPERPOSITION EFFECTS

15.0 Superposition Effects

Superposition is what happens when two or more waves are present at the same point. This might occur because they originate from different sources or because of reflection. If the waves are of the same type, the resultant disturbance at that point is the vector sum of the disturbances from each individual wave. This is called the **principle of superposition**. We can determine the effects of superposition graphically, by adding phasors or by calculation. Many important phenomena are linked to superposition, including interference and diffraction and the formation of standing (stationary) waves.

15.1 Two-Source Interference

If waves of the same type with equal wavelength, frequency, and amplitude are emitted from two sources placed a short distance apart (comparable to a few wavelengths), then a regular interference pattern is formed. The pattern consists of regions where the waves reinforce to produce maximum intensity (constructive interference) and regions where they cancel to produce minimum intensity (destructive interference).

The famous double slit experiment carried out by Thomas Young in 1801 provided strong evidence for the wavelength of light and enabled Young to calculate its wavelength. A similar setup with sound can be used to demonstrate superposition patterns and, in a modified form, to create noise-cancelling headphones.

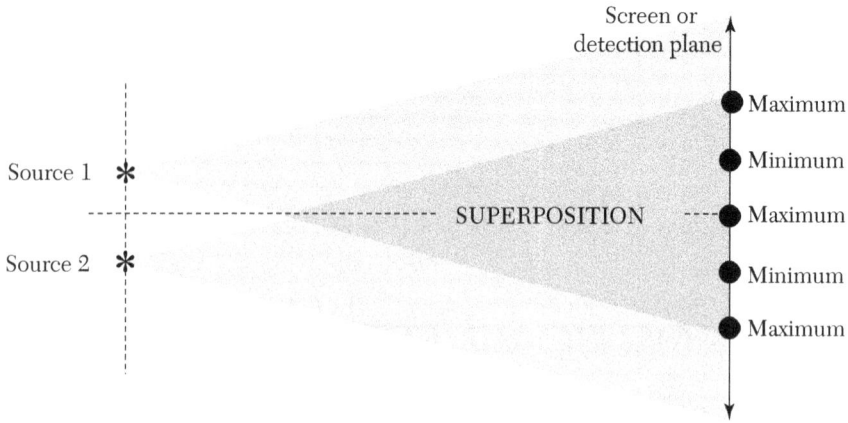

In order for stable clear interference effects to be created, the two sources must be **coherent**.

Coherent sources maintain a constant phase relationship.

This means that they must be the same type of wave and have the same wavelength and frequency. The sources do not have to be in phase, but the phase difference between them must be constant. They must also have comparable amplitudes; if one wave has a much greater amplitude than the other, then variations in intensity will be hard to detect.

15.1.1 Demonstrating Superposition Effects with Sound

If two loudspeakers are connected to the same signal generator, they will emit sound waves with the same amplitude, wavelength, and frequency, so an interference pattern will be formed where they superpose. The pattern can be explored by moving a microphone connected to an oscilloscope through the region of superposition. If this is done on a large enough scale, it is possible to walk through the superposition region and actually hear the variation in sound intensity.

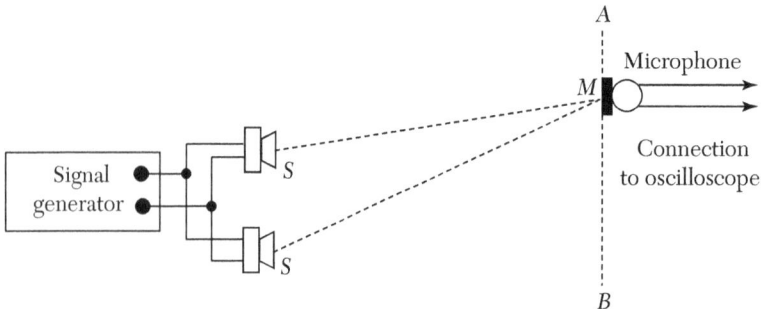

In the experiment above, the microphone is moved along the line *AB* and it detects a sequence of maxima and minima. When the microphone is in any position, the contributions from each speaker can be displayed separately by switching each one off in turn. The resultant disturbance depends on the phase difference between the waves as they reach the microphone. This depends on the path difference $\Delta x = S_2M - S_1M$. A path difference of one whole wavelength corresponds to a phase difference of 2π radians, so the relationship between path difference and phase difference is:

$$\Delta\phi = \frac{2\pi\Delta x}{\lambda}$$

For a maximum, the waves must reach the microphone in phase; this occurs when the path difference is a whole number of wavelengths. For a minimum, they must arrive in antiphase (an odd multiple of π phase difference); this happens when the path difference is an odd number of half wavelengths.

There will be a maximum at the center of the pattern because $S_2M = S_1M$ here, so the path difference and phase difference are both zero. Minima will occur when $x = \lambda/2, 3\lambda/2, 5\lambda/2$,etc., and these minima will be positioned symmetrically about the central maximum. In between the minima there will be further maxima when $x = \lambda, 2\lambda, 3\lambda$,etc. The intensity between maxima will vary continuously from maximum to zero and then back to a maximum. The path difference, phase difference, and effects at maxima and minima are tabulated below.

Path difference	Phase difference/rad	Resultant intensity
0	0	Central maximum
$\frac{1}{2}\lambda$	π	Minimum
λ	2π	Maximum
$\frac{3}{2}\lambda$	3π	Minimum
2λ	4π	Maximum
$\frac{5}{2}\lambda$	5π	Minimum

15.1.2 Demonstrating Superposition Effects with Light

Light from a filament lamp is polychromatic and incoherent, producing short wave trains from different atoms in the filament, so these sources are not well suited to producing interference effects. When incoherent light superposes, the interference effects average out so that we simply add intensities and do not get a pattern of maxima and minima. An example of this is the pool of light where a car's two head lamp beams overlap.

Lasers are ideal because they are intense monochromatic sources. However, the way in which laser light is produced means that the wave trains only remain coherent for a short time. If we wish to demonstrate interference, it is best to derive the two light sources from a single beam. This can be done by diffracting the beam with a single slit and then letting the light fall onto a double slit.

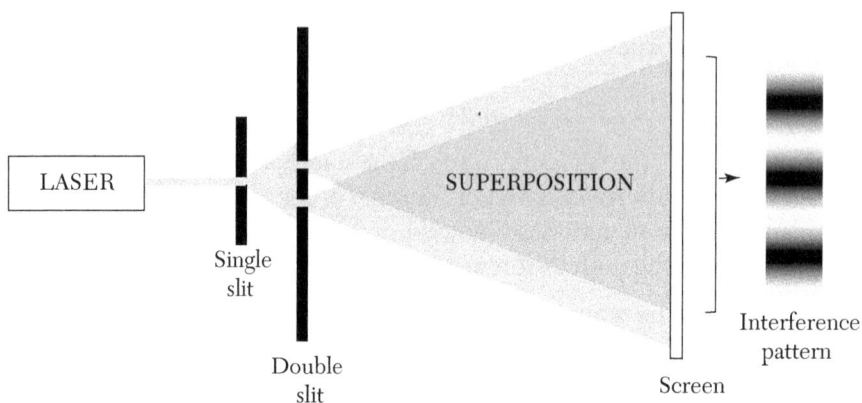

The diagram above has exaggerated the angle of spread from each slit. In practice the angles are small, and this allows us to use the small-angle approximation when deriving the relationship between the wavelength of the light and the structure of the pattern. It is also important to realize that the slit separation is very small compared to the distance between the double slits and the screen.

Here is an image of an actual double slit pattern formed using a green laser pen.

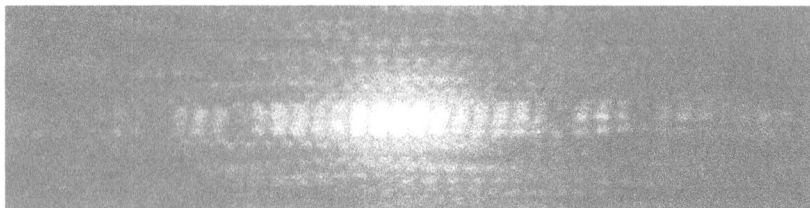

The small bright vertical patches are interference maxima (sometimes called interference fringes). Notice that the pattern itself fades in and out. This is a secondary effect caused by the diffraction pattern from individual slits rather than the interference pattern between different slits.

15.1.3 Using the Double Slit Experiment to Find the Wavelength of Light

Young's double slit experiment was used to measure the wavelength of visible light. We can derive an equation that relates the wavelength of light to three parameters: the separation of the double slits s, the distance from the double slits to the screen d, and the separation of adjacent maxima in the interference pattern y (fringe separation).

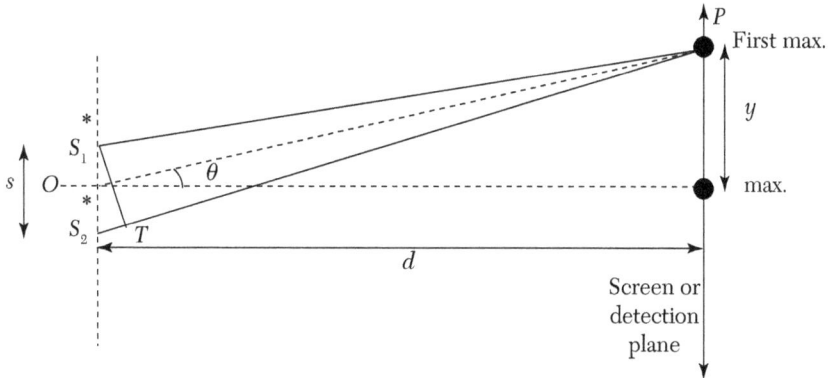

P is the position of the first maximum above the central maximum. A line drawn from O, half-way between the two slits, to P makes an angle θ with the central axis. If a line is now drawn from S_1 perpendicular to OP then distances S_1T and TP are equal. The distance S_2T is therefore equal to the path difference, and this must be equal to one wavelength.

$$\sin\theta = \frac{y}{d} = \frac{S_2T}{s} = \frac{\lambda}{s}$$

Rearranging,

$$\lambda = \frac{sy}{d}$$

The wavelength of light can be determined by measuring the slit separation, the fringe separation, and the distance to the screen.

15.1.4 Superposition of Harmonic Waves

When two harmonic waves superpose, the resultant effect is the sum of two sinusoidal oscillations. If the two waves have the same amplitude, this sum can be represented by adding two sine or cosine functions, y_1 and y_2, to give the resulting disturbance Y.

$$y_1 = A\sin(\omega t)$$

$$y_2 = A\sin(\omega t + \phi)$$

$$Y = y_1 + y_2 = A\{A\sin(\omega t) + \sin(\omega t + \phi)\}$$

where ϕ is the phase difference between the two oscillations. This will depend on the path difference x between waves from the two sources.

This can be simplified using a well-known trigonometric identity:

$$\sin R + \sin T = 2\,\sin\left(\frac{R+T}{2}\right)\cos\left(\frac{R-T}{2}\right)$$

to give:

$$Y = 2A\,\sin\left(\omega t + \frac{\phi}{2}\right)\cos\left(\frac{\phi}{2}\right)$$

This is most easily interpreted by grouping it differently so that there is a phase-dependent amplitude multiplied by a sinusoidal oscillation:

$$Y = \left[2A\cos\left(\frac{\phi}{2}\right)\right]\sin\left(\omega t + \frac{\phi}{2}\right)$$

The term in brackets represents the amplitude. When $\phi/2 = 0$, 2π, 4π, etc., the cosine term will be 1, representing a maximum in the interference pattern and the amplitude will be $2A$. When $\phi/2 = 0$, π, 3π, etc., it will be 0, representing a minimum.

The graphs below show the resultant oscillations when two waves, each of amplitude 2 units, combine with various phase differences. The two broken lines represent oscillations from waves 1 and 2, while the solid line is the sum of these.

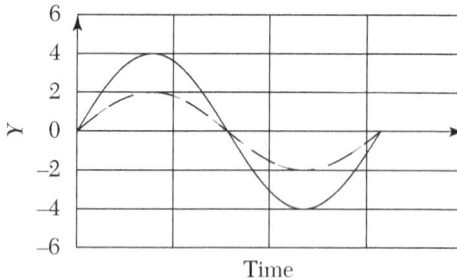

Phase difference = 0

Lines for waves 1 and 2 are on top of each other.

Resultant amplitude = 2 + 2 = 4

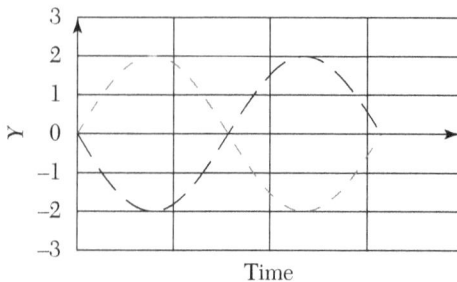

Phase difference = π

Lines for waves 1 and 2 are in antiphase.

Resultant amplitude = 0

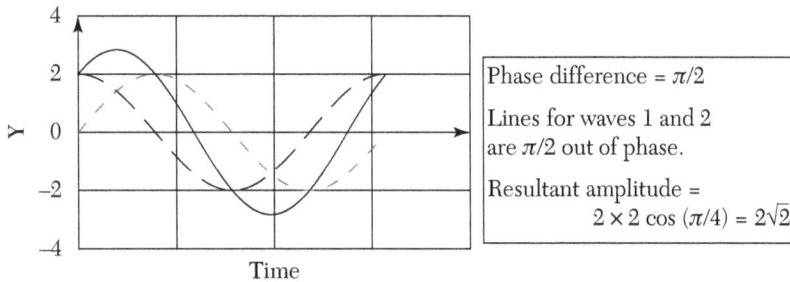

Phase difference = $\pi/2$

Lines for waves 1 and 2
are $\pi/2$ out of phase.

Resultant amplitude =
$2 \times 2 \cos(\pi/4) = 2\sqrt{2}$

Another way to determine the resultant of two superposed waves is to add phasors representing each wave. This approach is illustrated below for the final case where the amplitude of each wave is 2 units and the phase difference is $\pi/2$.

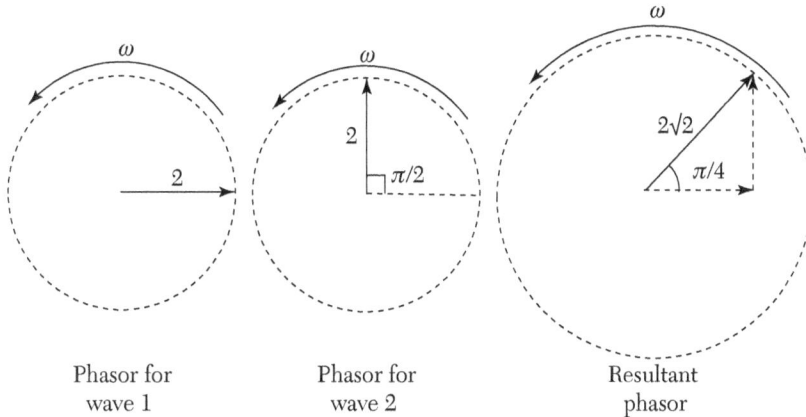

Phasor for
wave 1

Phasor for
wave 2

Resultant
phasor

Phasors are particularly helpful for visualizing how two or more waves will superpose.

15.2   Diffraction Gratings

A transmission grating consists of a large number of parallel, equally spaced narrow slits. When light passes through the grating, it is diffracted by each slit and light from a large number of sources overlaps to create a pattern of interference (often referred to as a diffraction pattern). Diffraction gratings produce intense well-separated and sharply defined maxima and are useful for analyzing the spectrum of light sources. This is called spectroscopy. The typical arrangement of apparatus used to measure wavelengths of light is shown below.

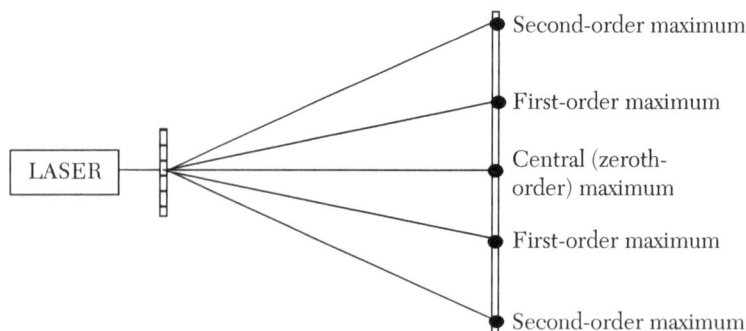

Here is an image of an actual experimental setup using a green laser pen as source.

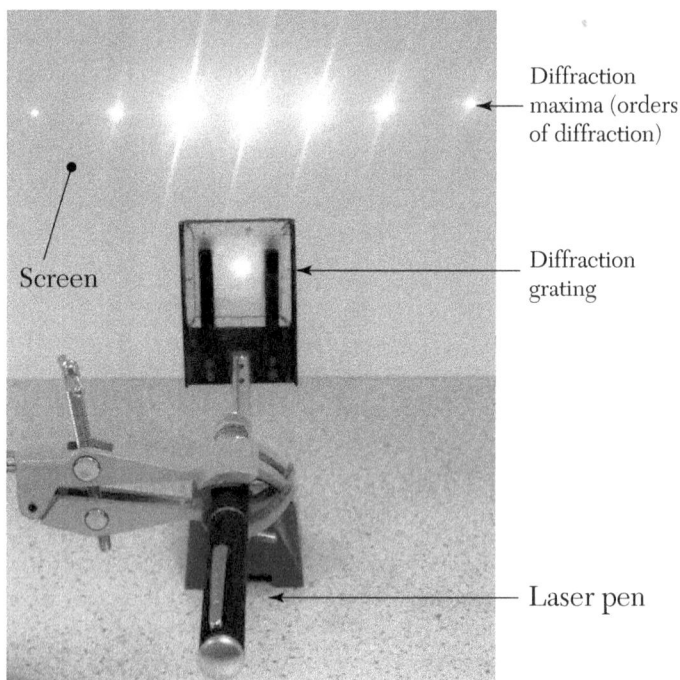

15.2.1 Diffraction Grating Formula

In practice the slit separation on a diffraction grating is very small compared to the distance from the grating to the screen, so that rays leaving adjacent slits and superposing at a point on the screen effectively travel parallel to one another. This simplifies the analysis because we can assume the rays are parallel as they leave the grating. In the analysis below, we will consider parallel rays traveling at an angle θ to the normal to the grating and superposing at a point P on the screen. The first diagram shows the large-scale situation and the second diagram is a highly magnified picture of some of the rays leaving slits in the grating.

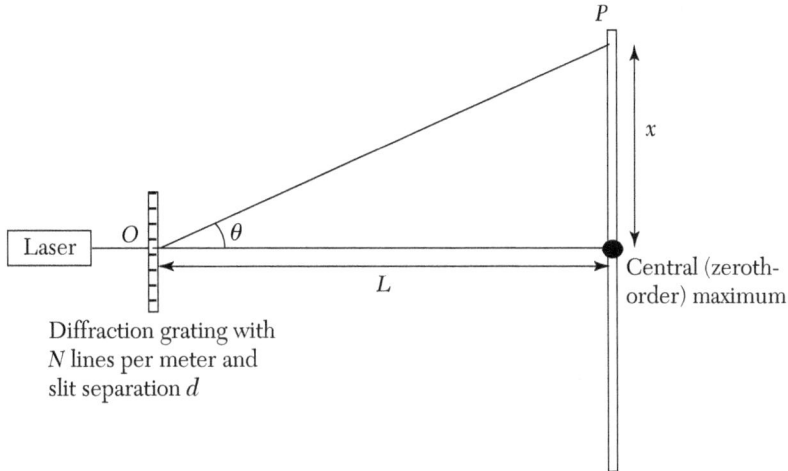

Diffraction grating with
N lines per meter and
slit separation d

Often the grating is described by the number of lines per meter(or per millimeter) N, rather than the slit separation d. These are related by $d = 1/N$.

The path difference between the ray from S_1 and the ray from S_2 is $S_2Q = d \sin \theta$.

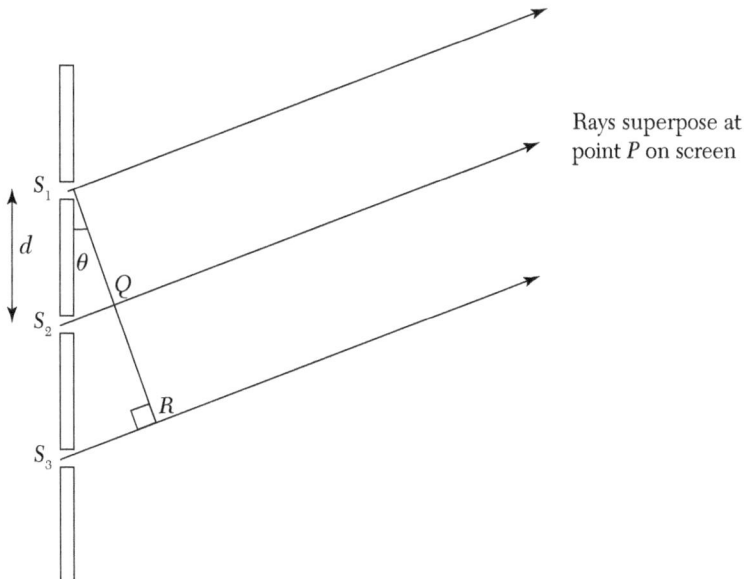

It is clear from the diagram that the path difference between the ray from S_1 and the ray from S_3 is just double this, and that the path difference to a slit m times further away is m times greater. It follows that if the rays leaving adjacent slits are in phase, then ALL rays across the entire grating will be in phase at that angle and an intense maximum will be created. These maxima are called orders of diffraction.

For rays from adjacent slits to be in phase, the path difference between them must be an integer number of wavelengths. The condition for a maximum is, therefore,

$$\sin\theta = \text{——} \quad \text{or} \quad n\lambda = d\,\sin\theta$$

where n is the order of diffraction (equal to the number of wavelengths path difference between rays from adjacent slits).

This is the diffraction grating equation. If we know d and can measure θ, we can calculate λ.

Intensity of maxima

Intensity I is proportional to amplitude-squared ($I \propto A^2$), so in the double slit experiment, the maxima will have double amplitude and four times the intensity of the light from a single slit. When we use a diffraction grating, N slits contribute to each maximum, so the amplitude is N times greater and the intensity is N^2 times greater. This makes the maxima much more intense than in the double slit experiment.

Sharpness of maxima

In the double slit experiment the intensity varies gradually between one maximum and the next minimum so that the maxima are not sharp. This is because the path difference between adjacent slits varies slowly as we move away from a maximum position. This is also true for adjacent slits in the diffraction grating, but the resultant effect on the screen is the sum of waves from slits all the way across the grating, and even a small displacement from the maximum position will result in a large change in path difference from a distant slit. The result is that the sum of rays across the grating falls to zero very rapidly as we move away from each maximum. This is best illustrated using phasors. The diagrams below compare the effect of moving the same short distance from a maximum for both the double slit and for a grating with just 8 slits (in practice gratings have many more than this).

As the angle moves away from a diffraction grating maximum, the phasors from slits across the width of the grating curl up more and more so that the maximum itself is sharp and has a series of closely packed secondary maxima on either side of it. The more slits, the sharper the maximum. This increases the precision with which wavelengths can be measured.

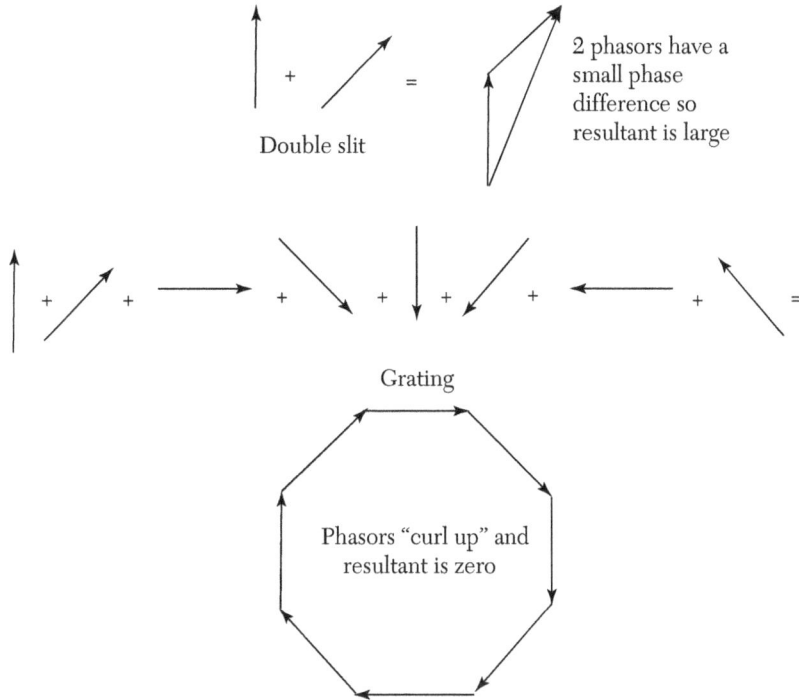

2 phasors have a small phase difference so resultant is large

Double slit

Grating

Phasors "curl up" and resultant is zero

Number of orders

The maximum possible path difference between rays leaving adjacent slits occurs when the rays leave parallel to the grating surface ($\theta = 90°$) and is then equal to the slit separation d. The maximum value of n is therefore d/λ. However, n can only be an integer, so the maximum number of orders must be the largest integer less than this; e.g., if $d/\lambda = 5.7$, then 5 orders of diffraction could be formed either side of the central maximum. In practice these might not all be visible. The intensity of light diffracted at large angles might be too low for them to be seen, or they might correspond to missing orders where there is a minimum of the single slit diffraction pattern (see Section 15.3.1).

15.2.2 Spectroscopy

Spectroscopy is the analysis of electromagnetic radiation to identify the wavelengths present in a source. This is one of the most important experimental techniques in science and is used in a range of different areas of research, from cosmology to atomic physics.

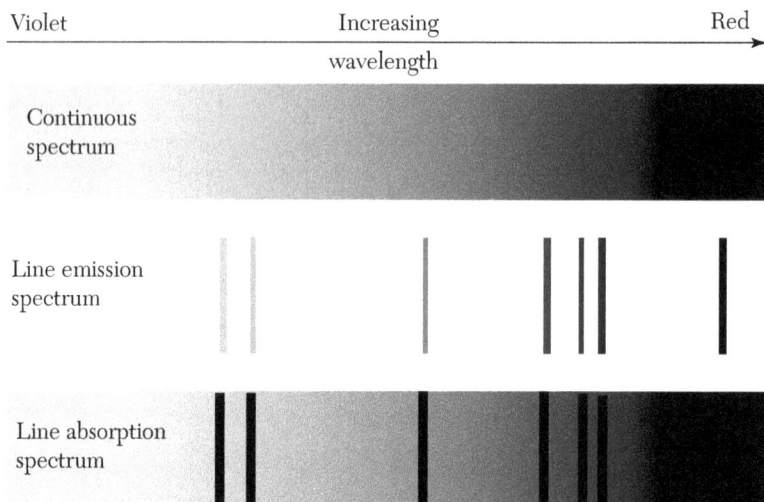

Violet   Increasing   Red

wavelength

Continuous spectrum

Line emission spectrum

Line absorption spectrum

When atoms or molecules are excited to higher energy states, the electrons can then return to lower energy states by making quantum jumps and emitting photons. The wavelength of the emitted photon is related to the energy change by the equation $\lambda = hc/E$ (see Section 27.2.1), so the greater the energy jump, the shorter the wavelength. For isolated atoms the energy levels are very distinct, so there is a discrete set of emitted wavelengths forming a **line emission spectrum**. The spectrum of each atom is unique, so an analysis of its spectrum can be used to identify the types of atom present in the source (e.g., a distant star). Atoms in molecules interact, producing a more complex spectrum that includes **bands** corresponding to small allowed ranges of energy. The atoms in solid materials are packed closely together, and electrons occupy wide energy bands so the characteristic emission from a hot solid is a **continuous spectrum**. When electromagnetic radiation is absorbed by atoms or molecules in lower energy states, photons whose energies correspond to allow quantum jumps can be absorbed and electrons are excited to higher energy states. This removes certain wavelengths from the radiation so that the transmitted radiation will contain dark lines or bands called an **absorption spectrum**. The wavelengths missing in the absorption spectrum are the same ones that would be present in the emission spectrum from the same element or compound.

A typical experimental setup to analyze light from a source (e.g., a laser or a discharge lamp) is shown below. The diagram shows the measurements that would need to be taken to calculate one wavelength present in the spectrum. If there are several, then each wavelength would produce its own set of orders and would need to be measured separately.

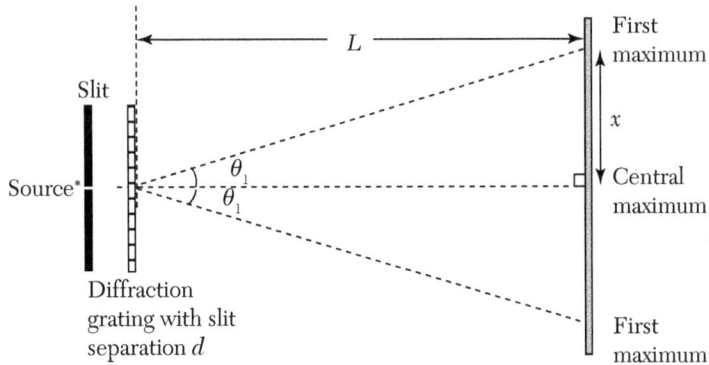

The angle θ_1 can be found by measuring x and L with a ruler and then using $\tan \theta_1 = x/L$.

In practice it is best to use an average value for θ_1 by measuring the first order on both sides of the central maximum. This helps to reduce errors due to alignment. The wavelength is then calculated from:

$$\lambda = d \sin \tau_1$$

If several orders are visible, each one can be used and wavelength can be calculated from:

$$\lambda = \frac{d \sin \tau_n}{n}$$

Another approach is to plot $\sin \theta_n$ against n. The gradient of this graph is λ/d.

15.2.3   Spectrometers

For more precise spectroscopy a specialized instrument called a spectrometer must be used. A traditional spectrometer consists of a collimating tube, which ensures that rays from the source arrive along a normal to the grating, and a telescope that is used to detect the orders of diffraction. These are all mounted on a rotating base so that the angle between the normal and each order can be measured from a Vernier scale engraved on the base.

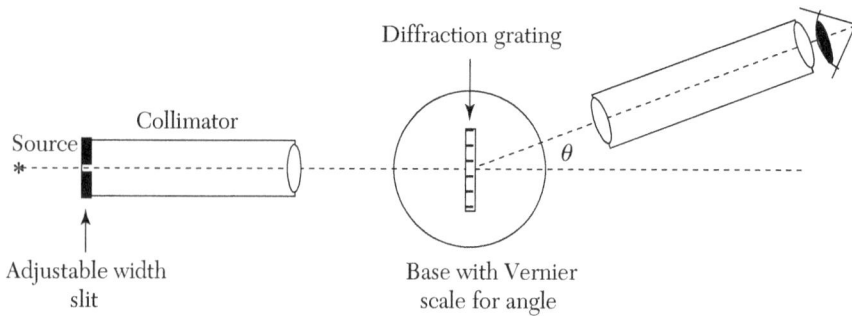

The eyepiece of the telescope has a built-in cross-hair so that the position of the image can be found precisely. The observer sees a fine vertical line in each wavelength at each order of diffraction. For greatest precision, the collimator slit must be adjusted so that it is very narrow.

Setting up a spectrometer requires great care to ensure that the rays leaving the collimator are parallel and the diffraction grating is in the plane perpendicular to these rays.

Digital spectrometers

Digital spectrometers are used to display spectra and to measure wavelengths but are rarely able to provide the level of precision achievable with a traditional setup. Most simple digital spectrometers use an optical fiber to direct light into the device where it then falls onto a reflection grating. This consists of a large number of parallel reflecting lines (like on the surface of a compact disc) and produces orders of diffraction by reflection. The light is detected by a CCD detector and the position on the detector corresponds to wavelength, so this can then be recorded digitally and used to generate a graph of intensity against wavelength.

15.3 Diffraction by Slits and Holes

When waves pass through a small gap or past the edge of an object, they spread out or diffract. This results in superposition effects that create interference patterns. For slits and holes the amount of diffraction depends on the ratio of the wavelength λ of the wave to the size d of the aperture (λ/d). The larger this ratio, the more significant the effect. This is illustrated below (simplified), showing plane waves passing through gaps of different size. As the ratio approaches 1, the waves spread out in all directions.

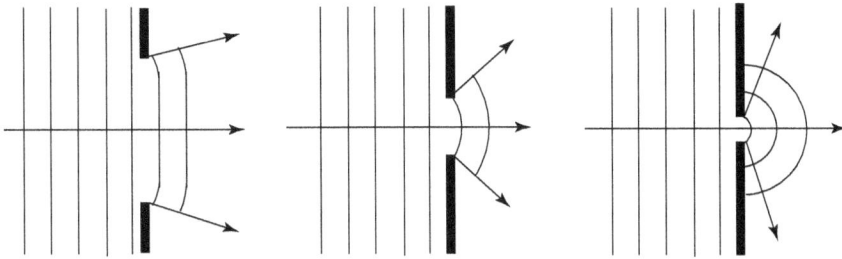

The intensity of the diffracted waves varies with angle and usually has maxima and minima.

15.3.1 Diffraction by a Narrow Slit

When waves pass through a narrow slit, a characteristic diffraction pattern is formed with a broad central maximum, and narrower dimmer secondary maxima spread out symmetrically on either side. A simple experimental arrangement is shown below:

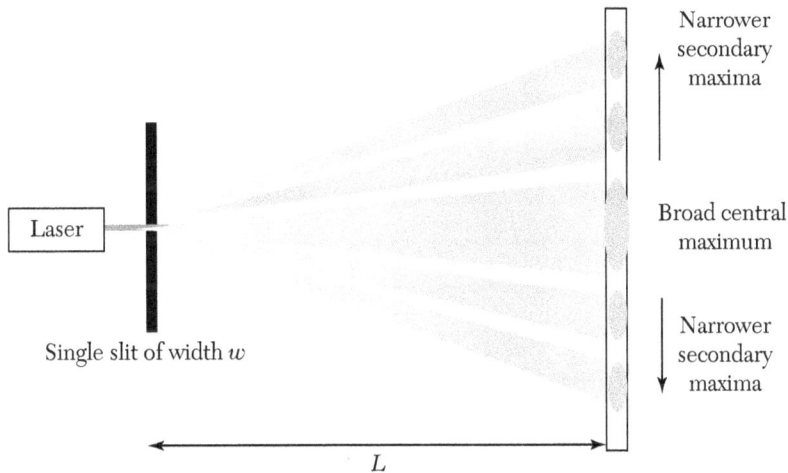

Here is a graph showing the intensity variation on the screen.

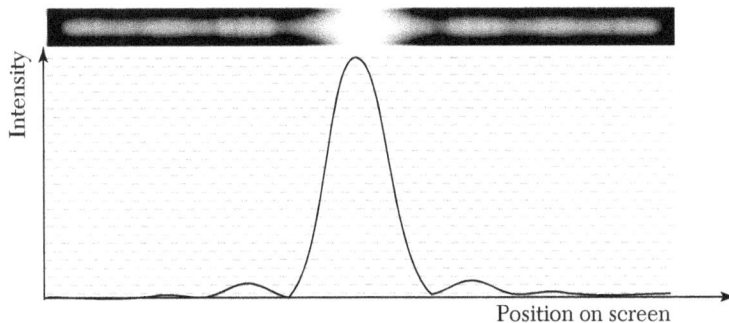

The pattern for a thin wire of diameter d is identical to that for a slit of width w when $w = d$.

15.3.2 Analysis of the Single Slit Diffraction Pattern

A complete mathematical analysis of this pattern is beyond the scope of this book, but we can derive a formula for the positions of the minima in the pattern by considering the angles at which light emitted from across the width of the slit interferes destructively. This is based on an idea from Christian Huyghens who realized that we can consider every point on a wave front as if it is an independent source of circular (or spherical) waves. We then need to add up the contributions from all such points. This leads to an integral that gives the resultant intensity at any point on the screen. To find the angles at which minima occur, we simply need to find the condition under which the waves from all points add to zero.

The diagram below shows the region close to the slit and has divided the slit up into a large number of point sources. The angle shown corresponds to the first minimum from the center of the diffraction pattern. Rays from the edges and center of the slit are shown. Rays from intermediate points have not been shown.

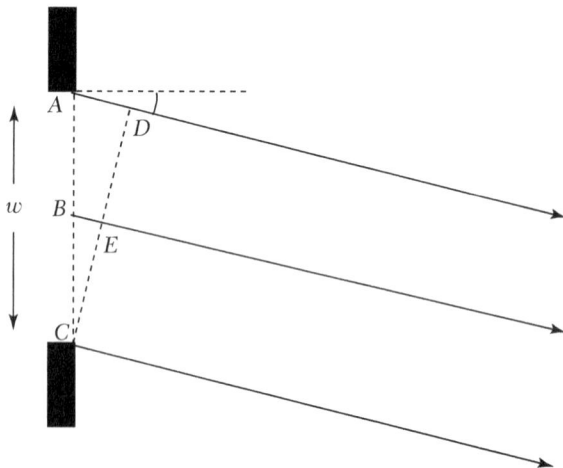

Rays from all points across the slit reach the same point on the screen and add to zero. Consider rays from A and B. If these two rays are π out of phase, the pair will add to zero. Now consider the two rays a small distance further down the slit from A and a small distance further down the slit from B. These will have the same small phase difference from A and from B but will also be π out of phase with each other and so will also add to zero. In fact, continuing this argument, all the rays will *cancel in pairs* across the slit and the resultant at the screen will be zero. The condition when this first occurs is when the ray at B is π out of phase with the ray at A and the path difference BE is $\lambda/2$. Consider triangle BCE when this occurs:

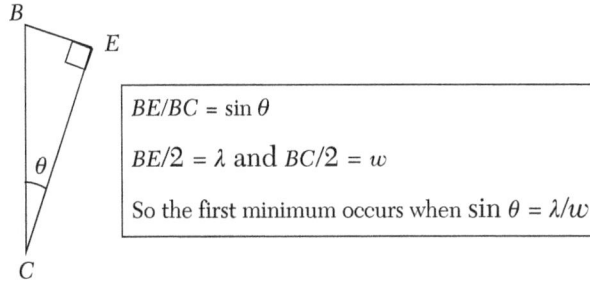

B

E

$BE/BC = \sin\theta$

$BE/2 = \lambda$ and $BC/2 = w$

So the first minimum occurs when $\sin\theta = \lambda/w$

θ

C

The condition for the nth minimum is when rays can cancel in pairs n times across the width of the slit. The path difference between the rays at each edge of the slit must therefore be $n\lambda$ so that:

$$\sin\theta_n = \frac{n}{w}$$

This has a similar form to the equation for the *maxima* of a diffraction grating—don't forget that this gives the positions of minima!

If the pattern subtends a small angle at the slit, then the minima are equally spaced and the central maximum, which goes from the minima on either side, is exactly double the width of the secondary maxima. This contrasts with the double slit pattern where the maxima have equal widths.

15.3.3 Diffraction Through a Circular Hole

The diffraction pattern through a circular hole or by a circular object is qualitatively similar to that for a slit, but the pattern consists of concentric rings rather than lines and the angular positions of the minima are not the same. The first minimum for a circular diffraction pattern occurs when:

$$\sin\theta = \frac{1.22\lambda}{D}$$

where D is the diameter of the hole or object.

It is quite easy to demonstrate this pattern in the laboratory. Lycopodium powder consists of fine pollen grains that are all roughly spherical and of about the same diameter. If a microscope slide is dipped into the powder and then held in front of a green laser pen, the diffraction pattern can be projected onto a screen. However, for this to be convincing you will need a completely blacked-out room. The pattern itself is the superposition of many patterns formed by the individual particles. However, the particles are very close together so they create a single pattern on the screen.

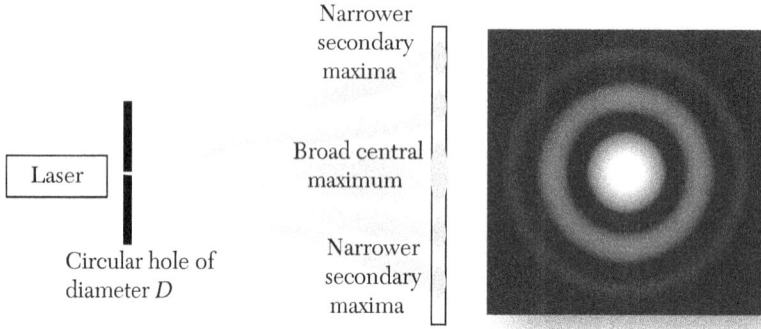

Diffraction patterns from slits, wires, holes, and circular objects can be used to measure the size of the diffracting object. There is an inverse relationship between the size of the object and the size of the pattern, so the smaller the object, the larger the angle at which the first minimum occurs. In practice the wavelength of the source needs to be known and the experiment consists of measuring the angle at which the first minimum occurs.

15.3.4 Resolving Power and the Rayleigh Criterion

When light enters the objective of an optical instrument, it diffracts. This limits how sharply the instrument can focus images. The amount of diffraction depends on the ratio λ/D, where D is the diameter of the objective, so instruments with large-diameter objectives working at short wavelengths have the greatest ability to resolve detail in the images they produce.

The ability of an optical instrument to resolve detail is called its **resolving power**, and the smallest angular separation between object points that can be distinguished as separate image points is called the **limit of resolution.** When instruments are compared, we use the Rayleigh criterion to give a value for the limit of resolution. Rayleigh suggested that the diffraction limit occurs when the central maximum of the diffraction pattern from one object falls onto the first minimum of the diffraction pattern from the other object. This would occur when the angular separation θ of the two objects at the objective is equal to $1.22\lambda/D$.

Object 1

θ

Objective
diameter D

Object 2

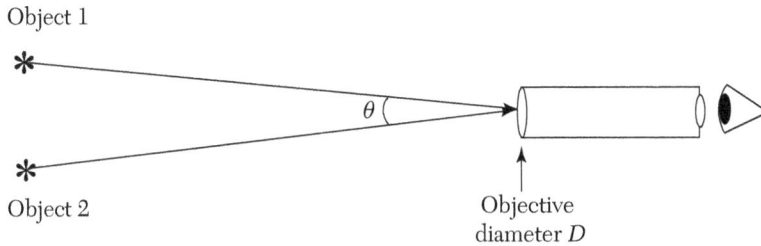

Rayleigh criterion:

$$\theta < \frac{1.22}{D} \quad \text{not resolved}$$

$$\theta = \frac{1.22}{D} \quad \text{limit of resolution}$$

$$\theta > \frac{1.22}{D} \quad \text{resolved}$$

In practice diffraction is only one factor that limits resolution, so these diffraction limits are rarely achieved. Other factors are aberration, atmospheric effects, sensitivity of the sensors, etc.

It is interesting to compare the limit of resolution of the human eye with that of a large optical and radio telescope. We will take 500 nm as a representative visible wavelength and 20 cm as a representative radio wavelength.

Human eye: pupil diameter 5.0 mm, wavelength 500 nm, limit of resolution 1.2×10^{-4} rad

Large optical telescope: objective diameter 10 m, wavelength 500 nm, limit of resolution 6.1×10^{-8} rad

Large radio telescope: objective diameter 500 m, wavelength 20 cm, limit of resolution 4.9×10^{-4} rad

The resolving power of the large optical telescope is, as expected, much greater than that of the unaided human eye, but it might be surprising to realize that the world's largest radio telescope "sees" the universe in less detail than our eyes, albeit seeing different features.

15.4   Standing (Stationary) Waves

Standing waves are formed when two similar waves traveling in opposite directions superpose. This can occur because of reflection or as a particular example of two-source interference. The resultant disturbance has positions called **antinodes**, where the waves always combine in phase to produce a large amplitude, and regions called **nodes,** where the waves always combine in antiphase and cancel out. Standing waves are responsible for the sound of musical instruments and the existence of energy levels in atoms.

15.4.1   Standing Waves on a String (Melde's Experiment)

One end of a string is attached to a vibrator and the other end is connected over a pulley to some hanging masses. At certain vibration frequencies, standing waves are set up on the string.

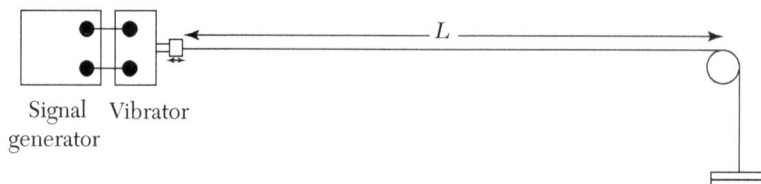

Signal generator   Vibrator

The ends of the string are effectively fixed. These are the **boundary conditions**, so there must be nodes at each end. Standing waves can then be formed at frequencies that "fit" the length of the string—in other words, patterns that have nodes at the ends. The lowest frequency, longest wavelength wave that sets up a standing wave is called the **fundamental**.

The sequence of different standing waves are **harmonics**. Since the speed v of transverse waves on a string is determined only by the tension and mass per unit length of the string, it is the same for all harmonics, and the frequencies form a simple sequence. The diagrams that follow show the first few harmonics for the string above.

It is clear that the harmonic frequencies are all integer multiples of the fundamental frequency f_1. The frequency of the nth harmonic will be nf_1. A similar series is obtained if the boundary conditions have an antinode at both ends (e.g., standing sound waves in a tube open at both ends).

If the boundary conditions are different at each end, so that there is a node at one end and an antinode at the other, then the sequence is f_1, $3f_1$, $5f_1$, etc., all the odd multiples of the fundamental frequency (e.g., standing waves in a tube open at one end and closed at the other).

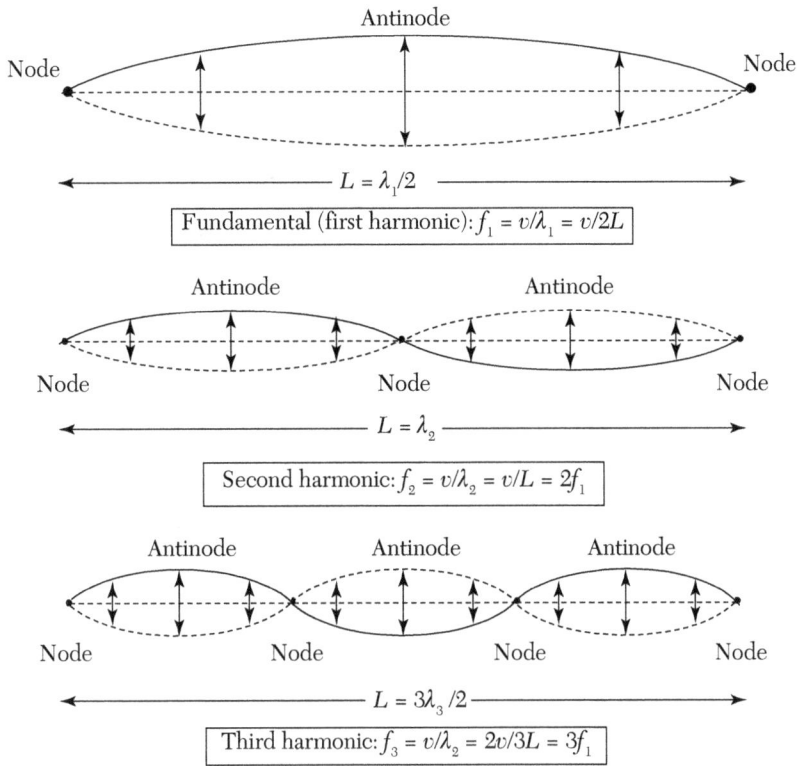

Fundamental (first harmonic): $f_1 = v/\lambda_1 = v/2L$

$L = \lambda_1/2$

Second harmonic: $f_2 = v/\lambda_2 = v/L = 2f_1$

$L = \lambda_2$

Third harmonic: $f_3 = v/\lambda_2 = 2v/3L = 3f_1$

$L = 3\lambda_3/2$

The speed of transverse waves on a string is given by:

$$v = \sqrt{\frac{T}{\mu}}$$

The frequencies of harmonics on the string above are, therefore,

$$f_n = \frac{n}{2L}\sqrt{\frac{T}{\mu}}$$

Here are some of the key features of standing waves:

- Nodes are regions where the waves always interfere destructively, giving zero amplitude.

- Antinodes are regions where the waves always interfere constructively, giving maximum amplitude.

- The distance between two adjacent nodes, or two adjacent antinodes, is equal to $\lambda/2$.

- All points between two nodes oscillate in phase.

▪ Points on either side of a node oscillate with a phase difference of π radians.

Standing waves on a stretched string produce sounds in stringed musical instruments. The player excites the string by plucking, strumming, hitting, or bowing it, and this generates waves with a wide range of frequencies that travel along the string and reflect from the fixed ends. Only wavelengths that satisfy the boundary conditions (nodes at each end) create standing waves, so these dominate the emitted sound, which will usually consist of a range of harmonics. The nature and quality of the sound we hear will depend on the amount of each harmonic present and the attack and decay rate of the sound.

15.4.2 Mathematics of Standing Waves

Since standing waves are formed by the superposition of two similar waves traveling in opposite directions, we can find an equation for a standing wave by adding the wave equations for these waves together.

$y_1 = A \cos(\omega t - kx)$ $\rightarrow$ wave traveling in $+x$ direction

$y_2 = A \cos(\omega t + kx)$ $\rightarrow$ wave traveling in the $-x$ direction

$Y = y_1 + y_2 = A \cos(\omega t - kx) + A \cos(\omega t + kx) \rightarrow$ equation of a standing wave

where $\omega = 2\pi f$ and $k = 2\pi/\lambda$

This can be rewritten using the trigonometric identity:

$$\cos A + \cos B = 2 \cos\left(\frac{A+B}{2}\right) \sin\left(\frac{A-B}{2}\right)$$

to give:

$$Y = 2A \cos kx \cos \omega t$$

where we have reversed the order of the cosines.

The first part of this, $2A \cos kx$, can be regarded as a position—dependent amplitude. This will be zero when $\cos kx = 0$, and this occurs when $kx = 0, \pi, 2^\pi$, etc.,i.e., $kx = n\pi$. These are the nodes, and the separation of two nodes is given by $k\Delta x = \pi$:

$$k = \frac{2\pi}{\lambda'},$$

so

$$\frac{2\pi\Delta x}{\lambda} = \pi$$

giving $\Delta x = \frac{\lambda}{2}$ (separation of nodes)

The maximum value of this term is when cos kx = 1. The amplitude is then 2A. This occurs at the antinodes, half way between nodes.

The second term, cos ωt, is a simple harmonic oscillation at all points in the standing wave.

15.5   Exercises

1. The diagram shows a two-source interference experiment using sound.

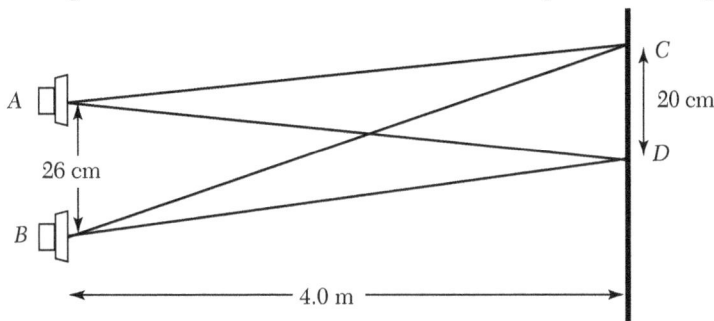

Speakers A and B are 26 cm apart along a line parallel to CD. They both emit sound waves of a single frequency and the same amplitude a, and they are in phase with one another. D is equidistant from A and B. When a microphone is placed at D it records a sound of maximum intensity. When it is moved along the line DC it gradually fades to become a minimum at C; immediately beyond C, the intensity increases.

(a) State the phase difference between the waves arriving at D.
(b) State the phase difference between the waves arriving at C.
(c) State the path difference (in wavelengths) between waves reaching C from sources A and B.
(d) State the amplitude and intensity of the resultant sound at D (assume $I = ka^2$ where k is a constant).
(e) State the amplitude and intensity of the resultant sound at C (assume $I = ka^2$ where k is a constant).
(f) What would happen to the amplitude and intensity of the sound at D if one of the speakers was switched off?
(g) Use the dimensions on the figure to work out the wavelength and frequency of the sound. The speed of sound is 340 ms⁻¹.
(h) How would CD change if the wavelength of the sound was doubled?
(i) How would the distance CD change if the separation of the speakers was doubled?
(j) Explain why the intensity of sound increases beyond D.

2. Monochromatic light of wavelength 589 nm is passed through a single slit, which diffracts the light onto double slits of separation 0.12 mm. The double slits act as coherent sources, and an interference pattern is formed on a screen 2.0 m beyond the double slits.

(a) Explain why it is important to use monochromatic coherent sources.

(b) Calculate the separation of maxima on the screen.

(c) Describe and explain the effect on the interference pattern if each of the following changes are made (independently):

 (i) The entire apparatus is immersed in water (refractive index 1.33).

 (ii) One of the two double slits is wider than the other.

 (iii) The separation between the double slits and the single slit is increased.

 (iv) The screen is moved further away from the double slits.

 (v) Light of longer wavelength is used.

 (vi) Polarizing filters are placed in front of each slit and one of them is slowly rotated through 360°, starting off with both polarizing directions parallel.

(d) Light from a car's headlamps overlaps. Explain why this does not form an interference pattern.

3. Explain in detail how and why the interference pattern from three slits compares with one from a double slit if they both have the same slit width and slit separation and if they are both illuminated with monochromatic light of the same frequency. Illustrate your answer with a graph showing how the intensity of the light varies with position on a screen.

4. A diffraction grating has 300 lines per mm and it is illuminated by light containing two strong emission lines at 480 and 520 nm. The width of each slit is 1.11 mm.

(a) What are the angular positions of the first-order maxima for each of these lines?

(b) What is the angular separation of lines in the second-order spectrum?

(c) How many orders of diffraction are there?

(d) Explain why the third order of diffraction will not be observable.

5. Describe an experiment using a diffraction grating to measure the wavelength of light from a laser pen. Your description should include:

- a labeled diagram of the apparatus

- what you will measure and how the measurements will be made

- how you will calculate the wavelength from the data you collect

- how you will maximize accuracy and precision

6. Red light of wavelength 525 nm passes along a normal to a narrow vertical slit of width 0.20 mm.
   (a) Calculate the angle to the normal at which the first minimum of the resultant diffraction pattern occurs.
   (b) Sketch a graph of intensity against position (in mm) for the diffraction pattern formed on a screen 1.2 m from the slit.
   (c) Explain how the minima in the pattern are formed.
   (d) How would the pattern change if the red light source was replaced with a blue light source of wavelength 450 nm?

7. Describe an experiment that could be carried out to measure the thickness of a human hair using a laser of known wavelength. Your description should include:

   - a labeled diagram of the apparatus

   - what you will measure and how the measurements will be made

   - how you will calculate the wavelength from the data you collect

   - how you will maximize accuracy and precision

8. The pupil of the human eye is about 5.00 mm in diameter.
   (a) Use the Raleigh criterion to calculate the theoretical minimum angular limit of resolution.
   (b) If the eye could reach this diffraction limit, calculate the maximum distance at which it could resolve two-point sources separated by a distance of 1.00 mm.
   (c) A pair of stars form a binary system 20 light years from Earth. It is just possible to resolve them into separate images using the naked eye. Estimate a lower limit for the separation of the stars.
   (d) Explain why your answer to (c) is a lower limit.
   (e) Explain why a small telescope, with aperture diameter 15 cm, can easily resolve these two stars.

9. (a) Estimate the maximum distance at which you could resolve a car's head lamps into separate sources.
   (b) It has been claimed that a spy satellite orbiting at a height of 600 km could resolve the letters in a car number plate on the surface of the Earth. Is this a realistic claim?

10. The speed of transverse waves on a string is given by $v = \sqrt{\dfrac{T}{\mu}}$, where T is the tension in the string (N) and μ is the mass per unit length of the string (kgm^{-1}).

A steel wire of diameter 0.12 mm is held at a tension of 26 N between two fixed points 0.75 m apart. The density of steel is 7800 kgm^{-3}.

(a) Calculate the mass per unit length of the steel string.

(b) Calculate the speed of transverse waves in the steel string.

(c) What is the frequency of the fundamental?

(d) List the frequencies of the first three harmonics.

(e) The temperature of the wire increases. State and explain how this will affect the frequencies of sound emitted by the vibrating wire.

CHAPTER 16

SOUND

16.1   Nature and Speed of Sound

Sound waves are longitudinal mechanical vibrations. Audible sound lies in the range 20 Hz to 20 kHz and travels at about 340 ms⁻¹ through the air. Sound waves with frequencies >20 kHz are called ultrasound, while those with frequencies <20 Hz are called infra sound. The wave itself consists of compressions and rarefactions of the medium through which it passes, so the speed of sound depends on mechanical properties of the medium such as its density and compressibility.

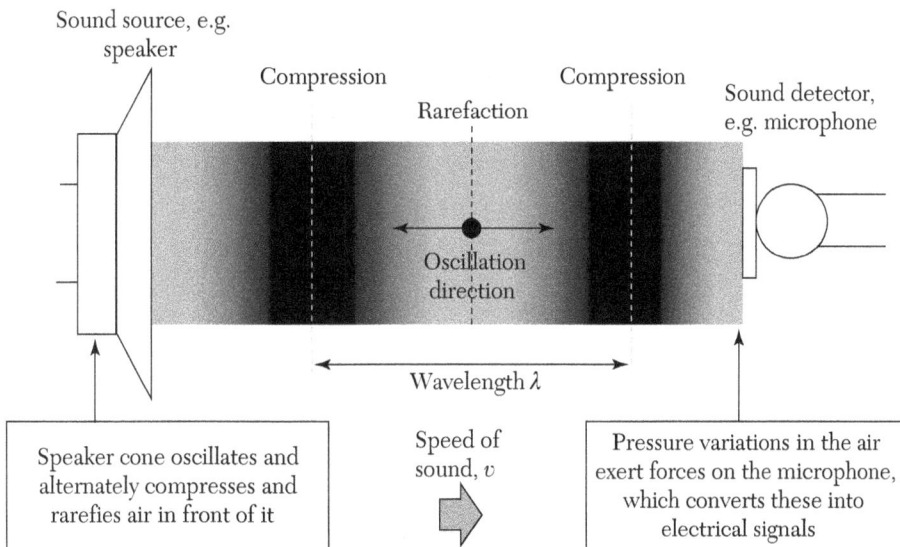

Sound source, e.g. speaker

Compression

Compression

Rarefaction

Sound detector, e.g. microphone

Oscillation direction

Wavelength λ

Speaker cone oscillates and alternately compresses and rarefies air in front of it

Speed of sound, v

Pressure variations in the air exert forces on the microphone, which converts these into electrical signals

The sound wave itself can be described either by the varying particle displacements at each point or by the varying pressure at each point.

The speed of sound in an ideal gas is given by:

$$v = \sqrt{\frac{\gamma RT}{M}}$$

where M is the molar mass of the gas and γ is the adiabatic gas constant.

The speed of sound in a solid material is approximately given by:

$$v = \sqrt{\frac{E}{\rho}}$$

where E is the Young modulus and ρ is the density of the solid.

The speed of sound in a liquid is given by a similar equation, but the Young modulus is replaced by the bulk modulus of the liquid, a quantity that measures the compressibility of the liquid.

Typical values for the speed of sound are:

Air, at atmospheric pressure and 20°C: $v = 343$ ms^{-1} (this falls to about 300 ms^{-1} at the altitude of a commercial jet)

Water: $v = 1482$ ms^{-1}

Steel: $v = 6000$ ms^{-1} (varies with type of steel)

16.2   Decibel Scale

The human ear has a logarithmic response to sound. This allows us to detect sounds over a very wide range of intensities (from about 1 pWm^{-2} to about 1 Wm^{-2}), but it also means that doubling the intensity of the sound does not double the apparent loudness of the sound. In fact, for sound to seem twice as loud, we have to increase its intensity by a factor of 10. The decibel scale takes this into account. Intensity levels are measured in bels (B), and 1 bel is equal to 10 decibels (dB). The intensity level in bels is related to the threshold of human hearing at $I_0 = 1$ pWm^{-2}. To calculate the intensity level in bels corresponding to an actual intensity I in Wm^{-2}, we use the equation:

$$\text{intensity level} (B) = \log_{10}\left(\frac{I}{I_0}\right)$$

$$\text{intensity level} (dB) = 10\log_{10}\left(\frac{I}{I_0}\right)$$

Here are some typical intensity levels:

Threshold of hearing: 0 dB

Breathing: 10 dB

Radio/TV: 70 dB

Live rock music: 110 dB

Jet take off (300 m away): 130 dB

Jet take off (25 m away): 150 dB (eardrum rupture)

Note that the decibel is certainly NOT an S.I. unit, and values of intensity level are comparative, not absolute.

16.3   Standing Waves in Air Columns

Many musical instruments (wind instruments) use standing waves in air columns to generate particular sounds. The length of the air column and the boundary conditions at its ends determine what harmonics will be present in the sound. A simple demonstration of these standing waves is shown on the left.

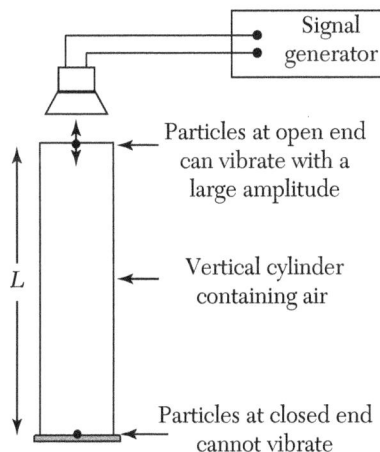

As the frequency of the signal generator is gradually increased, the sound intensity increases sharply at certain frequencies. These correspond to the standing waves formed when waves traveling down the column reflect from the base and the incident and reflected waves superpose.

The air column shown is closed at the bottom and open at the top. Particles near the bottom cannot undergo longitudinal oscillations, so this must be a node of displacement. The top of the column is open to the atmosphere, so particles are free to oscillate. This allows an antinode to form, but it actually occurs a small distance e above the top of the column. e is called the "end correction," and it is about half the radius of the tube. The effective length of the resonating column is therefore $L + e$.

The diagrams below (shown horizontally) show the standing waves that can be formed in a tube open at one end and closed at the other. Note that, while the diagrams look like transverse waves, they simply represent the amplitude of longitudinal vibrations at each position in the tube.

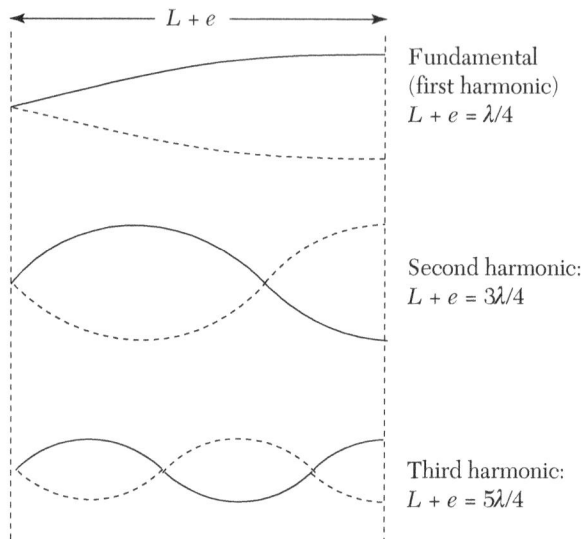

Fundamental (first harmonic)
$L + e = \lambda/4$

Second harmonic:
$L + e = 3\lambda/4$

Third harmonic:
$L + e = 5\lambda/4$

This harmonic series consists of odd multiples of the fundamental frequency:

$$L + e = \frac{(2n+1)_n}{4} = \frac{(2n+1)v}{4f_n}$$

$$f_n = (2n+1)f_1$$

If the air column is open at both ends, the boundary conditions will be displacement antinodes at both ends (i.e., a column of length $L + 2e$). If the column is closed at both ends, the boundary conditions will be displacement antinodes at both ends. In both cases the harmonic series will be integer multiples of the fundamental frequency.

It was mentioned previously that we can describe the sound wave in terms of particle displacements or variations of pressure. However, displacement nodes are actually pressure antinodes and vice versa. This can be understood by considering the particle motions on either side of a displacement node.

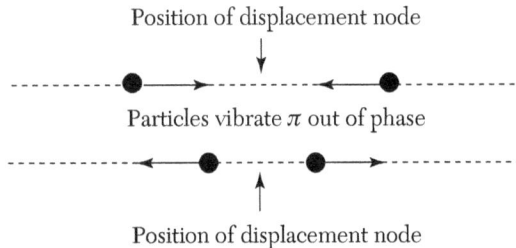

Position of displacement node

Particles vibrate π out of phase

Position of displacement node

The particles move toward the displacement node and increase the pressure and then move away from the displacement node and decrease the pressure. The variation of pressure has a large amplitude here, so it is a pressure antinode.

16.4   Measuring the Speed of Sound

Here are two simple ways to measure the speed of sound. The first one uses traveling waves and the second method uses standing waves.

Method 1

Sig gen.

Mic.

Mic.

To oscilloscope

To oscilloscope

x

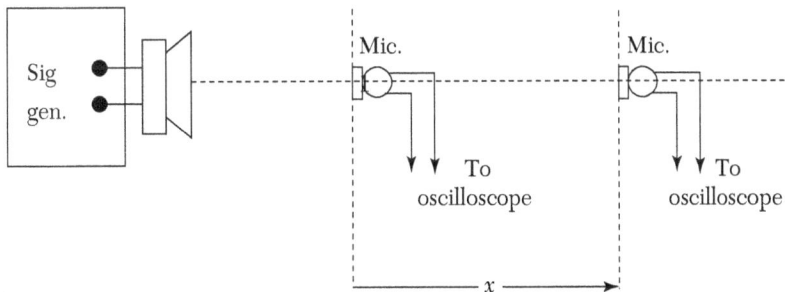

The signal generator is set to a single frequency (5–10 kHz is suitable) and the two microphones are connected to a dual-beam oscilloscope. The oscilloscope is triggered from the first microphone. The second microphone is then positioned close to the first one, and the two traces, which will be sinusoidal, are compared. Then x is changed until the signals are in phase. This position $x = x_0$ is recorded. Now the second microphone is moved away from the first and until the phase has changed by $2n\pi$ (i.e., has gone in and out of phase n times). The new position $x = x_n$ is recorded and the value of the wavelength calculated from $\lambda = (x_n - x_0)/n$. The frequency of the sound can also be measured from the oscilloscope (using $f = 1/T$), and the speed of sound is calculated from $v = f\lambda$.

Method 2

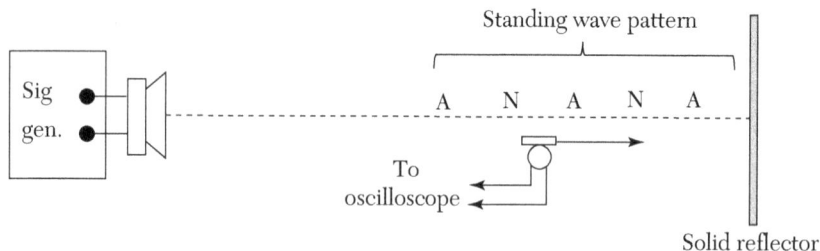

A standing wave is formed where the reflected sound superposes with incident sound. The microphone is moved along a line perpendicular to the reflector and the signal on the oscilloscope screen has periodic maxima every time the microphone passes through an antinode. The separation of adjacent maxima is $\lambda/2$, so an average value for the separation of adjacent nodes can be found and the wavelength can be calculated. The oscilloscope can also be used to find the frequency of the sound so that the speed can again be calculated from $v = f\lambda$.

16.5 Ultrasound

Ultrasound has frequencies >20 kHz, i.e., above the highest sound frequency audible to humans. Ultrasound scanning is used for medical imaging (e.g., in prenatal scans). Ultrasound pulses from a transmitter on the surface of the patient's skin partially reflect at each boundary inside the patient. The times of the returning pulses can be used to determine the depth of the boundary and to map out structure. This is done automatically so that the ultrasound scanner connects to a computer that displays a digital image of the organ or foetus being examined.

The fraction of the incident ultrasound intensity that is reflected is determined by the nature of the tissues on either side of the boundary. The key parameters are the density and speed of sound in the medium These are combined in the acoustic impedance Z of the medium:

$$Z = \text{(speed of sound)} \times \text{(density)}$$

The S.I. unit of the acoustic impedance is $\text{kgm}^{-2}\text{s}^{-1}$

The ratio of the reflected intensity I_r to the incident intensity I_0 is given by the expression:

$$\frac{I_r}{I_0} = \left(\frac{Z_2 - Z_1}{Z_2 + Z_1} \right)^2$$

The greater the difference in acoustic impedances between the two media, the greater the ratio. Here are some values of acoustic impedance:

Medium	Speed of sound/ms⁻¹	Density/kgm⁻³	Z/kgm⁻²s⁻¹
Air	340	1.2	410
Water	1480	1000	1.5×10^6
Muscle	1590	1070	1.7×10^6
Blood	1550	1060	1.6×10^6
Bone	4000	1500	6.0×10^6

The low value of Z for air compared to body tissues means that if there is an air gap between the ultrasound transmitter and the patient, then most of the ultrasound reflects from the surface and does not enter the body. To solve this problem, a gel is spread on the skin under the transmitter. The gel has an acoustic impedance similar to that of water or body tissues, so most of the ultrasound is transmitted rather than reflected.

16.6   Analysis and Synthesis of Sound

Most sounds that we hear contain a wide range of different frequencies, and it is possible to analyze the spectrum of sound in much the same way that we might use a spectrometer to analyze the spectrum of light. This is usually done using a microphone connected to a computer that is running an app that acts as a sound spectrum analyzer. This will then display a graph of sound intensity (usually in decibels) against frequency. Complex sounds can be analyzed into a sum or sequence of different sinusoidal components. This is called Fourier analysis.

In the same way we can add a sequence of harmonic sounds (sinusoidal functions) to synthesize a complex sound. This is called Fourier synthesis. The mathematics of Fourier analysis and synthesis are beyond the scope of this book, but the idea that complex sounds can be broken down into a sum of simple sinusoidal components is an important one, both theoretically and practically.

Excess pressure

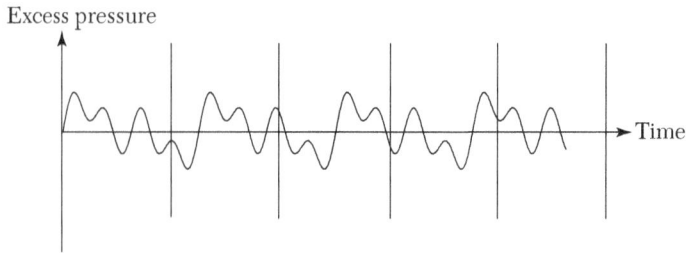

Time

The graph above shows the result of adding three sinusoidal sound waves of the same amplitude but with frequencies f, $2f$, and $4f$. This can also be represented by the equation:

$$y = A\sin(\omega t) + A\sin(2\omega t) + A\sin(4\omega t)$$

16.7   Exercises

1. Describe an experiment to measure the speed of sound. Your description should include:

▪ a labeled diagram of the apparatus

▪ an explanation of what you will measure and how the measurements will be made

▪ an explanation of how you will calculate the wavelength from the data you collect

▪ an explanation of how you will maximize accuracy and precision

2. A particular sound has an intensity of $1\mu\mathrm{Wm}^{-2}$ at a listener's ear.
   (a) Calculate the intensity level of this sound in decibels.
   (b) The intensity increases by a factor of 1000 to $1\mu\mathrm{Wm}^{-2}$. How many times louder does the sound seem to be?

3. A student sets up the apparatus below and records the amplitude of sound emitted by the air column as the tube is gradually filled with water. The end correction for the tube is 5.0 cm, and the frequency of the sound used is 1200 Hz. The speed of sound is 340 ms^{-1}.

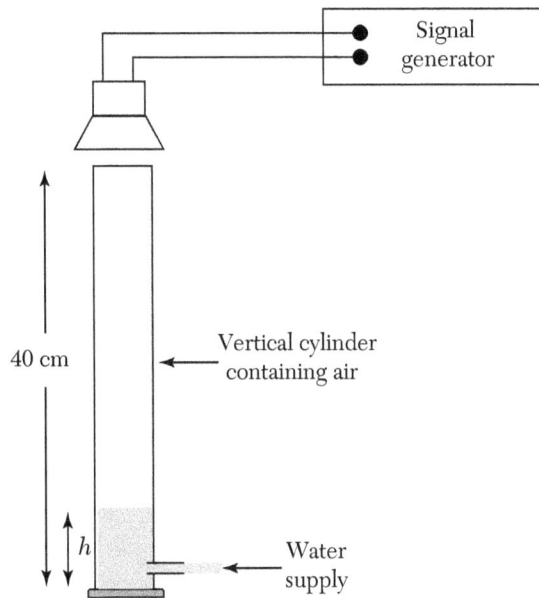

(a) Calculate the wavelength of the sound.
(b) Draw diagrams to show how standing waves of sound can be formed in the air column.
(c) Calculate the values of h at which the amplitude will be a maximum.

4. When a medical ultrasound scan is carried out, the doctor spreads a layer of gel onto the patient's skin before pressing the ultrasound transmitter/receiver against it.

(a) Explain why this is necessary.
(b) Calculate the fraction of the incident ultrasound intensity that is reflected when an ultrasound pulse reaches a boundary between muscle and bone. The acoustic impedance for muscle is 1.7×10^6 kgm^{-2}s^{-1} and that for bone is 6.0×10^6 kgm^{-2}s^{-1}.

17

ELECTRIC CHARGE AND ELECTRIC FIELDS

17.1   Electric Charge

Electric charge Q is a fundamental property carried by some fundamental particles. There are two types of charge, positive (e.g., the charge of the proton) and negative (e.g., the charge of the electron). Like charges repel one another, and unlike charges attract.

Charge is quantized. This means that the total charge on any object is always a multiple of a fundamental amount equal to the magnitude of the charge on an electron or a proton. The S.I. unit of charge is the coulomb (C), and the charge on an electron is:

$$e = -1.60217662 \times 10^{-19} \text{ C}$$

The charge on the proton has the same magnitude but opposite sign. Atoms contain equal numbers of protons and neutrons and are neutral. If an atom loses an electron, it becomes a positive ion; and if it gains an electron, it becomes a negative ion. You might have read that quarks, the fundamental particles inside protons and neutrons, have fractional charges. However, quarks are never found as individual particles; they are always combined in pairs (mesons) or triplets (baryons) and they always combine in a way that makes the total charge of the composite particle an integer multiple of e.

When charge flows it is called an electric current. Electric current I is defined as the rate of flow of electric charge:

$$I = \frac{dQ}{dt}$$

The S.I. unit of electric current is amp (A), and 1 A = 1 Cs⁻¹.

If the current is constant, the charge transferred in time t is given by:

$$Q = It$$

Some materials, e.g., metals, allow charge to flow through them. These are called conductors. Others such as plastics and rubber do not allow charge to flow and are called insulators. A third class of materials, with intermediate properties, is called semiconductors.

17.2   Electrostatics

Electrostatics deals with situations where objects become charged, remain charged, or lose charge. Most of these situations can be explained in terms of a transfer of electrons. The reason for this is because the electrons are on the outside of the atom, so when atoms interact (e.g., if two materials are rubbed together), the electrons can move from one place to another. Protons are locked inside the atomic nucleus, so proton transfer does not occur. When a neutral object loses electrons, it becomes positively charged; and when it gains electrons, it becomes negatively charged.

An object is said to be "earthed," when it is connected to the earth by a good conductor (e.g., an electrical wire). From the point of view of electrostatics, the Earth itself is a huge conducting sphere that can gain or lose any number of electrons while remaining neutral. Another way of looking at this is to say that the earth is always at zero potential so that any object connected to it is also at zero potential. The symbol to show an earth connection is shown on the right.

17.2.1   Charging by Friction

When a polythene rod is rubbed against a cloth, it becomes negatively charged. Electrons have been transferred from the cloth to the rod. Since polythene is a good electrical insulator, it can remain charged for some time and can be used to demonstrate electrostatic effects.

This process is called **charging by friction**. It is the same process that allows us to build up a charge when walking on a carpet. If the charge becomes large enough, we can experience a sharp electrical shock when we touch an earthed object. The shock is caused by the brief electric current that flows between our body and earth in order to neutralize us.

An acetate rod can be charged positively in the same way. However, in this case, electrons jump from the rod to the cloth and the rod retains

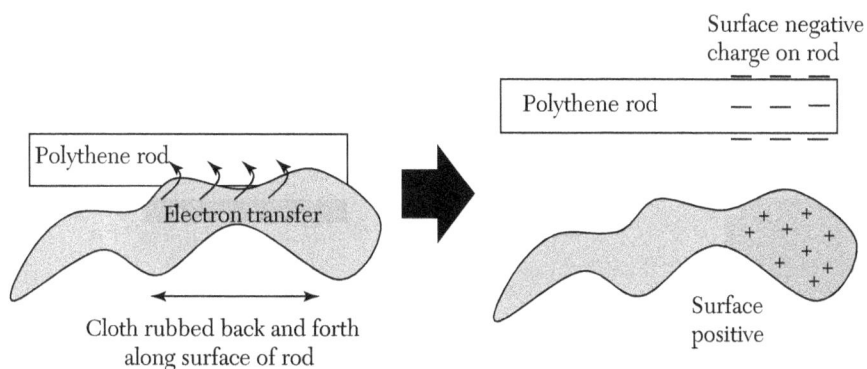

Polythene rod

Electron transfer

Cloth rubbed back and forth
along surface of rod

Surface negative
charge on rod

Polythene rod

Surface
positive

a positive surface charge. This is because atoms on the surface of the rod are no longer electrically neutral; some of them have fewer electrons than protons and so have a net positive charge. However, this positive charge has, once again, been brought about by the movement of the negatively charged electrons.

17.2.2   Gold Leaf Electroscope

The gold leaf electroscope is a useful device for detecting the presence of charge and for demonstrating simple electrostatic effects. Here is a diagram showing the structure of a gold leaf electroscope:

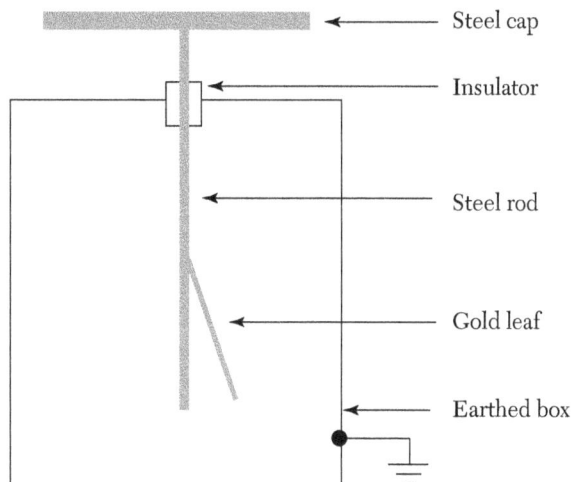

Steel cap

Insulator

Steel rod

Gold leaf

Earthed box

The cap rod and leaf are all metallic conductors, but they are isolated from the earth by an insulator. If a positively or negatively charged rod is held close to the cap but not touching it, the leaf will rise. This is because the charge on the rod exerts electrostatic forces on the free electrons making them move. The rod and leaf gain the same charge and repel one another.

The leaf is very thin, so it responds to this by rising. In the diagrams below, only the cap, leaf, and rod are shown.

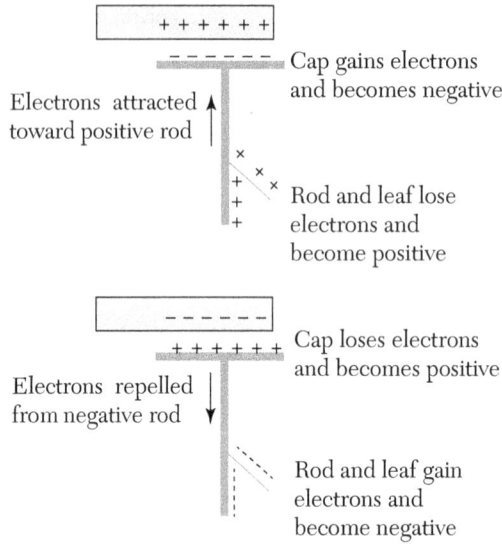

The process by which the cap gains a charge opposite to the charge on the rod is called electrostatic induction. In the examples above, there is no net charge on the cap leaf and rod; the electrons have just been redistributed. However, it is possible to use induction to give the electroscope a net positive or negative charge—this is called **charging by induction** and is explained in the sequence of diagrams below (resulting in a net positive charge).

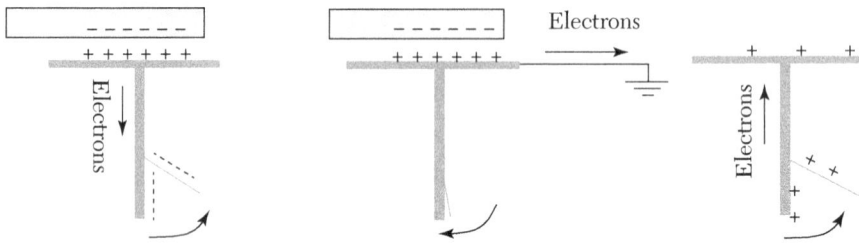

1. Hold a negatively charged rod close the cap. The leaf rises.

2. Keeping the negatively charged rod in place; momentarily earth the cap and then disconnect the earth. Electrons flow to earth and the leaf falls.

3. Remove the rod. The electroscope has lost electrons, so the cap, rod, and leaf have a net positive charge. When the rod is removed, electrons spread out and there is a positive charge on the cap, leaf, and rod. The leaf rises.

A similar process, starting with a positively charged rod, can be followed to charge the electroscope negatively. Note that the final charge is always opposite to the charge on the rod.

17.2.3 Using a Coulomb Meter

A coulomb meter is an electronic meter that can be used to measure charge. The charge to be measured must be transferred to the coulomb meter and then the amount can be read off from the display. Here is an image of a coulomb meter.

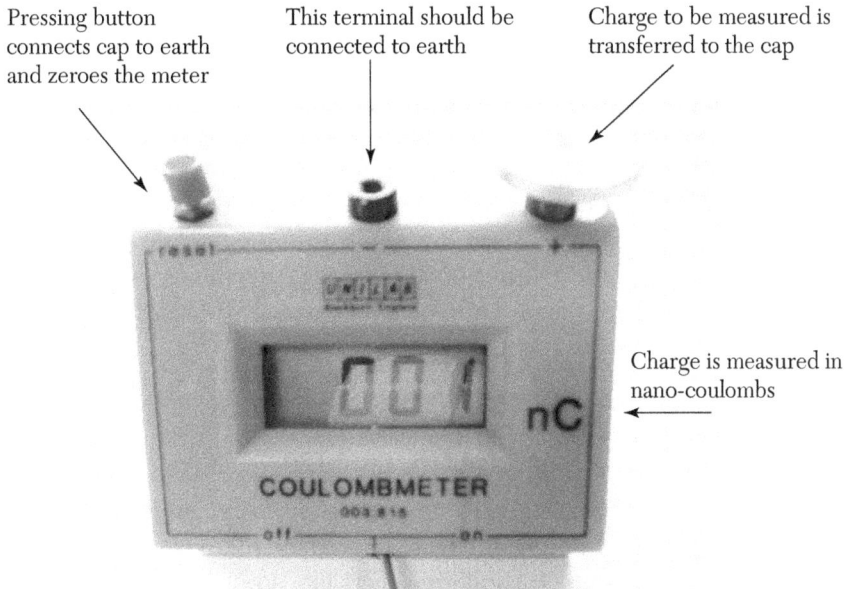

Pressing button connects cap to earth and zeroes the meter

This terminal should be connected to earth

Charge to be measured is transferred to the cap

Charge is measured in nano-coulombs

It is important that the meter is connected to earth before use.

The charge on an isolated conductor can be measured using a coulomb meter. For example, a conducting sphere could be isolated by suspending it from an insulating thread. It can then be charged by touching it momentarily with a lead connected to the high-voltage terminal of an H.T. supply (high-voltage supply). To measure the charge, it can then be connected momentarily to the cap of the coulomb meter (after the H.T. lead has been disconnected!). Charge flows from the sphere to the cap and the amount of charge transferred can be read from the display.

17.3 Electrostatic Forces

Charges obey a very simple qualitative force law: like charges repel and unlike charges attract. The magnitude of the force between two point charges depends on their size and separation and is given by Coulomb's law, an inverse-square law similar in form to the law of gravitation.

17.3.1 Coulomb's Law

The electrostatic force between two point charges Q_1 and Q_2 separated by a distance r in a vacuum is directly proportional to the product of the charges and inversely proportional to the square of their separation:

$$F = \frac{Q_1 Q_2}{4\pi \varepsilon_0 r^2}$$

where ε_0 is the **permittivity of free space**, a constant representing the ability of the vacuum to support an electric field. The S.I. unit of ε_0 is Fm^{-1} (F is the farad, a unit of capacitance equivalent to a coulomb per volt, CV^{-1}). If the charges are embedded in a different medium, the permittivity of free space is replaced by the permittivity of the medium ε, which is usually written as $\varepsilon = \varepsilon_r \varepsilon_0$ where ε_r is the relative permittivity of the medium, a dimensionless number. Here are some values of relative permittivity: vacuum 1 (by definition), air 1.0006 (at S.T.P.) so this is usually taken to be 1; polythene 2.25; paper 3.85; mica 3–6; water 80.1 (at 20°C); titanium dioxide 86–173; calcium copper titanate>250,000,

Capacitors (see Chapter 19) consist of two conductors separated by an insulating (or dielectric) material, and their capacitance is directly proportional to the permittivity of the insulator.

While Coulomb's law applies to point charges, it can be shown that if charge is distributed uniformly over the surface of a sphere or uniformly throughout the volume of a sphere, it acts like a point charge of the same total charge located at the center of the sphere.

17.3.2 Investigating Electrostatic Forces

It is possible to investigate Coulomb's law using simple apparatus, but care has to be taken to avoid effects due to induction, etc., with surrounding objects.

Two small isolated conducting spheres are charged from the same H.T. supply and then suspended by insulating strings.

The charges repel one another, so the system hangs in equilibrium with the strings making an angle θ to the vertical. θ and r can be determined using a digital camera to capture an image of the apparatus with a suitable scale placed behind. The force F can then be found using the condition for equilibrium of forces:

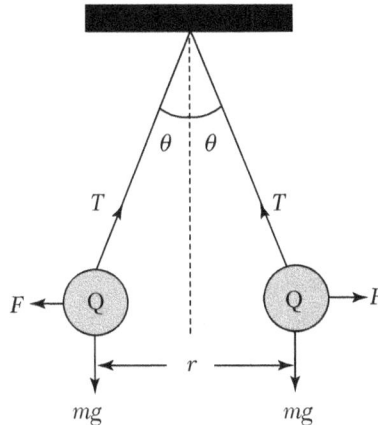

$T \cos \theta = mg$

$T \sin \theta = F$

So $F = mg \tan \theta$

Q can be found by discharging the sphere to a coulomb meter.

If Q is kept constant (charging from the same high voltage) and the mass of the spheres is varied, then F and r will change, so the inverse-square law can be tested.

Q can be varied by charging the spheres from a different voltage, but this will also change r unless a compensating change in mass is made.

17.4   Electric Field

Michael Faraday introduced the idea of an electric field to explain how distant charges can affect one another, even if there is a vacuum between them. Each charge sets up a field in the surrounding space and responds to fields from other charges. This makes electrostatic forces **local effects** (the field acts on the charged particle where the charged particle is) rather than an "action-at-a-distance" across space. It also implies that the electric field has physical properties at each point in space: it has a strength and direction. It is a **vector field**—there is an electric field vector at each point.

17.4.1   Electric Field Strength

Electric field strength E is defined as the force per unit positive charge at a point in space. If you were to place a small charge q at that point, it would experience a force F. The magnitude of the electric field strength at that point is given by:

$$E = \frac{F}{q}$$

The S.I. unit of electric field strength is NC^{-1}, which is equivalent to Vm^{-1}. The electric field is a **vector field**: the electric field points in the direction of the force on a positive charge.

This can be represented by drawing field lines. The direction of the field lines is the direction of the electric field, and the separation of the field lines represents the strength of the field. The two diagrams below show the shape of the electric field close to a point positive and a point negative charge.

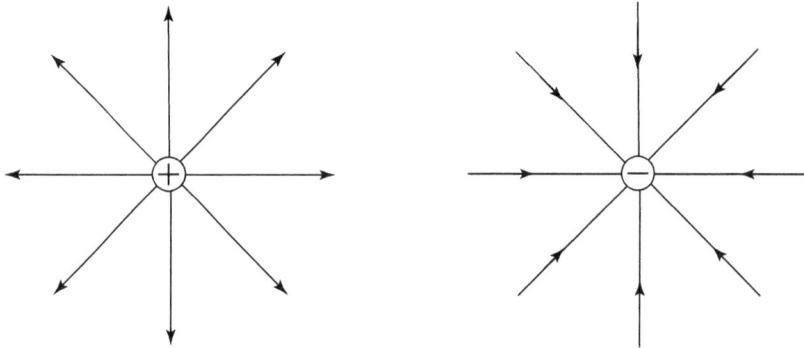

Electric field lines:

▪ start on positive charges and end on negative charges

▪ cannot cross

▪ point in the direction of the force that would act on a positive charge

Electric fields obey a **principle of superposition**. The resultant field at any point in space is the vector sum of all the electric fields at that point. In the example shown below, the resultant field at point C is the vector sum of the fields at that point due to charges A and B: $\underline{E}_C = \underline{E}_A + \underline{E}_B$

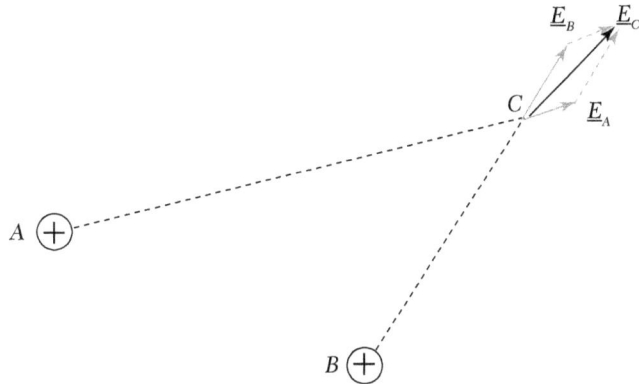

The diagram below shows the electric field close to a dipole—two charges of equal magnitude separated by a small distance. Many molecules have electric dipoles.

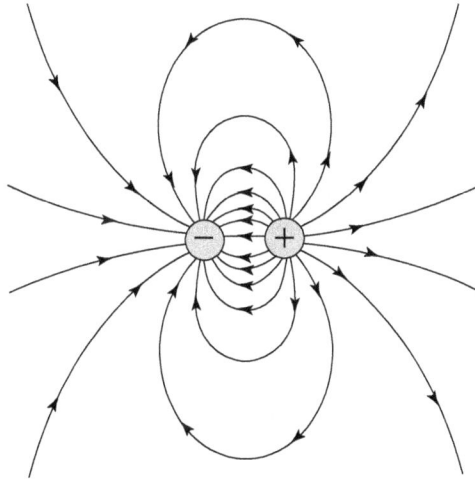

17.4.2 Electric Field Strength of a Point Charge

Imagine placing a small charge q a distance r from a point charge Q. The force exerted on q would be, from Coulomb's law,

$$F = \frac{Qq}{4\pi\varepsilon_0 r^2}$$

The electric field strength at that point is:

$$E = \frac{F}{q} = \frac{Q}{4\pi\varepsilon_0 r^2}$$

The electric field of a point charge obeys an **inverse-square law**.

This also makes sense from a geometrical point of view. Imagine the field lines from a point positive charge spreading uniformly out in all directions in 3D space. They would spread over an area $4\pi r^2$ at distance r from the charge, so the density of field lines would fall off as an inverse-square law.

This formula can be used to find the resultant electric field from a number or distribution of charges.

The diagram below shows an electric dipole consisting of two charges, $+Q$ and $-Q$, separated by a distance $2a$. Consider the electric field at a point P at distance x from the center of a dipole along its axis:

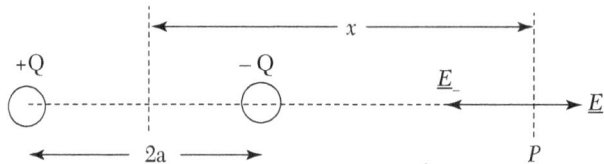

$$E_P = \frac{+Q}{4\pi\varepsilon_0 (x+a)^2} + \frac{-Q}{4\pi\varepsilon_0 (x-a)^2} = \frac{-axQ}{\pi\varepsilon_0 (x^2 - a^2)^2}$$

The negative sign indicates a resultant field to the left.

Here is another example at another point in the dipole field:

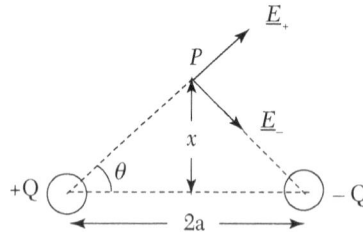

Point P is equidistant from the two charges, so the magnitude of fields E_+ and E_- are equal. The vertical components cancel. The resultant is therefore a horizontal field of strength:

$$E_P = \frac{2Q\cos\theta}{4\pi\varepsilon_0 (x^2 + a^2)} = \frac{2Qa}{4\pi\varepsilon_0 (x^2 + a^2)^{3/2}}$$

Sometimes the electric fields of two or more charged particles sum to zero at a point. This is called a **neutral point**. For example, half way between two charges of the same magnitude and sign:

$$E_P = \frac{Q}{4\pi\varepsilon_0 a^2} - \frac{Q}{4\pi\varepsilon_0 a^2} = 0$$

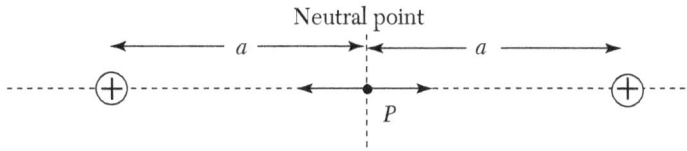

17.4.3 Gauss's Law

Consider the electric field that passes through a closed spherical surface of radius r centered on a point positive charge. Since lines of electric field cannot start or end in empty space, the flux of field lines through the surface is the same whatever the radius of the sphere.

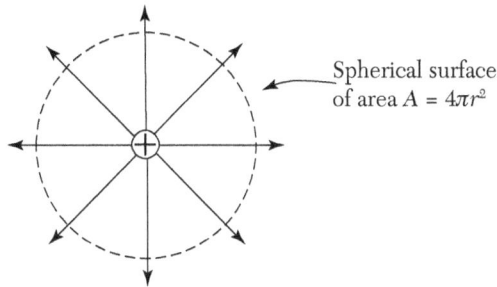

Spherical surface of area $A = 4\pi r^2$

The electric flux Φ_E is the product of the normal component of the electric field E and the area A of the surface, then:

$$\Phi_E = EA = E \times 4\pi r^2$$

When this is rearranged and compared with the expression for the field strength of a point charge, it is clear that the flux through the surface is directly related to the charge within that surface.

$$E = \frac{\Phi_E}{4\pi r^2} = \frac{Q}{4\pi\varepsilon_0 r^2}$$

so that:

$$\Phi_E = \frac{Q}{\varepsilon_0}$$

This is an example of a general result known as **Gauss's theorem**. This states that:

- The total flux through any closed surface is equal to the total charge contained within that surface divided by the permittivity of free space (or of the medium if the charge is not in a vacuum).

This is a powerful theorem that can be used to understand how the electric field behaves in a range of important situations. Gauss's theorem can be stated more precisely using an integral:

$$\Phi_E = \int\limits_{\text{surface}} E.dS = \frac{\sum\limits_{i=1}^{i=N} Q_i}{\varepsilon_0}$$

Where dS is an infinitesimal element of the surface area at some point and **E.dS** is the product of the perpendicular component of **E** at that point and the area dS. The integral sums these contributions to find the total flux (in the previous example, the field strength was constant and perpendicular to the surface, so we simply multiplied the values together). $\sum\limits_{i=1}^{i=N} Q_i$ is the sum of charges contained inside the closed surface. The next section shows how Gauss's theorem can be used to derive some important results.

17.4.4 Using Gauss's Theorem

(i) Electric field near the surface of a charged conductor

The electric field strength inside a perfect conductor must be zero. If it was not zero, charges would move until it became zero. This means that that when a conductor is charged the charge stays on the surface. The diagram below shows a small section of the surface of a charged conductor that has a surface charge density σ. The dotted lines indicate a Gaussian surface in the shape of a cylinder with the top surface of the cylinder (of area A) just above the surface of the conductor and the lower surface just below it.

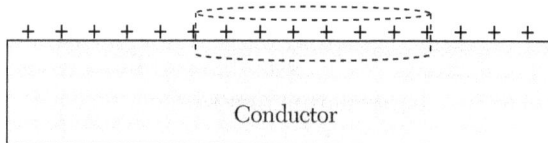

- The electric field through the lower surface must be zero because it is inside the conductor.

- The electric field through the sides of the cylinder must also be zero because there can be no component of electric field in the surface of the conductor.

- All of the electric field must pass through the upper surface.

Using Gauss's theorem

Flux through upper surface = EA = charge contained inside Gaussian surface divided by $\varepsilon_0 = \sigma A/\varepsilon_0$.

Conclusion: The electric field strength close to a conducting surface is $E = \dfrac{\sigma}{\varepsilon_0}$ and this is perpendicular to the surface.

(ii) Electric field strength inside a hollow conductor (Faraday cage)

Consider a closed hollow box made of conducting material and containing no free charges as shown below. The dotted line represents a closed Gaussian surface entirely within the conductor.

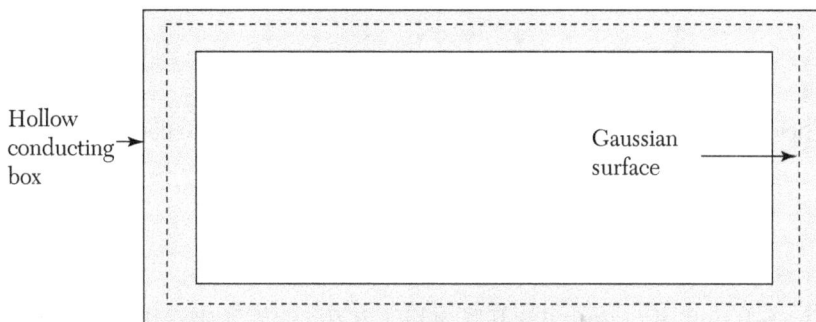

The electric field strength is zero inside the conductor, so the flux through the Gaussian surface is also zero. Using Gauss's theorem the total charge contained within the Gaussian surface and therefore inside the hollow box must also be 0. This remains the case even if other charged objects are brought close to the box or if electromagnetic waves are incident on the box. The field strength inside the box remains 0.

This is an example of a **Faraday cage**—a conducting box used to shield its contents from external electric fields and electromagnetic waves. Metal cars and aircraft act as Faraday cages protecting their occupants from lightning strikes. A room with conducting walls, floor, and ceiling can also be used to provide security against unwanted communications—e.g., mobile phone signals (which cannot enter or leave the room). The effect can be demonstrated by placing a mobile phone inside a metal cookie tin. When the tin is closed it is impossible to ring the mobile phone.

17.5   Electric Potential Energy and Electric Potential

Electric fields exert forces on electric charges, so when a charge moves from one point to another in an electric field, its electric potential energy changes. If the field does work on the charge, then the electric potential energy falls; and if work is done on the charge by an external agent, then the electric potential energy increases. This is shown in the diagram below where two positive charges are moved in a uniform electric field.

The electric field exerts a force downwards on A but A moves upwards so work is done on A by an external agent and the electric potential energy increases

The electric field exerts a force downwards on B and B moves downwards so work is done by the electric field and the electric potential energy decreases

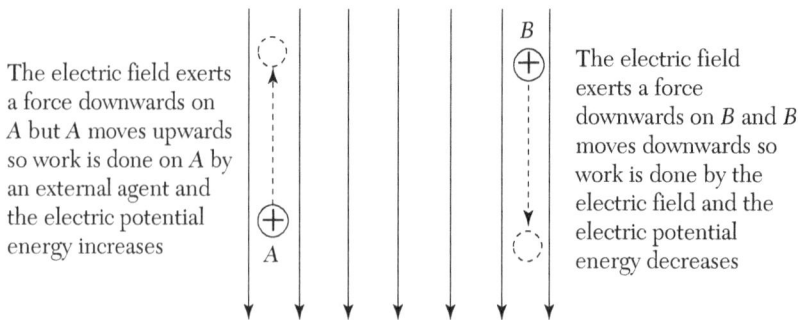

17.5.1 Electric Potential and Potential Difference

The electric potential V at a point in space is equal to the electrical potential energy (EPE) per unit charge (Q) if a small positive charge were placed at that point. This can be written as:

$$V = \frac{EPE}{Q}$$

The S.I. unit of potential is JC^{-1}, which is the volt, V. In other words, 1 V = 1 JC^{-1}. Electric potential, like energy, is a scalar quantity, so the potential at any point in space is the sum of potentials due to all fields at that point.

The electrical potential difference DV between two points is equal to the work that must be done per unit charge in moving the charge between the two points concerned.

$$\Delta V = W/Q$$

For example, if there is a potential difference of 3.0 V across an electrical component, then 3J of energy is transferred from electrical potential energy to other forms when 1 C of charge passes through the component.

(While we have distinguished the absolute potential V from the potential difference ΔV, here it is often the case that V is used for both.)

17.5.2 Electric Potential Gradient and Electric Field Strength

To derive the relationship between electric field strength and electric potential, we must investigate the work done on free charges placed in an electric field.

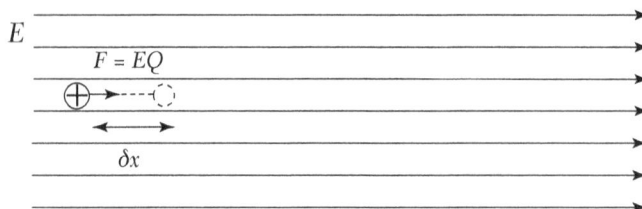

If a free point charge of value $+Q$ is moved a distance δx by an electric field of strength E, then the force on the charge is given by:

$$F = QE$$

and the work done on it is:

$$\delta W = F\delta x = QE\delta x$$

This work has been done by the electric field, so the electric potential energy of the system has fallen (in the same way that gravitational potential energy falls when an object is dropped):

$$\delta EPE = -QE\delta x$$

And the change in electric potential is:

$$\delta V = \frac{\delta EPE}{Q} = -E\delta x$$

$$\frac{\delta V}{\delta x} = -E$$

In the limit that $\delta x \rightarrow 0$, the relationship becomes:

$$E = -\frac{\mathbf{d}V}{\mathbf{d}x}$$

▦ Electric field strength = negative potential gradient

This can also be written as an integral to find the potential difference between two points A and B in an electric field:

$$\Delta V = \int_{V_A}^{V_B} dV = -\int_{x_A}^{x_B} E dx$$

Three immediate consequences of these equations are:

▦ There is no change in potential when a charge is moved in a direction perpendicular to the electric field lines. In this direction $E = 0$, so $dV/dx = 0$ too and the potential is constant.

▦ The potential changes more slowly with distance when the field is weak (dV/dx is lower because E is lower).

▦ The direction of the electric field is toward lower potential (this is the significance of the minus sign).

Equipotential surfaces can be drawn perpendicular to the field lines. The diagram below shows the equipotential surfaces (spherical surfaces) surrounding a point charge. The steps in potential between adjacent equipotentials is constant, so their separation increases farther from the central charge; this is because the field is weaker further out so the potential is changing more slowly.

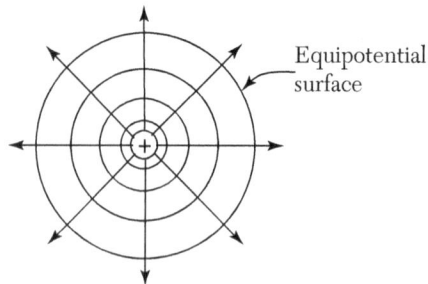

When a potential difference V is set up between two parallel conducting plates separated by a distance d, the electric field between them is approximately uniform so that the potential gradient $dV/dx = V/d$. The equation $E = -\dfrac{dV}{dx}$ can then be used to find the strength of the electric field.

$$= -\frac{dV}{dx} = -\frac{V}{d}$$

17.5.3 Accelerating Charged Particles in an Electric Field

Charged particles are often accelerated by an electric field. When a potential difference V is set up between two electrodes, a free particle of charge Q loses electric potential energy and gains kinetic energy as it moves between them. If there are no other energy transfers, then:

$$\frac{1}{2}mv^2 = QV$$

The velocity is then:

$$v = \sqrt{\frac{2QV}{m}}$$

Where m is the particle mass.

However, we must be careful when using this equation. If the velocity is a significant fraction of the speed of light (e.g., $v > 0.05\,c$), then we should use relativistic equations. The Newtonian equations above are useful for lower velocities but only give an approximate value for the final velocity.

Many electron tubes use an electron gun to accelerate electrons and form a beam. Here is a simplified diagram showing how the electron gun works.

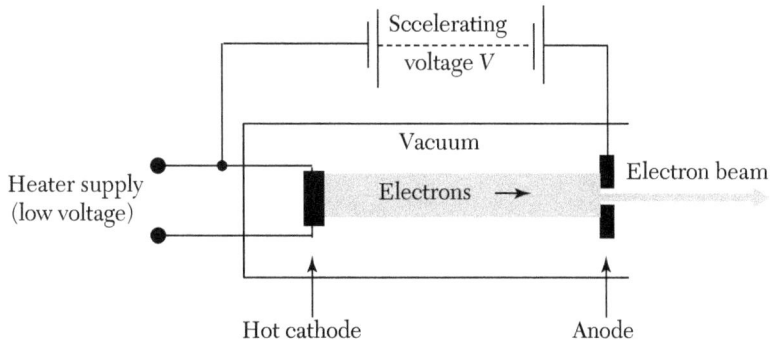

The cathode (negative terminal) is heated by passing a small current through it. This increases the kinetic of free electrons inside the cathode and allows them to leave the surface of the metal. The accelerating voltage V creates an electric field between the cathode and anode that accelerates the electrons. The electrodes are placed inside a vacuum tube so that the electrons are not scattered. The loss of electric potential energy eV is transferred to kinetic energy of the electrons as they reach the anode. A narrow beam passes through a small gap in the anode to create a beam. The electron velocity is:

$$v = \sqrt{\frac{2eV}{m}}$$

17.5.4 Deflecting Charged Particles in an Electric Field

When a charged particle is projected into an electric field, the electrical force on the particle acts parallel to the field. If a uniform electric field is perpendicular to the initial velocity of the charged particle, then the force is constant and deflects the charged particle into a parabolic path. This is analogous to the parabolic path followed by a massive particle projected horizontally in the Earth's gravitational field.

An electron deflection tube uses an electron gun to fire a horizontal beam of electrons into a region of uniform vertical electric field. The field is created by two parallel plates connected to a separate voltage supply.

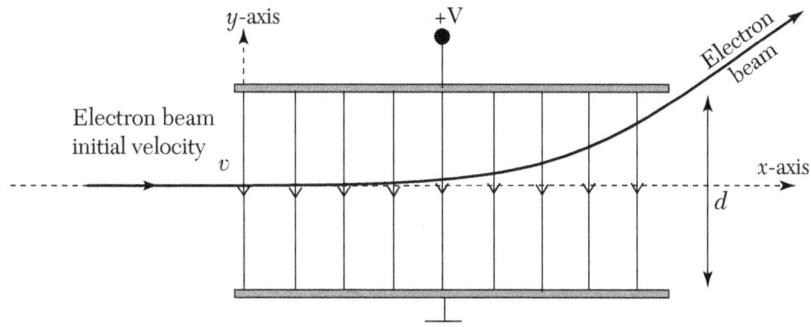

The resultant force on the electron is vertically upwards:

$$F = eE = \frac{eV}{d}$$

There will be a vertical acceleration:

$$a = \frac{eV}{md}$$

and the vertical deflection in the field is (using *suvat*):

$$y = \frac{eVt^2}{2md}$$

There is no horizontal force, so the horizontal component of velocity is constant and equal to v. The horizontal distance traveled in the field is therefore:

$$x = vt$$

Eliminating t from the equations for x and y gives:

$$y = \left(\frac{eV}{2mv^2}\right)x^2$$

This is the equation of a parabola.

17.5.5 Absolute Electric Potential of a Point Charge

The equation for potential difference in terms of field strength:

$$\Delta V = -\int_{x_A}^{x_B} E dx$$

can be used to find the absolute potential at point B if we know the potential at point A. In order to find the potential at any point in the universe, we must define the position where potential is zero. This is an arbitrary decision because forces and fields only depend on differences in potential and not on its absolute value. However, it makes sense to choose the position for the zero of potential in such a way that it is easy to use in calculations. The zero of potential in the electric field is taken to be at infinity. In other words, if all charges were separated so that they were at infinite distance from one another, then the total electrical potential energy would be zero.

Here are two qualitative examples of how this work:

- Consider two point positive charges a distance r from each other. These will repel each other, and if no other forces act on them, they will move toward infinity. While they are moving, electrical forces are doing work on them, transferring electrical potential energy to kinetic energy, so the electrical potential energy is *decreasing* all the time. However, we know that their electrical potential energy will be zero at infinity, so the initial electrical potential energy of two positive charges must have been positive. The same argument shows that the initial electrical potential energy of two negative charges a distance r apart must also be zero.

■ Now consider separating a positive and a negative charge that are initially a distance r apart. These two particles attract each other, so we would have to apply an external force to each of them to move them out to infinity. Work must be done by this external agent to separate them so the electric potential energy *increases* all the time and eventually becomes zero. The initial electric potential energy must have been negative. This is the case for the electron inside a hydrogen atom. It is attracted to the proton in the nucleus of the atom, and the system has a negative electric potential energy. The energy that must be put in to separate the two particles is the ionization energy for the hydrogen atom.

We can now derive an expression for the electric potential energy at a point in the electric field of a point charge. This is done by deriving an expression for the potential difference between that point and infinity and using the fact that the potential is zero at infinity.

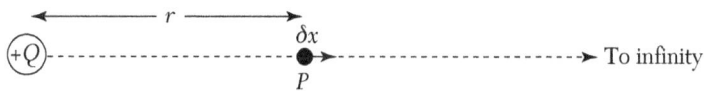

The electric field strength at a point distance r from a point positive charge Q is:

$$E = \frac{Q}{4\pi\varepsilon_0 r^2}$$

So the potential difference between P (at distance r from Q) and infinity is:

$$\Delta V = V_\infty - V_P = -\int_{x=r}^{x=\infty} E\,dx = -\int_{x=r}^{x=\infty} \frac{Q\,dx}{4\pi\varepsilon_0 x^2}$$

$$= -\left[\frac{-Q}{4\pi\varepsilon_0 r}\right]_{x=r}^{x=\infty} = \frac{-Q}{4\pi\varepsilon_0 r}$$

$$V_P = V_\infty - \frac{Q}{4\pi\varepsilon_0 r} = \frac{Q}{4\pi\varepsilon_0 r}$$

The potential varies as $1/r$.

Summary of equations for the electric field of a point charge

◽ Electric field (magnitude of vector) at distance r from a charge Q:

$$E = \frac{Q}{4\pi\varepsilon_0 r^2}$$

◽ Electric potential (scalar) at distance r from a charge Q:

$$E = \frac{Q}{4\pi\varepsilon_0 r}$$

17.6 Exercises

1. (a) A small electric cell can supply a continuous current of 2.0 A for 2 hours. How much charge passes through the cell in this time?

   (b) A large capacitor (component for storing charge) is charged using a steady current of 4.0 mA. It takes 0.75 s to reach full charge. How much charge does it store?

   (c) An electrostatic generator stores 8.0 C on a large metal dome. This charge leaks away through the air in 40 s. What is the average electric current during this time?

2. Write an instruction sheet explaining how to charge an electroscope positively using just a polythene rod and a cloth.

3. The diagram below is a model of an electric dipole.

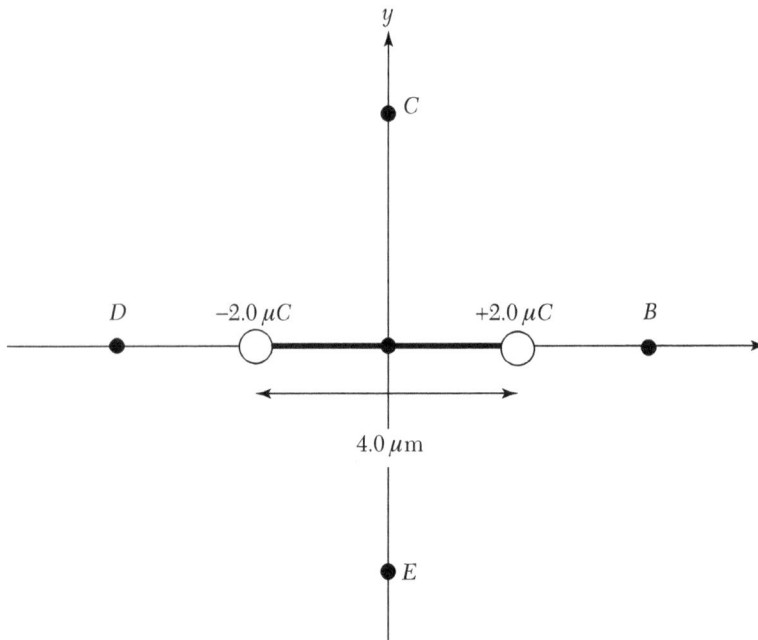

The dipole is located on the x-axis with its center at the origin.
(a) Calculate the electric field strength (and direction) and the electric potential at points A to E:

A $(0, 0)$
B $(+4.0 \, \mu m, 0)$
C $(0, +4.0 \, \mu m)$
D $(-4.0 \, \mu m, 0)$
E $(0, -4.0 \, \mu m)$

4. A hydrogen atom consists of a proton and an electron orbiting at a distance of 0.053 nm.
   (a) Calculate the force between the electron and the proton.
   (b) Assuming this force provides the centripetal force for circular motion of the electron, work out the electron's orbital speed and kinetic energy.
   (c) Calculate the electrostatic potential energy of the electron.
   (d) Calculate the total energy of the electron and explain the significance of its sign.
   (e) State the ionization energy of the hydrogen atom.

5. (a) Sketch the field lines and equipotentials round an isolated charged conducting sphere of radius 2.0 cm at a potential of +5000 V (include equipotentials at 1000, 2000, 3000, and 4000 V).
   (b) Calculate the electric field strength and potential at distances of 4.0 and 11 cm from the center of the positively charged sphere.

6. Copy the diagram below and sketch the electric field lines and equipotentials between the charged sphere and the earthed conducting plane shown below.

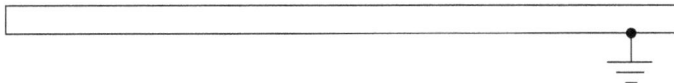

7. Use Gauss's theorem to show:
   (a) that the electric field strength inside a charged conducting sphere is 0
   (b) that the electric field strength immediately above a charged conductor is perpendicular to the surface and has a magnitude $E = \sigma/\varepsilon_0$ where σ is the charge density on the surface

(c) that the flux of electric field entering a volume of empty space is equal to the flux of electric field leaving that volume of space

8. (a) Prove that the uniform electric field between two parallel conducting plates separated by distance d and connected to a potential difference V is given by $E = V/d$.

   (b) Sketch the electric field lines and equipotentials between two parallel metal plates 2.0 cm apart with a potential difference of 5000 V (include equipotentials at 1000, 2000, 3000, and 4000 V).

   (c) Calculate the electric field strength between the plates.

   (d) Calculate the force on an alpha particle (charge +2e) half way between the plates.

   (e) How does this force vary if the alpha particle moves close to the positive or negative plate (ignore induction effects)?

   (f) An air molecule between the plates loses two electrons and becomes ionized. What is the ratio of the acceleration of the electron to the acceleration of this positive ion in the electric field? Assume that the mass of the ion is 60,000 times greater than that of the electron.

9. An electron in a vacuum tube is accelerated horizontally in an electron gun through a potential difference of 1500 V. It then enters a region of uniform vertical electric field of strength $2.0 \times 10^4 \, \text{Vm}^{-1}$ that extends 5.0 cm horizontally. It is deflected by the field and emerges into a field-free region.

   (a) Calculate the velocity of the electron as it enters the region of vertical electric field.

   (b) Describe the shape of the electron's path in the field. Assume the direction of the field is vertically upwards.

   (c) Calculate the time spent by the electron in the vertical field.

   (d) Calculate the vertical component of velocity gained by the electron in moving through the vertical field.

   (e) Calculate the angular deflection of the beam.

   (f) Calculate the work done by the vertical electric field on the electron.

18

D.C. ELECTRIC CIRCUITS

18.0 D.C. Circuits and Conventional Current

Electric current is the rate of flow of electric charge, and D.C. refers to circuits in which the current flows in one direction around the circuit. Since current can be carried by positive or negative charges, the direction of electric current needs a convention: *the direction in which a free positive charge would flow.*

There are three different microscopic ways in which current can flow.

Negative charge carriers (e.g., electrons in a metal or an n-type semiconductor): The electrons move in the opposite direction to the conventional current.

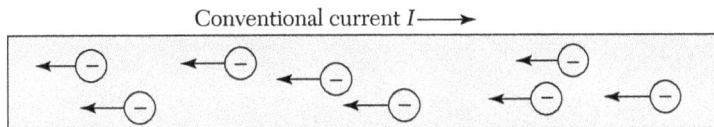

Positive charge carriers (e.g., positive ions in an ion beam or holes in a p-type semiconductor): The positive charge carriers move in the same direction as the conventional current.

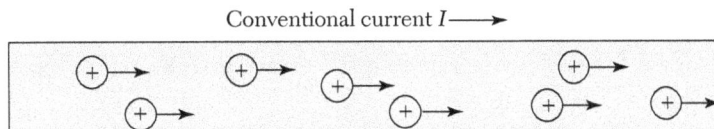

Positive and negative charge carriers (e.g., when current flows through an electrolyte): The positive charges move in the direction of conventional current and the negative charges move in the opposite direction.

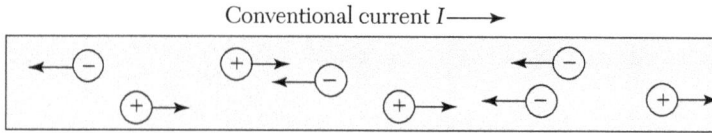

In all the examples above the velocities of the charge carriers represent their average **drift velocities**; the particles will also have random thermal motion, which might involve speeds that are several orders of magnitude greater than the drift velocities.

18.1 Charge and Current

Electric charge Q is measured in coulombs (C). Electric current I is the rate of flow of electric charge:

$$I = \frac{dQ}{dt}$$

The S.I. unit of electric current is A, and $1\ A = 1\ Cs^{-1}$.

If the current is constant, the equation above becomes $I = Q/t$ or $Q = It$, where Q is the charge passing a point in t seconds.

18.1.1 Charge Carriers and Charge Carrier Density

When current flows through a material, charge carriers move inside the material and we can derive an expression to link the microscopic movement of these charge carriers to the macroscopic current. In the diagram below, we assume that the charge carriers are positive, but the form of the relationship is not affected by this and it can be used for positive or negative charge carriers.

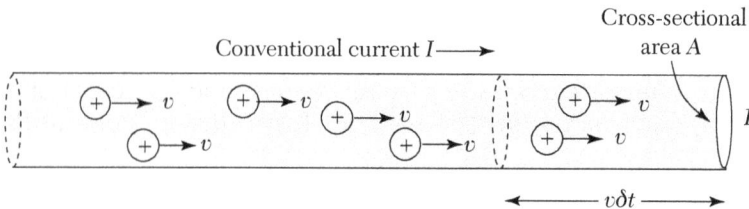

There are n charge carriers per unit volume in the conductor, and each charge carrier has a charge q.

Consider the charge leaving the right-hand end of the wire at P during a short time δt.

All charge carriers within a distance $v\delta t$ will leave in this time. The volume within this distance of point P is $Av\delta t$, so the number of charge carriers is $N = nAv\delta t$ and the total charge δQ passing P in time δt is:

$$\delta Q = qnAv\delta t$$

so the current is:

$$I = \frac{\delta Q}{\delta t} = nAvq$$

Very often the charge carriers have a charge equal to the electronic charge, e; so this equation becomes:

$$I = nAve$$

Typical values of the charge carrier density n for two different metals and two types of semiconductors are shown below. The values are for a temperature of 300 K.

Copper, 8.5×10^{28} m^{-3}

Silver, 1.1×10^{28} m^{-3}

Silicon, 1.5×10^{13} m^{-3}

Germanium, 2.4×10^{16} m^{-3}

Note that the carrier density in the pure semiconductors is very much smaller than for the metals, so the drift velocity in a semiconductor carrying the same current as a metal of the same dimensions will be *much greater* than in the metal. The carrier density is highly temperature-dependent in semiconductors and increases rapidly with temperature. Pure, or intrinsic, semiconductors contain both negative charge carriers (electrons) and positive charge carriers (holes) in equal numbers, but doped, or extrinsic, semiconductors have small quantities of other elements added to increase their carrier density of a factor of 10^6 or more.

18.1.2 Measuring Current

Electric current is measured using an ammeter.

The ammeter must be connected in series at the point where current is to be measured.

An ideal ammeter has zero resistance, so it has no effect on the current it is measuring. In practice ammeters do have a small internal resistance, so the current measured by the ammeter is slightly less than the current would be if the ammeter was not in the circuit. This difference is usually small enough to be neglected, but can be important, especially if the circuit resistance is particularly low.

Small currents are often measured in milli-amps (mA) or micro-amps (mA).

1 mA $= 0.001$ A $= 10^{-3}$ A

1 mA $= 0.001$ mA $= 0.000001$ A $= 10^{-6}$ A

18.1.3 Currents in Circuits—Kirchhoff's First Law

An electric circuit provides a complete conducting path for charge. Charge is conserved, so the amount of charge in the circuit cannot change. This leads to **Kirchhoff's first law**—at a junction in an electric circuit, the sum of currents entering the junction is equal to the sum of currents leaving the junction.

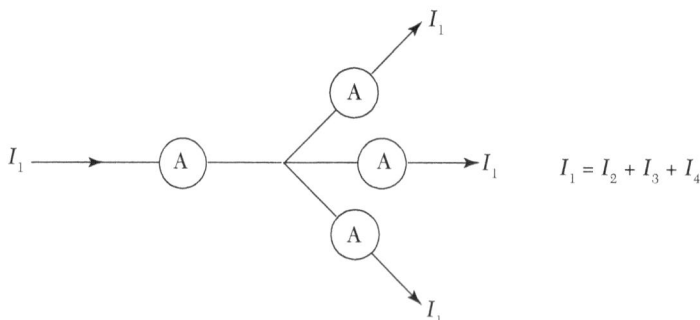

$$I_1 = I_2 + I_3 + I_4$$

If there are no junctions in the circuit, then the current is the same everywhere—i.e., at all points around a series circuit.

The current that flows in each branch of a parallel circuit is inversely proportional to the resistance of the branch.

18.2 Measuring Potential Difference

Potential difference is literally the difference in electric potential between two points in a circuit. This is measured using a voltmeter. The voltmeter must be connected in parallel between the two points. The diagram below shows how a voltmeter can be connected to measure the potential difference across a resistor.

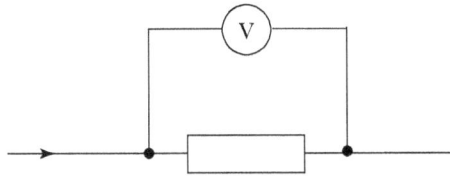

An ideal voltmeter has infinite resistance so that it does not draw any current from the circuit. In practice a real voltmeter will have a large but not infinite resistance, and this can affect readings if there are also very large resistors in the circuit.

18.2.1 EMF Potential Difference and Voltage

The potential difference across a source of electrical energy such as a battery or generator is called an emf (electromotive force), whereas the potential difference across a component that transfers electrical energy to other forms is called a potential difference or simply a "voltage."

Small voltages are often measured in milli-volts (mV) or micro-volts (mV).

Batteries are made up of cells, so the symbols for a battery and a single cell are different:

The longer line represents the positive end of the cell or battery. The emf of a battery consisting of cells in series is equal to the sum of the emfs of the individual cells. An ideal cell provides a constant emf and has no internal resistance. A real cell does have an internal resistance, so must be treated as an ideal cell of emf E in series with a resistor of resistance r and is usually represented as:

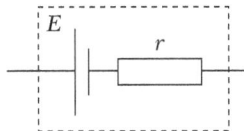

The behavior of real cells is analyzed in Section 18.5.

18.2.2 Kirchhoff's Second Law

When a charge carrier completes a closed loop around an electric circuit and returns to its initial position, it has exactly the same amount of energy at the end of the loop as it had at the beginning. If this was not the case, then energy would either have been created or destroyed, and this is impossible. This means that each charge must gain as much energy as it loses in completing a closed loop in the circuit. This leads to **Kirchhoff's second law**—the sum of emfs is equal to the sum of potential differences around any closed loop in an electric circuit.

$$\sum_{\substack{\text{closed} \\ \text{loop}}} \text{emfs} = \sum_{\substack{\text{closed} \\ \text{loop}}} \text{p.d.s}$$

If the potential differences are all across resistive components, this can be written as:

$$\sum_{\text{closed loop}} emfs = \sum_{i=1}^{i=N} I_i R_i$$

where there are N resistors in the loop and the ith resistor has a resistance R_i and current I_i.

The example below shows how Kirchhoff's second law can be applied to two different loops in an electric circuit:

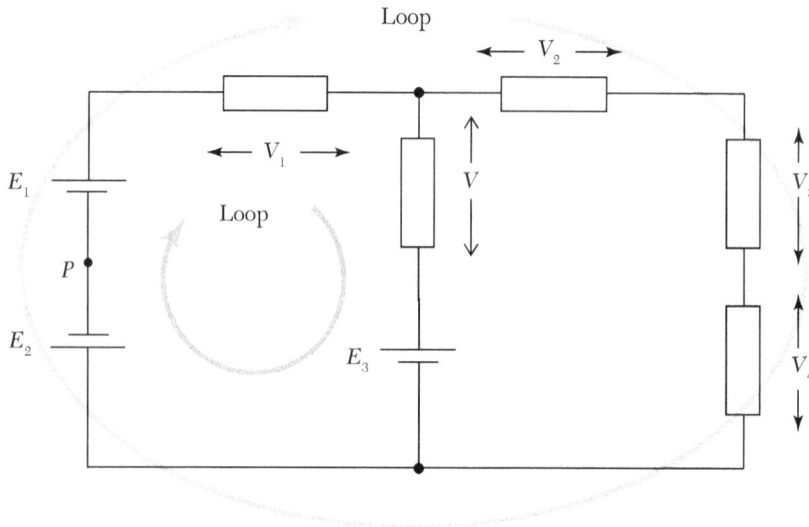

Each loop starts and ends at P, works in a clockwise direction, and provides a different equation:

Loop 1: $E_1 - E_3 - E_2 = V_1 + V_5$

Loop 2: $E_1 - E_2 = V_1 + V_2 + V_3 + V_4$

Care must be taken over the signs of the emfs and p.d.s.

While two loops are shown in the diagram, there is also a third loop in the right-hand section of the circuit. This is not included because it is not independent of the other two. The fact that all parts of this third loop are included in parts of the first two loops means that it would not provide additional information.

Kirchhoff's first and second laws generate a series of simultaneous equations that can be used to solve complex circuit problems (see Section 18.6.1).

18.3 Resistance

Resistance R is defined as the ratio of potential difference V across a component to current I through the component.

$$R = \frac{V}{I}$$

The S.I. unit of resistance is Ω, and $1\ \Omega = 1\ \text{VA}^{-1}$. Large resistances are measured in kΩ and MΩ.

$1\ \text{k}\Omega = 1000\ \Omega = 10^3\ \Omega$

$1\ \text{M}\Omega = 1000\ 000\ \Omega = 10^6\ \Omega$

18.3.1 Measuring Resistance

Resistance can be measured using an ammeter and voltmeter in the circuit shown below.

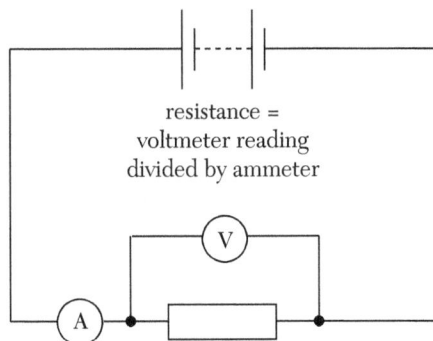

resistance =
voltmeter reading
divided by ammeter

While dedicated ammeters and voltmeters can be used, **multimeters** can be used instead. A multimeter can be used as a voltmeter or an ammeter depending on its settings. When using a multimeter, it is important to select appropriate settings before connecting the circuit to the power supply; otherwise there is a danger of damaging the meter or the circuit. For example, if the multimeter was set as an ammeter but connected in parallel like a voltmeter, it would short out the component and draw a large current.

Multimeters can also be used as an ohm-meter to measure resistance directly. When it is used like this, the component must first be isolated from the circuit, and then the meter must be connected directly across the component.

The image below shows a typical multimeter. The values shown on each range indicate the maximum value that can be measured on that range and shows the units in which it will be displayed. For example, a setting of 2 V D.C. would measure voltages from 0.00 to 2.00 V, whereas a setting of 200 mV would measure from 000 to 200 mV. When choosing a suitable scale, it is best to choose the most sensitive scale that has a maximum value greater than the value to be measured.

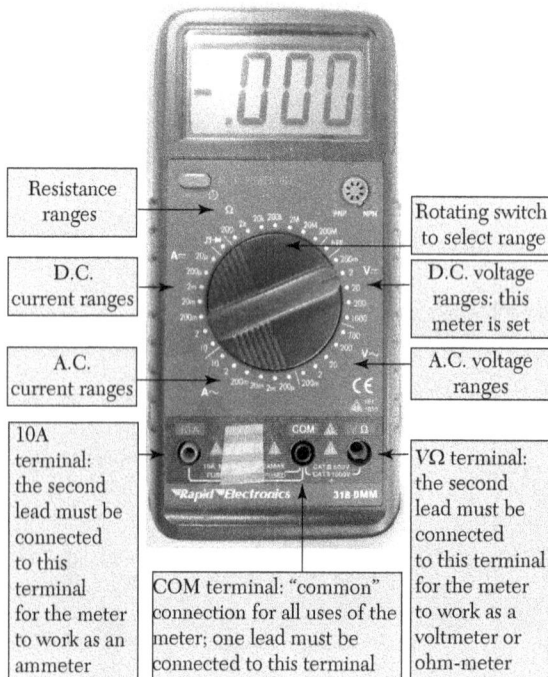

Resistance ranges

Rotating switch to select range

D.C. current ranges

D.C. voltage ranges: this meter is set

A.C. current ranges

A.C. voltage ranges

10A terminal: the second lead must be connected to this terminal for the meter to work as an ammeter

COM terminal: "common" connection for all uses of the meter; one lead must be connected to this terminal

VΩ terminal: the second lead must be connected to this terminal for the meter to work as a voltmeter or ohm-meter

The resistance of a component can also be found from its current–voltage characteristic. For example, if a particular component produced the characteristic shown below, its resistance at point P would be $R = V_p/I_p$.

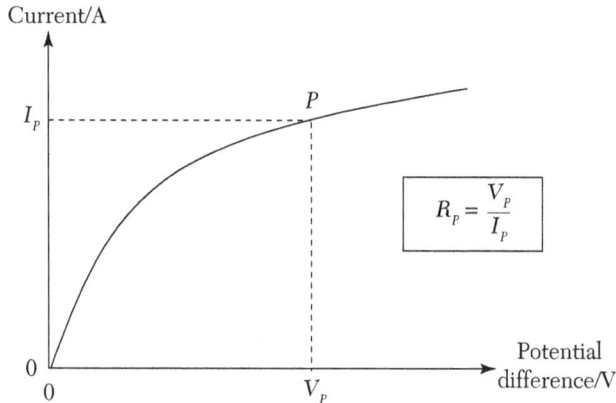

18.3.2 Current–Voltage Characteristics

The current–voltage characteristic of an electrical component can be determined using the circuit below. The part of the circuit in the grey box is called a potentiometer. As the moving contact goes from left to the right, the potential difference applied to the circuit containing the component R under test varies from 0 V to the maximum voltage of the battery.

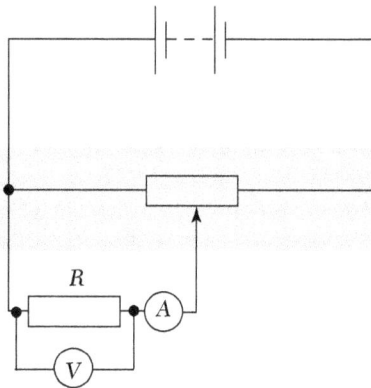

To reverse the current through the component, the battery or the component itself can be turned around. This allows both positive and negative values for I and V. The current–voltage characteristics for three components are shown below.

▦ A carbon resistor or a metal at constant temperature

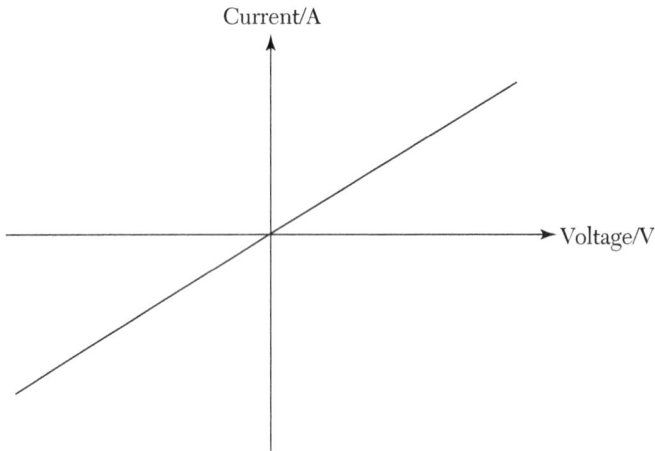

Current I is directly proportional to the potential difference V across the component:

$$I \propto V$$

$$\frac{V}{I} = \text{constant} = R$$

The resistance is constant.

Components that behave like this are described as "ohmic conductors" and are said to obey Ohm's law. Gustav Ohm investigated the electrical behavior of metals kept at constant temperature and discovered that the current passing through them and the potential difference across them are directly proportional, so Ohm's law really only applies to metals at constant temperature. Nowadays Ohm's law is invoked whenever current and voltage are directly proportional. The equation $V = IR$ is also sometimes referred to as Ohm's law, but this is not really correct—the equation defines resistance and only corresponds to Ohm's law when the resistance is constant.

▦ A metal filament (e.g., in a lamp)

Metal wires heat up when electric currents pass through them and their resistance changes. They are non-ohmic conductors.

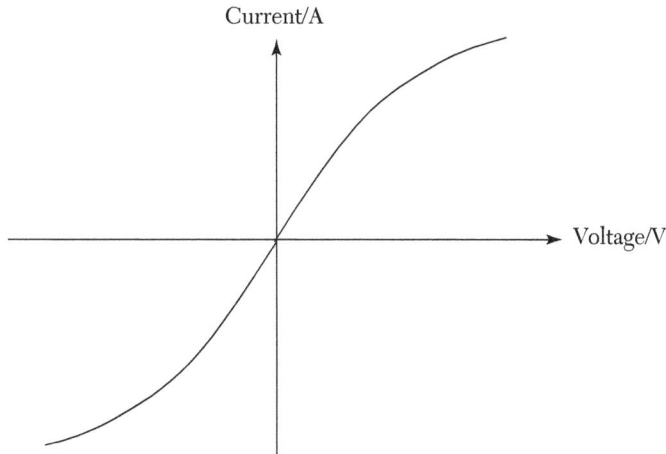

The fact that the graph is not a straight line through the origin shows that this is a non-ohmic conductor.

The ratio of V to I is increasing, so the resistance has increased as the current has increased.

The reason this happens is because the charge carriers passing through the metal transfer energy to the metal ions, making them vibrate more rapidly, and this in turn increases the scattering of charge carriers. More work has to be done to maintain the current (re-accelerate the charge carriers after scattering), so a higher voltage is needed for the same current. This increases the ratio V/I and increases the resistance.

Components whose resistance changes with temperature are called **thermistors**. The resistance of a metal increases with temperature, so it is a positive temperature coefficient (PTC) semiconductor. Platinum is commonly used as a PTC thermistor and can be used as a resistance thermometer. Semiconductors have a resistance that falls as temperature increases, so they are used as negative temperature coefficient (NTC) thermistors. The circuit symbol for a thermistor is:

Some semiconductors are light-dependent. Their resistance falls when they are illuminated. These are called light-dependent resistors (LDRs) and are useful both to measure light intensity and in sensing circuits that respond to changes in light intensity. The circuit symbol for an LDR is:

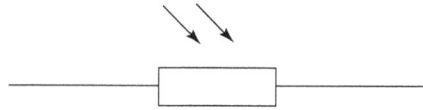

A semiconductor diode

The circuit symbol of a semiconductor diode is:

This is a conductor that conducts with very low resistance in one direction (forward bias) and acts as an insulator (infinite resistance) in the other (up to its breakdown voltage, at which point the diode resistance suddenly drops and it is likely to be destroyed by the current surge).

A small voltage is required in the forward direction before the diode begins to conduct. This "switch-on" voltage depends on the material from which the diode is made. For silicon diodes it is about 0.6 V; and for germanium diodes it is about 0.2 V. Once the forward voltage exceeds this value, the diode begins to conduct with a very low resistance, and care must be taken not to allow the current to grow so large that it melts the diode (e.g., by having a fixed resistor in series with it).

Diodes are important components in rectifier circuits, used to convert alternating current to direct current.

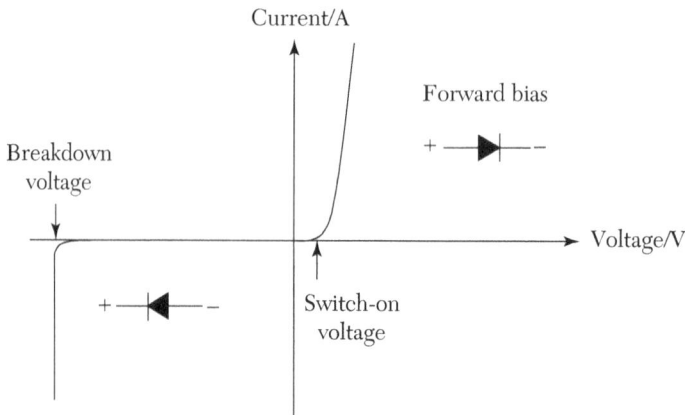

Current/A

Forward bias

Breakdown voltage

Voltage/V

Switch-on voltage

Some diodes emit light when they conduct. These are called light emitting diodes (LEDs). The circuit symbol for an LED is shown on the right.

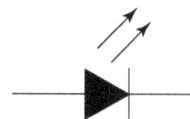

18.3.3   Resistors in Series and in Parallel

▣ Resistors in series

To replace several resistors in series by a single resistor R_S, we would need a resistor that has the same ratio of V to I as the set of series resistors. For the resistances to be equal, the total voltage across all the series resistors must equal the voltage across the single resistor. The same current I flows through all of the resistors, so:

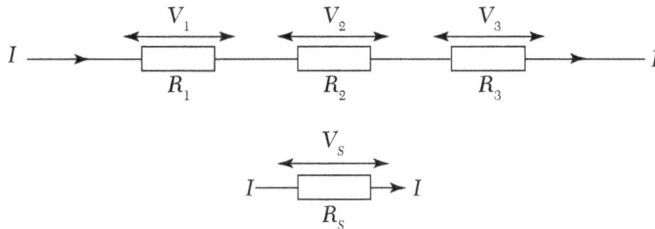

$$V_S = V_1 + V_2 + V_3$$
$$V_S = IR_S = IR_1 + IR_2 + IR_3$$
$$R_S = R_1 + R_2 + R_3$$

The total resistance of several resistors connected in series is the sum of the individual resistances. The general relationship for N resistors in series is:

$$R_{\text{series}} = \sum_{i=1}^{i=N} R_i$$

▣ Resistors in parallel

To replace several resistors in parallel by a single resistor R_P, we would need a resistor that has the same ratio of V to I as the set of parallel resistors. We start by using the fact that the same potential difference is across each arm of a parallel circuit.

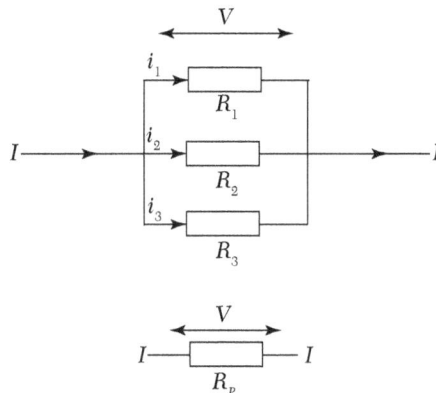

Using Kirchhoff's first law:

$$I = i_1 + i_2 + i_3$$

$$\frac{V}{R_P} = \frac{V}{R_1} + \frac{V}{R_2} + \frac{V}{R_3}$$

$$\frac{1}{R_P} = \frac{1}{R_1} + \frac{1}{R_2} + \frac{1}{R_3}$$

In general, the reciprocal of the equivalent resistance is equal to the sum of the reciprocals of all the resistances of the resistors connected in parallel. For N resistors in parallel this can be written more formally as:

$$\frac{1}{R_P} = \sum_{i=1}^{i=N} \frac{1}{R_i}$$

For the simple but very common situation where there are just two resistors in parallel, the equation can be rearranged to:

$$R_P = \frac{R_1 R_2}{(R_1 + R_2)}$$

This is easily remembered as "product over sum"; however, it only works for two resistors in parallel.

The formula for parallel resistors shows that the total resistance of several resistors connected in parallel is always *less* than the smallest resistance in the network.

18.3.4 Resistivity

The resistance of a component depends not only on the type of material from which it is made but also on the dimensions of the component. Resistivity is a property of the material alone and does not depend on its dimensions. For example, the resistances of copper wires of different lengths and cross-sectional areas differ, but the resistivity of the copper from which they are made is the same (at the same temperature).

For a cylindrical wire of length l, cross-sectional area A, and resistance R, the resistivity is given by:

$$\rho = \frac{RA}{l}$$

The S.I. unit of resistivity is Ω m.

Typical resistivities (at 20°C) for different conductors are shown below:

Metal	Resistivity
Silver	1.6×10^{-8} Ωm
Copper	1.7×10^{-8} Ωm
Aluminum	2.8×10^{-8} Ωm
Tungsten	5.8×10^{-8} Ωm
Platinum	1.1×10^{-7} Ωm
Constantan	4.9×10^{-7} Ωm
Steel	7.2×10^{-7} Ωm (varies depending on type of steel)
Nichrome	1.3×10^{-6} Ωm

The resistivity of a length of a resistance wire can be measured using the circuit below:

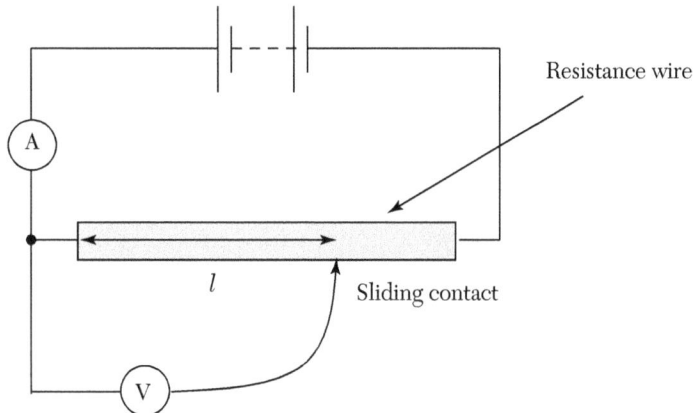

- There is a constant current I in the resistance wire. This is measured using the ammeter.

- The moving contact is used to measure the potential difference V for a range of lengths l.

- The diameter d of the wire is measured using a micrometer screw gauge. It is best to measure several different diameters at different points on the wire to obtain an average value. The cross-sectional area of the wire is ¼ πd^2.

- The length of wire l can be measured using a meter ruler.

The resistivity is then found from a graph of V against l:

$$V = IR = \left(\frac{4I\rho}{\pi d^2}\right)l$$

The gradient of the graph of V against l is equal to $\left(\frac{4I\rho}{\pi d^2}\right)$.

$$\rho = \left(\frac{\pi d^2}{4I}\right) \times \text{gradient}$$

18.4   Electrical Energy and Power

When a charge Q moves between points at different potential, energy is transferred to the charge from the electric field or vice versa. The energy transfer per unit charge is equal to the potential difference V, so the total energy transfer E is:

$$E = QV$$

The rate of transfer of energy is the electrical power P:

$$P = \frac{dE}{dt}$$

The S.I. unit for power is W.

For charge moving between points with a constant potential difference:

$$P = V\frac{dQ}{dt} \quad \text{or} \quad P = VI$$

Resistors transfer electrical energy to thermal energy. The power transfer in a resistor is given in various forms by using the equations $P = VI$ and $V = IR$:

$$P = VI = \frac{V^2}{R} = I^2R$$

If the current in the resistor is constant, then the energy transfer E in time t is given by:

$$E = VIt$$

18.4.1   EMF and Internal Resistance of a Real Cell

While an ideal cell is simply a source of constant emf, a real cell has an internal resistance. This arises because work must be done to move charge carriers through the material of the cell itself. A real cell is modeled as a constant source of emf E in series with a fixed internal resistance r.

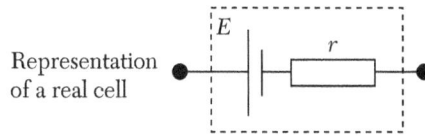

Representation of a real cell

When the cell is connected to an external circuit, current flows through it, and there is a voltage drop across the internal resistance. This results in the potential difference at the cell terminals, the terminal p.d. V, falling below the emf of the cell.

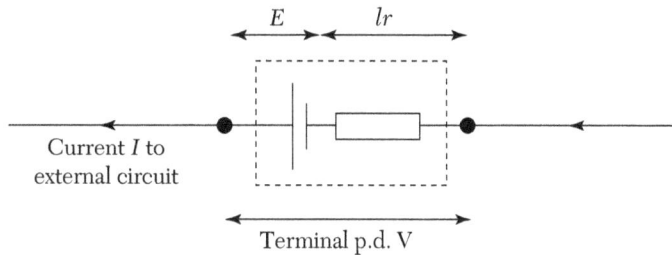

Current I to external circuit

Terminal p.d. V

The potential difference across the internal resistance is equal to the work that must be done per coulomb of charge to move charge carriers through the cell. The terminal p.d. is therefore equal to the difference between the cell emf and these "lost volts":

$$V = H - Ir$$

As the current drawn from the supply increases, the terminal p.d. falls.

The internal resistance also limits the maximum current that can be drawn from the cell. This occurs when the cell is short-circuited by connecting its terminals together with a conductor of negligible resistance. The only resistance in the circuit is the internal resistance, so the short-circuit (maximum) current from the cell is:

$$I_{SC} = \frac{H}{r}$$

The internal resistance of a typical 1.5 V alkaline AA cell is about 0.15 Ω, so the maximum current that could be drawn from it is 1.5/0.15 = 10 A. However, shorting the cell would drain it very quickly, so the working current for a device operated by AA cells must be much less than this. If a cell or battery is required to provide a very large current, it must have a very low internal resistance. Car batteries have an emf of 12 V but need to provide over 100 A when the ignition is switched on and the battery turns the starter motor. Typical car batteries have an internal resistance < 0.01 Ω.

18.4.2 Measuring the Internal Resistance and EMF of a Cell

The emf and internal resistance of a cell or battery can be determined using the circuit below.

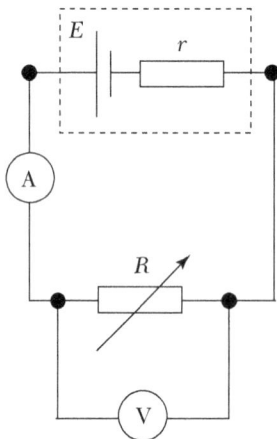

As the value of the load resistor is changed, the current I in the circuit and the terminal p.d. V change too. I and V are related by the equation:

$$V = H - Ir$$

comparing this to: $y = mx + c$

A graph of V against I is a straight line with a negative gradient and a positive intercept:

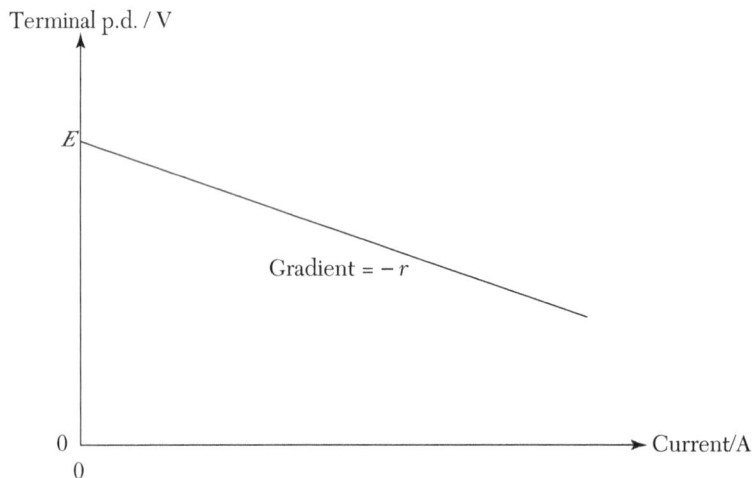

- intercept on y-axis = emf

- gradient = $-r$

The terminal p.d. is equal to the cell emf when no current is drawn, so a simple way to measure the emf is to connect a voltmeter with very high resistance across the cell terminals. The open-circuit terminal voltage is equal to the emf of the cell.

18.4.3 Power Transfer From a Real Cell to a Load Resistor

The power transferred to the load resistor in the circuit below is $P = I^2R$, but I depends on the load resistance R. How does P depend on R?

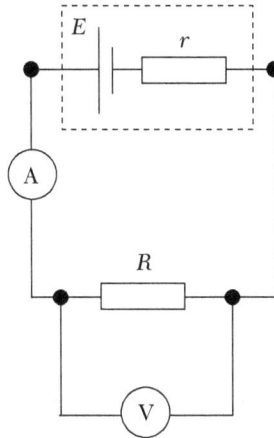

$$I = \frac{H}{(R+r)}$$

$$P = I^2R = \frac{H^2R}{(R+r)^2}$$

It is clear that P is 0 when $R = 0$.

P also becomes asymptotic to 0 as $R \to \infty$ since the denominator then dominates the expression.

The expression is positive for all values of R, so there must be a maximum value for P at some point. This occurs when $R = r$, i.e., the load resistor has a resistance equal to the internal resistance.

(The fact that maximum power transfer occurs when $R = r$ can be demonstrated using calculus. You need to find the condition for $dP/dR = 0$; this must be a stationary value of the function and, in this case, it is a maximum.)

Power transfer P

P_{max}

0

0          $R = r$ Load resistance R

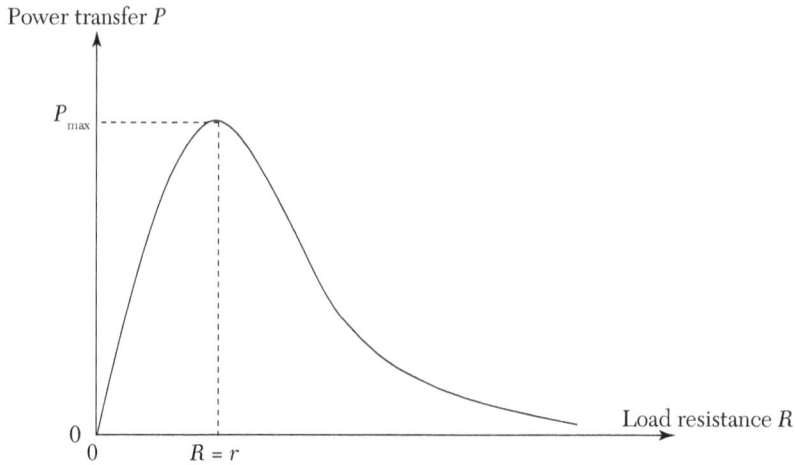

It might be tempting to think that maximum power transfer also corresponds to maximum efficiency for the system, but this is not the case.

$$\text{efficiency} = \frac{\text{power transferred to } R}{\text{power transferred from the emf}} \times 100\%$$

Power transferred to R is: $P_{out} = I^2 R = \dfrac{E^2 R}{(R+r)^2}$

Power transferred from the emf is:

$$P_{in} = EI = \frac{E^2}{(R+r)}$$

Efficiency is given by $\dfrac{P_{out}}{P_{in}} \times 100\% = \dfrac{R}{(R+r)} \times 100\%$

Surprisingly the maximum efficiency would be when $R \rightarrow \infty$, but this would also correspond to zero power transfer. At maximum power transfer, the efficiency is 50%.

18.5   Resistance Networks

Potential dividers

A potential divider is used to provide an output voltage V_{out} that is a fraction of the supply voltage V_{in}.

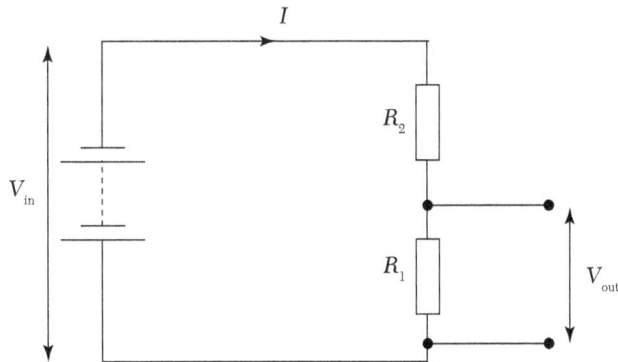

The current I in the circuit is the same through both resistors.

Using Kirchhoff's second law:

$$V_{in} = IR_1 + IR_2$$

$$V_{out} = IR_1$$

$$\frac{V_{out}}{V_{in}} = \frac{R_1}{(R_1 + R_2)}$$

If V_{out} is connected across a load resistor R_L then the fraction changes because the lower part of the potential divider now has two resistors in parallel $(R_1$ and $R_L)$ so that:

$$\frac{V_{out}}{V_{in}} = \frac{R_P}{(R_P + R_2)} \text{ where } R_P = \frac{R_1 R_L}{(R_1 + R_L)}$$

Potential divider circuits can also be used as sensing circuits, producing an output voltage that depends on external conditions such as temperature or light intensity.

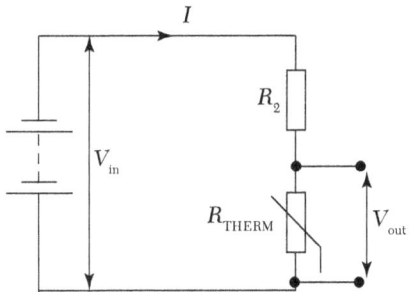

Temperature sensor: As temperature increases R_{THERM} falls so V_{out} falls. Swapping positions of R_2 and R_{THERM} makes R_{out} rise with temperature

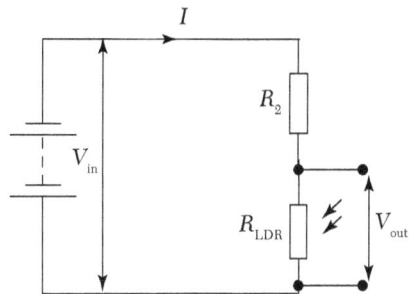

Light intensity sensor: As light intensity increases R_{LDR} falls so V_{out} falls. Swapping positions of R_2 and R_{LDR} makes R_{out} rise with light intensity

Using Kirchhoff's laws to solve resistance networks

Kirchhoff's two-circuit laws can be used to produce a set of simultaneous equations for currents and p.d.'s that can be used to find all the currents and p.d.'s in a network. Here is an example:

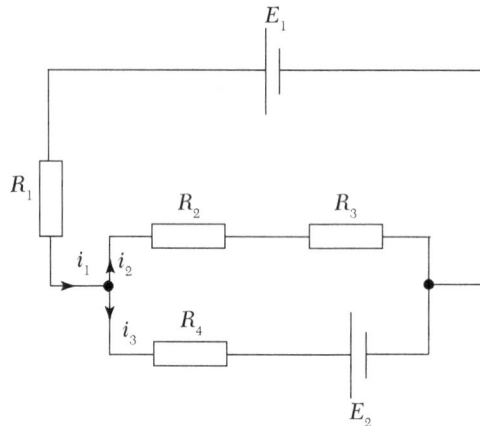

Applying Kirchhoff's first law,

$$i_1 = i_2 + i_3$$

Applying Kirchhoff's second law to the loop passing through resistors R_1, R_2, and R_3:

$$E_1 = i_1 R_1 + i_2 R_2 + i_2 R_3$$

Applying Kirchhoff's second law to the loop passing through resistors R_1, R_4, and E_2:

$$E_1 = i_1 R_1 + i_3 R_4 - E_2$$

There are three independent equations, so it is possible to solve for three unknowns. For example, if we know the two emfs and the values of all three resistors, we can find the current at any point in the circuit.

18.6   Semiconductors and Superconductors

18.6.1   Semiconductors

Semiconductors have a resistivity that lies between that of a conductor, such as copper, and an insulator, such as polythene. At room temperature the charge carrier density for a pure (intrinsic) semiconductor is typically 10^{13}–10^{16} m^{-3} compared to 10^{27}–10^{28} for a metal. However, the charge carrier density in a semiconductor is highly temperature-dependent and can be changed by the addition of impurity atoms (this is called "doping").

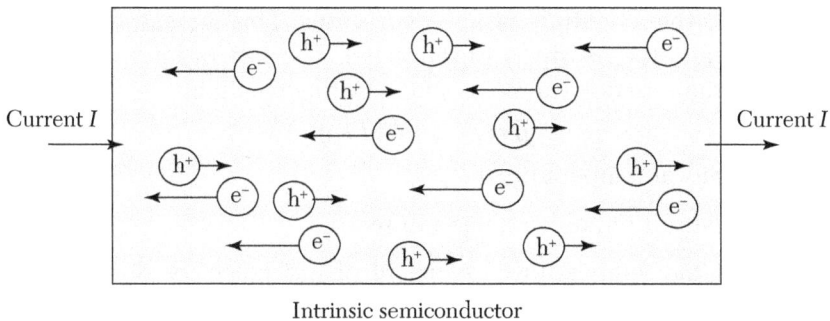

Intrinsic semiconductor

At absolute zero, semiconductors are pure insulators with no free charge carriers. However, as temperature increases, thermal agitation frees a small proportion of the valence electrons and allows them to become "free" or conduction electrons. Whenever an electron is freed, it leaves behind a "hole," which behaves very much like a particle carrying a positive charge +e. This means that a pure semiconductor contains equal numbers of electrons and holes and that both contribute to current flow. While the density of holes and electrons is equal, they do not contribute equally to the current because their mobility (ability to move through the material) differs.

The electrical properties of intrinsic semiconductors can be changed by introducing impurity atoms. Semiconductors such as silicon and germanium have four electrons in the outer shell, so adding atoms of an element such as boron, aluminum, or gallium, with three outer electrons, has the result of donating one additional hole per doping atom. Semiconductors doped in this way have an excess of holes (positive charge carriers) and are called **p-type semiconductors**. If phosphorus, antimony, or arsenic, with five electrons in the outer shell, is added, this will donate one electron per doping atom. Semiconductors doped in this way have an excess of electrons (negative charge carriers) and are called **n-type semiconductors**. Semiconductor devices such as diodes, transistors, and integrated circuits depend on doped semiconductors and are often constructed from junctions between n-type and p-type materials.

18.6.2 Variation of Resistance of a Metal with Temperature

When an electric current passes through a metal, electrons drift in the opposite direction and collide with ions in the metal structure. This scatters the electrons and dissipates energy, creating electrical resistance. The higher the temperature, the more random thermal energy in the metal and the greater the amplitude of ionic vibrations. This increases the electron scattering and the resistivity of the metal. The rate of increase is roughly linear, so the variation can be modeled by the equation:

$$\rho = \rho_0 (1 + \alpha(TT_0))$$

ρ = resistivity at temperature T

ρ_0 = resistivity at reference temperature T_0

Values for the temperature coefficient of resistance for different metals based on $T_0 = 293$ K are shown below:

Metal	Temperature coefficient of resistance
Aluminum	4.3×10^{-3} K^{-1}
Copper	4.0×10^{-3} K^{-1}
Silver	4.0×10^{-3} K^{-1}
Tungsten	4.5×10^{-3} K^{-1}

It might be expected that this linear dependence of resistivity on temperature would continue down to absolute zero and would approach zero at that value. In 1908 the Dutch physicist Kamerlingh Onnes investigated how the resistivity of mercury varied as he lowered the temperature toward absolute zero. At first the resistivity dropped approximately linearly, but at 4.2 K it suddenly dropped to zero. The mercury conducted electric currents *with zero resistance*—it had become a **superconductor**. Other metals exhibit similar properties, having a characteristic **transition temperature** below which they superconduct. Here are the transition temperatures for several different metals:

Metal	Transition temperature
Aluminum	1.2 K
Lead	7.2 K
Mercury	4.2 K
Tin	3.7 K
Zinc	0.9 K

These are called low-temperature superconductors. In the 1980s, several high-temperature superconductors were discovered. These tend to be ceramic materials with complex crystalline structures, such as yttrium barium copper chloride, but they become superconducting above the temperature of liquid nitrogen (77 K).

Superconductors are essential for the construction of extremely strong electromagnets (e.g., in particle accelerators such as the LHC at CERN) because huge currents can circulate in superconducting coils without generating any heat.

Physicists hope one day to discover or manufacture a room-temperature superconductor. Among other applications this would allow the transmission of electrical energy with no energy losses from ohmic heating.

18.7 Exercises

1. Explain in terms of the free electron model of a metal:

(a) how metals conduct electricity

(b) why the resistance of a metal increases with temperature

2. A copper wire has radius 0.20 mm and carries a D.C. current of 120 mA. Calculate the drift velocity of the electrons in the wire. The charge carrier density in copper is $n = 8.5 \times 10^{28}$ m^{-3}.

3. The resistivity of a sample of germanium is 6.0×10^{-2} Ωm. A cylinder of germanium 5.0 mm long and 0.50 mm in radius has a p.d. of 10 V applied across its ends.

(a) What current passes through the specimen?

(b) What is the current density (I/A) in the sample?

(c) What is the conductivity ($1/\rho$) of germanium?

(d) What is the conductance ($1/R$) of this sample?

(e) The density of charge carriers in germanium is about 10^{16} m^{-3}; calculate their average drift velocity (assume the charge on each charge carrier is 1.6×10^{-19} C).

(f) Name the majority charge carriers in each of the following:

(i) a metal

(ii) a p-type semiconductor

(iii) an n-type semiconductor

4. A is a 240 V 100 W filament lamp. B is a 12 V 6 W filament lamp.

(a) What happens when they are connected in parallel across a 12 V D.C. supply?

(b) What happens when they are connected in series across a 240 V A.C. supply?

5. A cell of emf E and internal resistance r is connected across a load resistor R, and the p.d. across R is measured using a high-resistance voltmeter.

(a) Draw a circuit diagram of this arrangement.

(b) Derive an expression for V in terms of E, R, and r.

(c) A set of values for V are obtained for a wide range of values of R. Rearrange the equation derived in (a) to show how a straight line graph can be drawn from this data and used to determine E and r.

(d) Sketch the graph, indicating its important features.

6. The table below gives the resistivities of three metals at two different temperatures.

Metal	Resistivity at 273K	Resistivity at 373K
Aluminum	2.45×10^{-8} Ωm	3.55×10^{-8} Ωm
Copper	1.55×10^{-8} Ωm	2.38×10^{-8} Ωm
Iron	8.70×10^{-8} Ωm	16.61×10^{-8} Ωm

(a) Which is the best conductor?

(b) An insulated cable contains a single metal core of diameter 0.50 mm. What is the resistance of 5.0 m of this cable at 273K if it is made of: (i) aluminum, (ii) copper, (iii) iron.

(c) By what percentage does the resistance of each cable increase when its temperature rises from 273 to 373K?

(d) Estimate the resistance of the copper cable at room temperature (about 293 K) and state any assumptions you had to make.

(e) A 5.0 m composite wire is made by joining 1.0 m of copper wire of diameter 0.50 mm to 4.0 m of iron wire of diameter 0.60 mm. What current would flow through this wire at 273 K if a cell of emf 6.0 V was connected across its ends?

7. (a) Calculate the readings on the ammeter and voltmeter in the circuit below assuming that the internal resistance of the battery is negligible.

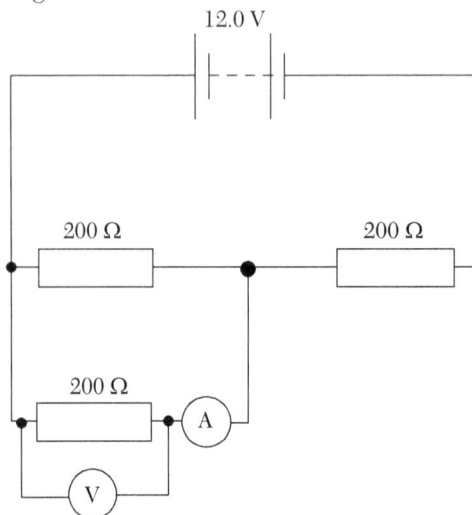

(b) In fact the internal resistance is 10 Ω. How does this affect your answers to part (a)?

8. You are provided with three identical resistors each of resistance R. Calculate and list the values of all the resistances you can make using one or more of these resistors stating which arrangement corresponds to which value.

9. A resistance network is constructed using 12 identical 1 Ω resistors, with each resistor placed so that it forms one edge of a cube. Calculate the resistance between two opposite corners of the cubic array.

10. The circuit below is used to detect changes in light intensity. The resistance of the LDR is 50 kΩ in the dark and 100 Ω in bright light.

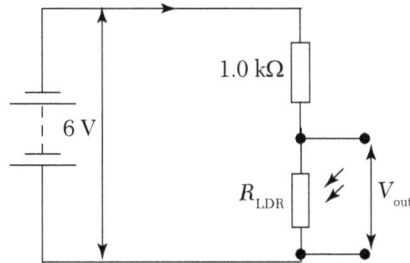

(a) Calculate the output voltage in the dark and in bright light.
(b) How does the value of the top-fixed resistor affect the range of the output voltages?

11. Use Kirchhoff's laws to determine the readings on the three ammeters and the voltmeter in the circuit below. Assume that the cell's internal resistance can be neglected.

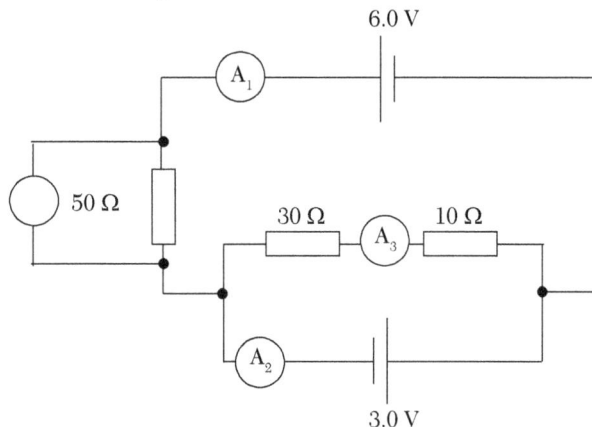

CHAPTER 19

CAPACITANCE

19.1   What is a Capacitor?

A capacitor consists of two conductors separated by an insulator. This might be two parallel metal plates with air between them or it could be a person standing on an insulator that separates them from the earth. When there is a potential difference between the plates, opposite charges gather on each plate and the capacitor is said to be charged. Capacitors are common components present in most electrical circuits.

The image shows a range of capacitors. The cylindrical capacitors are electrolytic capacitors that must be connected in a particular direction in the circuit (otherwise they might explode!), while the others are ceramic or paper capacitors. The insulator or dielectric material between the plates affects the amount of charge that can be stored on the plates.

19.1.1 Capacitors and Charge

The circuit below can be used to investigate what happens when a capacitor is charged.

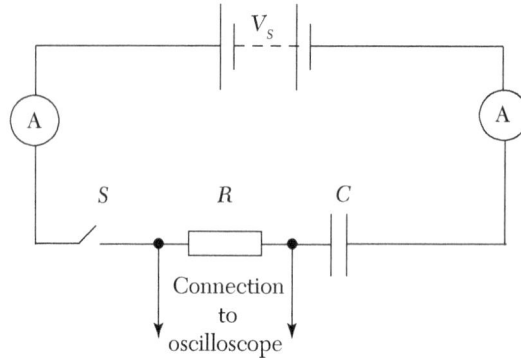

When switch S is closed, both ammeters jump to a positive reading that gradually falls back to zero. This shows that charge has moved around the circuit; however, it cannot cross the gap between the two plates, so what has happened is electrons have left one plate (giving it a net positive charge) and moved onto the other plate (giving it a net negative charge). The current in the circuit is monitored by connecting an oscilloscope across the resistor in series with the capacitor.

The oscilloscope trace actually displays the voltage across the resistor, but this is directly proportional to the current because the resistor is constant. The current decays exponentially to zero as the capacitor charges up. Once fully charged there is a charge of $+Q$ on one plate and ^-Q on the other plate. While it is true that the net charge on the capacitor is actually zero, we refer to the charge on the positive plate and say that the capacitor is "charged" or "stores a charge Q."

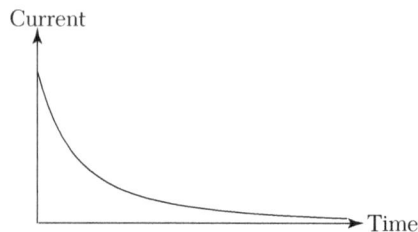

19.1.2 Capacitance

The charge stored on a capacitor is directly proportional to the potential difference between the two conductors. This can be verified using the arrangement shown below:

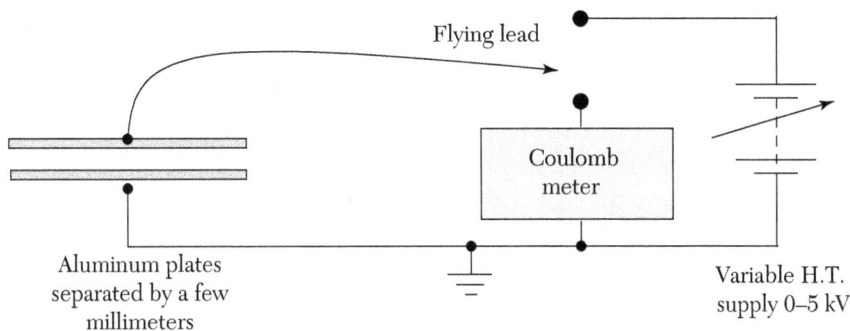

- Connect the flying lead momentarily to the positive terminal of the H.T. supply.

- Zero the coulomb meter.

- Move the flying lead to the terminal of the coulomb meter and measure the charge transfer.

- Increase the voltage and repeat.

Typical results would look like this:

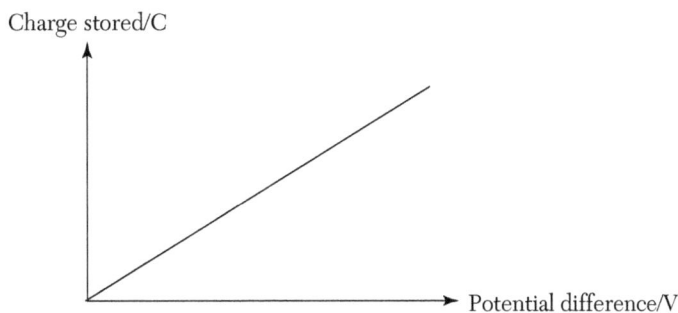

Charge stored on the capacitor is directly proportional to the p.d. between its plates:

$$Q \propto V$$
$$Q = CV$$

Where C is the capacitance.

The S.I. unit of capacitance is F; $1F = 1CV^{-1}$. A 1F capacitor would store 1 C of charge per volt of potential difference between its plates. This would be a very large capacitance; typical values used in electronic circuits are usually measured in nF or μF.

19.1.3   Energy Stored on a Charged Capacitor

When a capacitor is charged, the power supply forces charges onto the plates against the electrostatic repulsion of the charges that are already there. This increases the electric potential energy in an analogous way to how the compression of a spring stores strain energy. The charged capacitor is therefore a store of electrical energy that can be released if it is allowed to discharge through a suitable circuit.

When a capacitor with initial charge Q is discharged through a fixed resistor of resistance R, the work done by the capacitor in moving a charge δq around the circuit is equal to $V\delta q$, where V is the potential difference across the capacitor as the charge moves around the circuit. The total work done will be the area under a graph of V against Q or the integral of $V\delta q$.

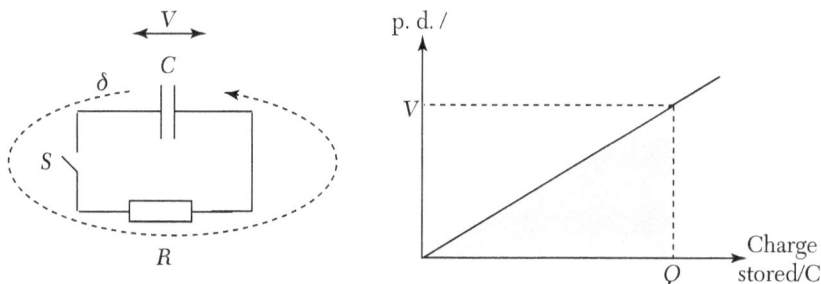

Energy stored = area up to charge Q

$$E = \tfrac{1}{2}\,QV$$

(It might seem strange that the graph shows charge increasing to the right when the capacitor is discharging—remember that the x-axis does not represent time!)

Using calculus,

$$E = \int_{q=0}^{q=Q} V dq = \int_{q=0}^{q=Q} \frac{q dq}{C} = \frac{Q^2}{2C}$$

The two expressions for E are equivalent since $C = Q/V$. There is also a third form of this expression in terms of just V and C (found by using $Q = CV$ to eliminate Q):

$$E = \tfrac{1}{2}\,CV^2$$

Here are the three equations for energy stored on a capacitor:

$$E = \frac{Q^2}{2C} = \frac{CV^2}{2} = \frac{QV}{2}$$

One of the key uses for a capacitor is as a temporary store of electrical energy that can be used when the capacitor is discharged—e.g., in a camera flashlight. One of the advantages of using capacitors as energy storage devices is that the energy can be accumulated slowly and released quickly so that the power delivered can be much greater than the power used to charge the capacitor.

19.1.4   Efficiency of Charging a Capacitor

When a capacitor is charged from a supply, the work done by the supply is always greater than the energy stored on the capacitor. This is because some of the work done by the supply is transferred to heat by resistance in the circuit.

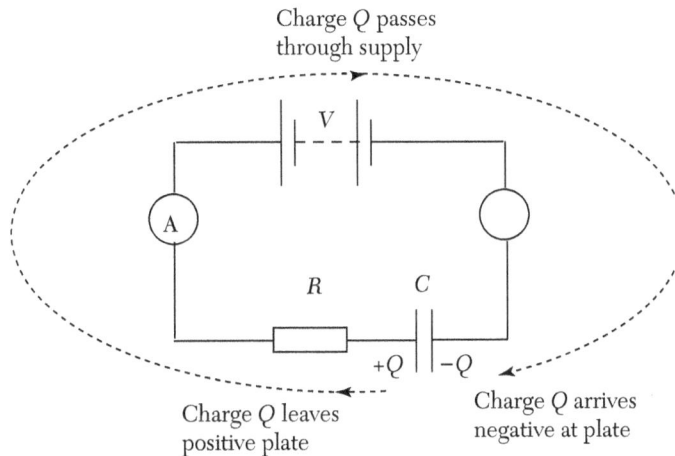

Charge Q passes through supply

Charge Q leaves positive plate

Charge Q arrives negative at plate

Work done by supply, $W = QV$

Energy stored on capacitor, $E = \frac{1}{2} QV$

Energy dissipated in resistor, $E = \frac{1}{2} QV$

The charging process has an efficiency of 50% regardless of the resistance.

19.2   Parallel Plate Capacitor

A simple capacitor can be constructed from two metal plates of area A separated by a distance d in air. A uniform electric field (ignoring edge effects) of strength E is set up between the plates when a p.d. V is placed across them.

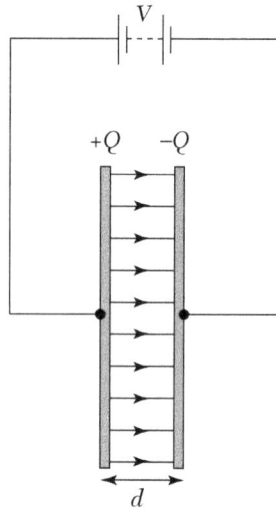

The electric field strength is related to the charge density by the equation (see Section 17.4.4):

$$E = \frac{\sigma}{\varepsilon_0} = \frac{Q}{A\varepsilon_0}$$

Where σ is the charge density on one plate.

The electric field strength is also equal to the (negative) potential gradient:

$$E = (-)\frac{V}{d}$$

Equating the two expressions for E (ignoring signs):

$$\frac{Q}{A\varepsilon_0} = \frac{V}{d}$$

The capacitance of a parallel plate air capacitor is, therefore,

$$C = \frac{Q}{V} = \frac{\varepsilon_0 A}{d}$$

And if the air gap is replaced by a dielectric material of relative dielectric constant ε_r, the capacitance is given by:

$$C = \frac{\varepsilon_0 \varepsilon_r A}{d}$$

To create large capacitance, we need a high dielectric constant, large area, and small separation.

19.3   Capacitor Charging and Discharging

In order to find a mathematical description for the variation of charge, current, and voltage during charging and discharging, we must start with a differential equation for the rate of flow on or off of the capacitor when it already stores a charge q and has a potential difference V across its plates. It is simpler to set this up for discharge than for charging, so we will begin with the equations for the discharge of a capacitor through a fixed resistor.

19.3.1   Equations for Capacitor Discharge

A capacitor of capacitance C is connected in parallel with a fixed resistor of resistance R. The capacitor has an initial charge Q_0 and an initial potential difference V_0. When the switch S is closed, the voltage across the capacitor and the resistor are equal, so current flows through the resistor and the capacitor gradually discharges. t seconds after S has closed, the charge has fallen to Q and the voltage across the plates is V. At this instance the discharge current is:

$$I = -\frac{dQ}{dt} = \frac{V}{R}$$

The negative sign indicates that the charge on the capacitor is falling.

However, $V = Q/C$, so we can form a first-order differential equation that can be solved by separation of variables:

$$\frac{dQ}{dt} = -\frac{Q}{RC}$$

$$\int_{Q=Q_0}^{Q(t)} \frac{dQ}{Q} = -\int_{t=0}^{t} \frac{dt}{RC}$$

$$\ln\left(\frac{Q(t)}{Q}\right) = -\frac{t}{RC}$$

$$Q(t) = Q_0 e^{-\frac{t}{RC}}$$

The charge on the capacitor decays exponentially. Since $V = Q/C$ and $I = V/R = Q/RC$, the current and voltage are both directly proportional to the charge and decay at the same rate. The term $e^{-\frac{t}{RC}}$ is equal to the fraction of charge remaining after time t.

The graph below shows how the charge, current, or p.d. across the capacitor changes as it discharges:

Percentage of Q_0, I_0 or V_0

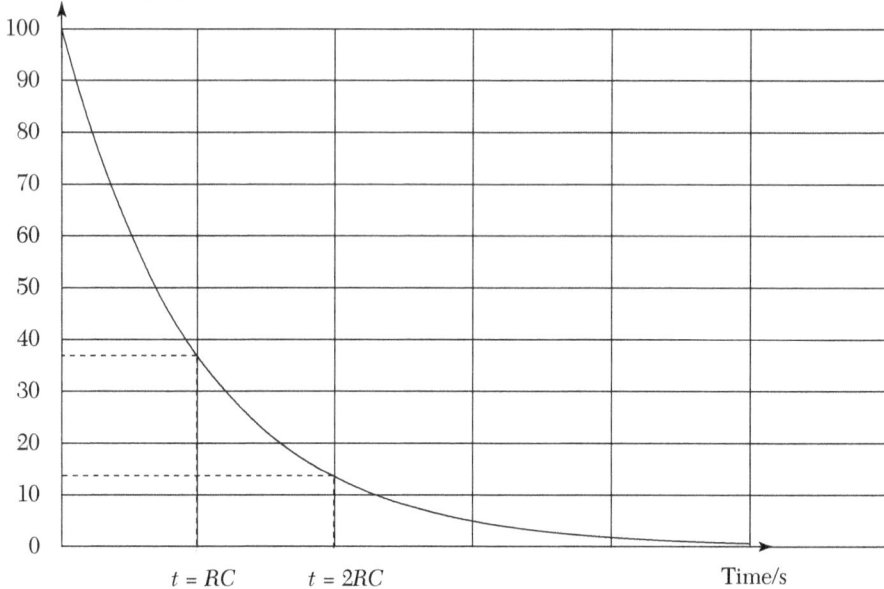

The quantity RC is called the **time constant** of the circuit. The dimensions of RC are the dimensions of time, so in S.I. units it is time in seconds. After one time constant $(t = RC)$, the fraction remaining is $e^{-\frac{RC}{RC}} = e^{-1} \approx 0.37$, so there will be 37% of the initial charge remaining after one time constant. After n time constants the fraction remaining is e^{-n}. This falls rapidly with n. After three time constants, there is 0.050 of the original charge, so the discharge is 95% complete. After five time constants, 0.0067 of the original charge remains, so the discharge is >99% completed. As a rule of thumb, discharging and charging processes are considered complete when a time $t = 5\,RC$ (five time constants) has elapsed.

The time constant controls the rate of charging and discharging.

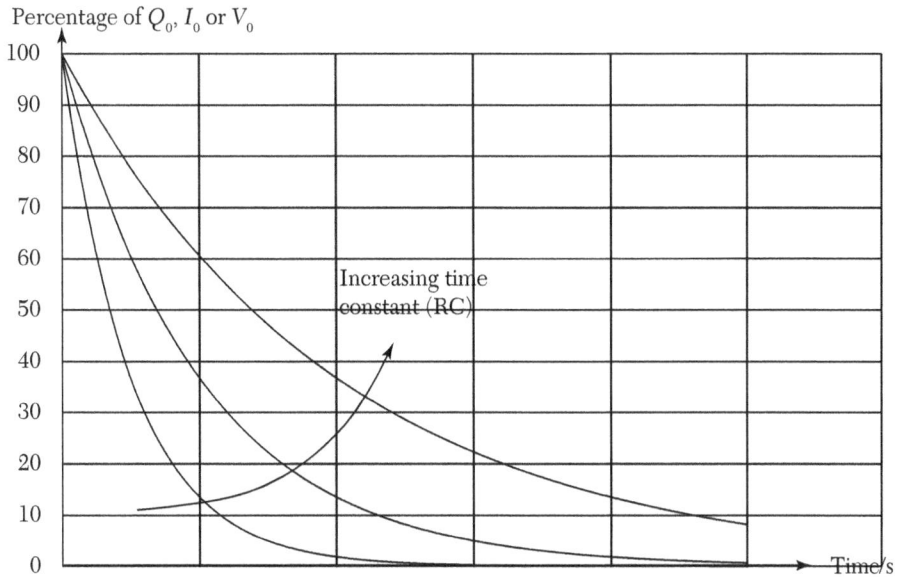

Percentage of Q_0, I_0 or V_0 vs Time/s, with curves showing Increasing time constant (RC).

19.3.2 Equations for Capacitor Charging

The diagram below shows a circuit used to charge a capacitor.

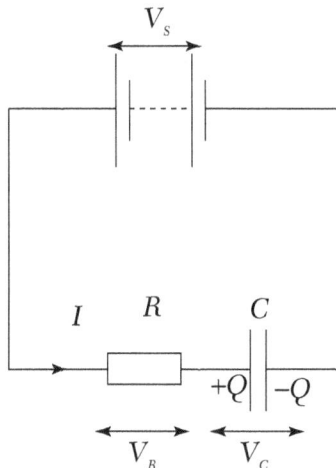

We can use Kirchhoff's second law to set up a differential equation for the charge on the capacitor:

$$V_S = V_R + V_C$$

$$V_S = IR + \frac{Q}{C}$$

$$V_S = R\frac{dQ}{dt} + \frac{Q}{C}$$

$$\frac{dQ}{dt} = \frac{1}{RC}(CV_S - Q)$$

$$\frac{dQ}{dt} = \frac{1}{RC}(Q_F - Q)$$

Where $Q_F = CV_S$ is the final charge when the voltage across the capacitor is equal to the supply voltage.

This is a first-order linear differential equation whose solution is:

$$Q = Q_F\left(1 - e^{-\frac{t}{RC}}\right)$$

The difference between Q and Q_F decays exponentially toward zero.

The voltage across the capacitor is directly proportional to the charge stored ($V = Q/C$), so the voltage rises in a similar way to the charge:

$$V = V_S\left(1 - e^{-\frac{t}{RC}}\right)$$

Current is the rate of change of charge, so the charging current decays exponentially from an initial value $I_0 = V_S R$

$$I = I_0 e^{-\frac{t}{RC}}$$

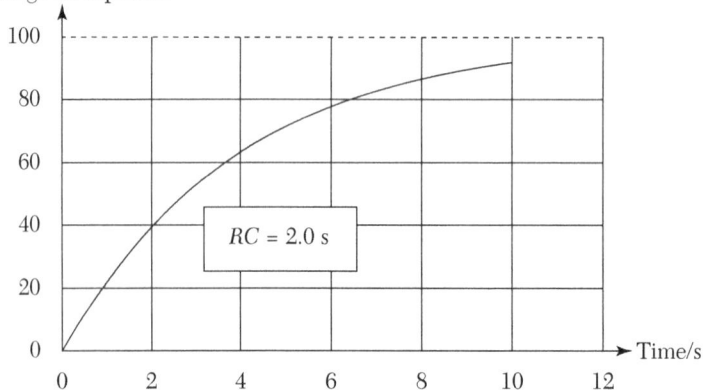

19.4 Capacitors in Series and Parallel

When capacitors are combined in series or in parallel, we can calculate the effective capacitance of the arrangement, i.e., the value of the single capacitor that could replace them.

19.4.1 Capacitance of Capacitors in Series

When several capacitors are connected in series, the wires connecting the inside plates together are isolated from the external supply. If the plate at one end of such a wire gains a positive charge, then the plate connected to the other end must gain an equal negative charge. As a consequence, all of the series capacitors must have a charge equal to that on the outermost plates that are connected to the external supply.

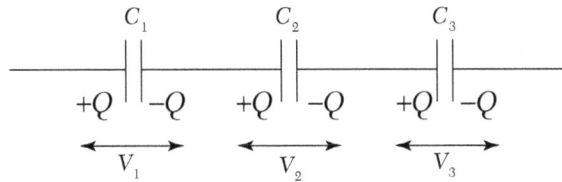

The total charge on the set of series capacitors is therefore Q, and the voltage across the set is equal to the sum of voltages across the individual capacitors.

$$V = V_1 + V_2 + V_3$$

$$\frac{Q}{C_{series}} = \frac{Q}{C_1} + \frac{Q}{C_2} + \frac{Q}{C_3}$$

$$\frac{1}{C_{series}} = \frac{1}{C_1} + \frac{1}{C_2} + \frac{1}{C_3}$$

For n capacitors in series,

$$\frac{1}{C_{series}} = \sum_{i=1}^{i=n} \frac{1}{C_i}$$

This has the same form as the equation for resistors *in parallel*, but here it applies when the capacitors are *in series*.

19.4.2 Capacitors in Parallel

When capacitors are connected in parallel, they all have the same voltage across them, and the total charge stored is the sum of the charges stored on each capacitor.

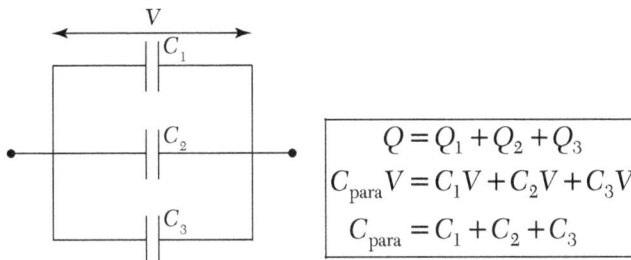

$$Q = Q_1 + Q_2 + Q_3$$
$$C_{para}V = C_1V + C_2V + C_3V$$
$$C_{para} = C_1 + C_2 + C_3$$

The total capacitance of several capacitors connected in parallel is the sum of their individual capacitances. For n capacitors in parallel,

$$C_{para} = \sum_{i=1}^{i=n} C_i$$

This has the same form as the equation for resistors *in series*, but here it applies when the capacitors are *in parallel*.

19.5   Capacitance of a Charged Sphere

When a charge Q is placed on a spherical conductor of radius a, it spreads evenly over the surface. The potential of the surface is given by:

$$V = \frac{Q}{4\pi\varepsilon_0 a}$$

So the capacitance of the sphere with respect to earth is

$$C_{sphere} = \frac{Q}{V} = 4\pi\varepsilon_0 a$$

19.6   Exercises

1. A $220\,\mu F$ capacitor is charged from a 6 V D.C. supply.
   (a) Calculate the charge stored on the capacitor.
   (b) Calculate the energy stored on the capacitor.
   (c) By what factors would the charge and energy stored on the capacitor change if the supply voltage was doubled 12 V?

2. A $470\,\mu F$ capacitor is connected in series with a $220\,\Omega$ resistor and charged from a 10 V D.C. supply.
   (a) Calculate the time constant for this circuit.
   (b) Roughly how long will it take to charge the capacitor?
   (c) How would your answer to (b) change if the capacitor was charged to 20 V? Explain your answer.
   (d) Once the capacitor is charged up (to 10 V), it is disconnected and discharged through a different resistor. If the discharge current falls to 50% of its original value in 10 s, what is the value of this resistor and what was the original value of the discharge rent?

3. When S in the circuit below is closed the capacitor charges.

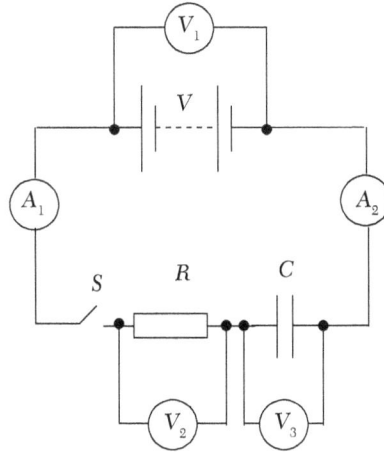

(a) Sketch a graph to show how the readings on ammeters A_1 and A_2 vary from the moment S is closed to the time that the capacitor is almost fully charged.

(b) Sketch a graph with three lines on it to show how the readings on voltmeters V_1, V_2, and V_3 vary from the moment S is closed to the time that the capacitor is almost fully charged.

4. The circuit below shows two capacitors that can be connected to a supply and/or to each other by closing one or both of two switches.

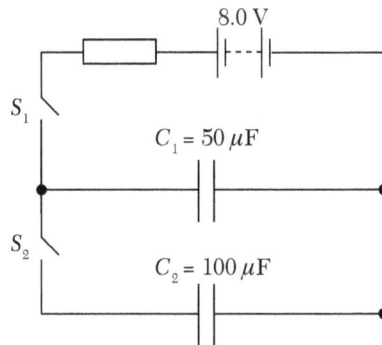

Initially both switches are open and both capacitors are discharged. S_1 is now closed and C_1 charges.

(a) Calculate the charge and energy stored on C_1.

S_1 is now opened and then S_2 is closed connecting the two capacitors together.

(b) Calculate the charge on each capacitor once they reach equilibrium.

(c) Calculate the total energy stored on both capacitors and compare this to the value you obtained in (a). Account for the difference. Now consider a different scenario. Both capacitors are discharged and then:

S_1 is now closed and C_1 charges. S_1 is kept closed and S_2 is also closed connecting the two capacitors together while still connected to the power supply.

(d) Calculate the charge on each capacitor once they reach equilibrium.

(e) Calculate the total energy stored on both capacitors and compare this to the value you obtained in (a). Account for the difference.

5. You are provided with three identical capacitors each of capacitance C. Calculate and list the values of all the capacitances you can make using one or more of these capacitors. State which arrangement corresponds to which value.

6. The diagram below shows a parallel plate air capacitor of area A and plate separation d connected to a power supply and a resistor.

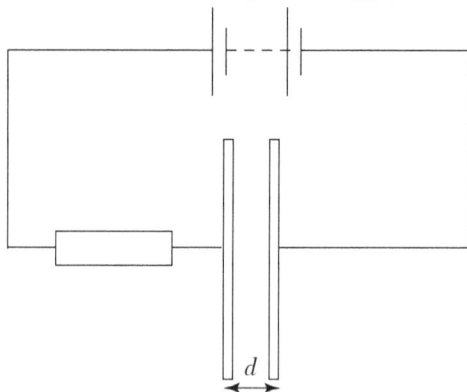

The plates are pulled apart by an external agent and their separation is doubled. Describe and explain what happens:

▪ to the charge stored on the capacitor

▪ to current flow in the circuit

▪ to the energy stored on the capacitor

▪ to the voltage across the capacitor

7. A 470 μF capacitor is charged to 10 V and then discharged through a 2200 Ω resistor. Calculate the charge and voltage remaining on the capacitor after 1.5 s.

20

MAGNETIC FIELDS

20.0   Magnetic Field

If a flat card is placed on top of a bar magnet and then iron filings are scattered on the card, they form a distinct pattern as shown in the image.

Each filing is like a small rod of iron that lines up with an invisible force field. If a magnetic compass is moved nearby, it too lines up with this force field and can be used to trace its pattern.

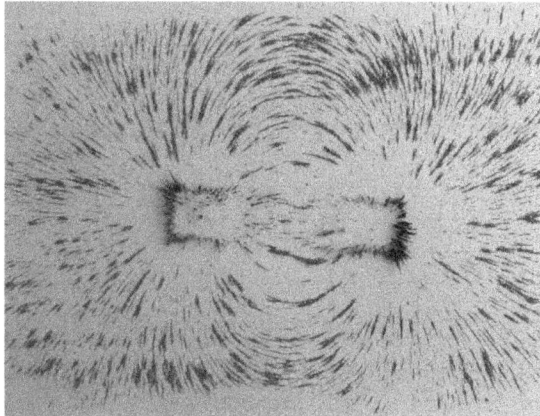

Michael Faraday explained these effects by saying that a bar magnet has two poles, north and south, and that these create a magnetic field in the space surrounding the magnet. The lines of magnetic field begin on north poles and end on south poles; they cannot start or end in space (although they can form closed loops). The direction of the field lines is the direction of force that would be exerted on a free north pole (even though free north poles do not exist in nature). This is all very similar to the description of the electric field and the link between electric field lines and electric charges.

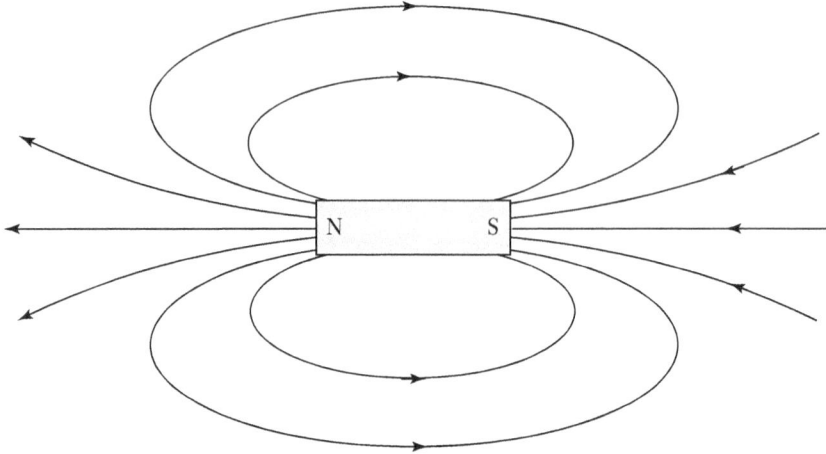

The behavior of a compass needle can be explained by considering the forces acting on its poles from the magnetic field. If the needle is not aligned with the field, then two forces of equal magnitude act along different parallel lines creating a couple. The couple is zero when the needle lines up the field.

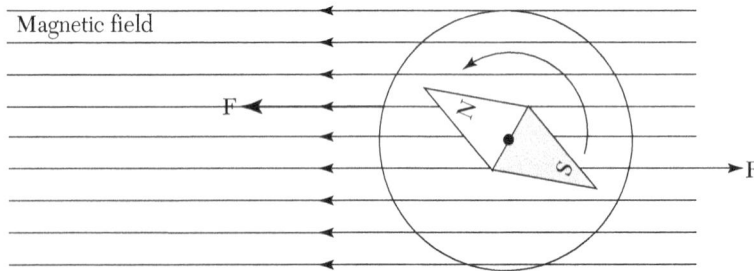

The Earth has its own magnetic field. The north pole of a compass points toward the Earth's North Pole. This implies that the geographic North Pole of the Earth is actually a south magnetic pole. Sometimes the poles on permanent magnets are called "north-seeking" or "south-seeking" poles, so that the north-seeking pole of a compass always points toward magnetic north.

From a distance, the magnetic field in space created by the Earth has a similar shape to that of a large bar magnet. However, the center of the Earth is very hot, and this high temperature would destroy any permanent magnetism (it is above the Curie point for iron and nickel). The Earth's field is generated by electric currents.

20.1   Permanent Magnets

Some materials can be permanently magnetized—e.g. iron, steel, cobalt, or nickel. These are described as ferromagnetic materials, and permanent magnets are constructed from these materials. If the material is easily magnetized and easily loses its magnetization, it is described as magnetically soft. If it is difficult to magnetize and demagnetize, it is described as magnetically hard. Pure iron is a soft magnetic material, and steel is a hard magnetic material. Alloys made of neodymium, iron, and boron are used to make the most powerful of all permanent magnets. These are called neodymium magnets.

The atoms of ferromagnetic materials are themselves magnetic, with each atom having its own dipole (like a tiny bar magnet). In an unmagnetized sample of the material, the atomic dipoles are in random orientations, so there is no net magnetic field. When the sample is magnetized, the atomic dipoles align and their fields add together in small regions called domains. The greater the magnetization, the larger the domains grow and the more they align with one another to create a stronger field.

Magnetization can be achieved by placing a ferromagnetic sample in an external field—e.g. from another permanent magnet or from an electromagnet. Molten rocks from volcanic eruptions often contain ferromagnetic minerals. As they solidify they become weakly magnetized in the direction of the Earth's own magnetic field. The alternating magnetization directions of volcanic rocks on the ocean floor has provided strong evidence that the Earth's magnetic field has changed polarity many times in the past and is likely to do so again in the future.

If a magnetized sample is heated, thermal vibrations can cause the atomic dipoles to change their orientation and destroy the alignment. There is a critical temperature for each ferromagnetic material above which it loses its permanent magnetic properties. This is called the Curie temperature or curie point. The Curie temperature for several materials is shown below:

Iron: 1043 K
Cobalt: 1400 K

Nickel: 637 K

Neodymium magnets: ~590–640 K

20.2 Magnetic Forces on Electric Currents and Moving Charges

In 1820, a Danish physicist, Hans Christian Oersted, demonstrated that magnetic fields are created by moving electric charges (electric currents). He did this by holding a magnetic compass close to a wire and switching on an electric current. The compass needle deflected from the Earth's field—a magnetic field had been created by the electric current. Further investigation shows that the magnetic field forms concentric rings around the current, getting weaker with distance. The magnetic field of a long straight current-carrying wire is shown below:

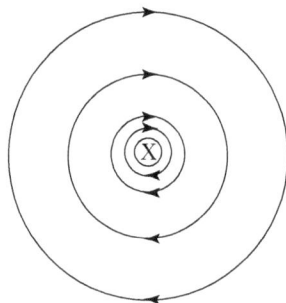

The direction of the magnetic field can be predicted using a **right-hand grip rule**. If you make a "thumbs up" with your right hand and align your thumb with the conventional current direction, then your fingers are curling in the direction of magnetic field lines.

Electric current is a flow of charge, so the source of the magnetic field is the moving charges inside the wire. A beam of charged particles moving through a vacuum would create a similar pattern of magnetic field.

It turns out that ALL magnetic fields originate from moving charges. Even the magnetic fields of permanent magnets originate from the movements of electrons inside the atoms of the material.

20.2.1 Magnetic Force on an Electric Current

Electric currents create magnetic fields and they also experience forces from them. This can be demonstrated using the apparatus below:

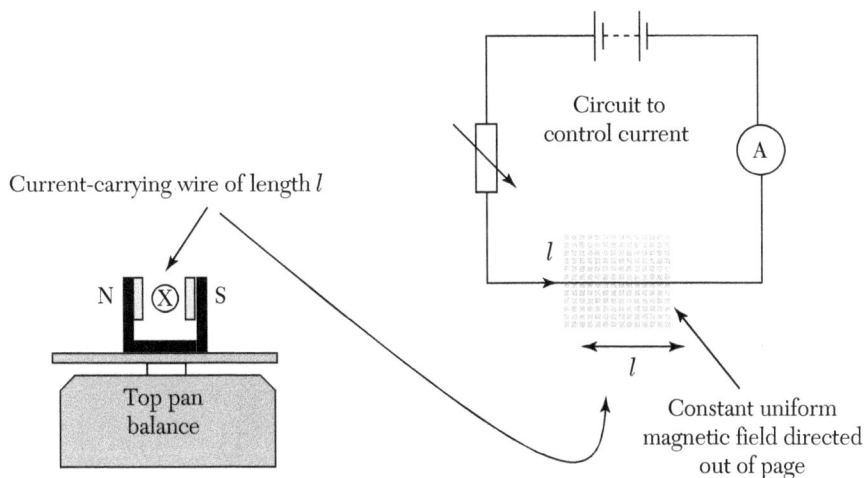

Current-carrying wire of length l

Circuit to control current

A

l

l

N | (X) | S

Top pan balance

Constant uniform magnetic field directed out of page

In the diagram above, the X on the end of the wire represents a current into the page.

A permanent magnet with flat pole pieces is placed on a top pan balance. The magnet creates a uniform horizontal magnetic field (directed from N to S). A separate circuit is used to control the current through a wire that passes horizontally between the poles perpendicular to the magnetic field lines. As the current is increased from zero, the reading on the top pan balance decreases, showing that the magnet is being pulled up and the wire (by Newton's third law) is being pulled down. The magnitude of the force increases as the current is increased. Experiments like this can show that the magnetic force F

- is directly proportional to current I

- is directly proportional to the length of wire in the field l

- is perpendicular to the current and the magnetic field

$$F \propto Il$$

The constant of proportionality depends on the strength B of the magnetic field, so we can use this effect to define the magnetic field strength:

$$B = \frac{F}{Il}$$

The S.I. unit of magnetic field strength is $NA^{-1}m^{-1}$. This is called the tesla (T); $1T = 1\ NA^{-1}m^{-1}$.

The magnetic force on a current-carrying conductor is often called the "motor effect" because this is used as the driving force in an electric motor. The direction of the force can be predicted using Fleming's left hand rule.

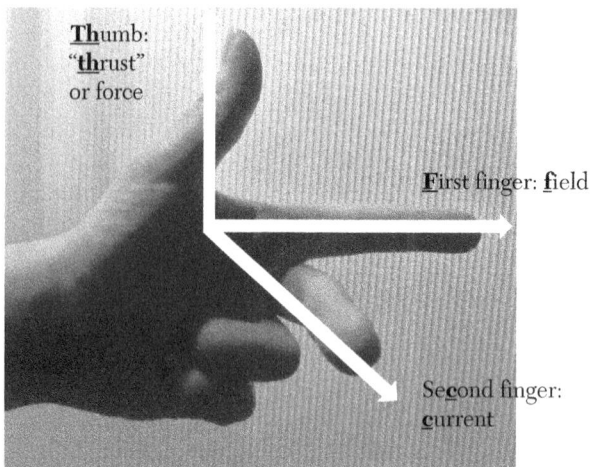

In the experiment above the current is perpendicular to the magnetic field. If the angle between the magnetic field and the current is varied, it is found that the magnetic force only acts on the component of current perpendicular to the field, so the more general equation for the magnetic force on a current-carrying conductor is:

$$F = BIl \sin \theta$$

where θ is the angle between the magnetic field and the current.

20.2.2 Force on a Moving Charge

The force on a current-carrying wire is really the sum of forces on the individual moving charges inside the wire. We can derive an expression for the force on a moving charge by using the microscopic equation for electric current.

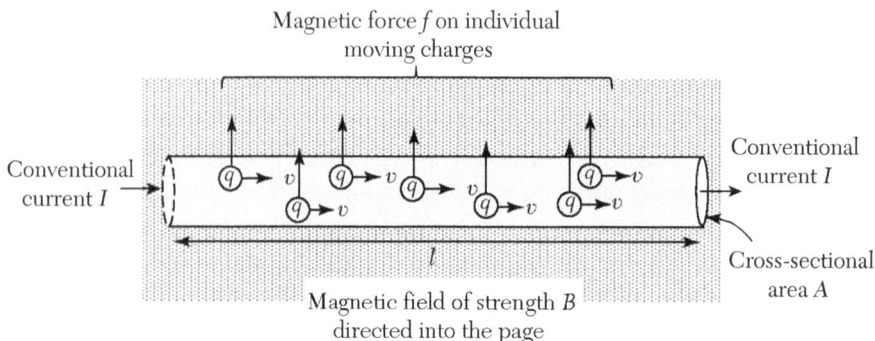

The total magnetic force on a length l of a conductor carrying current I at right angles to a uniform magnetic field of strength B is:

$$F = BIl$$

$$I = nqAv$$

$$F = BnqAvl$$

This force arises from the individual forces on N charge carriers where:

$$N = nAl$$

So the magnetic force on each charge carrier is:

$$f = Bqv$$

If the charge is moving in a direction at an angle θ to the magnetic field direction, the magnetic force on a moving charge is:

$$f = Bqv\sin\theta$$

This can be written using a vector cross-product: $\underline{f} = q\underline{v} \wedge \underline{B}$.

The direction of the magnetic force is given by Fleming's left-hand rule, but bear in mind that current direction refers to *conventional* current, so if the moving particle is negatively charged, the current is in the opposite direction to the motion.

Note that magnetic fields *only* affect moving charges. If $v = 0$ there is no magnetic force on the charge.

20.2.3   Path of a Moving Charged Particle in a Magnetic Field

The magnetic force is always perpendicular to the velocity; therefore it:

▫ cannot do any work on the charged particle, so it does not change the speed of the particle

▫ acts as a centripetal force, changing the direction of motion of the charged particle

Consider a particle with mass m and positive charge q moving into a region of uniform magnetic field of strength B with its velocity perpendicular to the magnetic field:

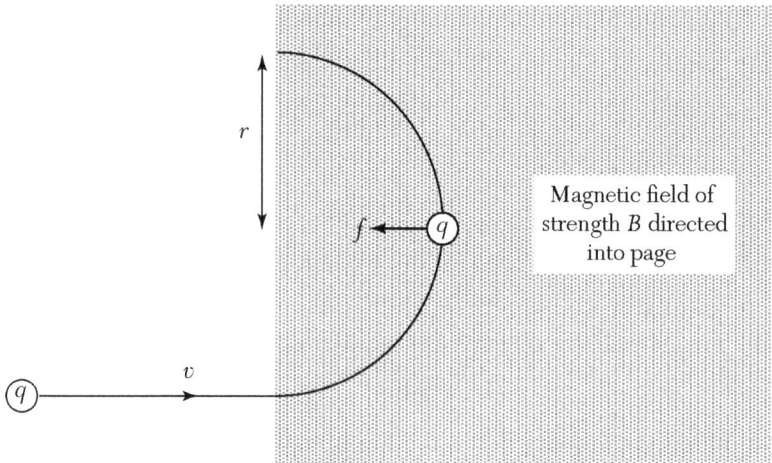

The magnetic field provides a constant centripetal force, so the charged particle moves in an arc of a circle of radius r:

$$f = Bqv = \frac{mv^2}{r}$$

$$r = \frac{mv}{Bq}$$

When particles are created in particle accelerators like the LHC at CERN, particle physicists use strong magnetic fields to deflect them so that they can determine their momentum and mass.

Lorentz force law

The total force on a charged particle in the electromagnetic field has two parts—the force from the electric field and the force from the magnetic field. This is expressed by the Lorentz force law:

$$f = f_{\text{electric}} + f_{\text{magnetic}}$$

$$f = q\underline{E} + q\underline{v} \wedge B$$

Note that the electric force is parallel to the electric field, whereas the magnetic force is perpendicular to the magnetic field (in a direction given by Fleming's left-hand rule).

20.2.4   Velocity Selector: Crossed Electric and Magnetic Fields

Moving charged particles can be deflected by both electric and magnetic fields. If the fields are perpendicular to one another and to the velocity of the incoming beam of charged particles, then the electric and magnetic force act along the same line and can be made to oppose one another. By adjusting the values of the fields, the two forces can be made to cancel out so that the beam is undeflected.

For the beam to be undeflected,

$$f_{\text{magnetic}} = f_{\text{electric}}$$

$$Bqv = Eq$$

$$v = \frac{E}{B}$$

Uniform electric field of strength E

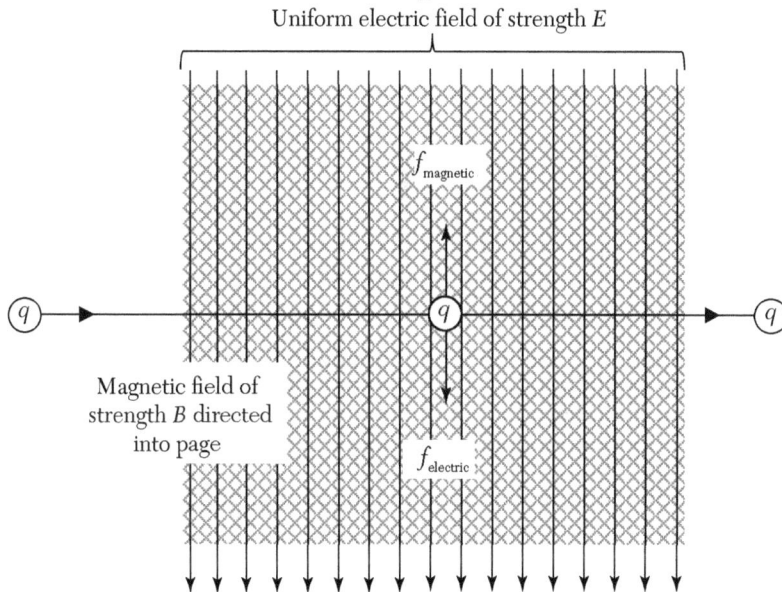

For any ratio of electric field strength to magnetic field strength, there is just one velocity that will be undeflected. This arrangement is called a velocity selector. If a stream of charged particles with a range of incident velocities enters the region of crossed fields, then only those satisfying the equation above go straight through.

Mass spectrometer

A velocity selector is used to send ion beams with a particular velocity into a mass spectrometer. The beams are then deflected in a constant magnetic field so that they move in a semicircular path. The radius of curvature of this path can be used to determine the masses of the ions. It is also possible to measure the amount of each type of ion that is present in the beam, so mass spectrometers are ideal for analyzing ionic ratios in samples.

A simplified diagram to illustrate the principle of the mass spectrometer is shown below. The entire apparatus is evacuated.

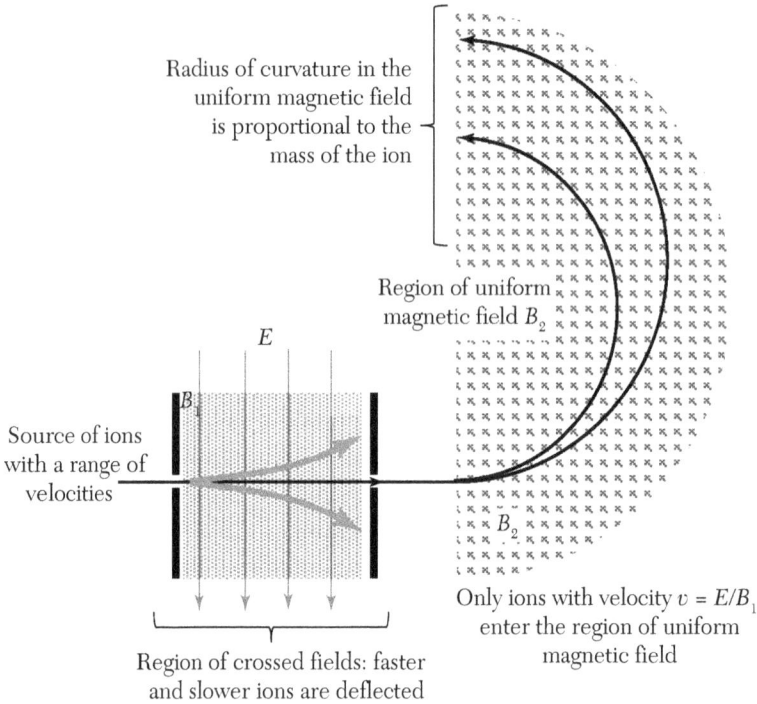

Radius of curvature in the uniform magnetic field is proportional to the mass of the ion

Region of uniform magnetic field B_2

E

B_1

Source of ions with a range of velocities

B_2

Only ions with velocity $v = E/B_1$ enter the region of uniform magnetic field

Region of crossed fields: faster and slower ions are deflected

The mass of the ion can be calculated using:

$$m = \frac{B_2 qr}{v} = \frac{B_1 B_2 qr}{E}$$

The greater the radius of curvature, the greater the mass of the ion.

20.3 Magnetic Fields Created by Electric Currents

20.3.1 Biot-Savart Law

The magnetic field created by a small element of an electric current is given by the Biot-Savart law. However, small current elements do not exist in isolation, so we must use integration to find the magnetic field at a point in space created by an electric current in a wire. This is analogous to the way we use Coulomb's law to find the resultant electric field at a point in space by adding up contributions from all the charges present.

The diagram below shows the contribution to the magnetic field strength δB at a point P from a small current element of length δl.

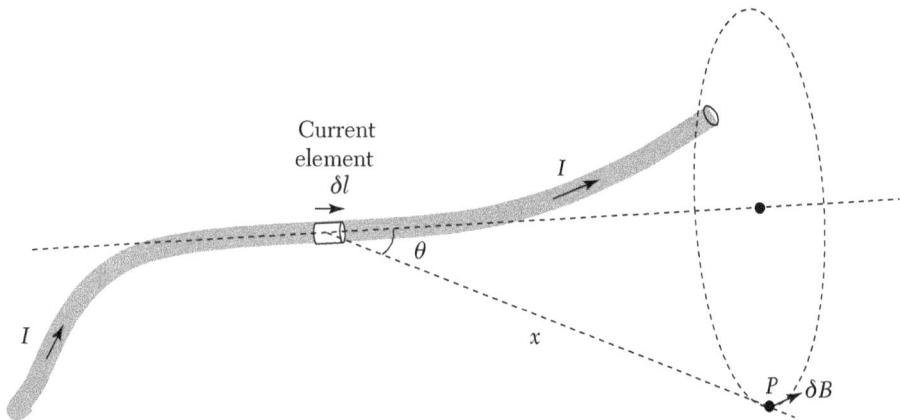

The direction of δB is found using the right-hand rule and using the continuation of the current element as the current direction. The magnitude of δB is given by:

$$\delta B = \frac{\mu_0 I \sin\theta}{4\pi x^2}\delta l$$

where μ_0 is the permeability of free space, a constant that determines the ability of a vacuum to support a magnetic field.

20.3.2   Magnetic Field at the Center of a Narrow Coil

This is the simplest example of the use of the Biot-Savart law because the angle θ is the same (and equal to $\pi/2$) for all current elements around the coil. The diagram has the axis of the coil into the page. The resultant magnetic field is also into the page, shown by the cross at the center.

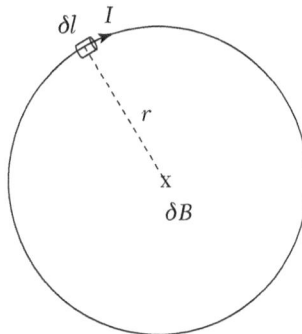

$$\delta B = \frac{\mu_0 I \sin\left(\pi/2\right)}{4\pi r^2}\delta l$$

$$B = \int_{l=0}^{l=2\pi r}\left\{\frac{\mu_0 I \sin\left(\pi/2\right)}{4\pi r^2}\right\}\delta l$$

All the terms in the curly brackets are constant, so this is a very simple integral resulting in:

$$B = \frac{\mu_0 I}{2r}$$

If the coil has N turns the magnetic fields add together:

$$B = \frac{\mu_0 N I}{2r}$$

An expression for the field strength at other points along the axis of the narrow coil can also be determined using the Biot-Savart law. The integration is a little more complicated, but the result is:

$$B = \frac{\mu_0 I r^2}{2\left(z^2 + r^2\right)^{3/2}}$$

where z is the distance along the axis from the center of the coil. When $z = 0$ the expression reduces to the previous equation (as it should!).

20.3.3 Magnetic Field of a Long Straight Current-Carrying Wire

For an infinitely long straight wire all of the current elements lie along the same line.

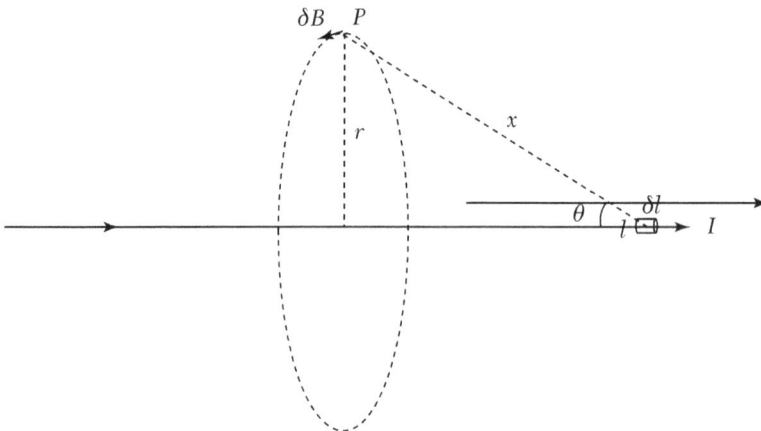

The resultant magnetic field forms concentric rings around the line of the electric current.

$$\delta B = \frac{\mu_0 I \sin \theta}{4\pi x^2} \delta l$$

The contributions from current elements to the left and right of P are the same, so the resultant magnetic field strength at P is:

$$B = 2 \int_{l=0}^{l=\infty} \frac{\mu_0 I \sin \theta}{4\pi x^2} dl$$

The three variables θ, x, and l are all related, so in order to carry out the integration we need to express two of them in terms of just one of them and the constant distance r.

$$x = \frac{r}{\sin\theta}$$

$$l = \frac{-r}{\tan\theta}$$

so
$$\frac{dl}{d\theta} = -\frac{r}{\sin^2\theta}$$

The integral now becomes:

$$B = -2 \int_{\theta=\frac{\pi}{2}}^{\theta=0} \frac{\mu_0 I \sin\theta\, d\theta}{r}$$

$$B = \frac{\mu_0 I}{2\pi r}$$

20.3.4 Magnetic Field Along the Axis of a Solenoid

A solenoid is a long coil, so the magnetic field along its axis can be found by integrating the contributions from narrow coils along the length of the solenoid. The result is that the field strength at the center of a long solenoid is:

$$B = \frac{\mu_0 NI}{l}$$

where N is the number of turns, I is the current in the solenoid, and l is the length of the solenoid.

The field strength drops to half of this value at each end and is small or negligible outside the solenoid (except near the ends).

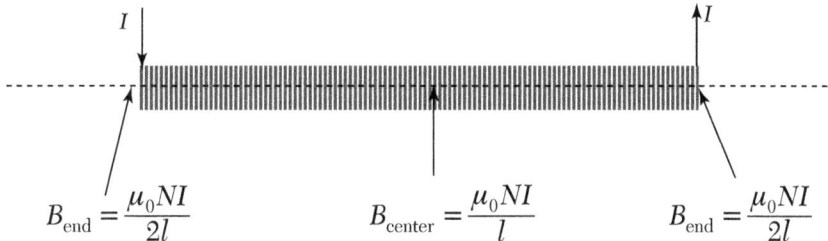

$$B_{\text{end}} = \frac{\mu_0 NI}{2l} \qquad B_{\text{center}} = \frac{\mu_0 NI}{l} \qquad B_{\text{end}} = \frac{\mu_0 NI}{2l}$$

The external field pattern is similar to that of a dipole bar magnet, one end of the solenoid acts like a north pole and the other acts like a south pole.

An electromagnet consists of a long coil wound around a soft iron core. The magnetization of the core increases the magnetic field strength, and the field at the center is then given by:

$$B = \frac{\mu_0 \mu_r NI}{l}$$

where μ_r is the relative permeability of the ferromagnetic core. Some typical values are shown below, but magnetic permeability depends strongly on the magnetic field strength, so these are representative values only.

Iron: 5000

Ferrite: 600

Nickel: 300

Carbon steel: 100

20.3.5 Ampère's Theorem

Ampère's theorem or "circuit law" states that, for any closed loop, the integral of the length elements multiplied by the component of magnetic field parallel to each element is proportional to the current enclosed by the path. This is really a consequence of the Biot-Savart law, but can be useful in situations with simple geometry.

B*.d*l is the component of ***B*** in the direction of ***dl*** multiplied by d*l* (i.e., the scalar product of ***B*** and ***dl***)

The constant of proportionality is the permeability μ_0.

$$\int_{\substack{closed \\ loop}} \textbf{\textit{B}}.\textbf{\textit{dl}} = \sum_{\substack{enclosed \\ by\ loop}} \mu_0 I$$

Ampère's theorem can be used to provide a simple derivation of the magnetic field strength a distance r from a long straight wire carrying a current I. By symmetry the magnetic field strength B is constant and forms concentric rings with the current at its center.

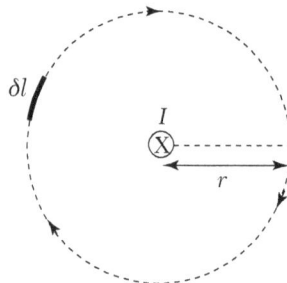

$$\oint_{\substack{\text{closed} \\ \text{loop}}} \boldsymbol{B}.d\boldsymbol{l} = 2rB = \sum_{\substack{\text{enclosed} \\ \text{by loop}}} \mu_0 I = \mu_0 I$$

$$B = \frac{\mu_0 I}{2\pi r}$$

20.4   Electric Motors

The "motor effect," where a force is exerted on a current-carrying conductor, is used in electric motors. This provides a way to transfer electrical energy to mechanical energy.

20.4.1   Turning Effect on a Coil in a Uniform Magnetic Field

A simple electric motor consists of a current-carrying coil in a magnetic field. Consider a rectangular coil of sides a and b, carrying a current I and lying in the plane of a uniform magnetic field of strength B.

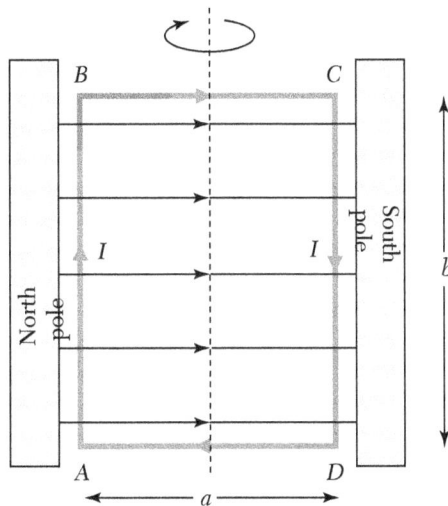

Sides AB and CD are perpendicular to the magnetic field, so magnetic forces act on these wires. On AB the force is into the page and on CD it is out of the page.

The magnitude of each force is $F = BIb$. These two forces create a turning effect about the central vertical axis (dotted line). The resultant couple or torque is:

$$\Gamma = 2 \times BIb \times \frac{a}{2} = BIab = BIA$$

where $A = ab$ is the area of the coil. If the coil has N turns, the torque is:

$$\Gamma = NBIA$$

20.4.2 A Simple D.C. Electric Motor

To make a working motor using a coil in a uniform magnetic field, the turning effect on the coil must always be in the same direction. For this to be the case, the direction of current in the coil must be reversed every half rotation. The reason for this is clear from the diagrams below, which show the end view of a motor coil as it rotates in the field. The end of the coil is shown in three different positions with side AB on the left and CD on the right. For the coil to continue to rotate in the same direction when AB moves past the vertical dotted line, the current direction must reverse. The thick black arrows represent the magnetic forces on the wires as the coil turns.

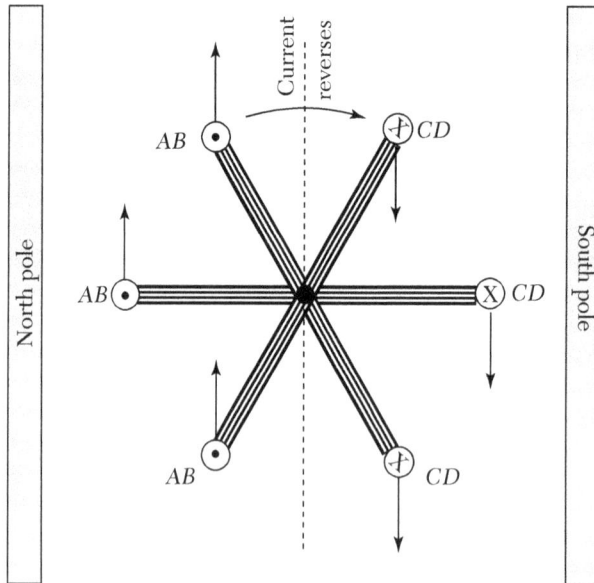

The device used to reverse the current is called a split-ring commutator. For a simple motor with a single coil, the commutator consists of a conducting cylinder split in half so that the halves are separated by an insulator. The commutator is attached to the axis of the motor and rotates with it. Current enters and leaves the coil via brushes that make a sliding contact with the surface of the commutator.

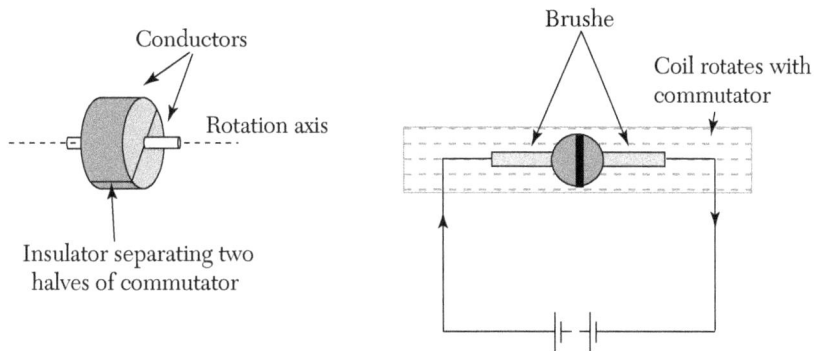

The commutator acts as a rotating switch ensuring that the direction of current in the coil stays the same as it rotates.

20.5   Exercises

1. The Earth's magnetic field is very similar to the field of a dipole bar magnet. However, geophysicists are sure that the field is not caused by a permanent magnet inside the Earth. Explain why not.

2. Two long straight parallel wires separated by 0.045 m each carry a current of 2.0 A in the same direction.
(a) Draw a diagram showing the magnetic field around one of the wires interacts with the other wire, and use this diagram to explain why the wires exert a force on one another. State the direction of this force.
(b) Calculate the magnitude of the force per unit length on each wire.
(c) Explain why coils of wire carrying a very large current must be able to withstand large stresses.

3. A mass spectrometer is used to measure the masses and abundancies of different isotopes. It does so by accelerating ions of each isotope to the same speed and then deflecting them into a semicircular path in a strong uniform magnetic field.
(a) Explain how a velocity selector, consisting of perpendicular electric and magnetic fields, can be used to select ions of the same speed from a group containing a wide range of different speeds.
(b) Ions of mass m_1 and m_2 and equal charge q enter the same uniform magnetic field of strength B at the same velocity v at right angles to the field lines and both are detected after moving through a complete semicircle in the field. Derive an equation for their separation.

4. The diagram below shows an end view of a rectangular coil in a uniform magnetic field of strength 0.05 T. The dot represents current out of the page and the cross represents current into the page.

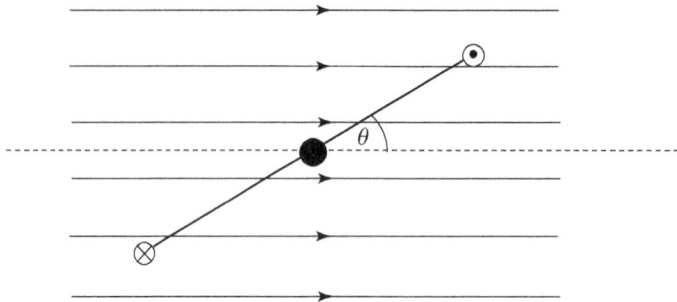

(a) Use a diagram to explain why there is a resultant moment on the coil and state the direction of this moment.

(b) Describe qualitatively how the moment on the coil changes as θ varies from 0° to 90°.

The coil has 80 turns and an area of 0.012 m². It carries a constant current of 0.65 A.

(c) Calculate the moment on the coil when $\theta = 0°$, 30°, 45°, 60°, and 90°.

(d) Explain how the coil would move if it was released from a horizontal position and allowed to move freely about a central axis directed into the page.

(e) Explain what has to be done to the current in the coil if it is to operate as a D.C. motor.

21

ELECTROMAGNETIC INDUCTION

21.1   Induced EMFs

21.1.1   What is Electromagnetic Induction?

An electric motor transfers electrical energy into mechanical work. This is a reversible process. If mechanical work is used to turn a motor, an emf is generated across its terminals; and if these are connected to an external load, electrical energy is generated. Motors and generators/dynamos are like mirror images of one another.

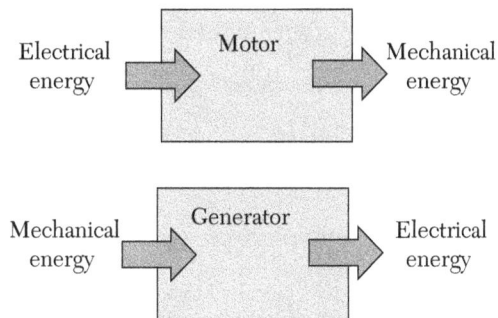

The underlying physics is electromagnetic induction—when a conducting wire cuts through a magnetic field or when the magnetic field passing through a coil changes, there is an induced emf in the conductor or coil. Electromagnetic induction was discovered by Michael Faraday in 1831.

21.1.2   Electromagnetic Induction Experiments

The "motor effect" is the force exerted on a current-carrying wire when it is perpendicular to a magnetic field. The force acts to push the wire in a direction perpendicular to the field and current. The inverse process is to use an external force to push the wire so that it cuts perpendicularly across the lines of the magnetic field. This results in an induced emf in the wire.

Here is a simple experiment that can be used to investigate electromagnetic induction effects when a wire cuts across the lines of a magnetic field.

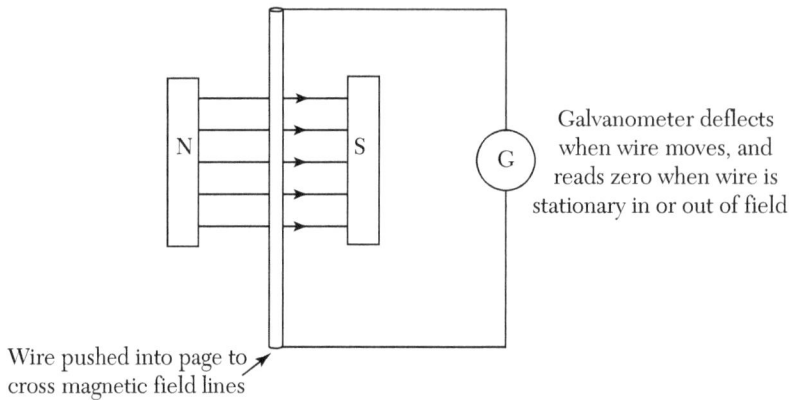

Galvanometer deflects when wire moves, and reads zero when wire is stationary in or out of field

Wire pushed into page to cross magnetic field lines

Several simple observations can be made:

- The deflection is greater if the wire is moved faster.

- There is no induction if the wire is stationary in or out of the field.

- The sign of the deflection changes if the direction of motion changes.

- If the wire is moved parallel to the field lines, there is no deflection.

We can explain all of these effects in the following way:

- When the wire cuts the lines of magnetic field, there is an induced emf across the ends of the wire.

- The magnitude of the emf is directly proportional to the rate of cutting field lines.

- The sign of the emf depends on the direction in which the field lines are cut.

Here is an experiment that can be used to investigate electromagnetic induction effects when the magnetic field through a coil is changed.

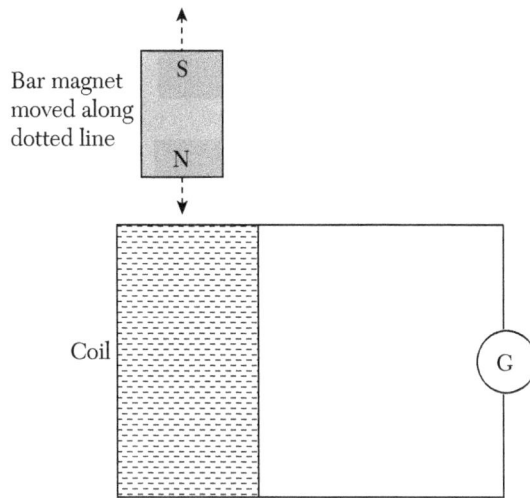

Several simple observations can be made:

- The deflection is greater if the magnet is moved faster.

- There is no induction if the magnet is stationary in or out of the coil.

- The sign of the deflection changes if the direction of motion changes.

- The effects are exactly the same if the coil is moved and the magnet is stationary.

We can explain all of these effects in the following way:

- Moving the magnet toward or away from the coil changes the amount of magnetic field passing through the coil.

- When the magnetic field in the coil changes, there is an induced emf across the ends of the coil.

- The magnitude of the emf is directly proportional to the rate of change of the magnetic field through the coil.

- The sign of the emf depends on whether the magnetic field through the coil is increasing or decreasing.

A third experiment shows that electromagnetic induction does not need relative motion; a changing magnetic field from an electromagnet can be used instead. This experiment is similar to Faraday's original experiments in the nineteenth century.

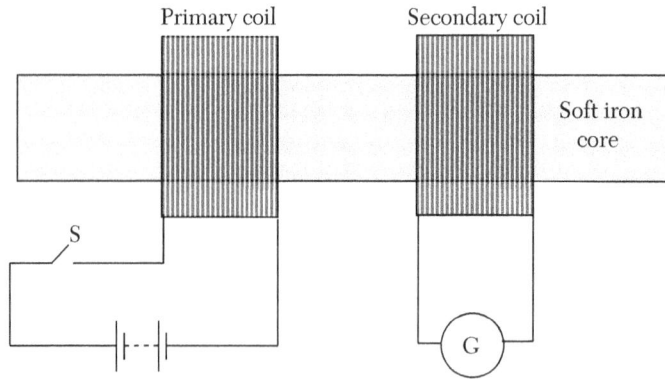

Several simple observations can be made:

▪ When S is closed there is a momentary deflection of the galvanometer and then it returns to zero.

▪ When S is open there is a momentary deflection of the galvanometer and then it returns to zero.

▪ If the supply voltage is increased, the deflections are larger.

▪ If the experiment is repeated without the iron core, the effects are similar but MUCH weaker.

We can explain all of these effects in the following way:

▪ When current flows in the primary coil, it becomes an electromagnet.

▪ The iron core increases the strength of the field and links the two coils magnetically.

▪ When the magnetic field in the secondary coil changes, there is an induced emf across the ends of the coil.

▪ The magnitude of the emf is directly proportional to the rate of change of the magnetic field through the secondary coil.

▪ The sign of the emf depends on whether the magnetic field through the secondary coil is increasing or decreasing.

▪ If the magnetic field passing through the secondary coil is constant or zero, there is no induction.

It is clear that electromagnetic induction only induces an emf in the secondary coil when the switch is opened or closed and the magnetic field through the secondary is changing. Opening or closing the switch causes a rapid change and results in a large induced emf. We can extend this experiment to show the effect of a continuously changing magnetic field by replacing the D.C. supply to the primary coil with an A.C. supply. The output from the secondary coil is now an A.C. emf that can be displayed on an oscilloscope. If a dual beam oscilloscope is used, the emf in the secondary can be compared with the current in the primary (which is in phase with the magnetic field).

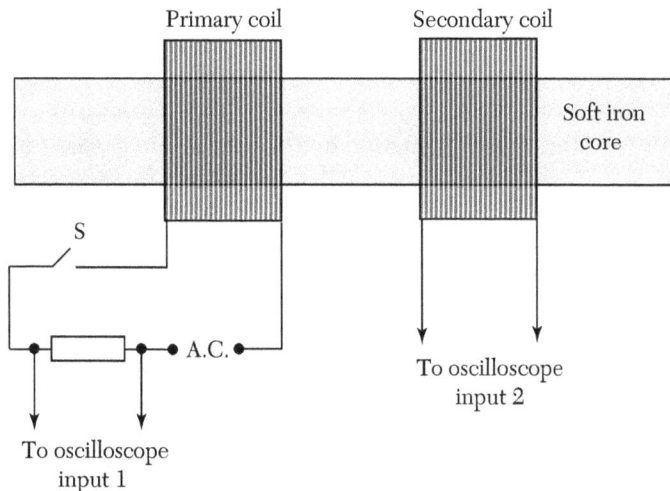

Note that the peaks of the induced emf occur at times when the rate of change (gradient) of the magnetic field through the secondary coil is greatest and that the induced emf is zero at times when the rate of change of the magnetic field through the secondary coil is zero (gradient is zero).

21.2 Laws of Electromagnetic Induction

The observations in the previous experiments can be explained using one simple equation—this is usually referred to as "Faraday's law,"but the mathematical formulation was actually first found by Neumann and the sign of the emf was explained by Lenz! This equation involves two new concepts—magnetic flux and magnetic flux linkage.

21.2.1 Magnetic Flux and Magnetic Flux Linkage

So far, we have given a qualitative explanation of electromagnetic induction in terms of changing the amount of magnetic field through a coil or changing the rate at which the magnetic field lines are cut. To give a quantitative formula for the induced emf, we need to define what we mean by "amount of magnetic field." This will depend on three factors—the strength of the field, its orientation, and the area through which it passes.

Magnetic flux

If a constant uniform magnetic field of strength B passes normally through a surface of area A, the flux Φ is defined as the product BA.

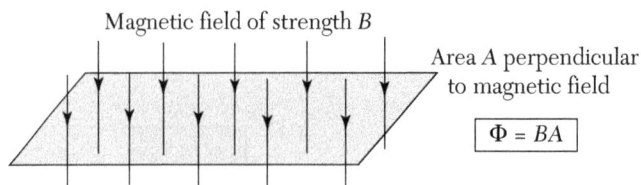

Magnetic field of strength B

Area A perpendicular to magnetic field

$$\boxed{\Phi = BA}$$

In general, the flux through an element of area δA that makes an angle θ to the normal to the surface element will contribute a flux element $\delta\Phi = B\delta A \cos\theta$ to the total flux through the surface.

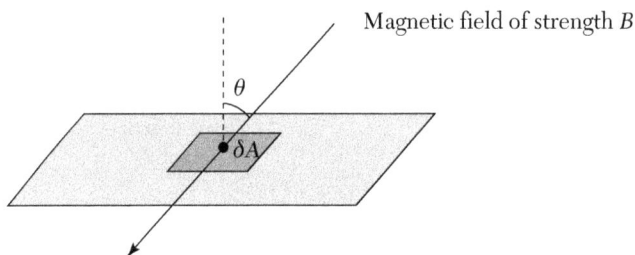

Magnetic field of strength B

θ

δA

The total flux is found by integrating these contributions across the surface:

$$\Phi = \int\limits_{\text{surface}} B\cos\tau\, dA$$

where B might vary from point to point.

The S.I. unit of magnetic flux is Tm^2, which is called Wb.

$$1\ Wb = 1\ Tm^2 \quad \text{or} \quad 1\ T = 1\ Wbm^{-2}$$

Writing the tesla in this way emphasizes the fact that it is the magnetic *flux density*.

Magnetic flux linkage

When magnetic flux passes through a coil of N turns, it links each turn in the coil, so it is convenient to define the magnetic flux linkage as the product of the magnetic flux and the number of turns linked by that flux:

$$\text{Flux linkage} = N\Phi$$

If the flux is caused by a constant uniform field of strength B along the axis of the coil, the flux linkage is simply **NBA**, where A is the area of the coil, N is a dimensionless number, so the S.I. unit of flux linkage is Wb.

21.2.2 Faraday's Law of Electromagnetic Induction

Faraday realized that the induced emf is directly proportional to the rate at which a conductor cuts the magnetic flux or the rate at which the magnetic flux linkage of a coil changes. This can be written mathematically as:

$$H = -\frac{d(N\Phi)}{dt}$$

We will discuss the significance of the minus sign later.

The term $\frac{dN\Phi}{dt}$ can be interpreted as the rate of cutting flux *or* the rate of change of flux linkage in a coil, depending on the context. The equivalence of these two descriptions can be seen by considering the emf induced in a straight conductor moving at constant velocity perpendicular to a magnetic field of strength B.

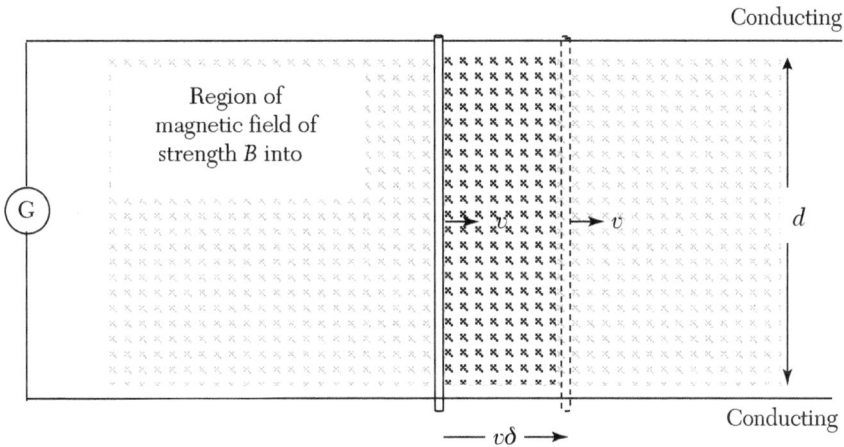

The conductor runs along two parallel conducting rails connected back to a stationary galvanometer. As the conductor moves, it cuts through lines of magnetic field. In time δt it cuts all the magnetic field lines in the darker shaded area, and the flux through the circuit increases by an amount equal to the flux through that additional area.

Flux cut in time δt, $\delta\Phi = Bdv\delta t$

Rate of flux-cutting:

$$\frac{d\Phi}{dt} = Bvd$$

Using Faraday's law, the induced emf in the moving conductor is:

$$H = (-)Bvd$$

Lenz's law

The negative sign in the equation of Faraday's law can be interpreted in the following way:

▪ The direction of the induced emf is such as to oppose the change that caused it.

This sounds a little obscure, but becomes clear when we apply it to the example above. In order to induce an emf, we needed to push the wire to the right. The existence of an emf in the circuit causes a current to flow. However, when a current flows in a conductor lying perpendicular to the magnetic field, there is a motor effect force on the conductor that is perpendicular to both the current and the field. In this case the force must lie either in the direction of v or in the opposite direction. If it was in the direction of v then once we had started the wire moving, the induced emf would create a current that experienced a force in the direction of motion and the wire would accelerate with no need for us to do any further work on it. Energy would be generated from nothing! This violates the law of conservation of energy, so it cannot occur. The force on the induced current must oppose the force moving the wire so that we do work to move it and "pay" for the electrical energy it generates. Lenz's law ensures that energy is conserved when an emf is induced. It is the reason we need to burn fuel to turn the generators in a power station.

21.2.3 Changing the Flux Linkage in a Coil

Consider a coil of N turns and area A initially perpendicular to a uniform magnetic field of strength B that is rotated through 90° so that it finally lies parallel to the field. The magnetic flux falls from a maximum value at the start to zero in a time δt. What is the average induced emf during this rotation?

Coil of N turns and area A

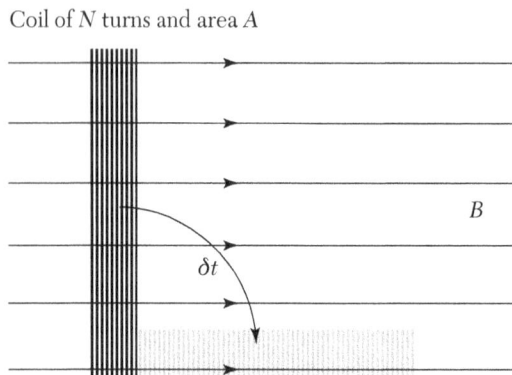

Change in flux-linkage: $\delta\Phi = NBA$

Rate of change in flux-linkage: $\dfrac{\delta\Phi}{\delta t} = \dfrac{NBA}{\delta t}$

Average induced emf: $= (-)\dfrac{NBA}{\delta t}$

The reason that this is an average value rather than a constant value is that the rate of change of flux linkage varies as the coil turns, being low at the beginning and a maximum near the end.

A more interesting situation is one where the flux linkage through the coil varies sinusoidally. This is the case in most electrical generators, where a coil rotates in a magnetic field.

Coil of N turns and area A

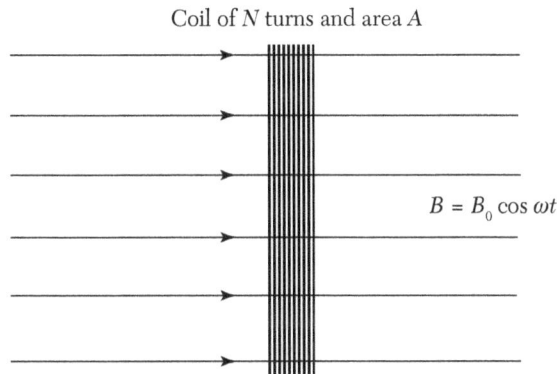

$B = B_0 \cos \omega t$

Flux-linkage: $\qquad \Phi = NB_0 A \cos \omega t$

Induced emf: $\qquad H = -\dfrac{d(N\Phi)}{dt} = \omega NB_0 A \sin \omega t$

This is an A.C. output with peak value $\omega NB_0 A$.

21.3   Inductance

When there is current in a coil, the coil becomes an electromagnet and creates a magnetic field. If the magnetic field changes, there is changing flux linkage inside the coil, so an induced emf is created that opposes the external supply that is changing the current. This is often called a "back emf." This opposition to changing current is called inductance. Coils act as inductors in electric circuits.

21.3.1 Self-Inductance

The tendency for a coil to oppose changes of current is called inductance. The greater the inductance, the stronger the opposition to changing currents. Inductance is defined by:

$$H = -L\frac{dI}{dt}$$

$$L = \frac{-H}{\left(\dfrac{dI}{dt}\right)}$$

Inductance is equal to the back emf per unit rate of change of current.

The S.I. unit of inductance is VsA^{-1} or the henry (H). $1H = 1\ VsA^{-1}$.

An expression for the self-inductance of a long solenoid can be determined by comparing the definition of inductance with Faraday's law:

$$H = -L\frac{dI}{dt} = -\frac{dN\Phi}{dt}$$

$$\Phi = NBA = \frac{\mu_0 N^2 IA}{l}$$

So

$$\frac{dI}{dt} \quad \frac{N\ A}{l}\frac{dI}{dt}$$

$$L = \frac{\mu_0 N^2 A}{l}$$

The symbol of an inductor is:

21.3.2 Rise of Current in an Inductor

When an inductor is connected to a D.C. power supply, the current through the inductor begins to increase, but as it does so, there is a back emf that opposes the supply and that limits the rate at which the current can rise.

$$\frac{dI}{dt} = -\frac{H}{L}$$

This effect means that it takes time for the current to reach its steady final value and that the supply does work against the back emf as this current is established. During this time, energy is being transferred from the supply to the magnetic field.

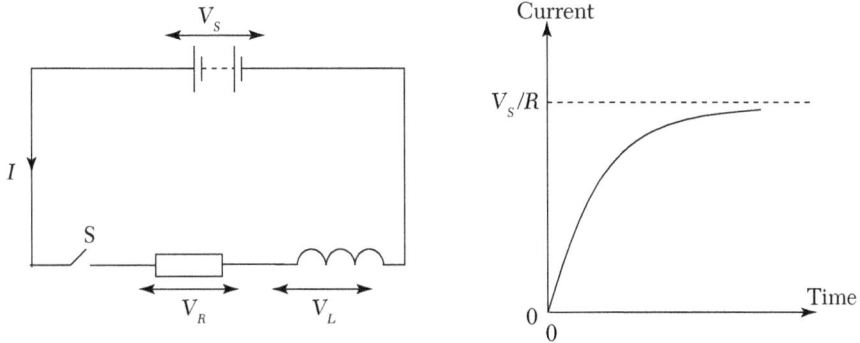

When S is closed, Kirchhoff's second law gives us:

$$V_S = V_R + V_L$$

$$V_S = IR + L\frac{dI}{dt}$$

This is a first-order differential equation that can be solved to give an expression for *I*.

$$I = I_0\left(1 - e^{-\left(\frac{R}{L}\right)t}\right)$$

This has a similar form to the equation for the charging of a capacitor, and (R/L) is the effective time constant.

If S is open the current falls to zero in a very short time so that dI/dt is very large. This results in a large back emf that can cause sparks across the switch terminals. Interrupting the current in a coil is a way to generate spikes of high voltage—e.g., for a car's spark plugs.

21.3.3 Energy Stored in an Inductor

When the current in an inductor is interrupted, there is a large back emf and the energy stored in the magnetic field is rapidly dissipated—e.g., by sparks and heating effects. We can derive an expression for the energy stored in an inductor of inductance L when a current I_0 flows in it by considering the work done to establish that current.

$$P = IV = IL\frac{dI}{dt}$$

Separating variables and integrating:

$$E = \int_{t=0}^{t=\infty} P dt = \int_{I=0}^{I=I_0} IL dI = \frac{1}{2}LI^2$$

21.3.4 Mutual Inductance

When two coils are placed close together, their magnetic fields can affect one another. If the current is changed in one coil, there will be a changing flux linkage in the other coil and an induced emf. The strength of this coupling is measured by the mutual inductance of the system.

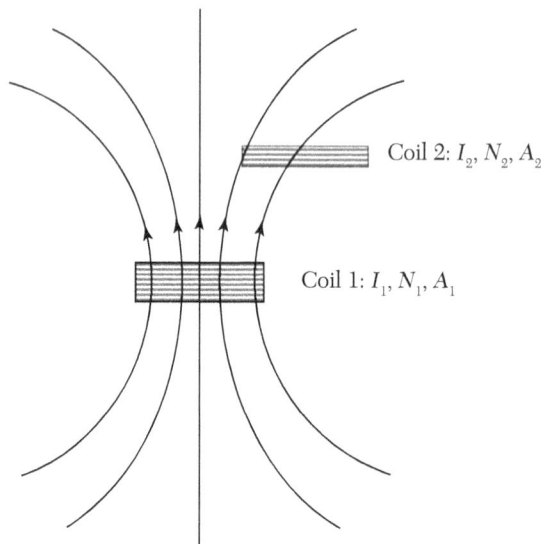

Coil 2: I_2, N_2, A_2

Coil 1: I_1, N_1, A_1

The mutual inductance of two coils is defined by:

$$H_2 = M \frac{dI_1}{dt}$$

$$H_1 = M \frac{dI_2}{dt}$$

The S.I. unit of mutual inductance is H. Mutual inductance depends on the self-inductance of each coil:

$$M_{12} = kL_1L_2$$

where k is a coupling constant that depends on how the coils are arranged and how they are magnetically linked.

21.4 Transformers

Electromagnetic induction is the key principle on which the generation and distribution of electrical energy depends. Thermal power stations use fossil or nuclear fuels to transfer chemical or nuclear energy to thermal energy and then use this to power generators that output A.C. electricity. Transformers step up the voltage so that electricity can be transmitted with low losses over large distances. More transformers are used to step down the voltage for consumers.

21.4.1 An Ideal Transformer

A transformer is a device that uses mutual induction to step up or step down an A.C. voltage.

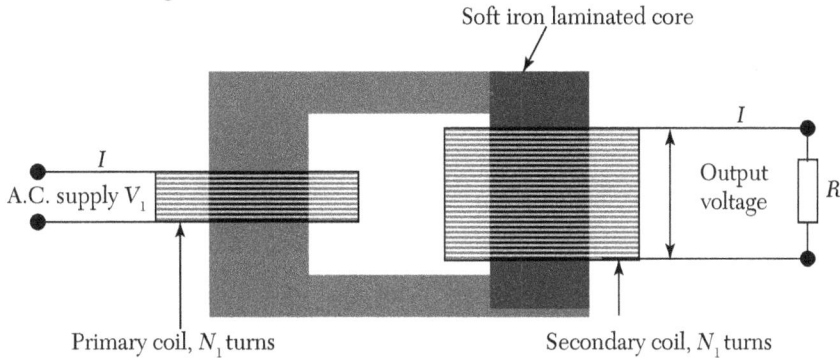

Soft iron laminated core

Primary coil, N_1 turns

Secondary coil, N_1 turns

The arrangement is very similar to that of Faraday's original experiment. A.C. current in the primary creates an alternating magnetic flux in the core. The soft iron core enhances the strength of the magnetic field and acts as a magnetic circuit, linking the primary with the secondary. The changing magnetic flux in the secondary induces an A.C. emf in the secondary. The secondary emf is given by:

$$H = -\frac{dN\Phi}{dt}$$

where $N\Phi$ is the flux linkage in the secondary coil and $\Phi \propto I_1$ (primary current)

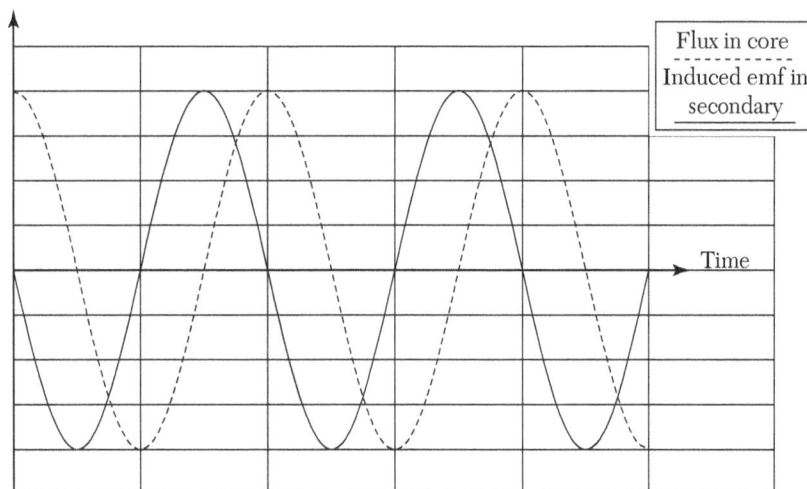

Flux in core
Induced emf in secondary

Time

The peaks of emf correspond to times when the flux in the core has its greatest rate of change.

For an ideal transformer, all of the flux created by the primary passes through the secondary. If there are equal numbers of turns on both primary and secondary, the back emf in both coils will be equal and will equal the supply voltage. However, the voltage on the secondary is also directly proportional to the number of turns on the secondary ($V_2 \propto N_2$), so changing the turns ratio (N_2/N_1) must change the voltage ratio (V_2/V_1) in a similar way. This leads to the transformer equation:

$$\frac{V_2}{V_1} = \frac{N_2}{N_1}$$

The voltage ratio is equal to the turns ratio.

▪ If $N_2 > N_1$ then $V_2 > V_1$ and it is a step-up transformer.

▪ If $N_2 < N_1$ then $V_2 < V_1$ and it is a step-down transformer.

While a transformer can increase voltage, it cannot create energy; the power supplied to the transformer must be equal to or greater than the power it delivers to a load. An ideal transformer has 100% efficiency, so:

$$P_{in} = P_{out}$$

$$I_1 V_1 = I_2 V_2$$

$$\frac{I_2}{I_1} = \frac{V_1}{V_2} = \frac{N_1}{N_2}$$

The current ratio is the inverse of the voltage and turns ratios.

21.4.2 Transmission of Electrical Energy

The electrical power generated at power stations is transmitted over long distances via transmission lines. These are conducting cables that have resistance, so there is ohmic heating and some of the power is dissipated. In order to reduce the power loss and to increase the efficiency of transmission, transformers are used to step up the voltage in the transmission lines. This reduces the current in the cables and therefore reduces the amount of ohmic heating, increasing the efficiency of transmission. Step-down transformers are then used to reduce the voltage for consumers (very high voltages are difficult to insulate).

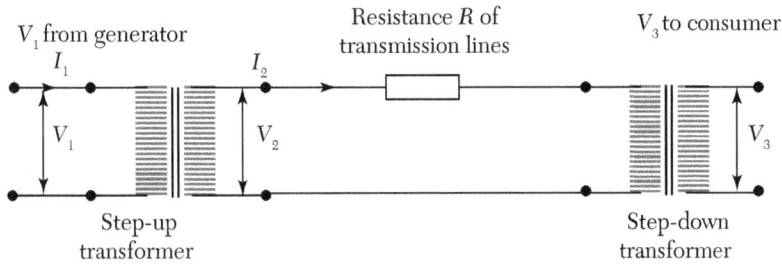

The input power from the generator is:

$$P_1 = I_1 V_1$$

The power transmitted is:

$$P_2 = I_2 V_2 = P_1$$

$$I_2 = \frac{P_1}{V_2}$$

The power loss in transmission is:

$$P_R = I_2^2 R = \frac{R P_1^2}{V_2^2} \propto \frac{1}{V_2^2}$$

So increasing the transmission voltage reduces the power loss.

Note that V_2 is the potential difference between the transmission lines and NOT the potential difference across resistor R. The potential difference across R depends on I_2 and falls as V_2 increases.

21.4.3   Real Transformers

The ideal transformer discussed above is 100% efficient. Real transformers can have high efficiency but do have some energy losses. There are four main causes of energy loss in a real transformer:

- Copper losses: ohmic heating of the conducting wires forming the coil (usually copper wires).

- Flux losses: the magnetic circuit provided by the transformer core might not confine all of the magnetic flux so that some of the flux created by the primary coil does not pass through the secondary coil.

- Eddy current losses: transformers work with A.C., so there is an alternating magnetic field in the core itself. If this is made from a conducting material (e.g., soft iron), there will be induced current loops in the core. These will dissipate heat. In order to reduce these losses, the core is usually laminated—cut into slices that are separated by thin layers of insulator. High-frequency transformers, e.g., in radio frequency circuits, use non-conducting cores (e.g., ferrite cores) to prevent these losses.

▪ Hysteresis losses: the continual magnetization and demagnetization of the core creates alternating stresses inside the core that dissipate energy as heat. Soft iron has low hysteresis losses, so, in addition to its high permeability, it is good material for low-frequency transformer cores.

Our discussion of transformers has assumed that the loads attached to them are purely resistive and ignores the phase relationships between currents and voltages in the two coils. In practice this is quite complex and the inductance and capacitance of the circuits needs to be taken into account.

21.5   A Simple A.C. Generator

A simple A.C. generator consists of a coil rotating with constant angular velocity in a constant uniform magnetic field. The diagram shows an end view of such a coil when it has rotated through an angle θ from the horizontal. The coil has area A and N turns.

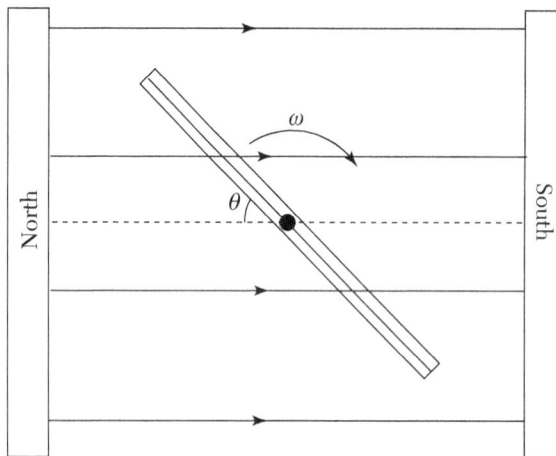

Flux linkage:

$$\Phi = NBA\sin\theta$$

The coil is rotating with constant angular velocity ω:

$$\theta = \omega t$$

Induced emf, $H = (-)\dfrac{dN\Phi}{dt} = (-)\omega NBA\cos\omega t$

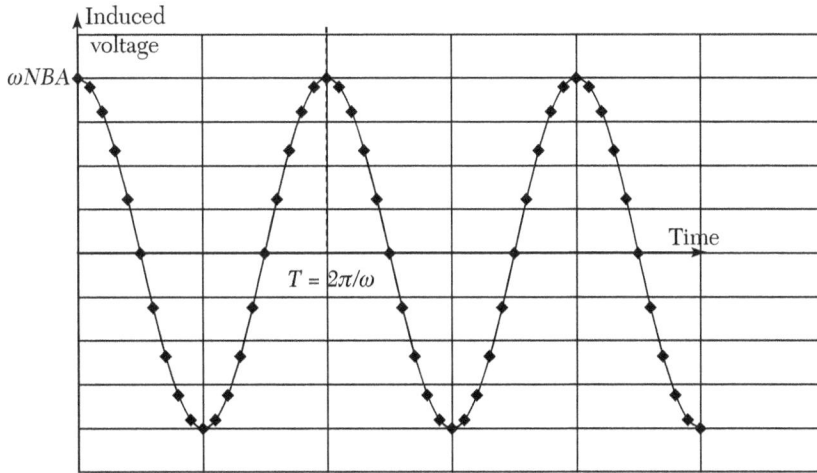

The negative sign in the equation has been ignored in the graph. The output is an A.C. voltage with peak value $E_0 = \omega NBA$.

21.6 Electromagnetic Damping

Lenz's law states that the induced emf is always in such a direction as to oppose the change that caused it. If the induced emf creates an induced current, the forces that act on this current oppose the motion causing the change in flux linkage. This acts like a braking force and can be demonstrated using the apparatus shown below.

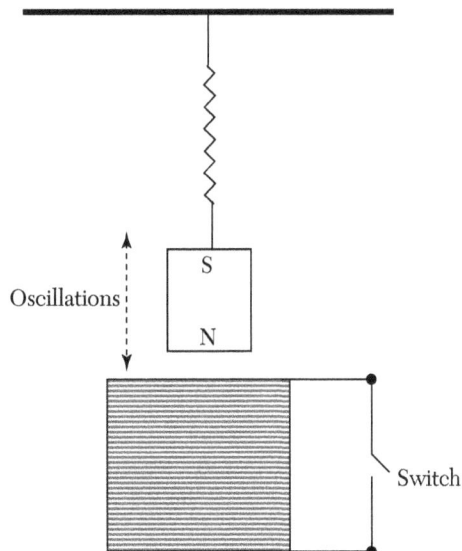

A bar magnet is supported from a spring. It is displaced and allowed to oscillate vertically, while the switch is left open. An alternating induced emf is created in the coil, but there are no induced currents because there is not a complete circuit. The oscillations are undamped (apart from air resistance,etc.) and the magnet oscillates for a long time.

When the switch is closed, the induced emf causes alternating currents in the coil. By Lenz's law these currents flow in a direction that opposes the change that caused them. As the North Pole approaches the top of the coil, the current moves in such a direction that the top of the coil is also a north pole, repelling the magnet. When the magnet moves away from the coil, the induced currents form a south pole at the top of the coil, attracting the magnet. It is clear that the electromagnetic forces act in the opposite direction to the motion, like friction, and provide additional damping. The oscillations decay more rapidly with the switch closed.

Another simple demonstration involves placing a cylindrical neodymium magnet on a thick sheet of aluminum or copper and then tipping the sheet until the magnet slides down. Aluminum and copper are not ferromagnetic metals, so when the magnet is stationary, there are no magnetic forces acting on it. However, when it begins to move, the magnetic flux cuts through the

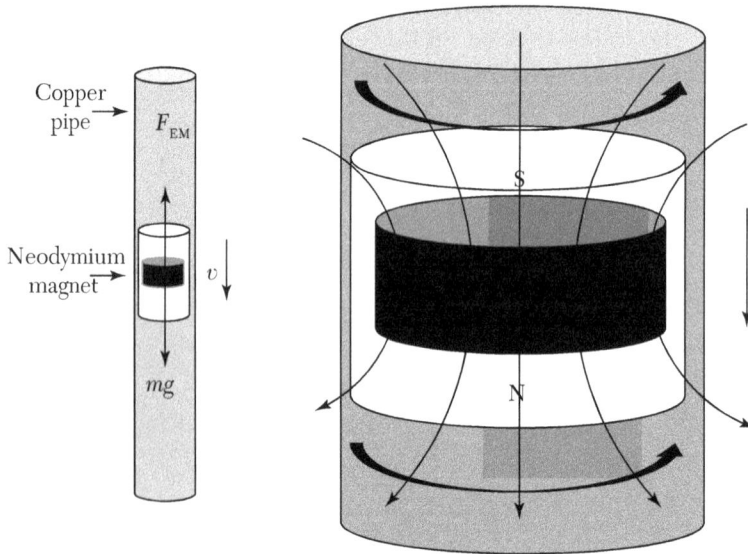

As magnet falls, its flux cuts the conductor surrounding it. This induces current loops above and below the falling magnet. These create magnetic fields that oppose the magnet's motion.

conductor inducing emfs. The induced emfs create current loops and the magnetic fields of these loops oppose the motion of the magnet. There is a magnetic drag force that slows the magnet down. This can be a surprisingly large effect if the metal sheet is titled at a large angle.

Another intriguing demonstration involves dropping a cylindrical neodymium magnet down a copper water pipe with an internal diameter slightly greater than the diameter of the magnet. The magnet falls at a slow terminal velocity. Electromagnetic damping forces balance its weight.

21.7   Induction Motors

Electromagnetic damping exerts a damping force on a magnet when it moves past a conductor. By Newton's third law there is an equal and opposite force on the conductor. This can be used to create a motor. If you move a neodymium magnet rapidly past a thin sheet of non-ferromagnetic metal (e.g., copper or aluminum), the sheet will begin to accelerate in the direction of the moving magnet.

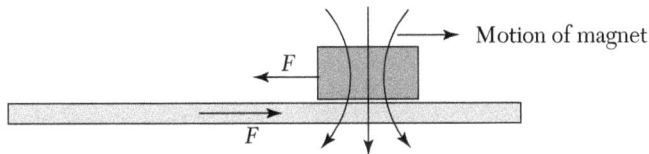

The moving magnetic field induces emfs in the metal that create current loops. By Lenz's law these create magnetic fields that oppose the relative motion, resulting in the forces on the magnet and conductor as shown above. The effect is that the conductor tends to follow the moving magnetic field.

In an induction motor, coils are used to create a rotating magnetic field. A non-ferromagnetic rotor is placed in the rotating field and it too rotates. This has the big advantage that there are no brushes. The idea was first suggested by Nikola Tesla, and induction motors are now widely used from DVD players to electric cars.

A rotating magnetic field can be created by superposing two alternating magnetic fields at right angles to one another and giving them a phase difference of $\pi/2$. If we represent each magnetic field by a phasor, then the resultant of the two phasors represents the rotating field.

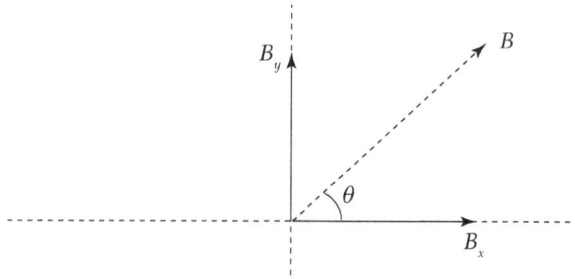

$$B_x = B_0 \cos \omega t$$

$$B_y = B_0 \sin \omega t$$

Resultant magnetic field strength is:

$$B = \sqrt{B_x^2 + B_y^2} = B_0 \sqrt{\cos^2 \omega t + \sin^2 \omega t} = B_0$$

The direction of the magnetic field is:

$$\theta = \tan^{-1}\left(\frac{B_y}{B_x}\right) = \tan^{-1}(\tan \omega t) = \omega t$$

In other words, this creates a magnetic field of constant strength B that rotates with constant angular velocity ω. A simple way to create such a field is to use two pairs of coils arranged as shown below:

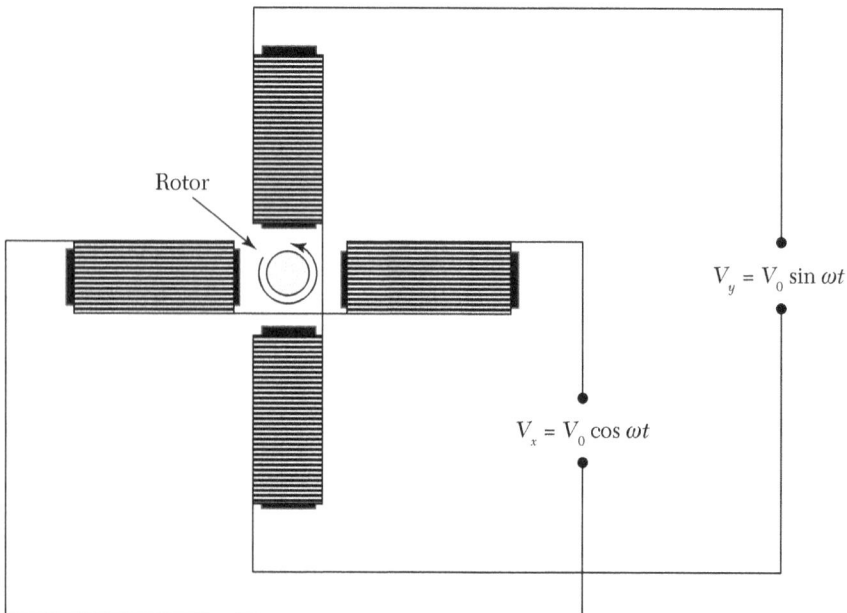

21.8   Exercises

1. The diagram below shows a bar magnet and a coil.

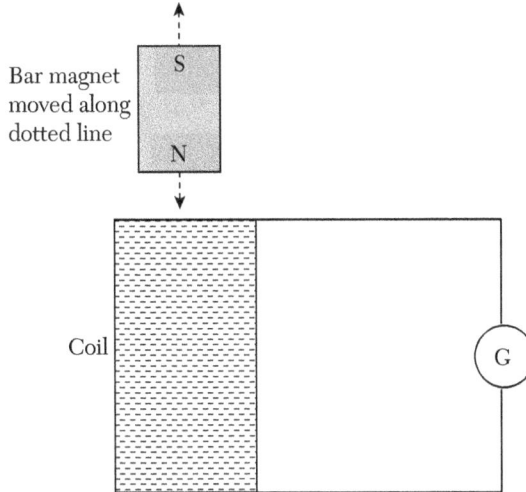

While the magnet is moving toward the coil, the galvanometer deflects to the right.
(a) Explain why this occurs.
(b) State and explain, using Faraday's law, what happens when:
  (i) the magnet is moved toward the coil at a higher speed
  (ii) the magnet is moved away from the coil
  (iii) the magnet is stationary inside the coil
(c) The experiment is repeated but with the galvanometer removed so that the ends of the coil are not connected to anything. Discuss whether the motion of the magnet near the coil has any effect now.

2. The Earth's field has a strength of about 50 mT. A square coil of side 20 cm having 200 turns is placed so that its plane is perpendicular to the Earth's field. The coil has a total resistance of 50 ohms.
(a) Calculate the flux through the coil.
(b) Calculate the flux linkage through the coil.
(c) The coil is quickly turned (about an axis in the plane of the coil perpendicular to the field) through 90° so that its plane is now parallel to the Earth's field. It takes 0.50 s to complete the rotation. Explain why there is a current in the coil as it rotates.
(d) Calculate the average current as the coil is rotated.
(e) Calculate the charge that has moved around the coil during the process.

(f) How (if at all) would your answers to parts (d) and (e) be affected if the coil was turned through 90° in 0.10 s instead of 0.50 s? Explain.

3. When a strong magnet is dropped through a copper tube, it falls at a constant velocity. Use your knowledge of electromagnetism to explain this as fully as you can. Use diagrams if this helps.

4. An aircraft is flying through due north at 200 ms^{-1} in a place where the vertical component of the Earth's magnetic field is 40 mT. Its wingspan is 50 m and its wings are made of a conducting material.
   (a) Explain why there is a potential difference between its wing tips.
   (b) Calculate the potential difference between its wing tips.
   (c) Discuss whether the voltage generated could be used to power an electrical device using wires connected to each wing tip.

5. (a) Draw a labeled diagram of a transformer that could step down a main supply of 240 V 50 Hz A.C. to 20 V 50 Hz A.C. The primary coil has 1200 turns.
   (b) Explain how the transformer works.
   (c) Explain the function of the core and explain why soft iron is a suitable material for it.
   (d) Explain how laminating the core reduces energy loss from the transformer.
   (e) Transformers are very efficient, but there are still losses. State three ways in which energy can be dissipated by a transformer.
   (f) A 20 V, 40 W lamp is connected to the secondary of the transformer. Calculate the current drawn from the primary (assume that the efficiency is 100% and neglect inductive effects).
   (g) Sketch a graph to show how the voltage in the secondary coil is related to the flux in the core. Explain how the graph illustrates both Faraday's and Lenz's laws.

6. A coil with inductance and resistance is connected via a switch to a power supply of 1.0 V. The graph below shows how the current in the coil increases with time from the moment the switch is closed.

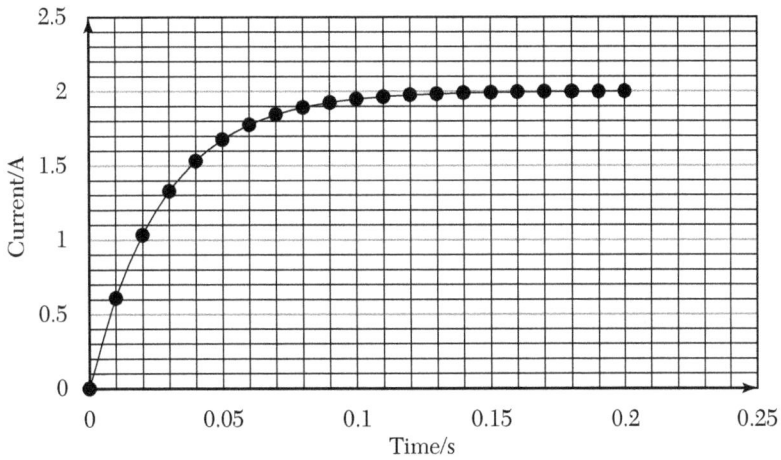

(a) Use the graph to calculate the inductance and resistance of the coil.
(b) Calculate the energy stored when there is a current of 2.0 A in the coil.
(c) When the switch is opened, a spark is observed to jump across its contacts. Explain why this occurs.

7. The diagram below shows an end view of a simple generator consisting of a coil rotating at a constant rate in a uniform magnetic field. The A.C. output is connected via slip rings to a purely resistive load.

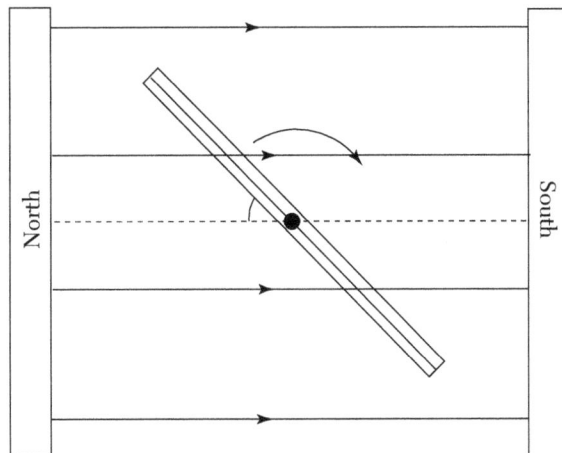

Here is some data about the generator:

Magnetic field strength   :   0.12 T
Area of coil   :   80 cm²
Number of turns on coil   :   250
Rotation frequency   :   60 Hz
Load resistance   :   50 W

(a) Calculate the peak output emf.
(b) Sketch a graph to show how the output emf varies with time as the coil completes one rotation.
(c) Calculate the peak current in the load.
(d) Calculate the peak power transferred to the load (assume that the resistance of the coil is negligible).
(e) Discuss how your answers to (b), (c), and (d) would be affected if the resistance of the coil was not negligible.

22

ALTERNATING CURRENT

22.1   A.C. and D.C.

D.C. stands for "direct current"and A.C. stands for "alternating current." D.C. flows in one direction around a circuit, so the supply polarity is constant. A.C. changes direction periodically, so the polarity of the supply alternates. Most A.C. supplies are sinusoidal and can be represented by:

$$V = V_0 \sin \omega t$$
$$I = I_0 \sin \omega t$$

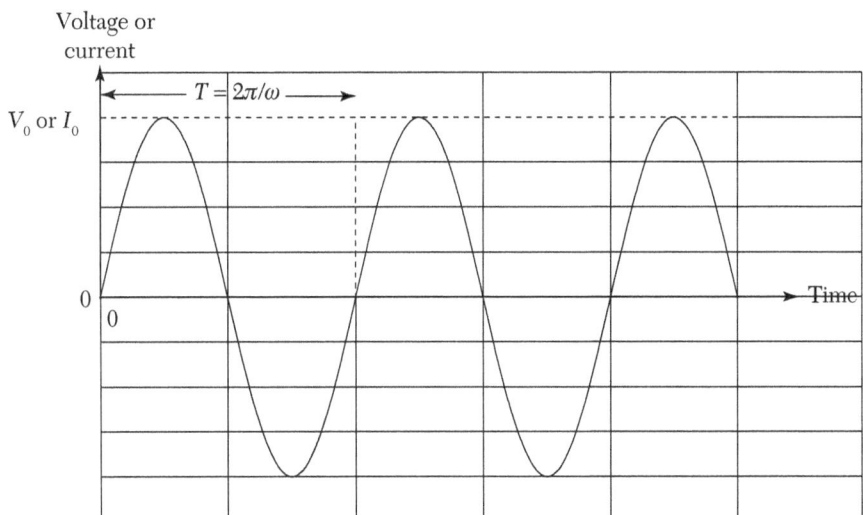

22.1.1   A.C. Power and RMS Values

The instantaneous power of an A.C. supply is given by $P = IV$ and varies throughout each cycle:

$$P = IV = I_0 V_0 \sin^2 \omega t$$

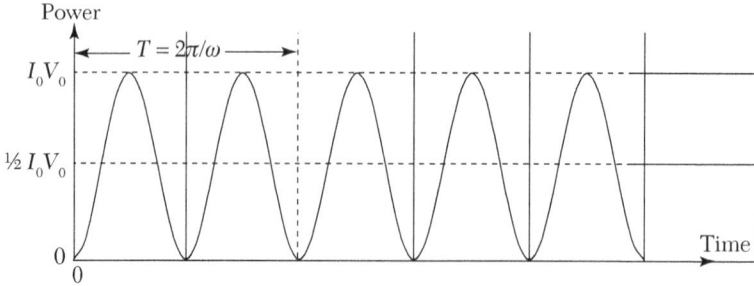

The power is always positive and peaks twice as frequently as the current or voltage. The peak power is $I_0 V_0$, and the average value of the power is $\frac{1}{2} I_0 V_0$.

This can be shown using the trigonometric relation:

$$\sin^2 A = \frac{1}{2} - \frac{1}{2} \cos 2A$$

$$\sin^2 \omega t = \frac{1}{2} - \frac{1}{2} \cos 2\omega t$$

The average value of a cosine term over a whole number of cycles is zero, so the average of a sine-squared term is ½.

The average A.C. power is, therefore,

$$P_{AC} = \frac{1}{2} I_0 V_0$$

The rms values of current and voltage (for sinusoidal variation) are:

$$I_{rms} = \frac{I_0}{\sqrt{2}}$$

$$V_{rms} = \frac{V_0}{\sqrt{2}}$$

So the average A.C. power is:

$$P_{AC} = I_{rms} V_{rms}$$

By using the rms values instead of peak values, we can express the formula for average A.C. power in the same way as we express the formula for D.C. power. This means that an A.C. supply voltage of 240 V rms would light a lamp to the same brightness as a D.C. supply of constant value 240 V.

In this sense the rms value is the "D.C. equivalent" value. However, an A.C. voltage of 240 V rms actually peaks at ±339 V. We can show where the factor of $1/\sqrt{2}$ comes from:

$$V = V_0 \sin \omega t$$

$$V^2 = V_0^2 \sin^2 \omega t$$

$$V^2 = V_0^2 \sin^2 \omega t = \frac{1}{2} V_0^2$$

$$V_{rms} = \sqrt{V^2} = \sqrt{V_0^2 \sin^2 \omega t} = \frac{V_0}{\sqrt{2}}$$

A.C. meters are usually calibrated to give rms values.

22.2   Resistance and Reactance

While current and voltage in a resistor are always in phase, this is not the case for capacitors or inductors. These components introduce a phase difference between the current and voltage, so it is not possible to define resistance in the same way as for a resistor.

22.2.1   Resistors in A.C. Circuits

The relationship between current and voltage for a pure resistance is defined by:

$$R = \frac{V}{I}$$

So if $V = V_0 \sin \omega t$,

$$R = \frac{V_0 \sin \omega t}{I_0 \sin \omega t} = \frac{V_0}{I_0}$$

The resistance is constant and independent of the frequency of A.C. The current and voltage are in phase with each other, and the average power dissipated by the resistor is:

$$P = \frac{V_{rms}^2}{R} = I_{rms} V_{rms} = I_{rms}^2 R$$

22.2.2   Capacitors in A.C. Circuits

The situation is different for a capacitor. The equation that defines the relationship between current and voltage for a capacitor derives from the defining equation for capacitance:

$$Q = CV$$

$$I = \frac{dQ}{dt} = C \frac{dV}{dt}$$

So the current depends on the derivative of the voltage:

$$V = V_0 \sin \omega t$$

$$I = \omega C V_0 \cos \omega t = I_0 \cos \omega t$$

$$I_0 = \omega C V_0 = 2\pi f C V_0$$

This introduces a $\pi/2$ phase difference between the voltage and the current with the current leading the voltage.

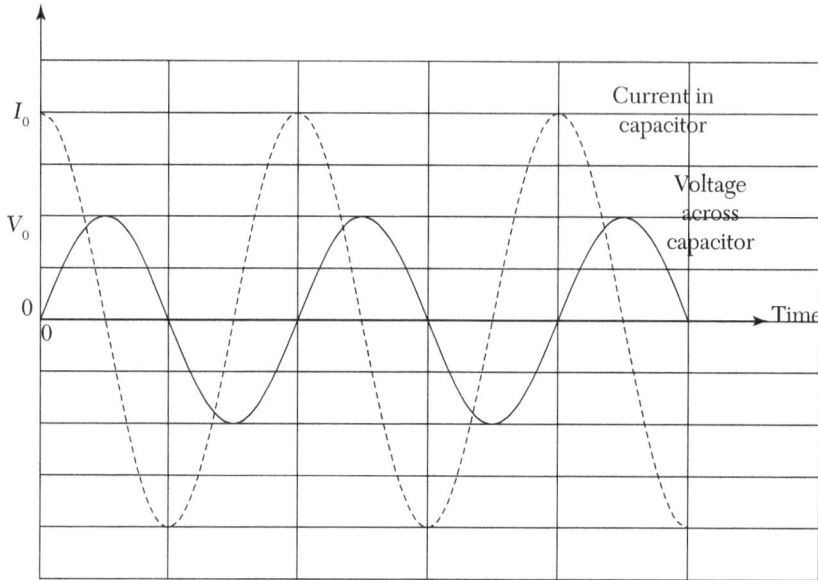

The ratio of V/I is of little use because it varies continuously (being zero at some points and infinite at others). However, the ratio of the peak values of voltage and current is constant; this is the reactance X_C ("kie-cee") of the capacitor.

$$X_C = \frac{V_0}{I_0} = \frac{1}{\omega C} = \frac{1}{2\pi f C}$$

The S.I. unit of reactance is Ω, but reactance is not the same as resistance because the peak values occur at different times.

Reactance can be used to find the peak or rms current in a capacitor if we know the peak or rms A.C. voltage across it. However, the interesting thing about reactance is its frequency dependence.

$$X_C \propto \frac{1}{f}$$

At low frequencies the reactance is high, becoming infinite at $f = 0$ (i.e., for D.C.). This makes sense because the capacitor is actually a break in the circuit so no D.C. can pass through it. At high frequencies the reactance becomes very small, so high-frequency signals can pass through a capacitor almost unimpeded. Capacitors (and inductors) are used in circuits designed to filter out or separate high or low frequencies.

It is also interesting to consider the energy flow when a capacitor is connected to A.C.

$$P = IV = I_0 V_0 \cos \omega t \sin \omega t = \frac{1}{2} \sin 2\omega t$$

The average value of a sine over an integer number of cycles is zero, so the average power delivered to the capacitor during one cycle of A.C. is also zero. For half the time, energy flows from the supply to the capacitor (as it charges); and for the other half, energy flows from the capacitor back to the supply (as it discharges). There is no net energy flow onto the capacitor.

22.2.3 Inductors in A.C. Circuits

The equation that defines the relationship between current and voltage for an inductor derives from the defining equation for inductance:

$$V = -- = L\frac{dI}{dt}$$

E is the back emf in the inductor, which opposes the supply voltage V, hence the change of sign.

If the current is $I = I_0 \cos \omega t$:

$$V = -\omega L I_0 \sin \omega t = V_0 \sin \omega t$$

$$V_0 = -\omega L I_0$$

Once again there is a $\pi/2$ phase difference, but this time the voltage leads the current by $\pi/2$.

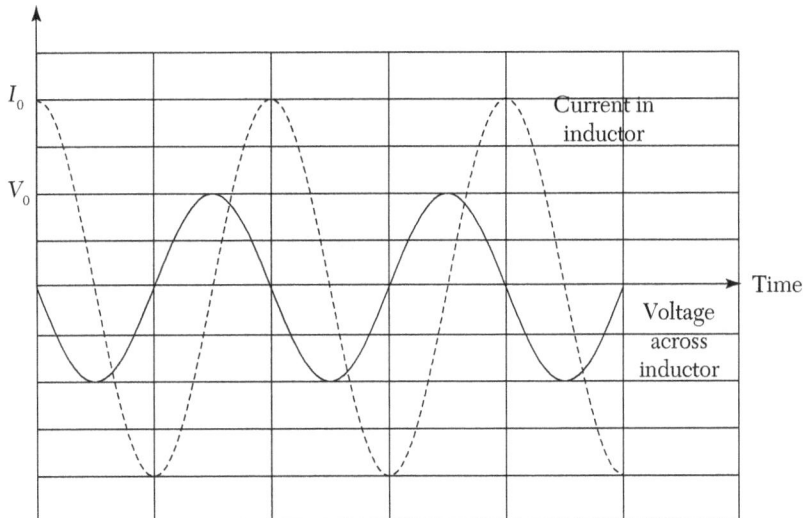

The reactance X_L("kie-ell") of the inductor is defined as the ratio of peak voltage to peak current:

$$X_L = \frac{V_0}{I_0} = \omega L = 2\pi f L$$

This is also measured in Ω and is frequency-dependent. At low frequencies the inductor has very low reactance; low-frequency signals pass through an inductor easily. At high frequencies the reactance becomes very great so that high-frequency signals are severely impeded. This frequency dependence is the opposite of that for the capacitor.

In common with the capacitor, however, there is no net energy flow into the inductor. During one cycle, energy is used to build up the magnetic flux in the inductor, but then energy flows out of the inductor as the field collapses.

In practice inductors consist of coils of conducting wire that has some resistance, so we do not usually encounter pure inductance. A real inductor is modeled as a pure inductor in series with a fixed resistor:

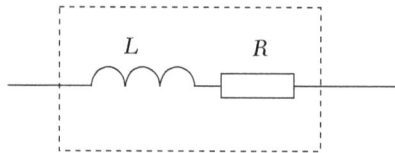

22.3   Resistance, Reactance, and Impedance

Resistors, capacitors, and inductors all respond differently to A.C. signals. The reason for this is that while current and voltage are in phase in a resistor, current leads voltage by $\pi/2$ in a capacitor and lags behind the voltage by $\pi/2$ in an inductor. To determine the relationship between the supply voltage and the voltages across individual circuit components, we must consider their relative phases.

22.3.1   Phasor Diagrams for A.C. Series Circuits

The simplest way to include relative phases is to use a phasor diagram. The phasors representing voltages across a resistor V_R, capacitor V_C, and inductor V_L connected in series are shown below (phasors rotate counterclockwise and current is used as the reference because it is the same in all components in a series circuit). The lengths of phasors representing V_C and V_L will vary with frequency.

$$V_L = IX_L$$

$$V_R = I_R$$

$$I$$

$$V_C = IX_C$$

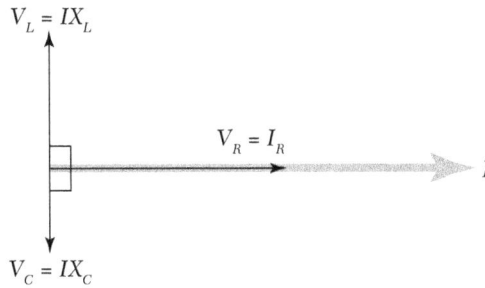

The supply voltage is equal to the sum of the phasors across the three components.

22.3.2   Impedance

The impedance Z of a load that contains resistive, capacitive, and inductive components is defined as the ratio of the peak voltage across the load to the peak current in the load:

$$Z = \frac{V_0}{I_0}$$

and is measured in Ω.

For a series circuit V_0 is found by adding the voltage phasors for the components, and except in special cases, the voltage and current in the circuit are not in phase, so the peak voltage and peak current occur at different times. The general expression for the impedance of a series A.C. circuit containing all three types of components is derived below:

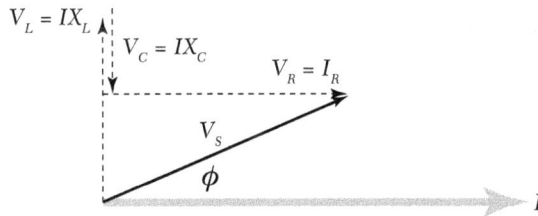

$$V_L = IX_L$$

$$V_C = IX_C$$

$$V_R = I_R$$

$$V_S$$

$$\phi$$

$$I$$

Pythagoras' theorem can be used to find the magnitude of V_S:

$$V_S = \sqrt{V_R^2 + V_L^2 + V_C^2} = I\sqrt{R^2 + X_L^2 - X_C^2}$$

and the impedance is:

$$Z = \frac{V_S}{I} = \sqrt{R^2 + X_L^2 - X_C^2} = \sqrt{R^2 + (\omega L)^2 - \left(\frac{1}{\omega C}\right)^2}$$

The phase angle ϕ is given by:

$$\tan\phi = \frac{V_L - V_C}{V_R} = \frac{X_L - X_C}{R}$$

For particular values of L and C, V_L and V_C are equal in magnitude and add to zero (because they are π out of phase with each other). Under these circumstances, the phase angle is 0 and the impedance is a minimum and purely resistive: the current has a maximum value and is in phase with the supply voltage. This is called resonance; and if resistance is small, it can result in a large increase in current.

The only component that dissipates energy is the resistor, so the power dissipated in this circuit is:

$$P = IV_R = IV_S \cos\phi$$

$\cos\phi$ is called the "power factor" for the circuit.

22.4 A.C. Series Circuits

A.C. electric circuits containing resistors, capacitors, and inductors have interesting frequency-dependent behavior. Some circuits can be made to oscillate and resonate at particular frequencies. These tuned circuits are extremely important in communications systems.

In series circuits, the amplitude and phase of the current through each component is the same, but the voltage can vary both in amplitude and phase.

22.4.1 RC Series Circuit

The behavior of an RC series circuit is quite straightforward. As frequency is increased, the reactance of the capacitor falls, so V_C also falls. This reduces the impedance of the circuit and reduces the phase difference between the supply voltage and the current. The amplitude of A.C. rises toward a maximum value determined only by the resistance, $I_{max} = V_S/R$. At high frequencies, the circuit behaves as if it is purely resistive and the supply voltage is in phase with the circuit current.

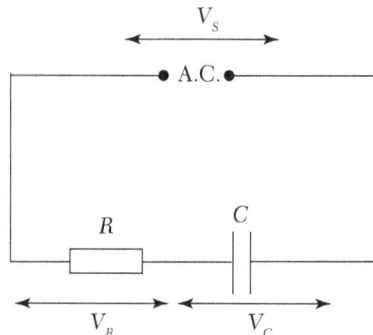

$$V_R = I_R$$

$$\phi$$

$$I$$

$$V_C = IX_C$$

$$V_S$$

$$V_S = \sqrt{V_R^2 + V_C^2} = I\sqrt{R^2 + X_C^2}$$

$$Z = \frac{V_S}{I} = \sqrt{R^2 + X_C^2} = \sqrt{R^2 + \left(\frac{1}{2\pi fC}\right)^2}$$

$$\tan\phi = \frac{\Phi_C}{R} = \frac{1}{2\pi fRC}$$

22.4.2 RL Series Circuit

The behavior of an *RL* series circuit is also straightforward. As frequency is increased, the reactance of the inductor increases, so V_L, the circuit impedance, and the phase difference between the supply voltage and the current also increase, and the amplitude of the A.C. falls toward zero. The maximum current is approached at low frequencies. In D.C., limit $I_{max} = V_S/R$ and the supply voltage and current are in phase.

$$V_S$$

$$\text{A.C.}$$

$$R \qquad L$$

$$V_R \qquad V_L$$

$$V_L = IX_L \qquad V_S$$

$$\phi \qquad V_R = I_R$$

$$I$$

$$V_S = \sqrt{V_R^2 + V_L^2} = I\sqrt{R^2 + X_L^2}$$

$$Z = \frac{V_S}{I} = \sqrt{R^2 + X_L^2} = \sqrt{R^2 + (2\pi fL)^2}$$

$$\tan\phi = \frac{\Phi_C}{R} = \frac{L}{2\pi fR}$$

22.4.3   RCL Series Circuit

In an *RCL* series circuit, the frequency dependence of the capacitor reactance and inductor reactance oppose one another—as frequency increases, the reactance of the capacitor falls and the reactance of the inductor rises. This leads to resonance behavior at the frequency when they are equal.

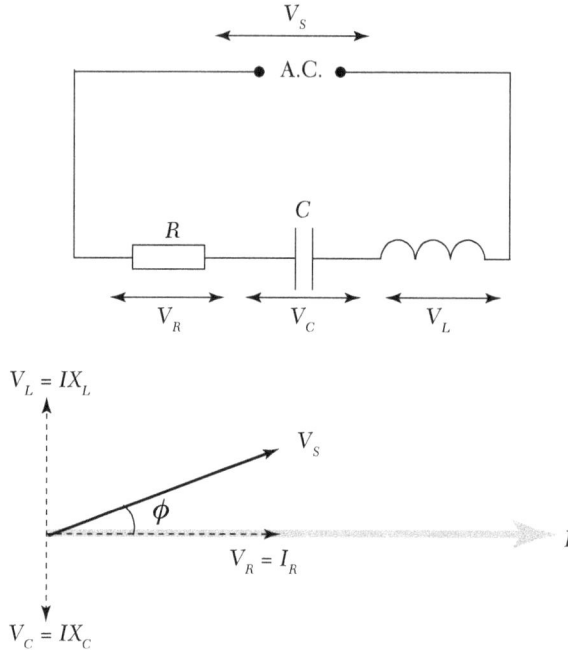

Previously we derived expressions for the impedance and phase of an *RCL* circuit:

$$Z = \frac{V_S}{I} = \sqrt{R^2 + X_L^2 - X_C^2} = \sqrt{R^2 + (\omega L)^2 - \left(\frac{1}{\omega C}\right)^2}$$

The phase angle ϕ is given by:

$$\tan \phi = \frac{V_L - V_C}{V_R} = \frac{X_L - X_C}{R}$$

Resonance occurs at a frequency f_0 when the reactance of the capacitor is equal to the reactance of the inductor.

$$X_L = X_C$$

$$\omega L = \frac{1}{\omega C}$$

$$2\pi f_0 L = \frac{1}{2\pi f_0 C}$$

$$f_0 = \frac{1}{2\delta}\sqrt{\frac{1}{LC}}$$

At this frequency the current in the circuit has its maximum value:

$$I_{max} = \frac{V_S}{R}$$

The circuit acts as a purely resistive load, and the current and supply voltage are in phase.

The graph below indicates how the resistance, impedance, and circuit current vary with frequency for an RCL series circuit.

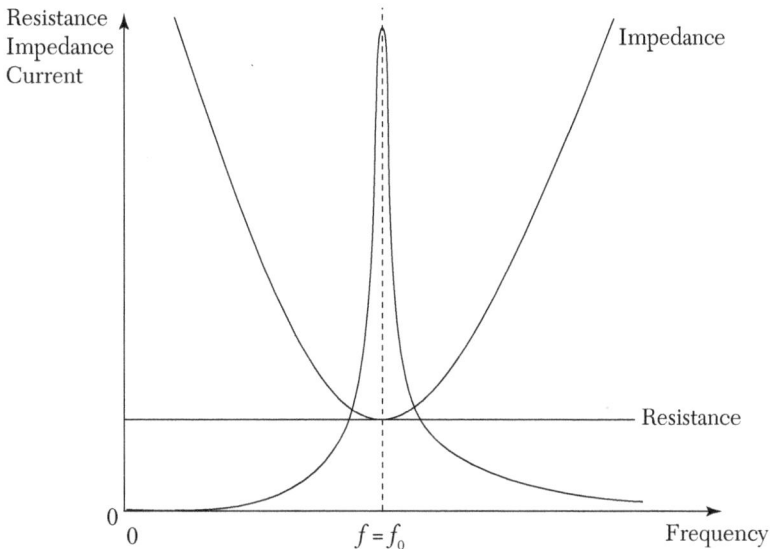

For $f < f_0$ the reactance of the capacitor is greater than that of the inductor, and the circuit is said to be capacitive. For $f > f_0$ the reactance of the inductor is greater than that of the capacitor and the circuit is said to be inductive.

22.4.4   Parallel Circuits Containing Resistors, Capacitors, and Inductors

The analysis of parallel circuits follows a similar procedure to the analysis of series circuits. However, the voltage across each component is now in phase and can be used as a reference phasor. The current in the resistor is in phase with the voltage; the current in the capacitor leads by $\pi/2$; and the current in the inductor lags by $\pi/2$.

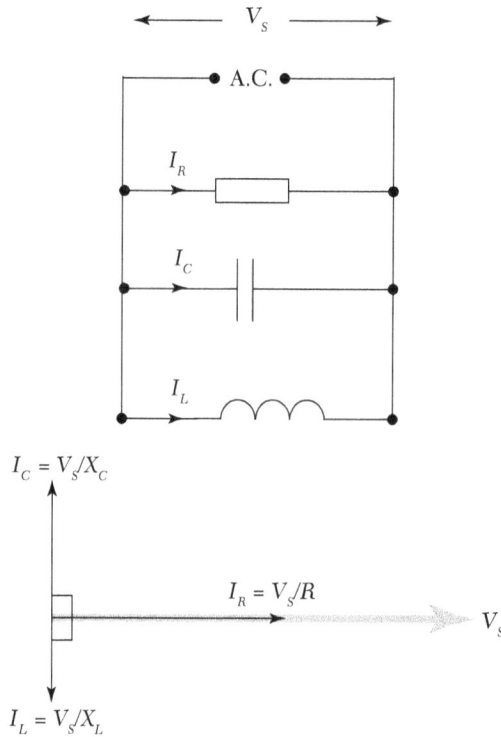

A voltage resonance occurs when $X_L = X_C$ when the circuit behaves as a purely resistive load and the impedance is R. The resonant frequency is again f_0:

$$f_0 = \frac{1}{2\pi}\sqrt{\frac{1}{LC}}$$

At resonance the currents in the capacitor and inductor are equal in magnitude but opposite in direction at every moment.

22.5   Electric Oscillators

A simple oscillator circuit can be constructed from a capacitor and an inductor as shown below. If the capacitor is charged from the D.C. supply and then connected across the inductor, currents in the circuit undergo a series of oscillations.

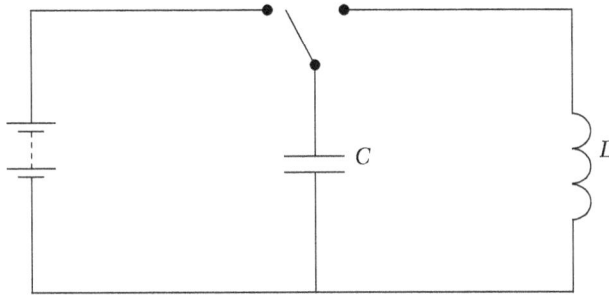

When the capacitor is connected to the inductor, the voltages across them must be equal, so that:

$$\frac{Q}{C} = -L\frac{dI}{dt}$$

Since $I = dQ/dt$ we can form a second-order differential equation for Q and t:

$$\frac{Q}{C} = -L\frac{d^2Q}{dt^2}$$

$$\frac{d^2Q}{dt^2} = -\left(\frac{1}{LC}\right)Q$$

This equation has exactly the same form as the equation of motion for simple harmonic motion, so the solutions will have the same form too:

$$Q = Q_0 \sin \omega t$$

where

$$\omega = \sqrt{\frac{1}{LC}}$$

The charge on the capacitor oscillates with a frequency,

$$f = \frac{1}{2\pi}\sqrt{\frac{1}{LC}}$$

which is the same as the resonant frequency for an *RCL* circuit.

The voltage across the capacitor oscillates at the same frequency:

$$V_C = \frac{Q_0}{C} \sin \omega t$$

and the current is

$$I = \frac{dQ}{dt} = \omega Q_0 \cos \omega t$$

In practice it is not possible to connect a capacitor to a perfect inductor. There is always some resistance in the circuit, so energy is dissipated as the oscillations occur. This leads to damping. The rate of decay of the amplitude of the oscillations depends on the resistance in the circuit; the more resistance, the greater the rate of decay.

22.5.1 Mechanical Analogy

The equations for electrical oscillations have the same form as those for the oscillation of a mass on a spring.

Compare the equations for electrical and mechanical oscillations:

$$\frac{d^2Q}{dt^2} = -\left(\frac{1}{LC}\right)Q$$

$$\frac{d^2x}{dt^2} = -\left(\frac{k}{m}\right)x$$

This identity of mathematic form allows us to make an analogy between mechanical oscillators and electrical oscillators using the following correspondences:

Mass spring oscillator	LCR electrical oscillator
Charge, Q	Displacement, x
Current, dQ/dt	Velocity, dx/dt
Mass (inertia), m	Inductor, L
Spring constant, k	Inverse capacitance, $1/C$

Analogies like this occur all over physics. They help us understand new phenomena in terms of ones we are already familiar with, but they are also useful in their own right. For example, this correspondence between the mechanical and the electrical means that we can model mechanical systems using electrical circuits—e.g., to test new designs for structures.

22.6 Exercises

1. (a) Discuss the advantages and disadvantages of using D.C. batteries compared to A.C. mains as sources of electrical energy.

(b) Explain the advantages of using A.C. for long-distance transmission of electricity.

2. The diagram below enables the same lamp to be connected to either a D.C. or an A.C. supply.

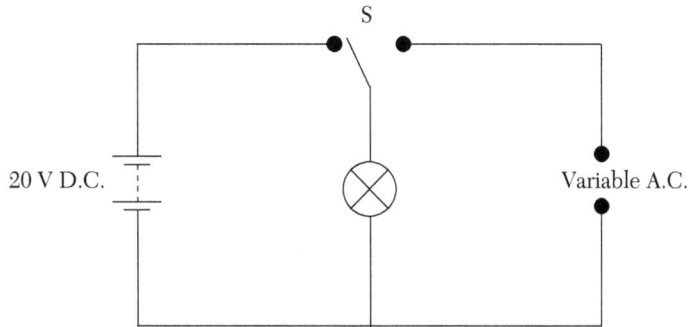

The A.C. supply is adjusted until the lamp lights equally brightly from both supplies.

(a) Calculate the peak value of the rms voltage.

(b) State the rms value of the A.C. voltage.

(c) Discuss whether the lifetime of the bulb (a filament lamp) will depend on the type of supply used to light it.

3. Copy and complete the table below by adding resistance or reactance values at each frequency:

Component/frequency	0 Hz (D.C.)	100 Hz	1000 Hz	10,000 Hz
100 Ω resistor				
100 μF capacitor				
100 μH inductor				

4. The circuit below contains a resistor and an inductor.

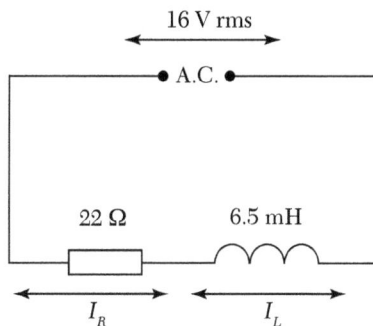

The frequency of the A.C. supply is 1000 Hz.

(a) Calculate the impedance of the circuit.

(b) Calculate the rms current in the circuit.

(c) Calculate the peak voltage across the inductor.
(d) Calculate the power dissipated by the resistor.
(e) Explain why no power is dissipated by the inductor.
(f) Calculate the phase difference between the supply voltage and the current.

5. The circuit below has a resistor, a capacitor, and an inductor connected in series.

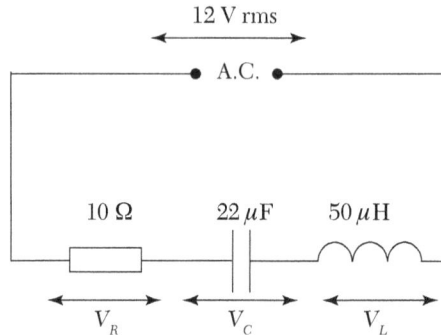

12 V rms

A.C.

10 Ω   22 μF   50 μH

V_R V_C V_L

(a) Calculate the impedance of the circuit at 0 Hz, 1.0 kHz, 5.0 Hz, 10 kHz, and 100 kHz.
(b) Explain what is meant by resonance and calculate the resonant frequency for this circuit.
(c) Sketch a graph to show how the rms current varies with frequency from 0 to 100,000 Hz.
(d) Sketch a graph to show how the phase difference between the supply voltage and the current varies as the frequency changes from 0 to 100,000 Hz.
(e) Describe the energy transfers that take place in the circuit at resonance.

CHAPTER 23

GRAVITATIONAL FIELD

23.1 Gravitational Forces and Gravitational Field Strength

Gravity is one of the four fundamental forces. It has infinite range and obeys a similar inverse-square law to electrostatics. All masses create gravitational fields, but, unlike the electrostatic forces between charges, which can be attractive or repulsive, gravitational forces are always attractive. The gravitational force acting on a mass close to the surface of the Earth is called weight.

23.1.1 Newton's Law of Gravitation

Newton stated that two point masses would exert an attractive force on each other that is directly proportional to the product of the masses and inversely proportional to their separation.

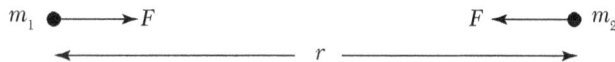

$$F = -\frac{Gm_1m_2}{r^2}$$

The minus sign indicates attraction.

G is the universal constant of gravitation, $G = 6.674 \times 10^{-11} \ \mathrm{Nm^2kg^{-2}}$.

Newton was also able to show that the force of attraction between spheres of uniform density is the same as the attraction between two point masses placed at their centers. This means that we can treat object like planets and stars as point masses when considering orbital motion. It is also important to note that, by Newton's third law, the forces on each mass have the same magnitude, even if the masses are different. For example, the weight of an apple in the Earth's gravitational field is the same as the weight of the Earth in the apple's gravitational field. It is also the case that the gravitational force exerted on the Earth by the Moon is equal in magnitude to the gravitational force exerted on the Moon by the Earth.

The resultant gravitational force on a body affected by the gravitational fields of several other objects (e.g., the Earth affected by the Sun, Moon, and other planets) is the vector sum of the gravitational forces from each of the other objects.

23.1.2 Gravitational Field Strength

The idea that gravitational forces arise from a gravitational field removes the difficulty of an action-at-a-distance explanation. The Moon is attracted to the earth because it experiences a force from the gravitational field *where it is*, i.e., a **local force**.

The gravitational field strength g at a point in space is defined as the gravitational force per unit mass at that point.

$$g = \frac{\text{gravitational force}}{\text{mass}} = \frac{F}{m}$$

The S.I. unit of gravitational field strength is Nkg^{-1}.

For a point or uniform spherical mass, the field strength at a distance r from the center of mass M can be determined by considering the force per unit mass acting on a small mass m placed at that distance:

$$g = \frac{F}{m} = -\frac{GM}{r^2}$$

This is an inverse-square law.

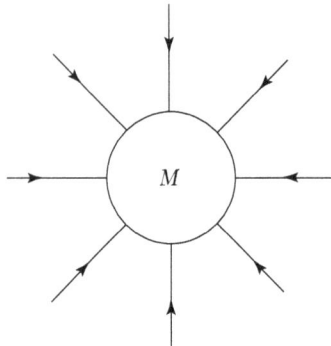

The gravitational field can be represented using field lines in a similar way to the electric field.

The fact that this is an inverse-square law and gravitational field lines can only begin on masses allows us to create a form of Gauss's theorem for the gravitational field.

Gravitational flux through a surface is defined as:

$$\Phi_G = \int_{\substack{\text{closed} \\ \text{surface}}} g.dA = -4\pi G \sum_{\text{enclosed}} m$$

Outer Outer Gaussian surface encloses mass M, so by symmetry:

$$4\pi r_1^2 g = -4\pi GM$$

$$g = -\frac{GM}{r_1^2}$$

Uniform spherical shell of mass M

Inner Gaussian surface encloses no mass, so by symmetry:
$$4\pi r_1^2 g = 0$$
$$g = 0$$

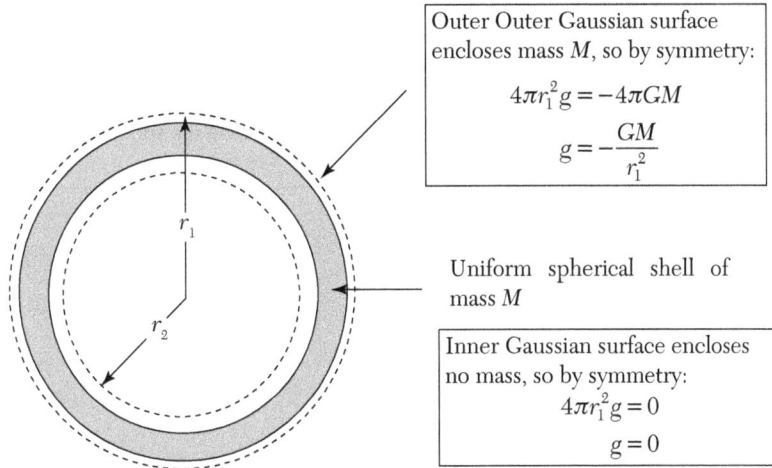

The negative sign again arises because of the attractive nature of the gravitational field; the flux entering through a closed surface is equal to $4\pi G$ times the total mass enclosed by the surface. In the same way as in electrostatics, this form of Gauss's law is particularly useful for situations with spherical symmetry. For example, we can use it to show that the gravitational field strength inside a hollow uniform spherical shell is zero everywhere.

23.1.3 Gravitational Field Strength of the Earth

We will model the Earth as a spherical mass of uniform density. This is not strictly correct—the Earth is actually an oblate spheroid, with a larger equatorial diameter than polar diameter and its density increases toward its core. However, this simple model is useful to give an idea of how the Earth's field varies outside and inside its surface. Assume that the Earth has mass M_E and radius R_E and uniform density ρ.

The field outside the surface is that of a point mass M_E located at the center of the Earth:

$$g(r > R_E) = -\frac{GM_E}{r^2}$$

We can determine an expression for the field strength inside the earth by recalling that the field strength inside a hollow sphere is zero. This implies that the field strength at distance $r < R_E$ from the Earth's center is that of the mass inside radius r.

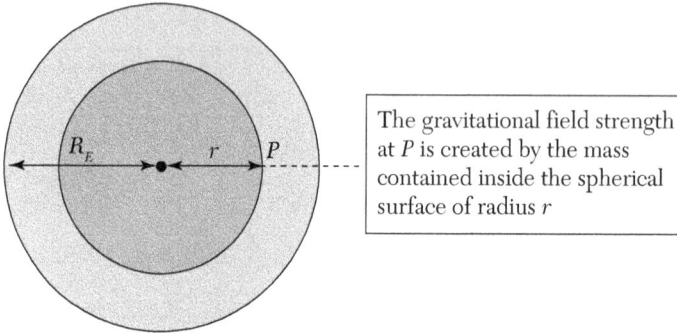

The gravitational field strength at P is created by the mass contained inside the spherical surface of radius r

Mass inside radius r is:

$$m = \frac{4}{3}\pi r^3 \rho$$

$$\rho = \frac{M_E}{\frac{4}{3}\pi R_E^3}$$

$$m = \frac{4}{3}\pi r^3 \times \frac{M_E}{\frac{4}{3}\pi R_E^3} = \frac{r^3}{R_E^3}M_E$$

$$g(r < R_E) = -\frac{GM_E r}{R_E^3}$$

The magnitude of g is zero at the center of the Earth and increases linearly to the surface. The gravitational field strength at the Earth's surface is:

$$g_{\text{surface}} = -\frac{GM_E}{R_E^2}$$

This is about 9.8 Nkg⁻¹ although it varies by a small percentage at different locations.

In many situations on Earth we assume that g is constant. This is a realistic assumption when vertical heights h are much smaller than the radius of the Earth. When this is not the case, we must use the inverse-square law. However, even at the height of the Hubble space telescope

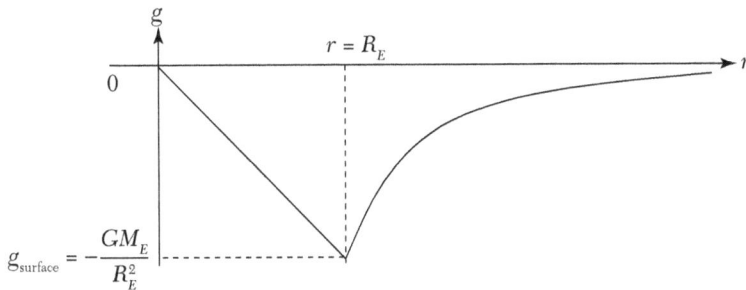

(559 km) the gravitational field strength is about 8.2 Nkg⁻¹, i.e., over 80% of its surface value. The apparent weightlessness of astronauts inside an orbiting spacecraft is not due to g being zero—it isn't. They are in free fall with the same acceleration as their spacecraft, so they do not experience a reaction force to their own weight. The very fact that they are orbiting is because of their weight!

23.2 Gravitational Potential Energy and Gravitational Potential

Gravitational potential energy is the energy a body has because of its position in the gravitational field. When it is moved from one position to another, energy is either transferred to or from gravitational potential energy. For example, lifting a case from the floor and placing it on a table requires an external agent to apply an upward force and to move this force; so work is done on the case and its gravitational potential energy increases.

If an apple falls from a tree to the ground, a gravitational force (its weight) acts on the apple and does work on it, so its gravitational potential energy decreases.

23.2.1 Change in Gravitational Potential Energy

Consider a particle of mass m moving from point A to B in a gravitational field as shown below.

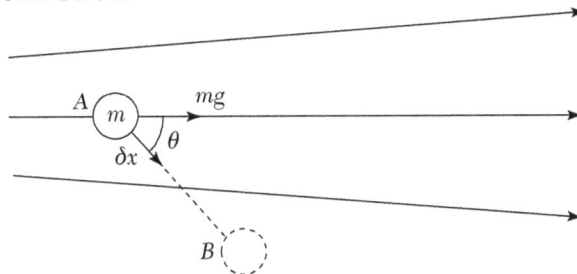

The work done by the gravitational field to move it a short distance δx along the line AB is:

$$\delta W = mg \cos \theta \delta x$$

So the total work done is:

$$W_{AB} = \int_A^B mg.dx$$

where **mg.dx** is the scalar product of the two vectors mg and $\delta\underline{x}$, i.e., the sum of the component of the gravitational force in the direction of motion along each line element. The gravitational forces do work on the particle, so its gravitational potential energy falls:

$$\Delta GPE_{AB} = -\int_A^B mg.dx$$

The differential form of this equation is:

$$\frac{dGPE_{AB}}{dx} = -mg$$

The rate of change of gravitational potential energy with distance is equal in magnitude but opposite in direction to the gravitational force

Uniform gravitational field

Changes of gravitational potential energy in a uniform field of strength g provide a simple (and familiar) formula.

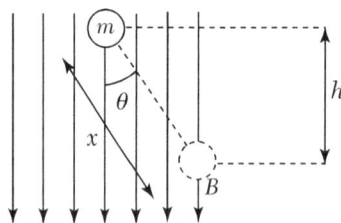

$$\Delta GPE_{AB} = -\int_A^B m\boldsymbol{g}.d\boldsymbol{x} = = mgx\cos\theta = -mgh$$

While this is only exact when g is constant, it is a good approximation for vertical displacements close to the surface of the Earth that are small compared to the radius of the Earth.

23.2.2 Gravitational Potential

The gravitational potential V_G at a point in the field is equal to the gravitational potential energy per unit mass at that point.

$$V_G = \frac{GPE}{m}$$

The S.I. unit of gravitational potential is Jkg^{-1}.

▪ The zero of gravitational potential is taken to be at infinity.

This is the same convention used to define the zero of electrical potential, and it makes sense because when particles are separated by very large distances, their interactions, under an inverse-square law, become negligible. Once we have defined the zero of potential, we can determine the absolute potential (and potential energy) at any point. This is equal to the work that must be done per unit mass to move a small mass from infinity and to place it at point P.

Gravitational potential at P a distance r from a point or uniform spherical mass M

A small mass m is moved from infinity and placed a distance r from a mass M.

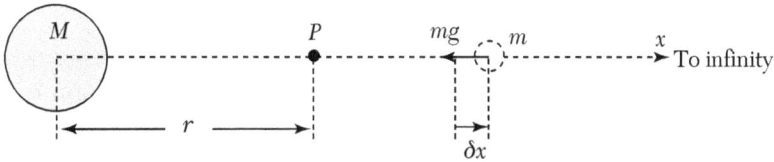

The change in gravitational potential energy of the small mass is:

$$\Delta\text{GPE} = \text{GPE}(\infty)\text{GPE}(r) = -\int_{x=\infty}^{x=r} mg.dx$$

where

$$g = -\frac{GM}{x^2}$$

$$\Delta\text{GPE} = \text{GPE}(\infty)\text{GPE}(r) = -\int_{x=r}^{x=\infty} m\left(-\frac{GM}{x^2}\right)dx = \frac{GMm}{r}$$

$$\text{GPE}(r) = -\frac{GMm}{r}$$

$$V_G(\boldsymbol{r}) = \frac{\text{GPE}(r)}{m} = -\frac{GM}{\boldsymbol{r}}$$

All gravitational potential energies are negative because work would have to be done to move any mass to infinity. This is a consequence of the attractive nature of gravitational forces.

23.2.3 Gravitational Field Lines and Equipotentials

When a mass moves in a direction perpendicular to the lines of the gravitational field, the component of gravitational field in the direction of motion is zero, so no work is done on or by the gravitational field. The gravitational potential energy of the mass is constant and it is moving along a gravitational equipotential. For a point or uniform spherical mass, the equipotentials are concentric spherical surfaces.

The field lines are perpendicular to the equipotentials.

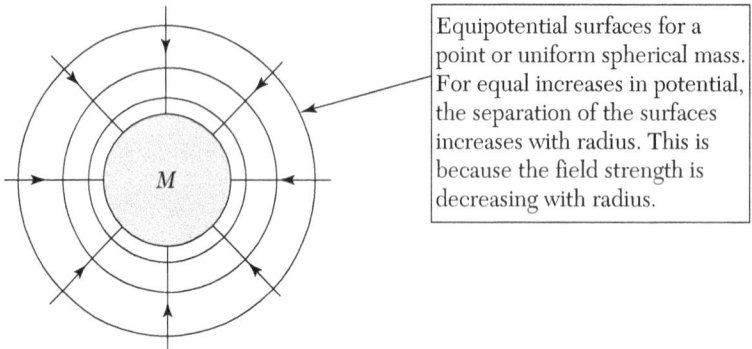

Equipotential surfaces for a point or uniform spherical mass. For equal increases in potential, the separation of the surfaces increases with radius. This is because the field strength is decreasing with radius.

Close to the surface of the Earth the gravitational field is approximately uniform. The equipotentials are planar surfaces parallel to the surface of the Earth and are equally spaced.

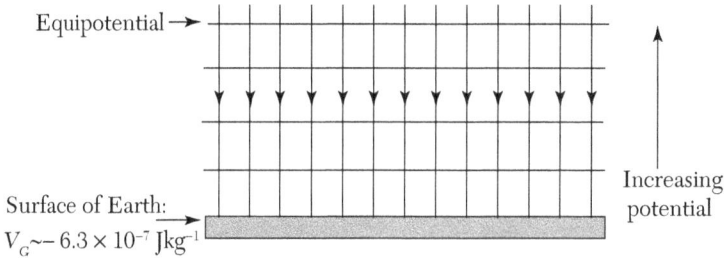

Equipotential →

Surface of Earth:
$V_G \sim -6.3 \times 10^{-7} \, \text{Jkg}^{-1}$

Increasing potential

23.2.4 Gravitational Potential Energy in the Earth's Field

The gravitational potential at the Earth's surface is:

$$V_G\left(R_E\right) = -\frac{GM}{R_E} = -6.2510^7 \, J\text{kg}^{-1}$$

The change in potential energy when a mass is moved from one place (A) to another (B) in the field is given by:

$$\Delta\text{GPE}(AB) = m\Delta V_G = m\left(V_G\left(B\right) - V_G\left(A\right)\right)$$

Consider a mass m moved through a vertical height h from the surface of the Earth:

$$\Delta\text{GPE}(AB) = m\Delta V_G = m\left(-\frac{GM}{\left(R_E + h\right)} - -\frac{GM}{R_E}\right) = m\left(-\frac{GM}{\left(R_E + h\right)} + \frac{GM}{R_E}\right)$$

This can be simplified to:

$$\Delta GPE(AB) = \frac{GMmh}{R_E(R_E + h)}$$

When $h << R_E$ this becomes:

$$\Delta GPE(AB) = \frac{GMmh}{R_E^2} = mgh$$

This is a familiar result for situations where g is constant.

As $h \to \infty$:

$$\Delta GPE(AB) \to \frac{GMm}{R_E}$$

This is equal to the work that must be done against gravitational forces to move a mass m from the surface of the Earth to infinity, i.e., to completely escape from the Earth's field.

23.2.5 Escape Velocity

The escape velocity is the minimum initial velocity required for a body projected from the surface of a planet or star to escape from the gravitational field. In order to escape, the total energy of the projectile must be positive.

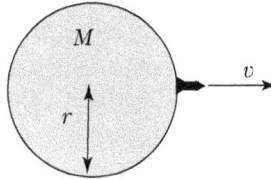

Total energy at surface, TE = KE + GPE

$$TE = \frac{1}{2}mv^2 - \frac{GMm}{r} \geq 0$$

The escape velocity is therefore given by:

$$\frac{1}{2}mv_{esc}^2 - \frac{GMm}{r} = 0$$

$$v_{esc} = \sqrt{\frac{2GM}{r}}$$

Note that this is independent of the mass m of the projectile.

The escape velocity from the surface of the Earth is:

$$v_{esc} = \sqrt{\frac{2GM_E}{R_E}} = 1.12\,10^4\,ms^{-1}$$

Black holes

The speed of light is a limiting speed in the universe, so if the escape velocity reaches this value, then nothing—not even light—can escape. Black holes are objects that are sufficiently massive and compact that the escape velocity at a certain distance from the object is equal to the speed of light. This distance is called the Schwarzschild radius, and while a thorough analysis of black holes requires general relativity, it is possible to derive some useful results from Newton's law of gravitation.

Consider a black hole of mass M. At a certain distance from the center, called the **Schwarzschild radius** R_S, the escape velocity is equal to the speed of light, c.

$$v_{esc} = c = \sqrt{\frac{2GM}{R_S}}$$

$$R_S = \frac{2GM}{c^2}$$

Another way to look at this is to realize that a body of mass M will become a black hole if all of its mass is compressed inside its Schwarzschild radius, R_S. For the Earth to become a black hole this radius is just under 1 cm! The Earth would have to be compressed to the size of a table tennis ball!

A sphere of radius R_S surrounding the black hole is called the **event horizon**. This is because no events inside this radius can communicate with the outside Universe. In a sense everything inside the event horizon has been cut off from the rest of the Universe.

23.3   Orbital Motion

Newtonrealized that circular motion requires a centripetal force and suggested that if an object was projected horizontally at a high enough speed (and drag was negligible) then the object would go into orbital motion around the Earth.

Newton derived his inverse-square law of gravitation by analyzing the motion of the Moon. He argued that the gravitational force of attraction to the Earth provides a centripetal force so that the Moon is in free fall. However, the centripetal acceleration of the Moon is about $1/R_E^2$ times smaller than the free fall acceleration (9.8 ms^{-2}) of a mass dropped near the surface of the Earth, so if both objects fall because of the Earth's gravity, its strength must get weaker with distance and obey an inverse-square law.

Objects projected horizontally from a
point above the Earth's surface

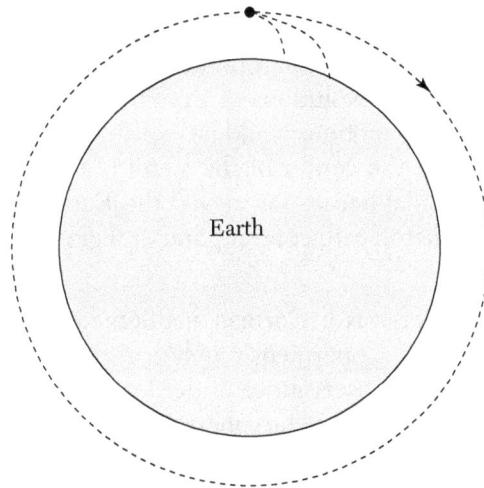

Earth

Newton's law of gravitation was published in *Principia Mathematica* in 1687 and provided astronomers and physicists with the mathematical tools to explain the motions of objects in the solar system and beyond. It is hard to overemphasize the importance of this work; it provided a single elegant explanation for motions across the universe. In our own solar system it explained the motions of all the planets and other bodies (such as comets and asteroids) using a single principle.

23.3.1 Early Ideas about Planetary Motion

The planets have been observed and identified since ancient times because they move against the background of "fixed stars." In many cultures the ability to predict their motions and interpret their meaning held great political and religious significance so that kings and emperors were prepared to support astronomers and astrologers.

The ancient Greeks proposed several models to explain their observations of planetary motion, but the most influential was Aristotle's geocentric (Earth-centric) model. This was developed by Ptolemy and became the basis for astronomical and astrological predictions for nearly 2000 years. The Ptolemaic system placed the Earth at the center of the universe with the Sun, Moon, and planets moving around it. The fixed stars were embedded in a crystal sphere surrounding the solar system. The details of the Ptolemaic system were refined over the years, and while it was physically incorrect it was still useful. The real problem with the Ptolemaic system was the fact that it did not *explain* the motions of the planets; each planet's motion had to be set up independently of the others, so there was no simple unifying principle for the system.

In 1542, Nicolaus Copernicus published his *De Revolutionibus Orbium Coelestiium*("On the Revolutions of the Heavenly Spheres") in which he argued that planetary motions could be explained more simply by assuming that the Sun was at the center of the system (a heliocentric model). This idea was controversial because it moved the Earth from the center of the universe and conflicted with contemporary religious views about the nature of the universe.

Johannes Kepler was a German mathematician and astronomer and strong proponent of Copernicus's heliocentric model. Kepler used the detailed astronomical observations of the Danish astronomer Tycho Brahe to develop three laws of planetary motion. He replaced the complex cycles and epicycles of Ptolemy and Copernicus with elliptical orbits. The fact that all of the planets could then be described by the same mathematical laws suggested that there was an as-yet-undiscovered law that governed all of their motions.

Kepler's laws of planetary motion

- All of the planets move in elliptical orbits with the Sun at one focus of the ellipse.

- A line drawn from the planet to the Sun sweeps out equal areas in equal times.

- The ratio r^3/T^2 is the same for all planets.
  r is the mean orbital radius and T is the period of the orbit.

Galileo Galilei used the newly invented telescope to make astronomical observations. In particular he discovered four moons of Jupiter. This was powerful evidence to suggest that the Earth is not the center of all orbital motion in the solar system. He was a strong supporter of the Copernican model, and Galileo's ideas in astronomy and physics were powerful influences on Newton.

When Newton published his theory of gravitation, he was able to show that an inverse-square law led to elliptical orbits and could be used to explain and derive all of Kepler's laws. For the first time there was a single simple theory that could explain planetary motions (and much more).

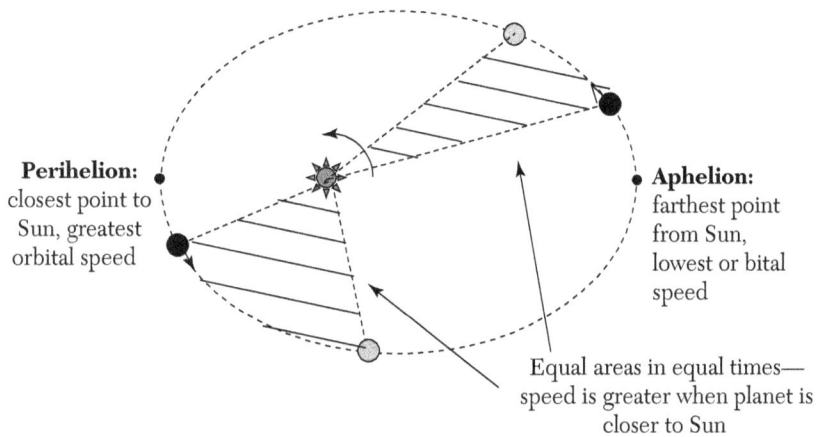

Perihelion: closest point to Sun, greatest orbital speed

Aphelion: farthest point from Sun, lowest or bital speed

Equal areas in equal times— speed is greater when planet is closer to Sun

23.3.2   Circular Orbits

A full mathematical analysis of elliptical orbits is beyond the scope of this book. However, most planetary orbits and many satellite orbits are approximately circular, so we will use Newton's law to analyze circular orbits. Gravity provides a centripetal force for circular motion.

Assume that the central body has a much greater mass than the orbiting body so that the motion of that central body can be neglected. The center of the orbit is then the center of the central mass (in fact both objects orbit about their mutual center of mass).

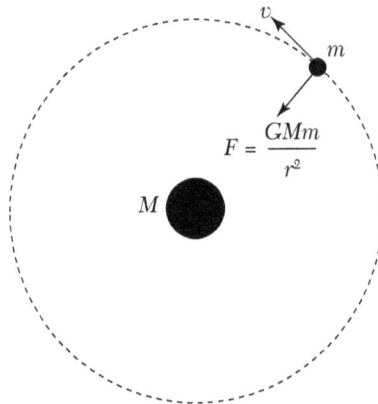

$$F = \frac{GMm}{r^2}$$

$$\frac{mv^2}{r} = \frac{GMm}{r^2}$$

$$v = \frac{2\pi r}{T}$$

$$\frac{r^3}{T^2} = \frac{GM}{4\pi^2}$$

This is Kepler's third law. Note that it is independent of m, so any body in a circular orbit of radius r about the same central mass will orbit with the same period. This law applies for any central body; all that changes is the constant $GM/4\pi^2$, which depends on the mass of the central body. For example, all of Jupiter's moons have the same ratio of r^3 to T^2 as each other, but this is different to the ratio for the planets around the Sun. All Earth satellites will also have the same ratio of r^3 to T^2 as the Earth's Moon.

The total energy of an orbiting body is constant. This is because gravity is a conservative field. When an object moves from one place to another (in the absence of any frictional forces), energy is transferred between potential and kinetic. For a circular orbit, both values are constant.

$$\text{Total energy} = \text{KE} + \text{GPE} = \frac{1}{2}mv^2 - \frac{GMm}{r^2}$$

So

$$\frac{mv^2}{r} = \frac{GMm}{r^2}$$

$$\text{KE} = \frac{1}{2}mv^2 = \frac{GMm}{2r} = -\frac{1}{2}\text{GPE}$$

$$\text{TE} = \frac{GMm}{2r} - \frac{GMm}{2rr} = -\frac{GMm}{2r}$$

The total energy is negative. This makes sense because an orbiting mass is in a bound state. It would require an external agent to do work to remove the mass to infinity.

23.3.3 Artificial Satellites

The first artificial satellite was Sputnik 1 launched by the Soviet Union in 1957. It moved in a slightly elliptical orbit at an altitude that varied between 200 and 900 km. Its orbital period was 96 minutes. In 2016 there are over 1000 operational satellites in orbit around the Earth; more than half of which were launched by the USA.

Orbital period of an artificial satellite is determined by its orbital radius:

$$\frac{r^3}{T^2} = \frac{GM}{4\pi^2}$$

$$T = \sqrt{\frac{4\pi^2 r^3}{GM}}$$

$$T \propto r^{3/2}$$

Satellites in low Earth orbits, like Sputnik, have periods of about 90 minutes. Low Earth polar orbits are useful for weather satellites and other satellites that need to scan the entire surface of the Earth because they pass over a different strip of the surface on each orbit and can observe the entire surface in a 24-hour period.

GPS satellites are further out, at an altitude of about 20,000 km and have an orbital period of about 12 hours. The GPS system creates a constellation of satellites around the globe and relies on several satellites being visible to a GPS receiver at any point on the Earth's surface at any time. The minimum number of operational satellites required to achieve this is 24, but the target number is 33. Over 70 GPS satellites have been launched, but not all of them are still operational.

Geostationary satellites are placed in an equatorial orbit at an altitude of about 35,800 km ($r = 42,200$ km) so that they have an orbital period of 24 hours. This ensures that they remain stationary above the same point on the Earth's equator because they complete one orbit as the Earth itself rotates once. This is called a **geosynchronous orbit** and is used for communications satellites. The fact that they remain stationary in the sky means that ground-based antennae do not have to track them.

23.4 Tidal Forces

Tidal forces occur because the strength of the gravitational field varies across an extended body. This has the effect of stretching or compressing the body parallel or perpendicular to the field. The ocean tides on Earth are an example of this, and their regularity can be explained by changing tidal forces as the Earth's orbital position changes with respect to the Moon (and, to a lesser extent, the Sun). However, tidal forces are a very common occurrence and can have very important consequences. Jupiter's Moon Io, for example, is compressed and stretched by tidal forces as it orbits close to the giant planet. These forces provide the energy to drive seismic activity on Io, and active volcanoes were observed on the surface of Io when the Voyager 1 spacecraft passed it in 1979.

23.4.1 Origin of Tidal Forces

The diagram below shows a spherical body (body 1) in the gravitational field of another body (body 2) of mass M.

The four squares A, C, D, and E represent small masses of size m on the surface of body 1 along diameters parallel and perpendicular to the line connecting it to the center of body 2. Body 1 is only affected by the gravitational field, so it can be considered to be in free fall toward body 2. In the free-falling reference frame, there will be a tension along line AC (C

is in a stronger field than B, and A is in a weaker field, so both are pulled away from mass B). There will also be a smaller compression along line DE because of the inward components of the gravitational field at the edges of body 1. The effect is to stretch the body 1 along AC and to compress it along DE. This is shown in the diagram below.

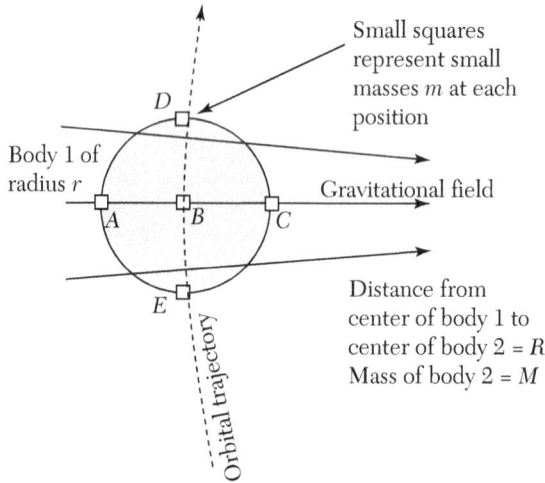

The tidal forces shown are those that act in the freely falling reference frame.

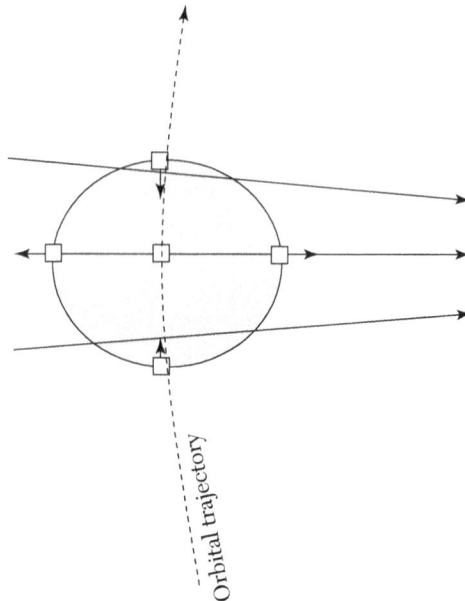

Tidal forces are differential forces and are defined as the difference between the actual gravitational force F_A on a small mass m and the force that would be exerted on it if it was at the center of the body, F_B. For example, the tidal force on the mass at A is given by:

$$\text{tidal force}(A) = F_A - F_B = \frac{GMm}{(R+r)^2} - \frac{GMm}{(R)^2} = \frac{-GMm(2Rr+r^2)}{(R+r)^2(R)^2}$$

For $r << R$, this simplifies to:

$$\text{tidal force}(A) = \frac{-2GMmr}{R^3}$$

The negative sign here indicates that the direction of the tidal force is away from the center of the body. This obeys an inverse-cube law with respect to distance, so tidal forces will only be significant for bodies that are relatively close together.

23.4.2 Earth's Ocean Tides

The ocean tides on planet Earth are driven by tidal forces. While both the Moon and the Sun affect the tides, the Moon's effect is significantly greater. This can be shown by comparing the maximum tidal force on a 1 kg mass due to each body.

Tidal force from the Moon

$M = 7.3 \times 10^{22}$ kg

$m = 1$ kg

$R = 3.9 \times 10^8$ m (mean distance to the Moon)

$r = 6.4 \times 10^6$ m (mean radius of Earth)

$$\text{tidal force}(A) = \frac{-2GMmr}{R^3} = 1.1 10^{-6} \text{N}$$

Tidal force from the Sun

$M = 2.0 \times 10^{30}$ kg

$m = 1$ kg

$R = 1.5 \times 10^{11}$ m (mean distance to the Sun)

$r = 6.4 \times 10^6$ m (mean radius of Earth)

$$\text{tidal force}(A) = \frac{-2GMmr}{R^3} = 5.1 10^{-7} \text{N}$$

The Moon's tidal effect is roughly double that of the Sun, but both are significant. As the Earth, Moon, and Sun change their relative positions, the tidal effects of the Moon and Sun on the Earth's oceans can reinforce or weaken one another. The largest or "spring tides"occur when the three bodies are aligned and the smallest or ""neap tides"occur when the lines from the Sun and Moon to the Earth are perpendicular. These effects are exaggerated in the diagrams below, which are not to scale!

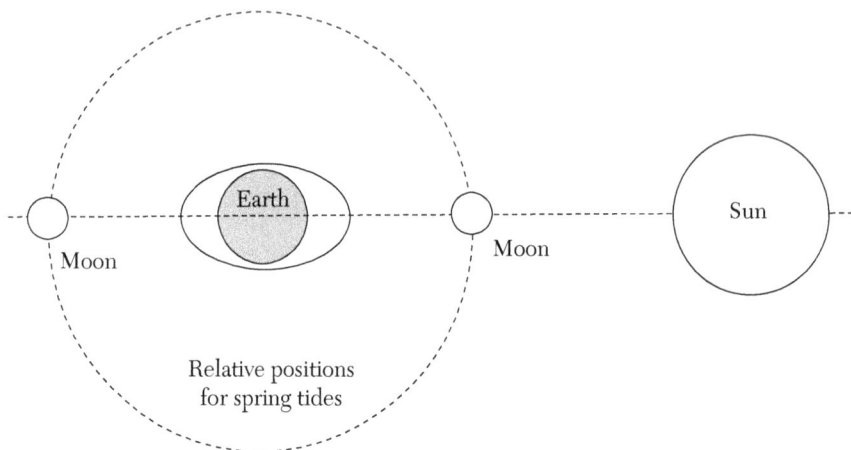

Relative positions
for spring tides

The period of the Earth's rotation is significantly less than the period of the Moon's orbit, so the Earth effectively rotates under the tidal bulges and there are approximately two high tides in each 24-hour period. However, the Earth's rotation pushes the bulge slightly ahead of the Moon's position.

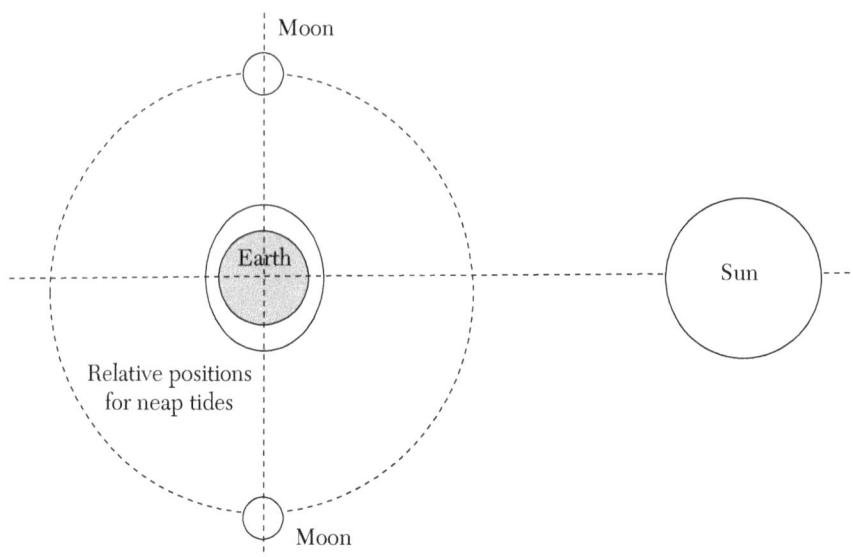

Relative positions
for neap tides

This has the effect of creating a retarding torque on the Earth that reduces its rotation period, making day length gradually increase and reducing its angular momentum. However, angular momentum for the system must be conserved, so the radius of the Moon's orbit also increases. The present rate of increase of day length is about 1.7 ms per century, and the Moon's orbital radius increases by about 3.8 cm per year. When the Moon first formed, the length of a day on Earth was only about 5 hours. The long-term effect of this interaction is that the Moon's orbital period and the Earth's rotation period will eventually become the same—they will be tidally locked.

23.5   Einstein's Theory of Gravitation

Newton's theory of gravitation is actually an approximation to a more fundamental theory discovered by Einstein in 1915. This is the general theory of relativity, and while the mathematics of that theory are well beyond the scope of this book, some of the key ideas are accessible without the mathematics.

23.5.1   Space–Time Curvature

The key idea in Einstein's theory is that gravity is not really a force field but a distortion of the geometry of space and time. What we experience as a gravitational field is actually the curvature of the space–time continuum where we are. This sounds rather obscure, but can be understood by analogy. Imagine stretching a rubber sheet out so that it is completely flat and then rolling a ball across its surface. In the absence of friction the ball would continue in a straight line at constant velocity. However, if a heavy mass is placed onto the surface and the experiment is repeated, the ball

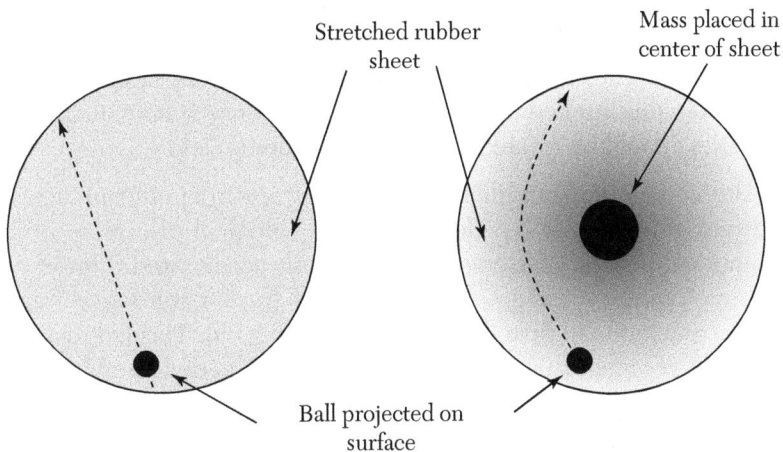

Stretched rubber sheet

Mass placed in center of sheet

Ball projected on surface

follows a curved path around the mass. The geometry of the rubber sheet has changed and the path followed by the ball has changed too.

Seen from above the presence of the central mass has caused the deflection of the ball. As far as the ball is concerned, it has simply followed the curvature of the surface. John Wheeler summarized Einstein's theory of general relativity by saying that:

- Matter tells space how to curve.

- Space tells matter how to move.

In Newtonian mechanics, an object continues to move in a straight line unless acted upon by a resultant external force. In Einstein's universe, objects move along the shortest paths in curved space–time. These paths are called geodesics. A good analogy is found in the paths followed by aircraft on long-haul flights around the Earth. The shortest path between two points on the Earth's surface is an arc of a great circle around the center of the Earth. According to Einstein's theory, the Earth's orbit around the Sun is actually a geodesic in space–time. The Earth is not being deflected by a gravitational force; it is simply following the local curvature of space–time and that curvature has been caused by the presence of the Sun.

23.5.2 Equivalence Principle

One powerful idea that led Einstein to the general theory of relativity is the equivalence principle. He realized that the effects of gravity and the effects of acceleration are actually equivalent. To understand what this means, imagine you are on a fairground ride that allows you to free fall vertically before bringing you safely to rest. While you are in free fall, you cease to feel the reaction from your seat and you feel "weightless." From an outsider's reference frame, you are falling because of your weight, but from your own reference frame, the gravitational field strength seems to be zero. To Einstein this implied that freely falling reference frames at each point in space–time could be treated as if the gravitational field was zero.

Alternatively it means that the physical effects in a uniformly accelerated reference frame must be identical to the physical effects in a uniform gravitational field. An observer in a uniformly accelerated reference frame with acceleration a cannot distinguish this from a reference frame in a uniform gravitational field of strength g where $g = a$. This led to a powerful thought experiment that showed that light must be deflected by gravity.

Imagine a laboratory inside a rocket that is accelerating vertically upwards. The acceleration of the rocket is a and observers inside the rocket feel exactly the same as if they were in a uniform gravitational field of strength $g = a$. Now imagine that a horizontal beam of light enters the rocket through a port hole on one side and leaves through a port hole on the other side. Since the rocket has constant acceleration the beam will follow a parabolic path inside the rocket. Now imagine that the same rocket is at rest on the surface of a planet with surface gravity g. If a beam of light enters horizontally through the same window, it must, according to the equivalence principle, follow the same parabolic path as before and deflect downwards toward the surface of the planet. If this was not the case, then an observer inside the rocket could tell whether the rocket was at rest in a gravitational field or accelerating simply by observing the motion of light beams.

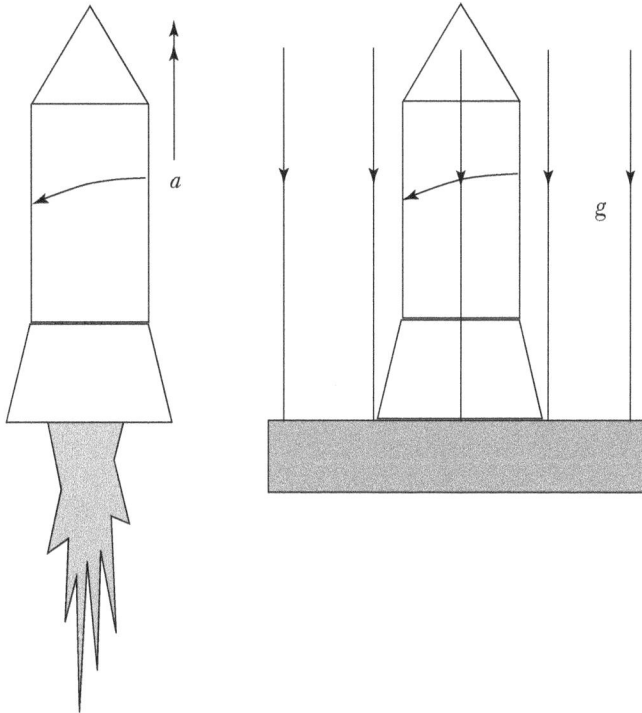

The British astronomer and mathematician, Arthur Eddington, tested this prediction in 1919. He looked for the apparent shift in the positions of stars close to the disc of the Sun during a total eclipse. He needed to do this during a total eclipse because stars close to the Sun's disc can only be observed when the bright disc of the Sun is eclipsed and effects on stars farther from the disc would be too small to measure. The measured shifts were in agreement with Einstein's theory. The theory has been tested many times and in many different ways since then, and results have all been in agreement with theory. General relativity is regarded as one of the fundamental theories of physics.

23.5.3 Gravitational Time Dilation

Einsteinrealized that the distortion of space–time by gravity would lead to time dilation effects. Time in a stronger gravitational field runs more slowly than time in a weaker gravitational field. This is important for GPS satellites because they are in a weaker gravitational field than the receivers on the surface, and a correction must be built in to retain location accuracy. (They must also be corrected for another time dilation effect caused by the relative motion between the satellite and the receiver.)

A simple formula for time dilation can be derived using Newton's theory of gravity. Consider a photon emitted vertically from a source on the surface of a star of mass M. As the photon travels away from the star, its gravitational potential energy increases, so photon energy and frequency fall. The change in time period of the photon in different positions in the gravitational field can be used to compare the rates of clocks at these positions. The time period of the photon at A is T and at B is T'.

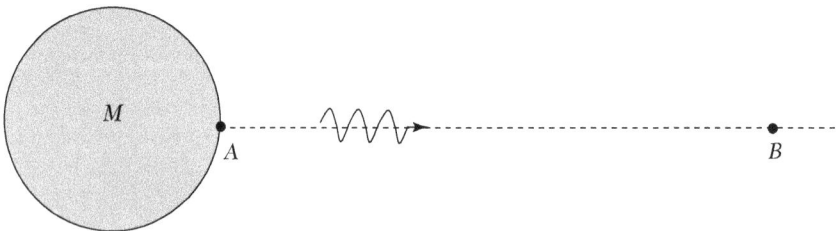

$$E_B - E_A = mV_B - mV_A = m\Delta V$$

m is the mass equivalent of the photon energy andDV is the change in GPE between A and B.

If the change in potential energy of the photon is small compared to its energy, $m\Delta V \ll hf$, then $T' \sim T$, so we can simplify the expression as follows:

$$hf' - hf = \frac{hf\Delta V}{c^2}$$

$$\frac{h}{T'} - \frac{h}{T} = \frac{h\Delta V}{Tc^2}$$

$$\frac{\Delta T}{T'T} = \frac{\Delta V}{Tc^2}$$

$$\Delta T \sim \frac{T \Delta V}{c^2}$$

$$T' = T\left(1 + \frac{\Delta V}{c^2}\right)$$

The greater the change in potential, the greater the difference in clock rates.

Gravitational time dilation becomes extreme in very strong gravitational fields (where the equation above is not valid). If we were to observe clocks close to the event horizon of a black hole, they would tick very slowly indeed, and time would stop at the event horizon. In this sense anything that actually enters the black hole has passed beyond the infinity of time in the external universe!

23.5.4   Gravitational Waves

Matter determines the curvature of the space–time continuum, so when the distribution of matter changes (e.g., an orbiting binary system), so does the local geometry. Einstein's field equations link these local changes to the rest of space–time, so that a disturbance travels outwards at the speed of light. This is a gravitational wave. The existence of gravitational waves was predicted by Einstein in 1916 as a consequence of his general theory of relativity.

When a gravitational wave passes, it causes periodic changes in the geometry of the objects in its path. The diagram below shows how a spherical object distorts along two perpendicular axes as a gravitational wave passes.

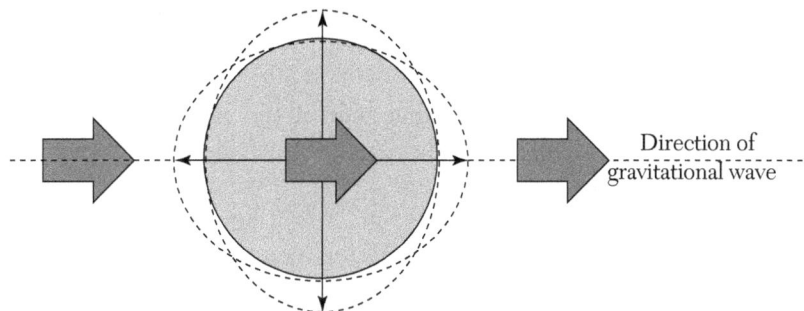

Direction of gravitational wave

However, the predicted amplitude of these vibrations is tiny, even from powerful astronomical events such as the collision of black holes. Typical amplitudes change the dimensions of an object by about 1 part in 10^{20}. For an object 1 m long, this is 10 billion times smaller than the diameter of an atom! This makes them extremely difficult to detect, and the first evidence for the existence of gravitational waves was indirect.

Gravitational waves transfer energy, so an orbiting binary system loses energy as it radiates and its period gradually changes. Alan Hulse and Joseph Taylor analyzed the period of a binary system that consisted of a pulsar and a normal star over a period of 30 years. They showed that the reduction in period was within 0.2% of the change predicted on the basis of gravitational radiation. The energy radiated is about 7.35×10^{24} W, about 2% of the Sun's luminosity. Their work won them the 1993 Nobel Prize in Physics and provided the first convincing experimental evidence for the existence of gravitational waves.

The most sensitive terrestrial detectors use an interferometer to detect small changes in light path over two perpendicular arms.

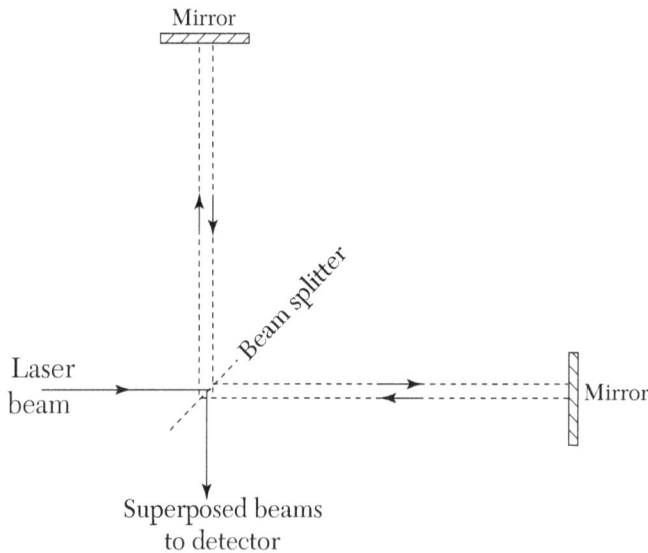

Monochromatic light is split so that half of the beam travels along the "vertical" arm and half along the "horizontal" arm (in reality, these will both be horizontal—e.g., NS and EW, and perhaps several kilometers long). The returning beams superpose and interfere. When a gravitational wave passes, the lengths of the arms fluctuate, introducing a periodic phase difference that makes the interference pattern change at the frequency of the wave. These fluctuations were finally discovered at the Laser Interferometer Gravitational Waves Detector (LIGO) in 2015 and announced in February 2016 (100 years after they were predicted). The signal detected by LIGO was consistent with the source being the inward spiral and coalescence of a pair of black holes of around 36 and 29 solar masses. This discovery opens the door to a new age of astronomy in which the most violent cosmic events are observed using gravitational wave "telescopes."

23.6   Exercises

1. Newton arrived at the law of gravitation by assuming that the centripetal force required to keep the Moon in a circular orbit was provided by gravity. In this question we will show how this leads to an inverse-square law. The first step is to compare the force on a 1.0 kg mass near the Earth's surface with the centripetal force on 1.0 kg of the Moon's mass.

    (a) Calculate the weight of a 1.0 kg mass close to the Earth's surface.
    $g = 9.8 \, \text{Nkg}^{-1}$.

    (b) Calculate the centripetal force required to keep 1.0 kg of the Moon's mass in its orbit around Earth. Moon's orbital period = 27.3 days. Mean radius of Moon's orbit = 3.84×10^8 m.
    Your answer in (b) is much smaller than in (c), but the Moon is further from the Earth, so if both forces arise because of gravitational attraction to the Earth, then the strength of this attraction must fall with distance.

    (c) Show that the forces in (a) and (b) are consistent with the idea that gravitational forces obey an inverse-square law with distance. Radius of the Earth = 6.4×10^6 m.

    (d) Give an argument to support the idea that gravitational attraction is proportional to the mass of the object attracted (think of the Moon in its orbit).

    (e) Give an argument to support the idea that gravitational attraction is proportional to the mass of the attracting object (think of Newton's third law).

    (f) Use your answers to (c), (d), and (e) to justify the form of Newton's law of gravitation.

2. Consider a mass m suspended at a point P just above the surface of a planet of mass M and radius R. Write down expressions for each of the following quantities:

    (a) the gravitational field strength and potential at P

    (b) the gravitational field strength and potential at a point Q a height $3R$ above the planet's surface

    (c) the work that must be done to raise mass m from P to Q

    (d) the final velocity of mass m, just before it strikes the planet's surface, if it falls freely from Q (ignore atmospheric drag).

(e) the escape velocity if the mass is projected away from the planet from rest at point P

(f) the escape velocity if the mass is projected away from the planet from rest at point Q

(g) the kinetic energy that must be supplied to m if it is to be projected tangentially from Q and then enter a circular orbit

3. (a) Derive an equation for g as a function of radius r from the center of the Earth assuming the Earth's density ρ is constant and that the Earth is a perfect sphere.

(b) The Earth is actually an oblate spheroid, flattened at the poles. Discuss how this affects the value of g at different points on the surface.

(c) The fact that the Earth rotates on its axis means that part of the gravitational force holding us to its surface must provide the centripetal force needed for us to rotate with the Earth. This affects our measured weight. Calculate the percentage change in weight of a man at the equator. $G = 6.7 \times 10^{-11}$ Nm^{2}kg^{-2}, $M_E = 6.4 \times 10^{24}$ kg, $R_E = 6.4 \times 10^6$ m, $T = 89,800$ s (rotation period).

4. Consider the Earth and Moon as an isolated system. At what distance from the Earth (along a line connecting it to the Moon) will a space craft experience no resultant gravitational force? $G = 6.7 \times 10^{-11}$ Nm^2 kg^{-2}, $M_E = 6.4 \times 10^{24}$ kg, $M_M = 7.2 \times 10^{22}$kg, $R_E = 6.4 \times 10^6$ m, $r = 3.8 \times 10^8$ m (separation of centers of Earth and Moon).

5. The Moon orbits the Earth in about 27.3 days at a distance of 380,000 km.

(a) State Kepler's three laws of planetary motion.

(b) Derive Kepler's third law for a planet in a circular orbit.

(c) Kepler's laws were originally stated about the planets in our solar system, but they can be applied to any similar system where satellites orbit a central body. Explain how the constant in Kepler's third law is affected when the law is applied to different systems.

(d) Use Kepler's third law to calculate the distance from the Earth at which an artificial satellite would be geostationary (i.e., have an orbital period of 24 hours or 1 day).

(e) Why must geostationary satellites be placed in equatorial orbits?

(f) Use Kepler's third law to derive a lower limit for the period of an artificial Earth satellite.

6. Ganymede is Jupiter's largest moon. Its orbit is circular with a radius of 1.07×10^9 m. The orbital period is 7 days, 3 hours,, and 43 minutes. Use Kepler's third law to find Jupiter's mass.

7. If the Olympics were to be held on the Moon, suggest how the shot putt record on the Moon would compare with the record on Earth. Explain your answer.

8. Both the Sun and Moon exert tidal forces on Earth's oceans.
   (a) Explain, qualitatively, how tidal forces arise.
   (b) Show that the tidal effects caused by the Sun are significantly less than those caused by the Moon.

9. Astronomers think that black holes may exist at the center of most galaxies. The evidence is from the high orbital speeds of stars near the center of the galaxies (including our own Milky Way). The star orbits because of attraction to the central object and that object must have a radius smaller than that of the star's orbit. Use this information to show that the object is likely to be a black hole if:

$$v \geq \frac{c}{\sqrt{2}}$$

where v is the orbital speed (assume the orbit is circular) and c is the speed of light.

The equation for the Schwarzschild radius is $R_S = \frac{2GM}{c^2}$

where M is the mass of the central object.

CHAPTER 24

SPECIAL RELATIVITY

24.1 The Postulates of Special Relativity

Classical physics consists of Newtonian mechanics (and gravity) and Maxwell's electromagnetic theory. However, by the end of the nineteenth century it became clear that mechanics and electromagnetism were in conflict, especially when physicists tried to understand what was meant by the speed of light. Up to that point the speed of a wave was measured relative to the medium through which it moved, but light is an electromagnetic wave that travels through a vacuum, so what exactly do we measure the speed of light against?

24.1.1 Absolute Space

Galileo realized that the laws of mechanics are exactly the same whether you are at rest or in a uniformly moving laboratory. In other words, if you find yourself inside a closed spacecraft with no windows, there is no mechanical experiment that you can carry out inside that spacecraft to determine whether it is at rest or moving with constant velocity. This is why we can comfortably drink coffee, read a book, and walk about inside a jet aircraft on a long-haul flight. This idea is called Galilean relativity:

- The laws of mechanics are the same in all uniformly moving reference frames.

These reference frames are called **inertial frames**.

If the laws of mechanics cannot distinguish rest from uniform motion, can anything? Is there a non-mechanical experiment that can distinguish motion from rest? This is an important question because if the answer is "yes" then there is one special reference frame that is at rest and all other reference frames are in motion. If the answer is "no" then the idea of being at rest is not special at all. The idea of a privileged reference frame at rest in the universe leads to the concept of **absolute space**, a fixed background or reference frame against which all other motions can be measured.

This is where the speed of light comes in. This speed emerges naturally from Maxwell's equations and represents the speed of a disturbance in the electromagnetic field. Other kinds of waves are all vibrations in some material medium; so physicists naturally thought that light must also have its own medium that fills the vacuum of space and is at rest in absolute space. They called this all-pervasive medium the **luminiferousether**. The speed of light would then be the speed relative to ether and to absolute space. They immediately began to think about how they might detect ether, or at least detect our motion relative to ether.

In principle, it should be relatively easy to detect the effect of the Earth's motion relative to ether. The Earth orbits the Sun, so its motion through ether ought to vary periodically during the year. If the speed of light is constant relative to ether, then the speed of light measured on Earth would be a relative velocity and should also vary—sometimes above and sometimes below the value given by Maxwell's equations. Several ingenious experiments were carried out to test this idea; the most famous of which was the Michelson–Morley experiment in 1887.

In the Michelson–Morley experiment a light beam is divided into two, and each half is sent on a round trip perpendicular to the other half of the beam. The two beams are then combined and form an interference pattern. Any difference between the time of flight on the two paths shows up as a shift in the interference pattern when the returning beams superpose. Since the Earth orbits the Sun, the experimenters expected the interference pattern to shift during the year as the Earth's motion relative to ether changes, and this changes the relative velocity of light, producing different time delays for each path. For example, if the Earth moves through ether in the same direction as light, then the relative velocity is lower; and if the Earth moves in the opposite direction, the relative velocity is higher.

The (simplified) arrangement of apparatus for the Michelson–Morley experiment is shown below. This is an example of an **interferometer** consisting of two arms of equal length. Motion of the laboratory through ether

Luminiferous ether at rest in absolute space

Earth's velocity v
relative to ether

Light beam velocity c
relative to ether

Light beam velocity $(c - v)$
relative to the Earth

Earth moving
relative to ether

Luminiferous ether at rest in absolute space

Apparatus at rest in laboratory

Mirror

Laboratory moves at
velocity v through ether

Half-silvered
mirror (beam
splitter)

Mirror

Monochromatic
light

Interference pattern

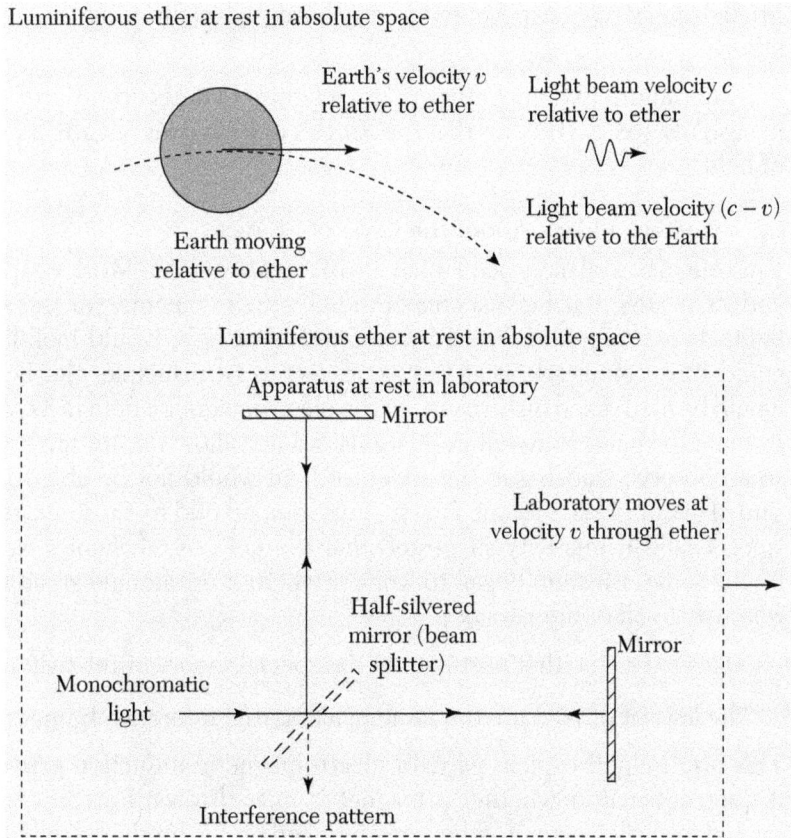

delays the light along both arms, but the delay is greater along the path parallel to motion through ether. If the apparatus is rotated through 90°, the greater delay should shift to the other arm of the interferometer and the interference pattern should shift. The apparatus used by Michelson and Morley was sensitive enough to detect shifts caused by relative velocities comparable to the Earth's orbital speed, so they expected to detect the Earth's motion through ether.

Despite carrying out the experiment carefully and repeating it at different times of the year, they could not detect any shift in the position of the interference fringes. Similar experiments have been carried out using more modern equipment but with the same results—i.e., no shift in interference fringes. This is called a **"null result"** and is probably the most famous null result in the history of physics. All attempts to detect ether have failed, and the fact that there is no delay in experiments such as the

Michelson–Morley experiment is very hard to explain using the concept of ether.

This failure to detect ether is also a failure to detect the Earth's motion in absolute space. It seems that the Earth's motion does not affect the speed of light.

24.1.2 Einstein's Ideas About the Laws of Physics

Einstein may well have been unaware of the Michelson–Morley experiment and its results, but he was concerned about the meaning of the speed of light. As a young man he had wondered what light would look like if he could travel alongside it at the same speed. In principle, the light wave would be like a fixed disturbance of the electromagnetic field in its reference frame. However, Maxwell's equations did not allow for any such solution, so an observer moving at the speed of light would not be able to use the same laws of physics as one at rest. This seemed odd to Einstein, especially since Galilean relativity suggested that the laws of mechanics would still be the same. Einstein began to wonder whether he should extend Galilean relativity to all of the laws of physics.

This led to the first postulate of the special theory of relativity:

■ The laws of physics are the same in all inertial reference frames.

This also helped explain why, in electromagnetic induction experiments, it does not matter whether a magnet is moved toward a coil or a coil is moved toward a magnet—the same induced emf is produced in both cases. Absolute motion is not important, but relative motion is.

Einstein thought that Maxwell's equations must be one of the fundamental laws of physics and so should apply in the same way in all inertial reference frames. Since Maxwell's equations led to a fixed value for the speed of light, Einstein proposed a second postulate:

• The speed of light is the same for all inertial observers.

If the laws of physics are the same for all inertial observers, then being at rest has no special significance to physics and the idea of an absolute space must be abandoned. If the speed of light is the same for all inertial observers, then it will be the same along each arm of a Michelson–Morley interferometer no matter how the apparatus is moving (as long as it is not accelerating), so no delay should occur and the null result is explained.

While these simple postulates explain the null result of the Michelson–Morley experiment and make the speed of light into a universal constant, they undermine the idea of absolute space and, as we shall see, the idea that there is an absolute time.

24.2 Time in Special Relativity

Newton thought that time was the same for all observers throughout the universe—an **absolute time**, as if one clock could be used to measure all time intervals. Einstein's thesis that the speed of light is the same for all inertial observers undermines this idea. For the speed of light to be the same for two observers in relative motion, they cannot agree on the distance traveled by the light and the time taken to travel from one point to another. Distances and time intervals must depend on the motion of the observer. This is the price that must be paid if observers in relative motion are to experience the same laws of physics.

24.2.1 Time Dilation

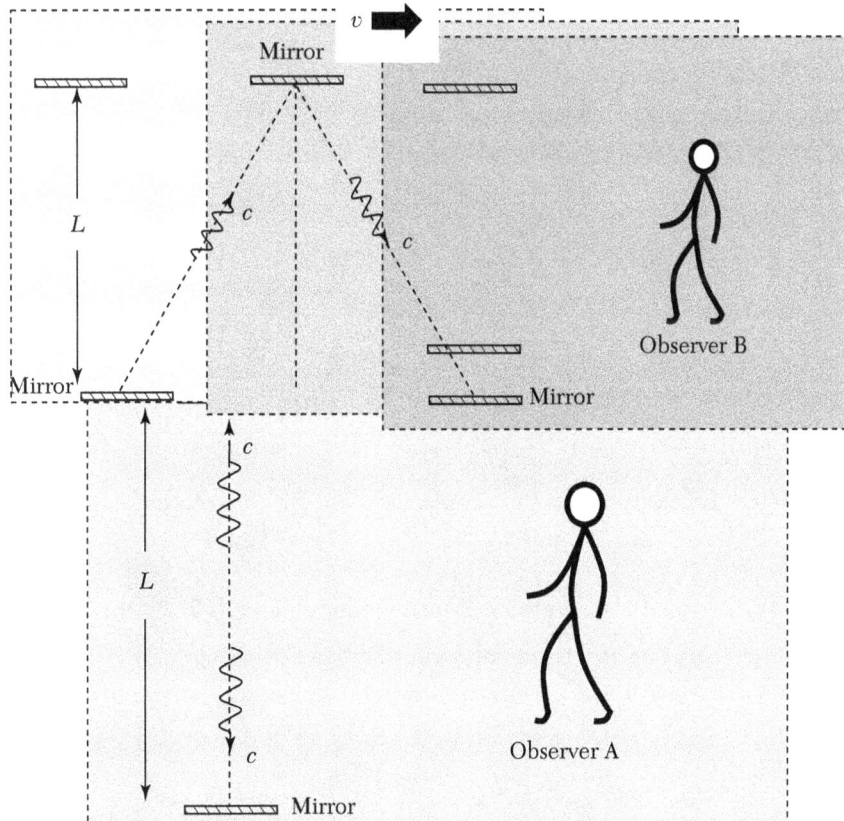

A simple thought experiment can be used to show how relative motion affects time intervals. It uses a hypothetical clock based on the constant speed of light. The light clock ticks as a light beam bounces up and down between two parallel mirrors. If the distance between the mirrors is L then the clock ticks once in a time $2L/c$ relative to an observer in the same reference frame. However, if the clock is in motion relative to that observer, the light path is longer. But the speed of light is the same (according to special relativity); so, for this observer, the clock takes longer to tick. Time in the moving relative frame has slowed down. The diagrams below show everything from the point of view of observer A who is at rest with respect to a light clock. He sees another observer, B, in a laboratory who moves past at velocity v relative to A's reference frame.

We can use Pythagoras theorem to calculate the length of the light path in B's moving laboratory as seen by observer A. It is very important to state which observer we are using because, for B—who is at rest with respect to his own light clock—the clock will appear to tick at its normal rate with the light pulses moving up and down at the speed of light.

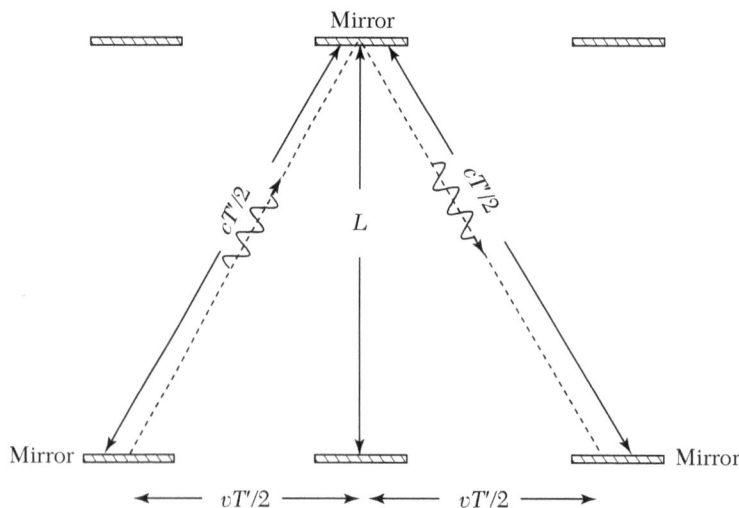

For A, the time between ticks of his own light clock is T where $T = \dfrac{2L}{c}$

Also for A, let the time between ticks on the moving (B's) clock be T'. Now use Pythagoras theorem to find the relationship between T and T'.

Considering either one of the two right-angled triangles above:

$$\frac{c^2 T'^2}{4} - \frac{v^2 T'^2}{4} = L^2$$

For A's own light clock:

$$L = \frac{cT}{2}$$

Therefore:

$$\frac{c^2 T'^2}{4} - \frac{v^2 T'^2}{4} = \frac{c^2 T^2}{4}$$

which can be rearranged to give:

$$T' = \frac{T}{\sqrt{1 - \dfrac{v^2}{c^2}}} = \gamma T$$

where

$$\gamma = \frac{1}{\sqrt{1 - \dfrac{v^2}{c^2}}}$$

This is Einstein's **time dilation** formula, showing how time slows down in a moving reference frame. That is, the time between "ticks" on the "moving" clock increases relative to the "rest" clock.

The "gamma factor"determines the significance of relativistic effects. For $v \ll c$ it approaches 1 and $T' = T$, which is consistent with an idea of absolute time. However, as v increases, the factor grows so that time in the moving reference frame (B's frame) slows down when observed from the rest frame (A's frame). As v approaches the speed of light, the gamma factor increases without limit so that time would slow to a halt at $v = c$. Note that we are not just talking about clocks here. As far as B is concerned, time in his reference frame is measured correctly by his own light clock, so when observed by A it is not just that the clock in B's frame slows down, so does B's aging process—*time* slows down.

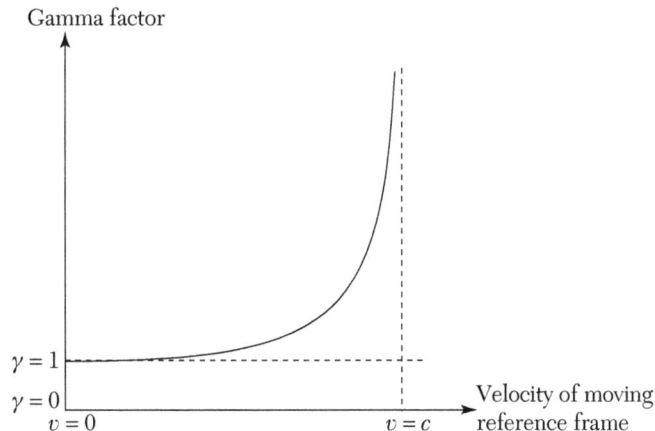

For relative velocities small compared to the speed of light, the gamma factor is negligible, so relativistic effects can be neglected. This allows us to assume that we all share the same time and space and can agree over measurements of time intervals and distances. For example, even at the speed of a jet airliner (about 300 ms⁻¹) the value of the gamma factor is just 1.0000000000005!

If A sees B's time slow down, then what does B see when he looks back at A? Both of the observers are in inertial reference frames, so the laws of physics are the same for both of them and the speed of light is constant in their reference frame. B will see A moving in the opposite direction at speed v and will see the light path in A's clock stretched out. B will see A's time run slow by the same time dilation factor that A saw B's clock run slow! The effect is the same for both observers. At first sight this seems to lead to a contradiction, but while the clocks are in motion, they cannot be placed at rest next to one another to compare them to find out which one has gained or lost time. To do this we would have to take at least one of them on a round trip.

24.2.2   The "Twin Paradox"

What happens when two clocks move relative to each other and then come back together so that the clocks can be compared? An observer moving with either clock ought to see the other clock running slow so that less time passes in the moving reference frame than in their own. This would lead to a paradox—how can clock A record less time than clock B AND clock B record less time than clock A? If the theory really predicts this, then there is a problem with relativity.

This is illustrated by the **twin paradox**. Imagine two twins A and B who are together on their 21st birthday. Twin A remains on Earth, while twin B undertakes a high-speed round-trip journey to a distant star. A sees B travel away and return and B's time runs slow on both the outward and return journeys. A concludes that B should be younger than A when they are reunited.

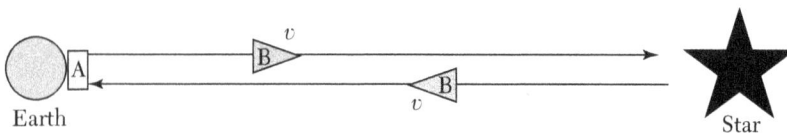

However, all inertial reference frames are equivalent, so B might argue that, from her point of view, A and the Earth move away and return and the star approaches and recedes. During the motion she sees A's time run slowly and so concludes that *A* should be the younger one when they reunite. Is this a valid way to describe the journey? In fact, it is not. If we ignore the relatively low velocity of the Earth, then A stays in the same inertial reference frame throughout and so *A*'s point of view is valid. *B*, however, undergoes three separate periods of acceleration: as she leaves Earth; as she turns around and accelerates back toward the Earth at the distant star; and then again as she comes to rest back on Earth. To describe the journey from *B*'s point of view, we must take into account the effects of these periods of acceleration. From *B*'s point of view she will experience inertial forces during these periods that are just like increases and decreases in the local gravitational field. These changes introduce gravitational time dilation effects (see Section 23.5.3) that are not present for *A*; so *B*'s view is not equivalent to *A*'s.

The traveling twin (*B*) will be younger when they reunite and the apparent paradox is resolved.

24.2.3   Relativity of Simultaneity

If time is absolute, then two distant events that occur at the same moment for one observer will also occur at the same time for any other observers wherever they are and however they are moving in the universe (once they have corrected for the time of flight of light to reach them from each event). Once we abandon the notion of an absolute space, as we have had to do, we cannot assume that simultaneous events in one reference frame will be simultaneous in all reference frames. It turns out that simultaneity is relative.

Consider an experimental method to synchronize two different clocks. They are both zeroed and then a flash of light is sent from a point midway between them. When the light hits a sensor on the side of each clock, the clock starts. In this way the clocks start simultaneously and then keep the same time in that reference frame.

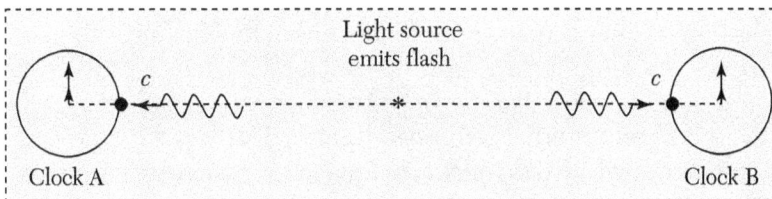

This method could be extended throughout a 3D space so that the times of all events in that space could be measured. If two events occur at the same time according to local clocks in this space, then those events are simultaneous in this reference frame.

The question is: will clocks synchronized in one inertial reference frame be synchronized in *all* inertial reference frames, regardless of their motion? Consider the process above carried out in a moving laboratory and observed from another inertial reference frame. For simplicity we will assume that the motion is parallel to the separation of the clocks and that the flash occurs at the moment the light source passes the stationary observer. The dotted clocks show their initial positions for the observer.

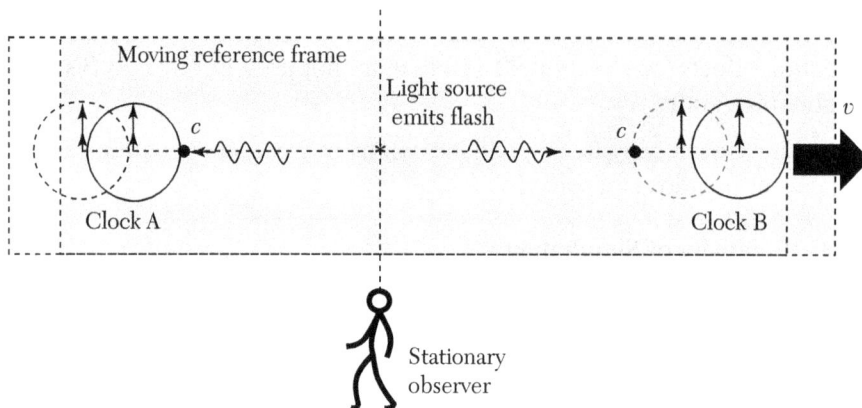

For the stationary observer the flash of light leaves the source and travels away in both directions at the speed of light. However, clock A is moving toward the source position and clock B is moving away from it. Light reaches clock A first so that clock starts first. Clock B starts later. The two clocks are not synchronized in the stationary observer's reference frame even though they are synchronized in the moving reference frame. For the stationary observer the synchronization "error" increases with the separation of the clocks.

This is a profound result. Two events that occur near A and B when the clocks show the same time are simultaneous in the moving reference frame but occur at different times for the stationary observer. This means that one moment in a particular inertial reference frame is spread over different times in different inertial reference frames. This destroys the concept of absolute time, which assumes a unique progression of moments for all observers.

It should again be noted that there is nothing special about the so-called "stationary observer." He is just in a different inertial reference frame and if he were to synchronize two distant clocks in his own reference frame, then an observer in the "moving" reference frame would think that there was a synchronization error.

24.3   Length Contraction

Our galaxy is over 100,000 light years in diameter. A space craft traveling at close to the speed of light would take more than 100,000 years to cross it as measured by an observer who remains on the Earth. However, for space travelers, the time that passes in their reference frame would be much less than this because of time dilation. Let's assume they travel at a constant velocity v that is so fast that the gamma factor is 20,000. The time that passes in the space craft is then only $100,000/20,000 = 5$ years!

If the observers on board the space craft work out the distance they have traveled, it will be $5(v/c)$ light years rather than $100,000 (v/c)$ light years as observed from the Earth. The distance traveled has contracted by the same gamma factor as the time. This is an example of relativistic length contraction.

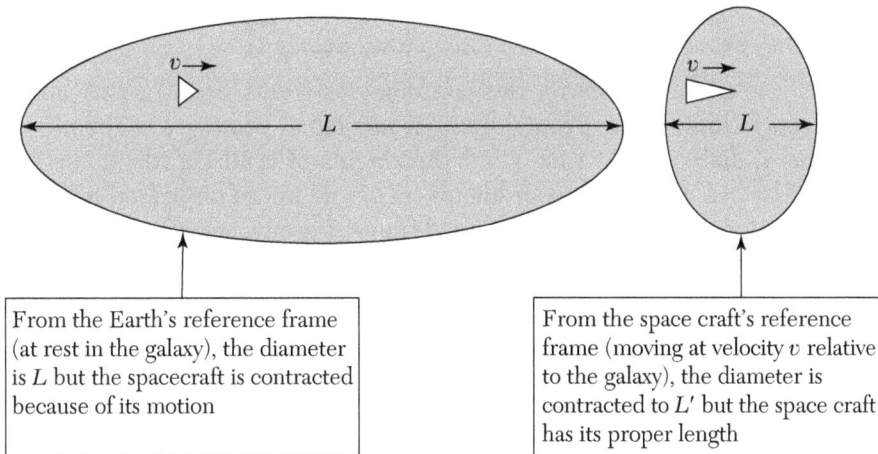

From the Earth's reference frame (at rest in the galaxy), the diameter is L but the spacecraft is contracted because of its motion

From the space craft's reference frame (moving at velocity v relative to the galaxy), the diameter is contracted to L' but the space craft has its proper length

The length contraction formula is:

$$L' = \frac{L}{\gamma}$$

$$\gamma = \frac{1}{\sqrt{1 - \dfrac{v^2}{c^2}}}$$

L is the length of an object measured in its rest frame: its **proper length** L' is the length of the same object when it moves past the observer at velocity v.

24.4   Lorentz Transformation

Length contraction, the relativity of simultaneity, and time dilation, all show that measurements of time and space in one inertial reference frame will not agree with measurements in another inertial reference frame in relative motion. Lorentz transformation equations relate the coordinates of an event in one inertial reference frame to the coordinates of the same event in another inertial reference frame moving at constant velocity with respect to the first.

24.4.1   Lorentz Transformation Equations

For simplicity we will define our axes so that the relative motion is along the x axis, so that y and z coordinates are not affected by the motion.

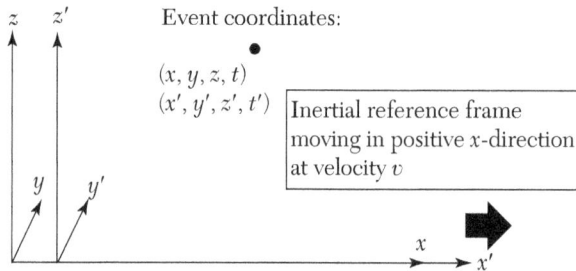

Event coordinates:

(x, y, z, t)
(x', y', z', t') Inertial reference frame moving in positive x-direction at velocity v

The diagram above shows the coordinate systems used in two inertial reference frames. The primed frame is moving at velocity v in the positive x- and x'-direction. At time $t = 0$ the origins of both coordinate systems coincides. An event with coordinates (x, y, z, t) in the unprimed frame has coordinates (x', y', z', t') in the primed frame. Lorentz transformation allows us to transform from one frame to the other.

▪ Lorentz transformation—to transform from the primed to the unprimed reference frame:

$$x' = \gamma(x - vt)$$
$$y' = y$$
$$z' = z$$
$$t' = \gamma\left(t - \frac{vx}{c^2}\right)$$

▪ Inverse Lorentz transformation—to transform from the unprimed to the primed reference frame:

$$x = \gamma(x' + vt')$$
$$y = y'$$
$$z = z'$$
$$t = \gamma\left(t' + \frac{vx'}{c^2}\right)$$

These equations can be used to derive the formulae for time dilation, length contraction, and synchronization differences.

We will use them to derive an equation for relativistic velocity addition.

24.4.2 Velocity Addition Equation

A simple example involving light shows that we cannot simply add velocities. Imagine a car passing you with its headlights switched on. The car is moving at velocity v and the light leaves the headlamps at velocity c relative to an observer inside the car. If space and time were absolute, then the light would have a relative velocity $v' = c + v$ relative to you. This cannot be the case because the speed of light is the same for all inertial observers, so it must still be c. This is the same problem that we ran into when discussing the Michelson–Morley experiment.

Consider an object moving at velocity u relative to an observer in a laboratory that is itself moving at velocity v relative to a second inertial observer. What is the velocity of the object w relative to this second observer?

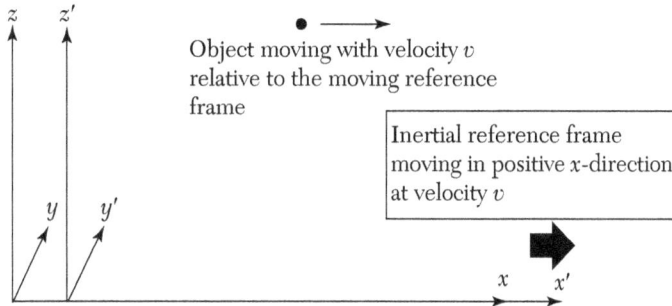

Object moving with velocity v relative to the moving reference frame

Inertial reference frame moving in positive x-direction at velocity v

We are trying to find the relationship between $w = dx/dt$, $u = dx'/dt'$, and v.

$$u = \frac{dx'}{dt'}$$

$$w = \frac{dx}{dt} = \gamma\left(\frac{dx'}{dt} + v\frac{dt'}{dt}\right) = \gamma\frac{dt'}{dt}\left(\frac{dx'}{dt'} + v\right) = \gamma\frac{dt'}{dt}(u + v) =$$

$$\frac{dt}{dt'} = \gamma\left(1 + \frac{v}{c^2}\frac{dx'}{dt'}\right) = \gamma\left(1 + \frac{uv}{c^2}\right)$$

$$w = \frac{\gamma}{\gamma\left(1 + \dfrac{uv}{c^2}\right)}(u + v)$$

$$w = \frac{(u + v)}{\left(1 + \dfrac{uv}{c^2}\right)}$$

If u and v are small compared to the speed of light, then this reduces to the familiar equation for velocity addition: $w = (u + v)$. The relativistic velocity addition equation has the interesting property that no two sub-light speeds can be added together to produce a speed greater than the speed of light. However long something accelerates, it will approach—but never quite reach—the speed of light. Consider the limiting case where $u = v = c$:

$$w = \frac{(u+v)}{\left(1 + \dfrac{uv}{c^2}\right)} = \frac{(c+c)}{\left(1 + \dfrac{c^2}{c^2}\right)} = \frac{2c}{2} = c$$

24.5   Mass, Velocity, and Energy

24.5.1   Mass and Velocity

When an object is accelerated at a constant rate, as measured in reference frames momentarily at rest with respect to the object, its acceleration measured by a stationary observer will gradually fall. This is because the time taken for the same increase in velocity is longer for the stationary observer as a result of time dilation.

Consider a particle of charge q and mass m_0 (as measured in a reference frame at rest with respect to the particle) accelerated by a constant electric field of strength E. The particle increases its velocity from v to $v + \delta v$ in a time $\delta t'$ in a reference frame moving with the particle at speed v relative to a stationary observer. The same increase in speed will take a longer time $\delta t = \gamma \delta t'$ for the stationary observer.

In the moving reference frame:

$$\delta v = \frac{Eq\delta t'}{m_0}$$

In the reference frame of the stationary observer:

$$\delta v = \frac{Eq\delta t}{m} = \frac{Eq\gamma\delta t'}{m}$$

Combining these two equations:

$$m = \gamma m_0$$

This shows that the mass will increase with velocity. The mass measured in the rest frame of the particle is m_0, and this is the fundamental mass of the particle. The additional mass is related to the kinetic energy of the particle as we shall see.

24.5.2 Mass and Energy

The equation for relativistic mass can be expanded as a series in terms of (v^2/c^2) using the binomial theorem:

$$m = \gamma m_0 = \frac{m_0}{\sqrt{\left(1 - \dfrac{v^2}{c^2}\right)}} = m_0\left(1 - \frac{v^2}{c^2}\right)^{1/2}$$

$$= m_0\left(1 + \frac{1}{2}\left(\frac{v^2}{c^2}\right) + \frac{\left(\dfrac{1}{2}\right)\left(\dfrac{-1}{2}\right)}{2!}\left(\frac{v^2}{c^2}\right)^2 + \cdots\right)$$

If this is multiplied throughout by c^2, then every term has the dimensions of energy:

$$E = mc^2 = \gamma m_0 c^2 = m_0 c^2 + \frac{1}{2}mv^2 + \cdots + \text{terms converging to zero}$$

The second term on the right is the classical kinetic energy and the converging series of terms beyond it (which are all negligible for $v \ll c$) are the relativistic corrections to the equation for kinetic energy.

The term $E = mc^2$ on the left-hand side of the equation must represent the particle's total energy when it is moving at velocity v. This becomes $E = m_0 c^2$ when the particle is at rest, so this is called the rest energy of the particle and it suggests that mass and energy are equivalent in some sense.

The kinetic energy of the particle is the difference between its total energy and its rest energy:

$$\text{KE} = mc^2 - m_0 c^2$$

This approximates to the classical formula KE = ½ mv² when the velocity is small compared to the speed of light, as we have seen.

The equivalence of mass and energy implies that all energy transfers are also mass transfers. However, the constant that relates energy to mass is the speed of light squared.

$$\Delta E = c^2 \Delta m$$

c^2 is so large (9×10^{16} m²s⁻²) that most energy transfers in everyday life make negligible difference to the mass. The only place where we see measurable mass changes because of energy transfers is in nuclear physics (see Section 26.1).

In contrast, the conversion of even a tiny amount of matter would release a huge amount of energy. The only process that converts mass completely to energy is the annihilation of matter and antimatter, e.g., the annihilation that occurs when an electron meets a positron (anti-electron).

24.6   Special Relativity and Geometry

Einstein published the special theory of relativity in 1905. In 1908 Hermann Minkowski, Einstein's mathematics teacher, published a paper in which he showed that the Lorentz transformations of special relativity are equivalent to rotations in a 4D continuum he called space-time. This effectively made special relativity into a geometric theory and ultimately led to Einstein's general theory of relativity in which gravitation is explained as a disturbance of the geometry of space-time. Here is a simple example to show how a geometric approach works.

24.6.1   Invariants

The fact that measurements of lengths and times made in different inertial reference frames differ from one another can be disconcerting. Physicists like to discover quantities that are the same for all observers. Such quantities are called **invariants**.

Consider two points on a 2D surface. The location of each point can be described by stating the coordinates of each point relative to a fixed coordinate system consisting of two perpendicular axes. However, there is an arbitrary choice about which set of perpendicular axes to use, and the coordinates will be different for different choices, as shown in the diagram below.

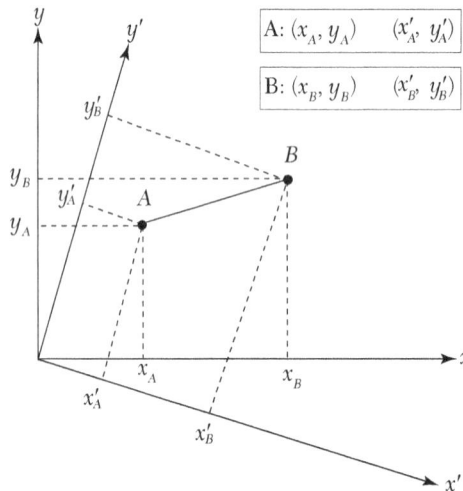

While the coordinates change when the axes are rotated, the distance AB does not. It is an invariant under rotation of the axes:

$$AB = \sqrt{\left(x_B - x_A\right)^2 + \left(y_B - y_A\right)^2} = \sqrt{\left(x'_B - x'_A\right)^2 + \left(y'_B - y'_A\right)^2}$$

It is also possible to write down a set of transformation equations to convert measurements made relative to the unprimed axes into measurements relative to the primed axes. It was the similarity between these transformation equations and the Lorentz transformation equations that alerted Minkowski to the idea that changing from one inertial reference frame to another is like a geometric rotation.

24.6.2 Space-Time

Space-time is a 4D continuum where the fourth dimension is related to time. Whereas a point in space is defined by three coordinates (x, y, and z) a point in space-time, called an event, has four (x, y, z, and t). In the S.I. system, distances and times are measured in different units—meters and seconds—so instead of simply using the time in seconds in a space-time diagram, we multiply the time by the speed of light so that the units on the time axis are also meters. Minkowskirealized that while measurements of time intervals and distances will differ for observers in different inertial reference frames, the Lorentz transformation ensures that the 4D "distance" between two events is the same for all inertial observers, regardless of their velocity. This 4D distance is calculated by Pythagoras theorem, but by treating the time differences as if they are mathematically imaginary quantities;in other words, the fourth dimension is actually the ict dimension where i is the square root of -1.

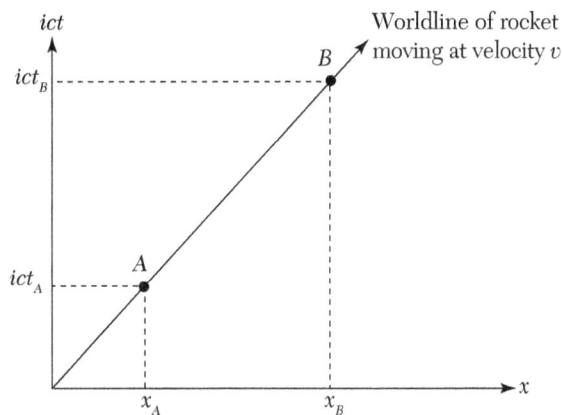

The above illustration is a space-time diagram. Only one spatial dimension is shown (x). Each point on the diagram represents an event—something that happens at a particular moment at a particular point in space. Lines on space-time diagrams are called **worldlines**; they represent a connected series of events. In this diagram the world line has constant gradient, so it represents an object moving at constant velocity in the positive x-direction relative to the origin of the reference frame, e.g., a rocket. Two events, A and B, lie on the world line, so these represent two separate moments when the rocket is in two different locations.

The interval AB is calculated in this reference frame by:

$$AB = \sqrt{\left(x_B - x_A\right)^2 + \left(ict_B - ict_A\right)^2}$$

Since the interval is an invariant, it must be the same using the coordinates of events A and B in any inertial reference frame, including that of the rocket. However, in the rocket's reference frame both A and B occur where the rocket is—in other words, at x' = 0 but at different times t_A' and t_B'. We cannot assume that the times are the same as those in the other reference frame. The space-time diagram in the rocket's reference frame looks like this:

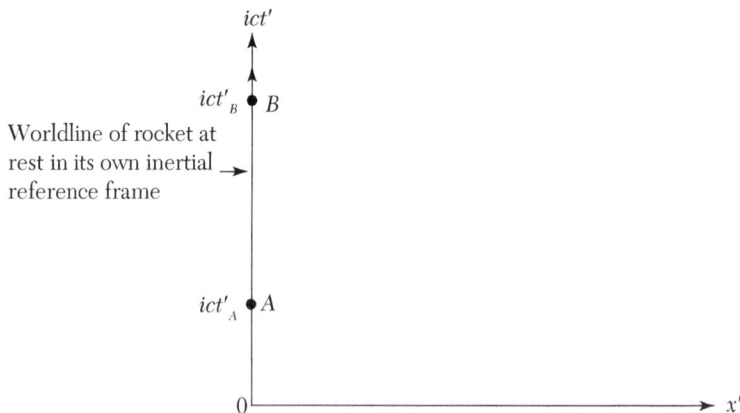

The interval AB in the rocket frame is:

$$AB = \sqrt{\left(ict'_B - ict'_A\right)^2}$$

We can equate the two expressions for the interval (because it is an invariant):

$$\sqrt{\left(ict'_B - ict'_A\right)^2} = \sqrt{\left(x_B - x_A\right)^2 + \left(ict_B - ict_A\right)^2}$$

We can also express the distance traveled in the unprimed frame in terms of velocity:

$$\left(x_B - x_A\right) = v\left(t_B - t_A\right)$$

After some algebraic rearrangement, we can express the time elapsed in the rocket frame (primed frame) in terms of the time elapsed in the rest frame (unprimed frame)

$$-c^2 \left(t'_B - t'_A\right)^2 = v^2 \left(t_B - t_A\right)^2 - c^2 \left(t_B - t_A\right)^2$$

$$\left(t'_B - t'_A\right) = \sqrt{\left(1 - \frac{v^2}{c^2}\right)}\left(t_B - t_A\right) = \frac{\left(t_B - t_A\right)}{\gamma}$$

This is time dilation. The time elapsed in the moving reference frame is less than the time elapsed in the rest frame by the gamma factor.

This illustrates how a geometrical approach in flat space-time reproduces the standard results of special relativity.

24.6.3   Mass, Energy, and Momentum

The interval is just one of the important invariants in relativity. We have already seen that when an object moves past us, its mass increases with velocity according to the gamma factor. We can also form an invariant quantity from momentum and energy:

Energy momentum invariant XE "invariant" $= E2 - p^2 c^2$

E is the total energy of the object and p is the momentum of the object.

Since this is an invariant, it will have the same value in the rest frame of the moving object, but in this frame, the object's mass is its rest mass m_0 and its momentum is zero:

$$E^2 - p^2 c^2 = \left(m_0 c^2\right)^2$$

This is a particularly useful relationship in particle physics.

24.7   Exercises

1. (a) Explain what is meant by:

▪ Absolute space

▪ Absolute time

▪ Inertial reference frame

▪ Galilean relativity

▪ Luminiferous ether

(b) Describe the Michelson–Morley experiment and explain why the null result could not be explained in terms of absolute space and absolute time.

(c) State the postulates of special relativity and show how the null result of the Michelson–Morley experiment is consistent with these postulates.

2. A spacecraft is manufactured on Earth and has a total length of 200 m when measured at rest on the surface of the Earth. During space trials it passes the Earth at a velocity of 0.90 c. The Earth can be assumed to be a sphere of diameter 13,000 km.

(a) Calculate the length of the space ship as seen by an observer on Earth.

(b) Calculate the diameter of the Earth as seen by an observer on the space ship.

(c) Calculate the time it would take for the spacecraft to reach a planet orbiting a star 20 light years from Earth:

(i) as measured by an observer on the Earth

(ii) as measured by an observer on the space craft.

(d) Calculate the distance traveled by the space ship as measured by an on-board observer during the journey.

(e) The astronauts spend 10 Earth years on the planet and then return to Earth at the same speed as on their outward journey. One of the astronauts has a twin brother who remained on Earth. Both were 21 at the start of the mission. How old will each twin be when the space ship returns to earth and they are reunited?

3. (a) Write down an equation for the relativistic "gamma factor."

(b) At what speeds does the gamma factor become 1.01, 1.10, 1.50, 5.0?

(c) Sketch a graph to show how the "gamma factor" varies with velocity.

(d) What is the significance of the case $v \ll c$?

The inertial mass of a moving object increases with velocity according to the equation:

$$m = \gamma m_0$$

(e) Show that as the velocity of a particle approaches the speed of light, the following approximation for its momentum can be used: $\sim \dfrac{E}{c}$, where p is the momentum of the particle.

(f) High-energy electrons have very short wavelengths and can be used to probe the internal structure of nucleons. Use the equation in (d) and the de Broglie relation to calculate the approximate accelerating voltage needed for an electron to be able to resolve details on the femtometer scale (10^{-15}m).

4. Two atomic clocks are synchronized and one of them is taken on a round trip lasting 10 hours at a steady velocity of 300 m/s in a jet aircraft. At the end of the journey it is placed beside the other clock and compared. (Ignore the effects of gravity and assume the "stay-at-home" clock is at rest throughout the experiment.)
   (a) Explain why there is a time difference between the two clocks at the end of the experiment and say clearly which clock will be fast or slow with respect to the other.
   (b) Calculate the expected time difference.

5. A space craft carrying an observer A has clocks, X and Y, at each end and is passing an observer B at speed v.

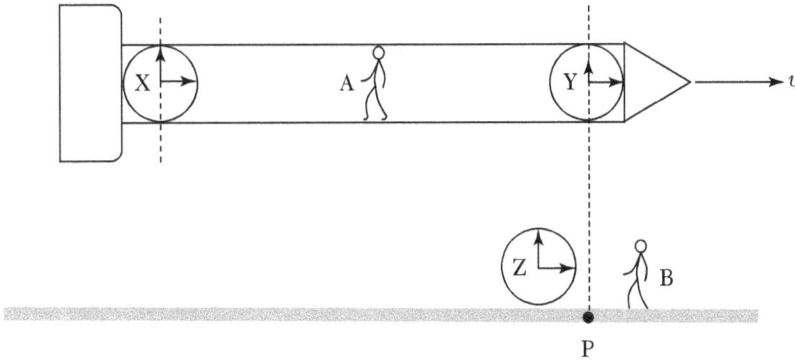

   (a) Explain what is meant by "relativity of simultaneity."
   (b) Observer A synchronizes clocks X and Y. Suggest a method by which he could do this.
   (c) Explain why the clocks will not be synchronized for observer B.
   (d) Both observers measure the distance between the two clocks. Observer A obtains a result l_0. What is the distance between the clocks according to observer B?
   A's method to measure the distance was to fire a pulse of light from Y (at time t_Y) and to record when it arrives at $X(t_X)$. The distance according to A is then $d = c\,(t_X - t_Y)$.
   B's method for measuring the length was as follows. He used clock Z and recorded the time at which each clock passed immediately above P. The distance according to B is then $d = v(t'_X - t'_Y)$.
   Explain carefully why these two methods must give different results and explain why both measurements are equally valid for the observers who made them.

CHAPTER 25

ATOMIC STRUCTURE AND RADIOACTIVITY

25.1 Nuclear Atom

The concept of an "atom" comes from the ancient Greeks and derives from the idea of an uncuttable smallest part of each element. While we still consider each atom to be the smallest part of an element, it is certainly not "uncuttable." At the end of the nineteenth century J.J.Thompson showed that electrons were small parts of all atoms; and at the start of the twentieth century, Ernest Rutherford probed the atom with high-energy alpha-particles and showed that all atoms have a tiny, massive core or nucleus. Later, in 1919, he managed to split a nitrogen atom. Rutherford's work resulted in the nuclear model of the atom.

25.1.1 Rutherford Scattering Experiment

The famous scattering experiment was carried out by his assistants, Geiger and Marsden, under Rutherford's direction. They used a radioactive source to fire a narrow collimated beam of alpha-particles at very thin gold foil. Rutherford expected the alpha-particles to penetrate the foil and to hardly be deflected because any charge present in the atom would be spread out and the forces on alpha-particles, which are positively charged, would be too small to cause large deflections. The experimental arrangement is shown below:

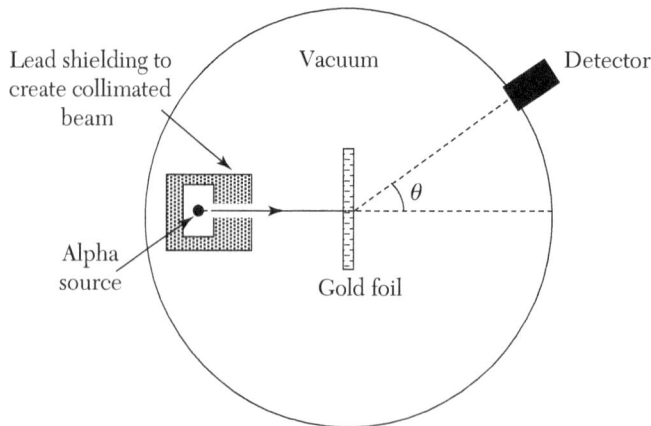

A moveable detector was used to record how many alpha-particles scattered through each angle.

The results were surprising:

- Most alpha-particles passed through the gold foil with little or no deflection (as expected).

- Some alpha-particles were scattered through large angles.

- A small number (about 1 in 10^4) scattered through angles >90° (back-scattered).

Rutherford assumed that any deflections must be electrostatic in nature and explained the results in the following way:

- The atom is mainly empty space, so most alpha-particles do not pass close to any concentrated charge centers.

- There is a tiny electrostatically charged central nucleus. The small proportion of alpha-particles that pass close to a gold nucleus will be deflected through large angles and those that make an almost direct hit will back scatter.

- The nucleus must contain most of the mass of the atom; otherwise the nucleus would recoil strongly and the alpha deflection would be less significant.

Rutherford used the scattering data and Coulomb's law to calculate an upper limit for the radius of the nucleus and found that the nuclear radius was of the order of 10^{-14} m or smaller, about 10^4 times smaller than the atomic radius.

Later work showed that the charge on a nucleus is equal to +Ze where Z is the atomic number, equal to the position of the atom in the periodic table of elements. The forces acting on the scattering nucleus and the scattered alpha-particles are shown below.

The diagram below shows typical alpha-particle trajectories close to a nucleus. On this scale the outside of the atom would be about 10 m away! Therefore, the vast majority of alpha-particles, which pass much further away from the nucleus than those shown in this diagram, experience weak electrostatic forces and suffer small deflections.

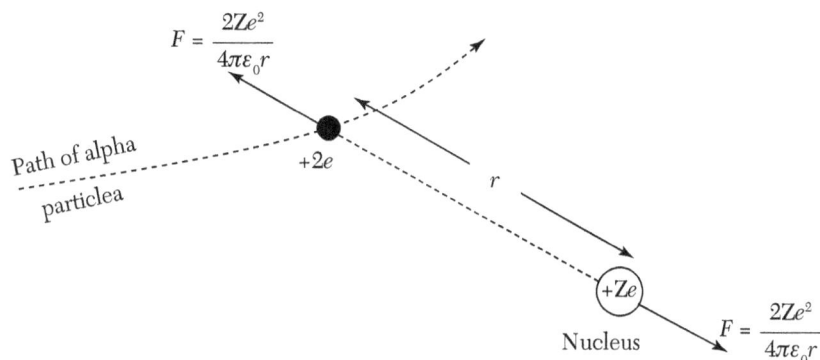

The greater the impact parameter (the perpendicular distance between initial path of the alpha-particle and the nucleus), the greater the minimum distance between the alpha-particle and the target nucleus and the smaller the angle of deflection. An alpha-particle traveling directly toward the center of the target nucleus makes the closest approach to it and is deflected back along its original path (deflection angle 180°).

25.1.2 Closest Approach and Nuclear Size

At closest approach the incident alpha-particle momentarily comes to rest as it deflects back along its original path. At this point all of the incident kinetic energy E_α of the alpha-particle has been stored as electric potential energy.

If the distance of closest approach is d, then:

$$E_\alpha = \frac{2Ze^2}{4\pi\varepsilon_0 d} \text{ and } d = \frac{2Ze^2}{4\pi\varepsilon_0 E_\alpha}$$

So the radius of the nucleus, r_n, must be less than this value:

$$r_n < \frac{2Ze^2}{4\pi\varepsilon_0 E_\alpha}$$

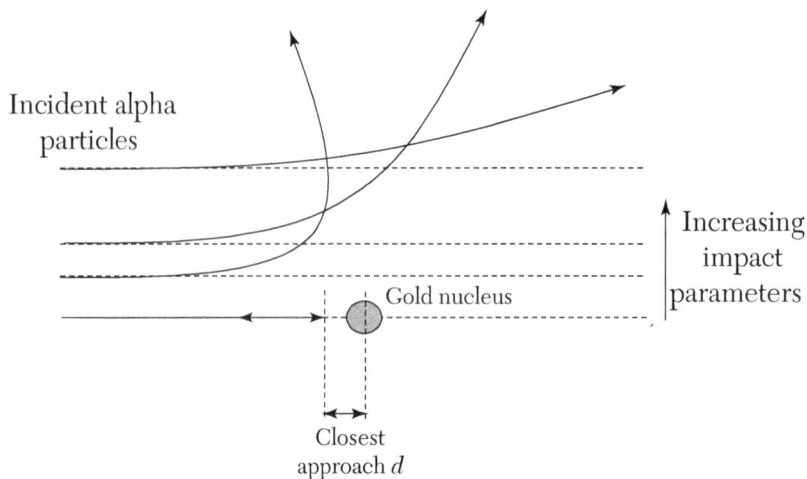

Closest
approach d

The alpha-particle energy in Rutherford's experiment was 4.7 MeV, and the atomic number of gold is 79. This gives an upper limit for the radius of gold nucleus:

$$r_n < 4.8 \times 10^{-14} \, \text{m}$$

25.1.3 Using Electron Diffraction to Measure Nuclear Diameter

Increasing the energy of incident alpha-particles reduces the distance of closest approach. However, alpha-particles are actually helium nuclei consisting of two protons and two neutrons. Neutrons and protons interact by the short-range, strong nuclear force, so when the alpha-particles get very close to a target nucleus, the forces are no longer simply electrostatic and Rutherford's analysis begins to break down. This limits the usefulness of alpha-particles for precise measurements of nuclear size. However, it is possible to use electrons instead. A high-energy electron beam behaves like waves of very short wavelength, and these are diffracted by the target nuclei just like light waves are diffracted by small spherical particles (e.g., lycopodium powder).

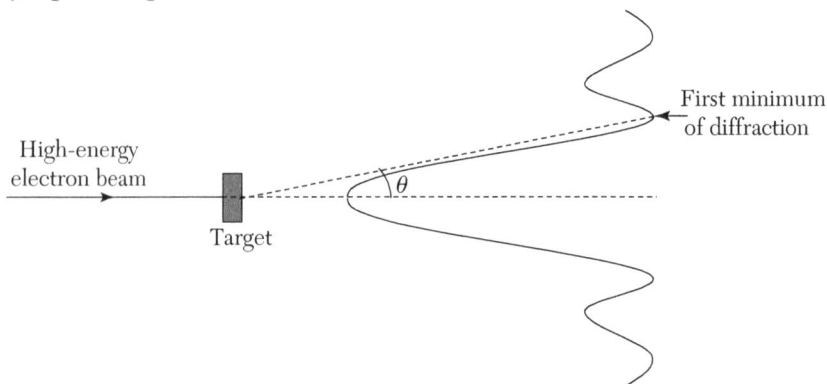

The first minimum of the diffraction pattern for a spherical object is at:

$$\sin\theta = \frac{1.22\lambda}{D}$$

Where D is the diameter of the target nucleus and λ is de Broglie wavelength of the electrons.

$$D = \frac{1.22\lambda}{\sin\theta}$$

The wavelength of electrons is calculated from de Broglie equation:

$$\lambda = \frac{h}{p}$$

However, in order to obtain a wavelength comparable to nuclear dimensions, the electron must be accelerated to very high energy, so relativistic equations must be used to determine electron momentum. In particular,

$$p^2c^2 = E^2 + \left(m_0c^2\right)^2$$

the energy required E is much greater than rest energy m_0c^2, so the equation simplifies to:

$$p = \frac{E}{c}$$

giving:

$$D = \frac{1.22hc}{E\sin\theta}$$

25.1.4 Nuclear Atom

The Rutherford nuclear atom consists of a small positively charged nucleus containing protons and neutrons (collectively called **nucleons**) and most of the mass of the atom surrounded by orbiting electrons. The number of protons in the nucleus is called the atomic number and corresponds to the position of the element in the periodic table. The atomic number is also equal to the number of orbiting electrons in the neutral atom. If an atom gains or loses an electron, it becomes an ion. The ratio of the atomic radius to the nuclear radius is about 20,000 in most atoms.

Nuclear nomenclature

▣ Z—the atomic number, equal to the number of protons in the nucleus and the position of the element in the periodic table

▣ A—the atomic mass number or nucleon number, equal to the number of protons plus the number of neutrons in the nucleus

▣ $N = A - Z$—the neutron number, equal to the number of neutrons in the nucleus

A nucleus of element X can be represented in the following way:

$$_Z^A X$$

For example, $_6^{12}C$ represents the nucleus of carbon-12 with 6 protons and 6 neutrons.

Elements are determined by their chemical behavior, and this depends on the arrangement of their outer electrons. Since the electronic configuration is determined by the charge on the nucleus, atoms with the same atomic number are chemically identical even if the number of neutrons in the nucleus varies. Atoms with the same atomic number and different neutron numbers (and mass numbers) are isotopes of the same element. For example,

$$_6^{12}C, \ _6^{13}C, \ \text{and} \ \ _6^{14}C$$

are all isotopes of carbon. They are chemically identical but have slightly different physical properties because of their differing masses. Carbon-14 is also unstable and, in common with many neutron-rich nuclei, it decays by beta-minus emission.

25.2 Ionizing Radiation

In 1896, the French physicist Henri Becquerel discovered that certain compounds of uranium emitted penetrating radiation that could cause ionization. He realized that this radiation was different from X-rays (discovered by Roentgen a year earlier) because the new ionizing radiation could be deflected by electric fields. Marie Curie coined the term "radioactivity"to describe the emission of this new type of radiation. She also discovered two more elements that were more radioactive than uranium—radium and polonium.

25.2.1 Types of Ionizing Radiation Emitted by Radioactive Sources

Three distinct types of ionizing radiation are emitted by radioactive sources—alpha, beta, and gamma. The nature and properties of each type are summarized below:

	Charge	Rest mass	Range in air*	Stopped by**	Nature of emission
Alpha	+2e	4u	Few cm	Card/skin	Helium nucleus
Beta°	−e	1/1840 u	~ 1 m	Few mm of Al	Electron
Gamma	Neutral	0	Indefinite	Several cm of lead	Photon

*There are two types of beta emission—beta-minus (electrons) and beta-plus (positrons). Here we are describing beta-minus.

**These depend on energy and so vary with different sources and are given here as a typical indication only.

The diagrams below show how each type of radiation is affected by electric and magnetic fields. Alpha-particles deflect less and in the opposite direction to beta-particles because they have more momentum and opposite charge.

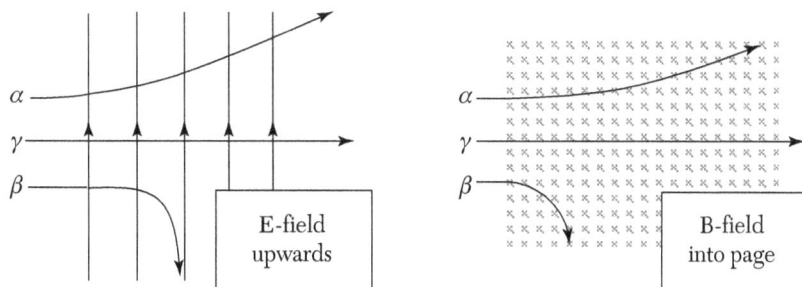

25.3   Attenuation of Ionizing Radiation

When ionizing radiation passes through matter, it transfers energy to the material and the ionizing beam is attenuated. When ionizing radiation is emitted in a vacuum, it spreads out and the intensity of the radiation falls.

25.3.1   Inverse-Square Law of Absorption

Radioactive emission is a random process, so ionizing radiation is emitted equally (on average) in all directions from a source. If the radiation is not absorbed by the medium into which it is emitted, then its intensity will fall as an inverse-square law. This will apply to all radioactive emissions in a vacuum but also to gamma-rays in air (since they are only weakly ionizing).

Consider a radioactive source that emits ionizing radiation at a rate R.

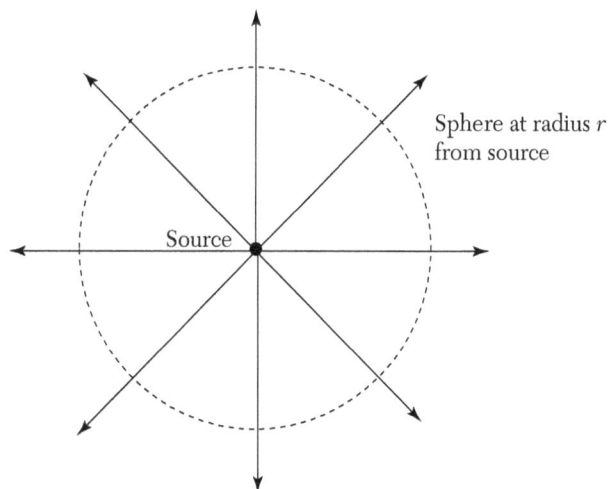

The intensity I of radiation at radius r will be:

$$I = \frac{R}{4\pi r^2}$$

This is an inverse-square law.

25.3.2 Exponential Absorption and Attenuation Coefficient

When gamma-ray photons travel through matter, the probability per unit length of their path is a constant. This means that the proportional change in beam intensity is always the same for the same thickness of material. For example, if 1.2 cm of material reduces the beam intensity from I to 0.5 I, then the next 1.2 cm of the same material will reduce it to 0.25 I; there is a constant half-thickness.

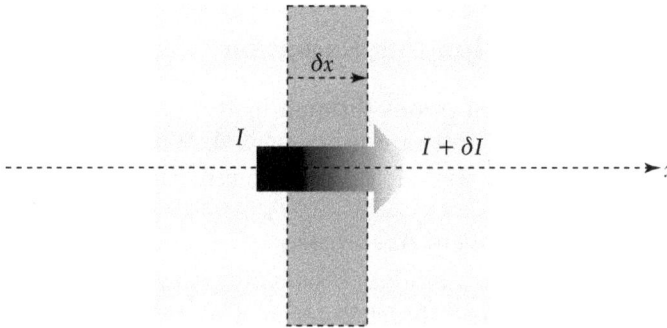

The diagram above shows part of the path of a gamma-ray beam through matter. Then intensity of the beam changes by an amount δI as it passes through a short thickness δx of material. Since the beam is being absorbed, δI is negative. The proportion absorbed per unit length is constant; so:

$$\frac{\delta I}{I \delta x} = -\mu$$

where μ is the absorption coefficient (a constant) for the medium with units m^{-1}.

In the limit that $\delta x \to 0$ this becomes the first-order differential equation:

$$\frac{dI}{dx} = -\mu I$$

whose solution is:

$$I = I_0 e^{-\mu x}$$

Intensity falls exponentially from its initial value I_0. As with all exponential changes, it has a constant proportion property in that the intensity will always fall by the same fraction in the same distance. We can therefore derive an expression for the half-thickness $x_{1/2}$ of the material, that is, the thickness of material that will reduce any initial intensity of the radiation by 50%.

$$e^{-\mu x_{1/2}} = \frac{1}{2}$$

$$x_{1/2} = \frac{\ln 2}{\mu}$$

The absorption coefficient and half-thickness depend on the energy of the gamma-rays and the nature of the medium. Half-thickness (or "half value layer," HVL) is a useful quantity to use in radiological protection when comparing the effectiveness of different shielding materials or the penetrating power of radiation.

X-ray penetration also decays exponentially with distance, and the table below gives half-thicknesses of human tissue, aluminum, and lead for X-rays of three different energies.

Medium	Half-thickness/mm		
	30 keV photons	**60 keV photons**	**120 keV photons**
Human tissue	20	35	45
Aluminum	2.3	9.3	17
Lead	0.02	0.13	0.15

25.3.3 Absorption of Beta-Radiation

Beta-radiation has a continuous range of energies even from the same source, so while the transmission of monoenergeticbeta-particles does obey an exponential decay law, the behavior of continuous beta-ray spectrum is more complex. There are also two distinct ways in which the electrons lose energy—by collision and scattering and by radiating photons as they decelerate (bremsstrahlung). In practice we usually refer to the range of beta-particles as a function of their energy. However, instead of giving the range as a distance, it is usually given in terms of a surface density. This is the density of the material multiplied by the range in that material. The reason for this is that the range depends on the material of the absorber, being shorter in denser media, but the surface density is equal to the product of range and density, so it is approximately the same for a variety of different media.

Here are some values of surface density for beta-particles of various energies:

Beta-particle (electron) energy/MeV	Surface density/kgm^{-2}	
	Aluminum	**Copper**
0.1	0.188	0.221
1.0	5.55	6.29
10	585	615

The relationship between range R, surface density σ, and density ρ is:

$$R = \frac{\sigma}{\rho}$$

While the surface densities for aluminum and copper are similar, the range of a 1.0 MeV beta-particle is very different because the two elements have quite different densities.

Range in aluminum:

$$R = \frac{\sigma}{\rho} = \frac{5.55}{2700} = 2.1 \times 10^{-3}\,\text{m}$$

Range in copper:

$$R = \frac{\sigma}{\rho} = \frac{6.29}{8960} = 0.70 \times 10^{-3}\,\text{m}$$

Copper is roughly three times as dense as aluminum, so the range in copper is roughly a third of the range in aluminum.

25.3.4 Absorption of Alpha-Particles

Alpha-particles lose their energy incrementally as they ionize atoms in the material, so alpha-particles of a particular energy will have a particular range. The nature of alpha-particle emission (as we shall see) results in each source emitting alpha-particles of one or a small number of distinct energies. This means that all alpha-particles from a particular source will have one or more distinct ranges in each medium.

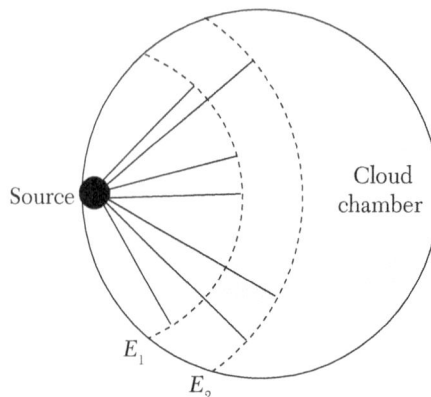

Alpha-particle tracks can be shown up using a device called a cloud chamber. A cloud chamber contains a super-cooled vapor that is on the point of condensing. When an alpha-particle passes through the vapor, tiny droplets condense around the ions it creates. This leaves a visible track rather like the vapor trail behind a jet aircraft. The diagram below shows tracks from an alpha source that emits alpha-particles of two distinct energies.

The high positive charge and large mass of an alpha-particle make it interact strongly with matter so that alpha-particles transfer energy quickly and have very short ranges in anything other than a gas. Even in air they are stopped within a few centimeters.

25.4 Biological Effects of Ionizing Radiation

When ionizing radiation is absorbed by matter, it transfers energy to the atoms and molecules with which it interacts. This can cause ionization and the breaking of atomic and molecular bonds. If ionizing radiation is absorbed by living tissue, it can cause cell damage. However, living things have evolved in a low-level radioactive environment, and damaged cells are continually repaired and replaced. This suggests that the effects of low doses of radiation may be small or even negligible, and some scientists have suggested there might be a threshold level of background radiation below which there are no harmful effects, but this has not been proven. In the absence of certainty we have to assume that low levels of radiation exposure pose a low level of risk to the health of living cells, and we judge what is a low level by comparison with levels of natural background radiation. If an experiment or a medical procedure increases your annual dose by an amount that is small compared to the dose you would receive naturally, then it is acceptable. If it increases your annual dose by much more than natural background radiation, then it may be dangerous and cause a significant increase in the likelihood that you suffer from a radiation-induced illness, such as cancer.

25.4.1 Natural Background Radiation

Radioactive isotopes are present in minerals in the Earth's rocks, soil, and atmosphere, in building materials, in the food we eat, and in our own bodies. Ionizing radiation is also produced by cosmic rays coming in from space and from artificial sources such as medical and dental X-rays, nuclear weapons testing, and the nuclear industry. We have evolved and live in a naturally radioactive environment, and our annual radiation dose varies depending on where we live and what we do. Taking a transatlantic flight, for example, increases our dose because there is less atmosphere above us to absorb incoming cosmic rays.

The pie chart below shows the origins of the typical annual radiation dose for a UK citizen. By far the largest contribution comes from radon gas, a natural product of the uranium decay series that seeps into the atmosphere from the ground and can accumulate in cellars and ground-floor rooms if they are not properly ventilated.

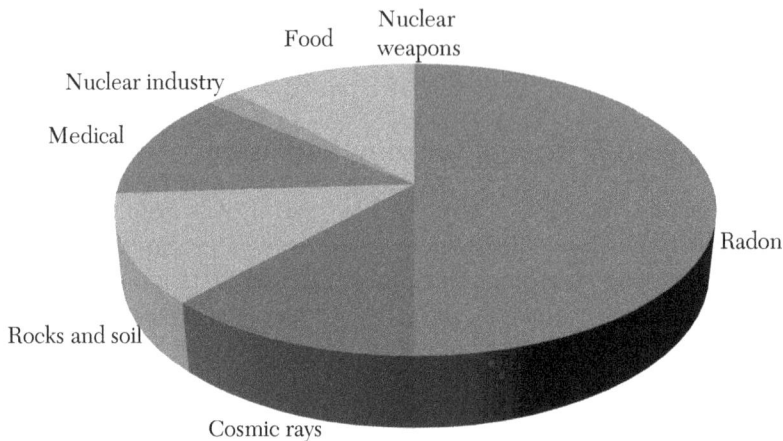

25.4.2 Measuring Radiation Dose

Several different units are used to measure the absorbed radiation dose. The gray is used to measure the total energy absorbed per kilogram of living tissue:

- Radiation dose in gray = energy absorbed (J)/mass of tissue (kg)

  1 gray (Gy) = 1 Jkg^{-1}

This does not take into account the nature of the radiation absorbed. The same dose in gray can produce very different effects depending on whether the tissue has absorbed alpha, beta, or gamma radiation.

The equivalent dose takes into account the nature and energy of the radiation absorbed by multiplying by a weighting or quality factor W_R determined by the type and energy of the radiation absorbed. Equivalent dose is measured in Sievert (Sv):

- Equivalent dose in Sieverts = radiation dose in gray × weighting factor

The Sievert is also in Jkg^{-1} because the weighting factor is dimensionless.

Typical weighting factors

X-rays, γ-rays, β-particles : $W_R = 1$

Neutrons : $W_R = 2\text{--}5$ (depending upon energy)

Protons : $W_R = 2$

α-particles, fission products : $W_R = 20$

When considering radiation hazards we must also bear in mind penetrating power. Alpha-particles have a very high weighting factor but are strongly absorbed and stopped by the outer layers of our skin. These layers are mainly dead cells, so alpha radiation is not particularly dangerous from outside the body. However, if an alpha source gets inside the body, e.g., breathing in a radioactive gas, it is particularly dangerous.

An older unit for equivalent dose is the rem (Roentgen equivalent man). This is still used in many textbooks (especially in the USA). 1 rem = 0.01 Sv

25.4.3 Effect of Radiation Dose on Human Health

There are many different factors that need to be considered when estimating the risks to human health from ionizing radiation. These include the nature and energy of the radiation, the tissues that are exposed to the radiation, the radiation dose, and the rate at which that dose is received. Most of the information about the effects of large absorbed doses comes from studies on the survivors of the nuclear attacks on Hiroshima and Nagasaki; and whilewe have a pretty good idea about the immediate effects of very high exposures, the long-term and random effects of lower doses are not well understood. The table below gives an idea of the effects of different radiation doses.

Equivalent dose (in 1day)		Effects
/milli-sievert	/rem	
0–250	0–25	No observable damage
250–1000	25–100	Mild symptoms, damage to bone marrow and spleen; short illness is likely—the higher the dose, the more serious the effects
1000–3000	100–300	Nausea, radiation sickness, suppression of the immune system and susceptibility to disease, more severe damage to spleen, lymph nodes and bone marrow—recovery often possible if treated
3000–10,000	300–1000	More severe than above but also hair loss, skin burns, diarrhea, hemorrhaging, damage to the central nervous system, sterilization … death likely (and almost certain if not treated)

According to UK government data, the annual average equivalent radiation dose for a UK citizen is 2.7 mSv (0.27 rem),while in the USA, this is 6.2 mSv (0.62 mSv).

Patients who survive high radiation doses have a significantly increased risk of developing cancers such as leukemia in the future. Those who have not been sterilized by the radiation also risk passing on genetic damage to their children, resulting in still births and birth defects. Approximately 50% of people exposed to 5000 mSv (500 rem) will die.

The annual exposure limit for nuclear industry employees in the UK is 20 mSv, about double the dose from a CT scan of the spine.

25.4.4 Reducing Risks in the Laboratory

There are several simple precautions that should always be followed when carrying out experiments with radioactive sources in a laboratory. The first and most obvious is that a risk assessment should be carried out before starting! This should include an estimate of the likely dose that will be received during the duration of the experiment and a list of actions that will be taken to minimize this dose. A judgement must also be made—do the benefits of carrying out the experiment (e.g., educational or medical benefits) outweigh the increase in the risk of damage to your or other people's health? In a school laboratory, sources are usually weak and well protected so that their use will result in a negligible increase in dose compared to the natural background; however, it is always important to be confident that this is the case!

Here are some additional procedures that should be used:

- Keep the source inside its shielded container except when carrying out the experiment

- Maximize the distance between the source and the experimenters

- Minimize the time of the experiment

- Do not direct a collimated beam toward anyone

- Always handle sources remotely, e.g., using tongs

- Include shielding between the experiment and experimenters

- Do not eat or drink while carrying out the experiment

- Wash your hands thoroughly after the experiment

▨ Cover any open cuts or scratches with a plaster

▨ If you drop the source or suspect it is damaged in any way, report this immediately

25.5 Radioactive Decay and Half-Life

Radioactive decay takes place inside the nucleus. It is a **random process**; we cannot predict when a particular nucleus will decay or in which direction the ionizing radiation will be emitted. It is also a **spontaneous process**;no external factors (e.g., temperature or pressure) affect when a decay will occur. Even though the process is random, it is still possible to predict the pattern of decay for a sample containing a large number of unstable nuclei. This is the probability of decay per unit that time remains constant so that the same proportion of nuclei decay in the same time.

Radioactive decay can be modeled quite simply using dice. Each die has 6 faces so the probability of landing with a "6"showing upwards is 1/6. If a large number of dice are rolled together, we would expect about 1/6 of them to show heads. The larger the number of dice, this prediction becomes better. In a similar way, the fraction of nuclei that decay in a particular time interval is always the same for any particular type of unstable nucleus (particular nuclide), and the rate of decay will be directly proportional to the number of nuclei in the sample. This leads to a differential equation:

$$\frac{dN}{dt} \propto -N$$

$$\frac{dN}{dt} = -\lambda N$$

N is the number of nuclei present at time t and the minus sign indicates that the number is falling with time.

λ is the decay constant and depends on the nuclide being considered. It has an S.I. unit s^{-1} and represents the probability of decay per unit time in the limit of small time intervals.

This is a first-order differential equation that can be solved by separation of variables. Its solution represents how the number of nuclei in the sample varies with time:

$$N = N_0 e^{-\lambda t}$$

This is exponential decay. The term $e^{-\lambda t}$ is equal to the fraction of the initial number of nuclei remaining after time t.

The **half-life** of the nuclide is the time taken for the number of nuclei in the sample to halve.

$$e^{-\lambda t_{1/2}} = \frac{1}{2}$$

$$t_{1/2} = \frac{\ln 2}{\lambda}$$

The graph below shows the decay curves for three nuclides, A, B, and C. Their half-lives are shown on the time axis. A has a half-life of about $0.35y$, B has a half-life of about 0.69 y, and C has a half-life of about $1.4\,y$.

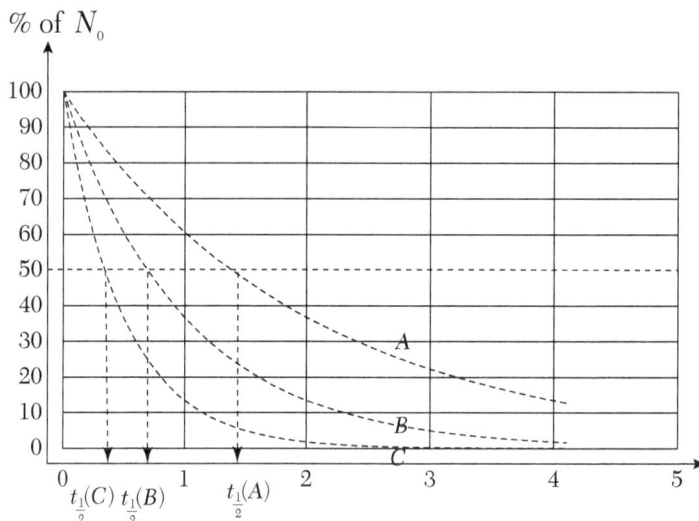

The fraction remaining after n half-lives is 1/2n. While the mathematical model suggests that the sample never completely decays, the number of nuclei remaining will eventually become so small that the model does not apply and the random nature of radioactive decay will result in significant fluctuations in decay rate until the last nucleus eventually decays.

The **activity** of a radioactive source is defined as the number of disintegrations (decays) per second taking place inside the source. This is not the same as the count rate in a detector because many emissions will miss the detector or not be detected and some will be absorbed before even leaving the source.

Activity = number of decays per second inside the source

$$A = -\frac{dN}{dt}$$

The S.I. unit of activity is Becquerel (Bq). 1 Bq = 1 decay per second.

This can be rewritten in terms of the number of nuclei in the source:

$$A = -\frac{dN}{dt} = -\lambda N_0 e^{-\lambda t} = -\lambda N$$
$$A = -\lambda N$$

This is a useful relationship that can be used to determine the half-lives of long-lived radioisotopes such as uranium. N can be determined from the mass of the sample, and A can be calculated from measurements of ionizing radiation emitted by the sample—i.e., a detector is placed close to the sample and the count rate is measured and scaled up to find the activity. Once A and N are known, the decay constant can be calculated and used to find the half-life.

25.6   Nuclear Transformations

When a nucleus decays, it changes to become a nucleus of a different element. The nuclear transformations equations for alpha, beta-minus, and gamma-decays are shown below along with two additional decays: beta-plus and electron capture. The numbers at the top of the equation are nucleon numbers (protons or neutrons). The numbers at the bottom of the equation are charges. Both nucleon numbers and charge numbers must balance on either side of the equation.

25.6.1   Alpha Decay

An alpha-particle is a helium nucleus consisting of 2 protons and 2 neutrons:

$$_2^4\text{He} \, or \, _2^4\alpha$$

Here is the transformation equation for the alpha decay of uranium-238:

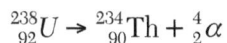

$$_{92}^{238}U \rightarrow _{90}^{234}\text{Th} + _2^4\alpha$$

The general equation for the alpha decay of X to Y would be:

$$_Z^A X \rightarrow _{Z-2}^{A-4}Y + _2^4\alpha$$

A has reduced by 4 and Z has reduced by 2.

25.6.2   Beta-Minus Decay

A beta-particle is a high-energy electron created inside the nucleus during beta-decay.

$$_{-1}^0 e \, or \, _{-1}^0 \beta$$

This is a subtle process involving the weak nuclear force and occurs when a neutron decays to become a proton inside the nucleus. In addition to the creation of an electron, conservation of lepton number (see Section 26.4.1) demands that an anti-neutrino is also emitted. This is an anti-particle of the neutrino; it is neutral, has a tiny rest mass, and is like an uncharged electron. Its symbol is $_0^0\overline{v}$.

The bar over the symbol indicates that this is an anti-neutrino.

Here is the transformation equation for beta-minus decay of carbon-14:

$$_6^{14}C \rightarrow \, _7^{14}N + \, _{-1}^0\beta + \, _0^0\overline{v}$$

The general equation for the beta-minus decay of X to Y would be:

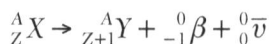

$$_Z^AX \rightarrow \, _{Z+1}^AY + \, _{-1}^0\beta + \, _0^0\overline{v}$$

There is no change in mass number A because one type of nucleon, a neutron, has changed into another type, a proton. Z increases by 1. The underlying nucleon transformation is:

$$_0^1n \rightarrow \, _1^1p + \, _{-1}^0\beta + \, _0^0\overline{v}$$

Beta decay results in three particles moving away from the position of the original nucleus. This can occur in a variety of ways with almost all the energy released being shared by the electron and anti-neutrino. The way in which this energy is shared is random, so the beta-particles emitted from a source have a continuous range of kinetic energies up to some maximum value.

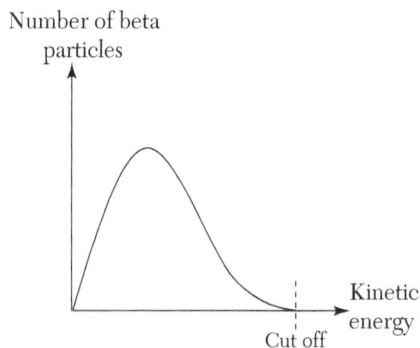

This differs from alpha decay where, with only two emerging particles, the energy is shared in a definite way and the alpha-particles have a discrete energy spectrum.

25.6.3 Gamma Emission

Gamma-rays are high-frequency electromagnetic photons. Gamma-ray emission occurs after some alpha or beta decays when the resultant nucleus is left in an excited state. The excited nucleus loses energy by making discrete quantum jumps to lower states until it reaches its lowest or ground state. Each quantum jump results in the emission of a gamma-ray photon whose energy is equal to the difference in energy between the nuclear energy levels.

$$\Delta E = hf$$

This results in a discrete energy spectrum for the gamma-rays.

Cobalt-60 decays to nickel-60 by emitting a beta-minus particle. However, the decay can occur to one of two excited states of the nickel nucleus (indicated by asterisks below) each of which then decays by emitting a gamma-ray.

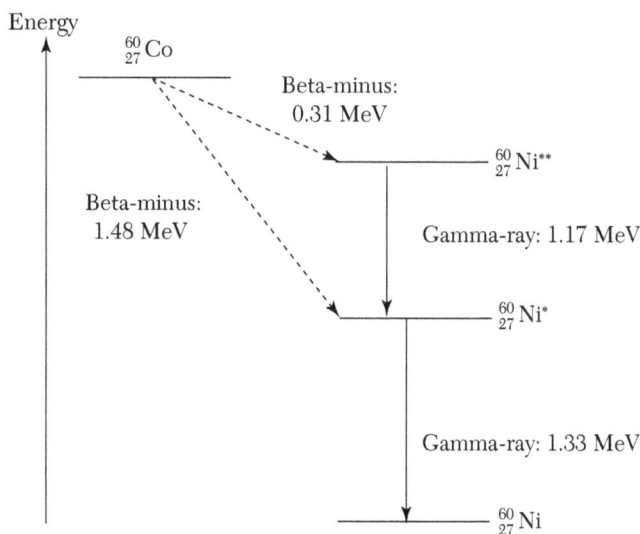

25.6.4 Beta-Plus Emission

Some proton-rich nuclei can decay by emitting a positron, an anti-electron. This converts a proton in the nucleus into a neutron. The emission of the positron is accompanied by the emission of a neutrino, and the positron and neutrino share the energy of the decay in a random way. This results in a continuous spectrum of positron energies. The nuclide oxygen-15 is a beta-plus emitter.

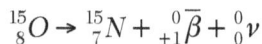

$$^{15}_{8}O \rightarrow {}^{15}_{7}N + {}^{0}_{+1}\overline{\beta} + {}^{0}_{0}\nu$$

The general equation for beta-plus decay is:

$$^{A}_{Z}X \rightarrow {}^{A}_{Z-1}Y + {}^{0}_{+1}\overline{\beta} + {}^{0}_{0}\nu$$

and the underlying nucleon decay is:

$$_1^1p \rightarrow {}_0^1n + {}_{+1}^0\overline{\beta} + {}_0^0\nu$$

The mass number A is unchanged, but atomic number Z decreases by 1.

25.6.5 Electron Capture

Electron capture is another way that a proton-rich nucleus can decay. The nucleus captures one of the atom's inner electrons and combines it with a proton to create a neutron and emit a neutrino, so the net effect on the nucleus is the same as beta-plus decay. However, the loss of an inner electron allows other electrons to cascade down to lower energy levels inside the atom emitting characteristic X-rays.

The general equation for electron capture is:

$$_Z^AX + {}_{-1}^0e \rightarrow {}_{Z-1}^AY + {}_0^0\nu$$

and the underlying nucleon transformation is again:

$$_1^1p + {}_{-1}^0e \rightarrow {}_0^1n + {}_0^0\nu$$

Electron capture is an alternative mode of decay for nuclei that undergo beta-plus decay.

25.7 Radiation Detectors

Alpha, beta, and gamma emissions are all forms of ionizing radiation, so ionization is the key to detection. A single alpha-particle, for example, can create tens of thousands of ion pairs before it is stopped in the air. Air is usually an insulator, but when ionizing radiation passes through the air, the ion pairs created make the air slightly conducting. This change in its electrical properties, or in the electrical properties of other gases, can be used to make a radiation detector.

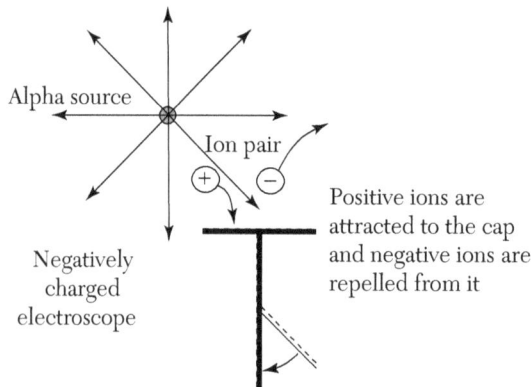

Alpha-particles are very strongly ionizing, much more so than beta-particles or gamma-rays; and if an alpha source is held close to a charged electroscope, the electroscope is discharged. Ions created in the air close to the cap create a weak conducting path to earth. A more sensitive detection method is required for beta-particles and gamma-rays unless the source is particularly intense.

25.7.1   Spark Counter

If the electric field strength in air is high enough, the air becomes ionized and it begins to conduct—a spark is formed. The breakdown field strength in air is about $3 \times 10^6 \, Vm^{-1}$. In the spark counter the field strength between a fine wire and a metal grille is adjusted so that it is just less than this value. When an alpha source is brought nearby so that ionizingradiation enters the gap between the electrodes, electrons are accelerated in the field, collide with air molecules, and cause further ionization. This results in an avalanche of charge that forms a spark. The greater the intensity of the ionizing radiation, the higher the spark rate. Spark detectors only work well with strongly ionizing radiation such as alpha-particles, but the principle behind the spark counter is used in many particle and radiation detectors.

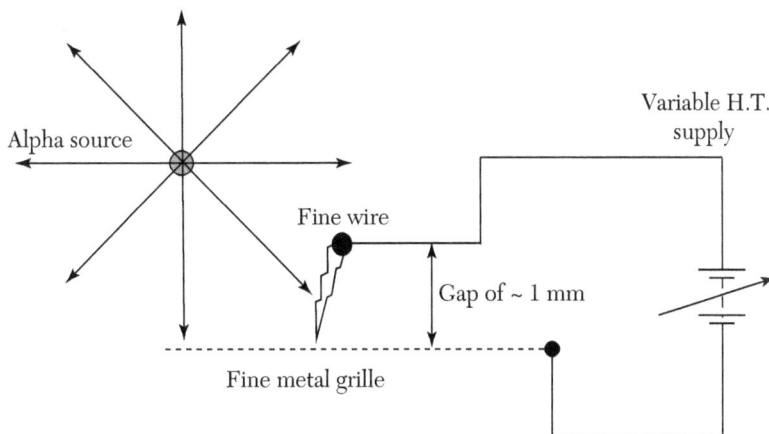

25.7.2   Geiger Counter

The Geiger counter is the most familiar radiation detector. It is really a more sophisticated version of the spark counter. However, it is much more sensitive than the basic spark counter and it counts individual ionization events inside a Geiger-Müller tube.

The tube itself contains an inert gas at low pressure, and a fine wire anode lies along the axis of the tube. When a potential difference is applied between this anode and the surrounding cylindrical cathode, there is a radial electric field inside the tube. When ionizing radiation passes through the low-pressure gas, some of the gas molecules become ionized. The electric field between the anode and cathode accelerates these ions and there is a small pulse of current in the external circuit. This generates a voltage pulse that can be detected and counted.

Alpha- and beta-radiation enters the Geiger-Müller tube through a thin end window, but gamma-radiation can also enter through the walls of the tube. All of the pulses are identical, regardless of the type of radiation, so a Geiger counter does not indicate the type of radiation detected. There is also a "dead time" following each voltage pulse. During this time the detector will not register any further ionization events. This limits the maximum count rate that can be measured accurately.

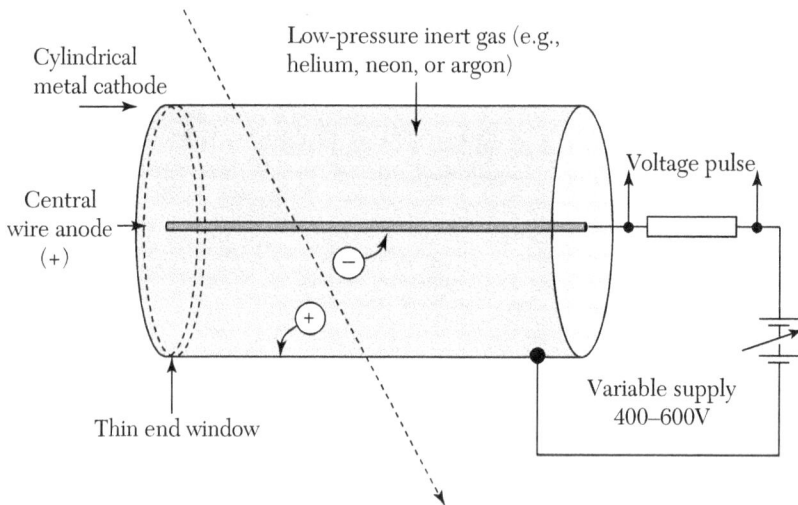

The count rate from a Geiger counter should not be confused with the activity of a source. The latter is the number of disintegrations per second inside the source. The count rate on the Geiger counter is related to this but is usually only a small fraction of the activity because of the emissions that miss the counter, are not detected by it, or are absorbed inside the source or between the source and the detector.

Count rates are usually recorded as counts per minute (cpm) and often have to be corrected for the average background count where the experiment is being carried out.

25.7.3   Using a Geiger Counter to Measure Count Rates

Before measuring the count rate from a radioactive source we must measure the average background count from the surroundings. This is done by setting up all of the apparatus except the source and then taking a series of measurements using the Geiger counter. Typically we might make five readings, each over 5minutes, and then use the average cpm as the average background count. After this has been done, we can carry out the experiment with the Geiger counter in the same position and the source present and measure the total cpm in a similar way (based on an average of five readings of 5minutes each). The corrected count is then the difference between the total count rate and the average background count rate.

Corrected count rate (ccpm) = total count rate (tcpm) − average background count rate (bcpm)

The diagram below shows a possible arrangement of apparatus to measure the penetrating power of beta-rays through aluminum:

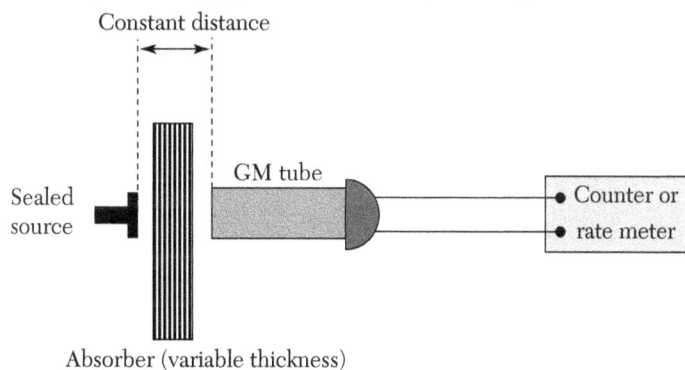

25.8   Using Radioactive Sources

All of the properties of ionizing radiation have useful applications.

▧ The exponential decay and constant half-life can be used like a clock to date archeological samples and minerals on Earth and elsewhere in the solar system (as described below).

▧ The penetrating power of ionizing radiation can be used to measure the thickness of materials and to locate medical tracers inside the body.

▧ The energy of ionizing radiation can be used to sterilize equipment and to kill cancer cells.

▧ The absorption of ionizing radiation can be used to form images of pipelines and structures and to locate cracks and defects.

25.8.1 Radiological Dating

The best-known dating technique uses the isotope carbon-14, a beta-minus emitter with a half-life of 5730 years. Approximately 1 in 10^{12} atoms of carbon in the Earth's atmosphere is carbon-14. The rest are carbon-12 and carbon-13, both of which are stable.

Atmospheric carbon-14 is continually replenished by cosmic rays from space that bombard the atmosphere resulting in a flux of neutrons. When neutrons strike nitrogen atoms (78% of the Earth's atmosphere is nitrogen), carbon-14 is created:

$$_0^1 n + {}_7^{14}N \rightarrow {}_6^{14}C + {}_0^1 p$$

This maintains a roughly constant proportion of carbon-14.

Living things continually exchange carbon with the atmosphere, so while they are alive, they contain the same proportion of carbon-14 as the atmosphere. When a living thing dies, the exchange ceases so the proportion of carbon-14 decays exponentially. Radiocarbon dating measures the fraction of carbon-14 in the carbon content of a sample of once-living material and uses this to estimate the age of the sample.

If the fraction in the atmosphere is f_0 and the fraction in the sample is f:

$$f = f_0 e^{-\lambda t}$$

where λ is the decay constant for carbon-14.

The age of the sample is then:

$$t = \frac{\ln\left(\dfrac{f}{f_0}\right)}{-\lambda} = \frac{\ln\left(\dfrac{f}{f_0}\right)}{-\ln 2}$$

$f/f_0 \times 100\%$

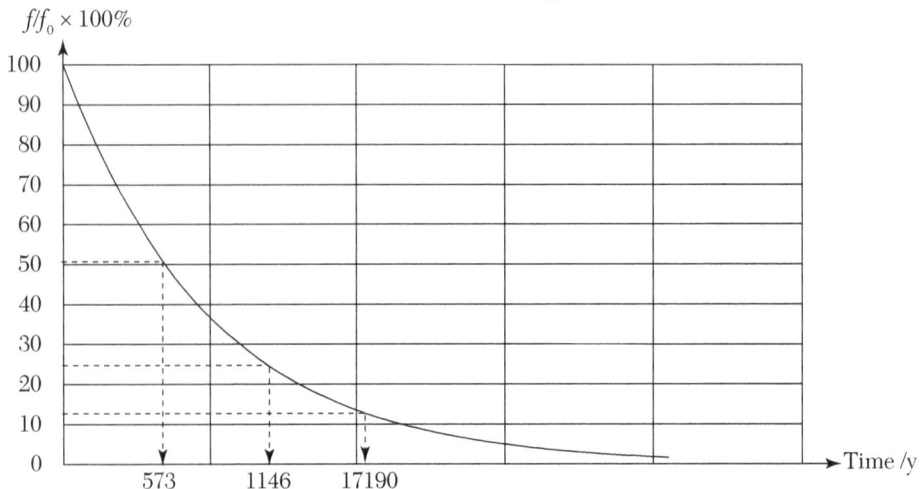

This method assumes that there has been no contamination of the sample and that the fraction of carbon-14 in the atmosphere has remained constant over the time period being measured. Radiocarbon dating methods are calibrated against other methods such as dendrochronology (tree ring counting).

25.8.2   Radiological Dating of Rocks

Radiocarbon dating cannot be used to date non-organic material unless there is some organic material known to be of the same age associated with it. However, radioactive dating methods (radiometric dating) using other isotopes can be used to date rocks. One method uses two unstable isotopes of uranium—uranium-238 (with a half-life of 4.5 billion years) and uranium-235 (with a half-life of 700,000 years). Both decay, via many steps, to stable isotopes of lead (lead-207 and lead-206, respectively), so the ratio of lead to uranium grows as time goes on and this ratio can be used as a clock. Another method uses the decay of potassium-40 to argon-40 (with a half-life of 1.3 billion years). All such methods rely on assumptions about the initial state of the rocks, e.g., that there was no argon present in a rock when it solidified so that all of the argon now present must have come from the decay of potassium, and about the isolation of the rock since its formation, i.e., so that no potassium or argon has been added from or lost to external sources.

25.8   Exercises

1. The discovery of the electron by J.J.Thomson in 1897 showed that atoms are not fundamental particles. Thomson thought that atoms were like "plum puddings" with negatively charged electrons embedded in a sphere of positive matter. However, Rutherford's interpretation of the alpha-particle scattering experiment carried out by Geiger and Marsden showed that Thomson's model was not correct and led to the nuclear model of the atom.

   (a) Use a diagram to describe Rutherford's alpha-particle scattering experiment.

   (b) List the main observations from the experiment.

   (c) Explain how evidence from the experiment supports the idea that:
      (i) atoms are mainly empty space
      (ii) then ucleus is charged
      (iii) then ucleus contains most of the mass of the atom.

    (d) Explain how, if we know the incident energy of alpha-particles, it is possible to estimate an upper limit for the size of a nucleus.

    (e) Calculate the closest approach of 5.0 MeV alpha-particles to a gold nucleus (Z = 79).

    (f) State an assumption that was made to carry out the calculation in (e).

2. State and explain two advantages of using high-energy electron beams from an accelerator rather than alpha-particles from radioactive sources to measure nuclear diameters.

3. (a) Complete the table below for neutral atoms.

Isotope	Symbol	Atomic number	Nucleon number	Protons	Electrons	Neutrons
Hydrogen			1			
Carbon-12	$^{12}_{6}C$					
Carbon-13						
Carbon-14						
Oxygen-16			8			
Iron-	$^{56}_{26}Fe$					
Gold-		79	197			
Uranium-235					92	
Uranium-238						

    (b) Use examples from the table to explain what is meant by an "isotope."

    (c) The heavier nuclei, such as iron, gold, and uranium, have an excess of neutrons. Suggest a reason for this.

4. A radioactive rock is tested in a school laboratory. Here are the results:

Setup	Number of counts in 5 minutes
No rock present	100
Rock alone present	900
Rock behind card	402
Rock behind 2mm aluminum sheet	398
Rock behind 1 cm lead	123
Rock alone present 24 hours later	197

    (a) What is the average background count (in cpm)?

    (b) Why must we use the term "average"?

    (c) Roughly what is the half-life of the source?

    (d) What is/are the main type(s) of radiation emitted by the rock? Give reasons for your answer.

5. Carbon-14 has a half-life of 5700 years. While alive, organisms exchange carbon with their surroundings and maintain a constant small fraction of the isotope carbon-14. When a living creature dies, its carbon-14 content is not replenished, so this fraction falls. If the activity due to carbon-14 is measured, the time since the organism died can be calculated.

   (a) Calculate the decay constant of carbon-14.

   (b) A sample taken from an archeological relic contains only 1/16 of carbon-14 that would be present in a sample of the same mass taken from a living creature. How old is the relic? What assumptions have been made?

   (c) How long does it take after death for the ratio of carbon-14 to carbon-12 to fall to 10% of its original value?

6. Two radioisotopes—X and Y—have half-lives of 1 and 2 hours respectively. N atoms of X have an activity equal to that of M atoms of Y. What is the ratio M/N?

7. Nuclide P decays to nuclide Q with half-life 1000 years. Q is stable. At $t = 0$ there are N atoms of P and none of Q present.

   Work out the number of nuclei of P and Q after 1000, 2000, 3000, and 5000 years. Tabulate your results and use them to sketch a graph showing how the number of atoms of both nuclides varies with time.

8. 8. Gamma-ray photons of a particular energy from a radioactive source have a 98% probability of penetrating 0.10 mm of lead.

   (a) A lead shield is placed in front of the source. Show that the intensity of transmitted radiation falls off exponentially with the thickness of the lead shield.

   (b) Calculate the half-thickness of lead for this radiation.

   (c) For safety reasons the intensity of gamma-rays from this source must be reduced to 1% of its original value. Calculate the thickness of lead shielding required to do this.

9. Discuss the advantages and disadvantages of using radioactive sources as tracers in medical diagnosis.

10. There are four naturally occurring radioactive series, all of which end with a stable isotope of lead. The uranium series starts with uranium-238 (Z = 92) and ends with lead-206 (Z = 82).

   (a) All of the decays in the series are either alpha decays or beta-minus decays. State the number of each type of decay in the entire series.

(b) The first three nuclides in the series are: uranium-238 (atomic number 92), thorium-234, palladium-234, and uranium-234. Write down balanced nuclear transformation equations for the first three decays.

(c) The half-lives of the four nuclides above are: 4.51×10^9 years, 24.1 days, 77s, and 2.47×10^7 years. Suggest how the relative abundances of each nuclide compare.

(d) The isotope bismuth-214 ($Z = 83$) can decay by either alpha or beta-minus decay, but whichever decay it undergoes, the series reaches lead-210 (atomic number 82) in two steps. Write down the nuclear transformation equations to show how this occurs by each route. (The element with $Z = 81$ is thallium and the element with $Z = 84$ is polonium.)

11. An experiment was carried out using a Geiger counter to monitor the activity of a radioactive source. The table below gives the average cpm, corrected for background, recorded during the experiment.

Time/s	Activity/10^{13} s^{-1}
0	1.64
50	1.36
100	1.13
150	0.935
200	0.775
250	0.643
300	0.533
50	0.422
400	0.366
450	0.304
500	0.252

Use a log-linear graph to determine the decay constant and half-life of this source.

CHAPTER 26

NUCLEAR PHYSICS

26.1 Nuclear Energy Changes

Nucleons inside nuclei are in bound states. This means that their total energy is negative and work would have to be done to separate them. The work that must be done is called the binding energy (B.E.) of the nucleus. Note that this is not energy *in* the nucleus, it is energy that must be supplied to break it apart or that would be released if the nucleus formed from separate nucleons. When nuclear transformations such as radioactive decay, nuclear fission or nuclear fusion occur, the total B.E. of the system changes and energy can be released, e.g., in the form of kinetic energy of the products of reaction. Einstein's mass–energy relationship, $E = mc^2$, links these energy changes to mass changes.

26.1.1 Nuclear Binding Energy

The B.E. of a nucleus is equal to the work that must be done to separate all of the nucleons in the nucleus so that they are no longer interacting. In theory, this would mean separating them to infinite distances but the strong nuclear force that binds them together is extremely short range and effectively falls to zero at distances greater than a few times 10^{-15} m:

The energy of a nucleon is lower when it is inside the nucleus than when it is a free particle so its mass is also lower. The total mass of a nucleus is therefore less than the sum of the rest masses of the free nucleons from which it is made. The difference in mass between the mass of the free nucleons and the mass of the nucleus is called the mass defect Δm for the

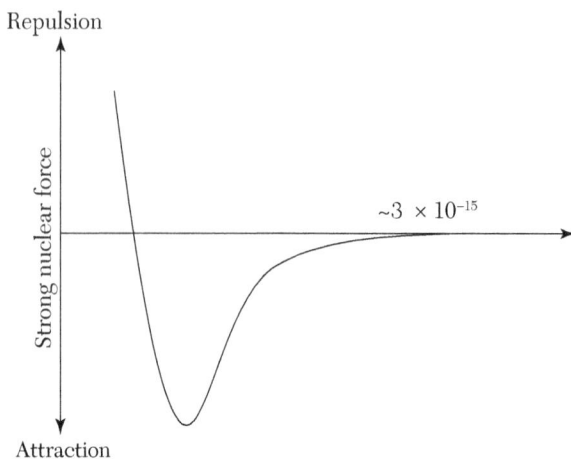

nucleus. The B.E. of the nucleus is equal to the mass defect multiplied by the speed of light squared.

Mass defect = (sum of masses of free nucleons) − (mass of nucleus)

$$\Delta m = \left(Zm_p + Nm_n\right) - M_{\text{nucleus}}$$

$$\text{B.E.} = c^2\Delta m$$

B.E. per nucleon:

$$\text{B.E.}/A = c^2\Delta m /A$$

When carrying out calculations of nuclear B.E. it is important to use nuclear mass and not atomic mass. Most tables of data give atomic masses so the mass of Z electrons must be subtracted from this. The data below has been used to calculate the B.E. and B.E. per nucleon of an oxygen-16 nucleus.

Particle	Mass/kg
Electron	9.1094×10^{-31}
Proton	1.6726×10^{-27}
Neutron	1.6750×10^{-27}
Oxygen-16 atom	26.5676×10^{-27}

Mass of oxygen-16 nucleus = (mass of oxygen−16 atom)
$$- 8 \times (\text{mass of electron})$$
$$= 26.5603 \times 10^{-27} \text{ kg}$$

Mass of nucleons = 8 × (mass of neutron) + 8 × (mass of proton)
$$= 26.7808 \times 10^{-27} \text{ kg}$$

Mass defect = Δm = (mass of nucleons) − (mass of nucleus)
$$= 0.2205 \times 10^{-27} \text{ kg}$$

B.E. = $c^2\Delta m$ = 1.9845×10^{-11} J = 124.0 MeV
B.E. per nucleon = B.E./16 = 7.752 MeV/nucleon

26.1.2 Atomic Mass Units (A.M.U.)

Atomic masses are usually stated in terms of unified atomic mass units, u.

$1 \text{ u} = 1/12 \times$ (mass of an unbound carbon-12 atom in its ground state)

$1 \text{ u} = 1.660\ 539\ 040 \times 10^{-27} \text{ kg}$

The mass of a carbon-12 atom in these units is 12 u and 1 u is approximately equal to the mass of a nucleon (proton or neutron).

The energy equivalent of 1 u is $1.494 \times 10^{-10} \text{ J} = 932.9 \text{ MeV}$

Nuclear energy calculations are often carried out by finding the mass defect in atomic mass units and then multiplying by the energy equivalent to 1 u.

26.1.3 Energy Released by Nuclear Decays

To calculate the energy released in a radioactive decay process we must find the mass defect for the reaction. Here are some examples.

Alpha decay of uranium-238 is:

$$^{238}_{92}U \rightarrow\ ^{234}_{90}\text{Th} + ^{4}_{2}\alpha$$

The relevant atomic masses are:

Atom	Mass
$^{4}_{2}\text{He}$	4.002603 u
$^{234}_{90}\text{Th}$	234.04364 u
$^{238}_{92}\text{U}$	238.05082 u

The mass defect for the reaction is the difference between the nuclear masses on each side of the equation. To calculate the nuclear masses we need to subtract 92 m_e from the left-hand side and $(90 + 2)m_e$ from the right-hand side. These electron masses cancel so we can work directly with the atomic masses:

$\Delta m = 238.05082 \text{ u} - (234.04364 + 4.002603) \text{ u} = 0.00458 \text{ u}$

The energy released is $E = 932.9 \times 0.00458 = 4.27 \text{ MeV}$

This is shared between the alpha particle and the recoiling nucleus and linear momentum must be conserved so (in the reference frame of the original uranium-238 atom) the two must travel in opposite directions and:

$$m_{nucleus} v_{nucleus} = m_\alpha v_\alpha$$

so that:

$$v_\alpha = \left(\frac{m_{\text{nucleus}}}{m_\alpha}\right)v_{\text{nucleus}} = \left(\frac{234}{4}\right)v_{\text{nucleus}}$$

The alpha particle travels much faster than the recoiling nucleus and carries away most of the kinetic energy.

$$E_\alpha = \frac{1}{2}m_\alpha v_\alpha^2 = \frac{1}{2}m_\alpha\left(\frac{m_{\text{nucleus}}}{m_\alpha}\right)^2 v_{\text{nucleus}}^2$$

$$= \left(\frac{m_{\text{nucleus}}}{m_\alpha}\right)E_{\text{nucleus}} = \left(\frac{234}{4}\right)E_{\text{nucleus}}$$

The alpha particle gets 234/238 E.

Beta-minus decay of carbon-14 is:

$$^{14}_{6}C \rightarrow {}^{14}_{7}N + {}^{0}_{-1}\beta + {}^{0}_{0}\bar{v}$$

The relevant atomic masses are:

Atom	Mass
$^{0}_{-1}\beta$	0.000549 u
$^{14}_{6}C$	14.003242 u
$^{14}_{7}N$	14.003074 u

The neutrino has negligible rest mass.

The mass defect for the reaction is the dfifference between the total rest masses on each side of the equation. To calculate the nuclear masses of carbon-14 and nitrogen-14 we must subtract 6 me from the atomic mass of carbon-14 and 7 me from the atomic mass of nitrogen-14.

$$\text{Mass defect} = (14.003242 - 6 \times 0.000549)\text{ u}$$
$$- ((14.003074 - 7 \times 0.000549) + 0.000549)\text{ u}$$

When the beta-minus is included, the electron masses cancel:

$$\Delta m = (14.003242 - 14.003074)\text{ u} = 0.000168\text{ u}$$

The energy released is $E = 932.9 \times 0.000168 = 157$ keV

This energy is shared between the recoiling nitrogen-14 nucleus, the beta-particle and the neutrino. The mass of the neutrino and beta-particle is negligible compared to the mass of the nucleus so virtually all the energy goes to the two light particles. However, this energy is shared randomly between them so beta-particles from a carbon-14 source are emitted with a continuous range of kinetic energies up to a cut-off value of 157 keV.

26.2   Nuclear Stability

Inside an atomic nucleus the nucleons are bound together by the strong nuclear force. However, the protons are all positively charged and repel one another so the strong nuclear forces and the electrostatic forces act in opposition. These two forces are very different—the nuclear force becomes very strong at short distances, binding nucleons together when they are close, but it has a very short range so it only strongly affects nearest neighbors. The electrostatic force however, obeys an inverse-square law, so all the protons exert significant repulsive forces on all other protons in the nucleus.

For very large nuclei the cumulative effect of coulomb repulsion outweighs the attractive force between nearest neighbor nucleons and the nucleus becomes unstable. Ultimately this sets a limit to the size of stable nuclei and explains why the periodic table contains only about 100 elements.

26.2.1   Nuclear Configuration and Stability

If a graph of neutron number against proton number is drawn the stable nuclei form a narrow band. On each side of this band there are unstable nuclei with specific decay modes that result in product nuclei lying closer to the band of stability.

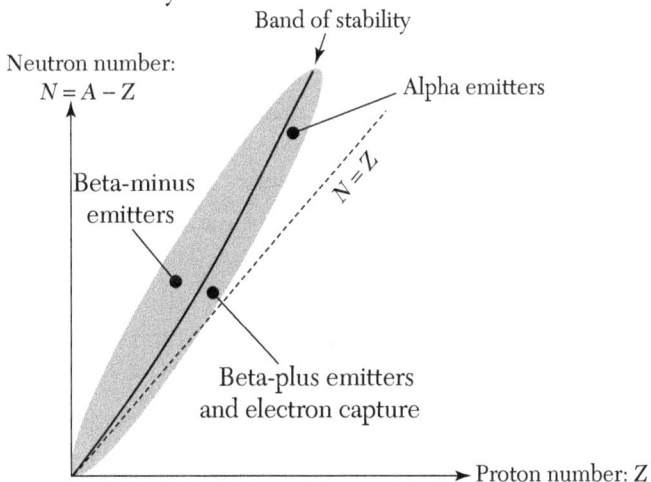

- Beta-minus emitters: N/Z too high—neutron-rich nuclei that approach stability by converting a neutron to a proton.

- Beta-plus emitters and electron capture nuclei: N/Z too low—proton-rich nuclei that approach stability by converting a proton to a neutron.

- Alpha emitters: N/Z too low—heavy proton-rich nuclei that approach stability by reducing both N and Z by 2. Since $N > Z$ this reduces the ratio of N to Z.

- Nuclear fission: some heavy nuclei close to the top of the band can undergo induced or spontaneous nuclear fission to create pairs of neutron-rich daughter nuclei which lie about half-way down the band and are beta-minus emitters.

The diagram below shows the effect of alpha and beta decays on a plot of proton number against neutron number:

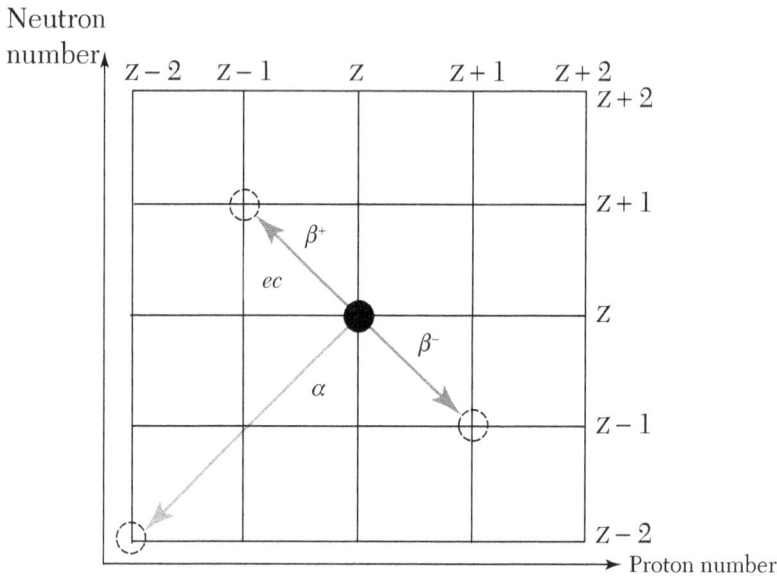

26.2.2 Nuclear Binding Energy and Stability

The diagram below shows the variation of nuclear B.E. per nucleon with nucleon number across the periodic table.

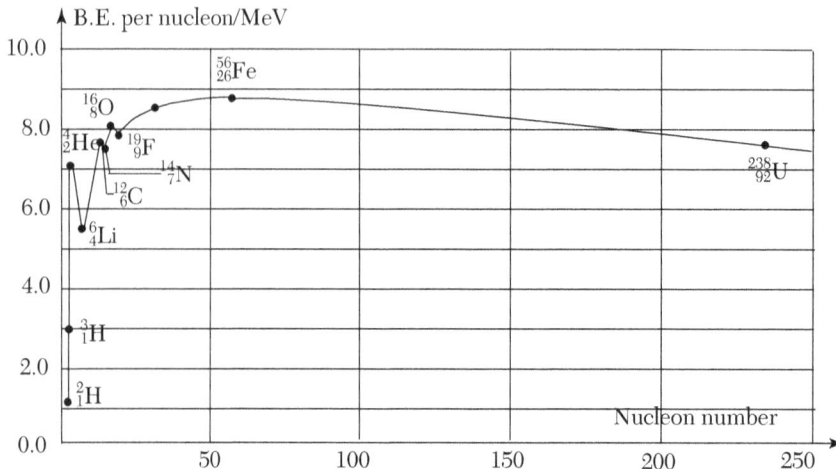

It is better to compare B.E. per nucleon rather than total nuclear B.E. because the latter depends on the number of nucleons in the nucleus so that a large value does not necessarily mean that the nucleus is particularly stable.

- B.E. per nucleon increases rapidly with nucleon number for light nuclei.

- The curve has a peak value that occurs for iron-56. This is the most stable nuclide.

- For nuclides heavier than iron-56 the B.E. per nucleon gradually falls.

- Most nuclides (from Oxygen to Uranium) have a B.E. per nucleon between 7.5 and 8.5 MeV/nucleon.

- Some light nuclides, such as helium-4, carbon-12 and oxygen-16 have particularly large B.E. per nucleon compared to other nearby nuclides.

26.3 Nuclear Fission and Nuclear Fusion

Einstein's equation $E = mc^2$ explains where the energy released in radioactive decays comes from but it also raised the question of whether there were other ways to release nuclear potential energy. There are two important processes that can be used to do this, nuclear fission and nuclear fusion. Nuclear fission is the splitting of a heavy nucleus to create two lighter daughter nuclei and nuclear fusion is the combination of two light nuclei to create a more massive nucleus. The B.E. per nucleon curve shows that both processes will release energy.

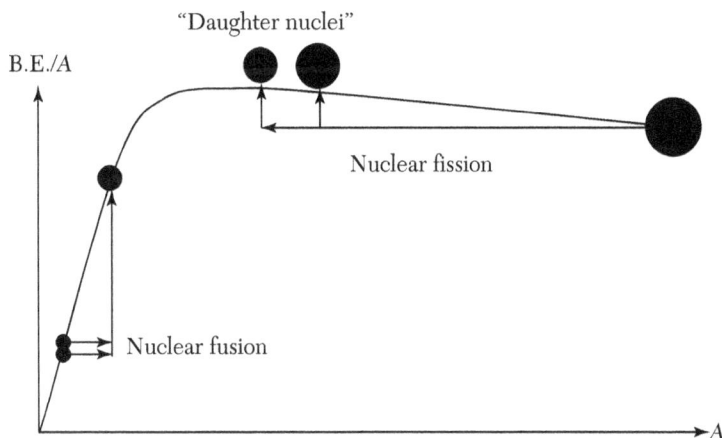

The initial steepness of the curve shows that nuclear fusion releases more energy per kilogram of fuel than nuclear fission. However, the high nucleon number of the fissioning nucleus shows that nuclear fission releases more energy per reaction than nuclear fusion.

26.3.1 Nuclear Fission

Nuclear fission was discovered by Otto Hahn and Lise Meitner in 1938. They were bombarding uranium with neutrons and noticed that lighter nuclei with mass numbers approximately half that of uranium were being formed. They came to the correct conclusion that the neutrons had induced fission reactions in nuclei of uranium. However, natural uranium consists almost entirely of two isotopes, uranium-238 (99.3%) and uranium-235 (0.7%). The fissionable isotope is uranium-235 while uranium-238 is more likely to absorb neutrons than fission.

There are many ways that the uranium-235 nucleus can split when it absorbs a neutron. Here is one nuclear equation for the induced fission of uranium-235:

$$^{1}_{0}n + ^{235}_{92}U \rightarrow ^{144}_{56}Ba + ^{90}_{36}Kr + 2\,^{1}_{0}n$$

The relevant atomic masses are:

Atom	Mass
$^{1}_{0}n$	1.009 u
$^{235}_{92}U$	235.044 u
$^{144}_{56}Ba$	143.923 u
$^{90}_{36}Kr$	89.920 u

We need to work with nuclear masses, but since there are equal numbers of electrons to subtract from each side of the equation we can, once again, simply use the atomic masses.

Mass defect = $(235.044 + 1.009)$ u $- (143.923 + 89.920 + 2 \times 1.009)$ u = 0.192 u

The energy released is $E = 932.9 \times 0.192 = 179$ MeV

The fact that additional neutrons are emitted could lead to further nuclear fission reactions. The Hungarian physicist Leo Szilardrealized that it would be possible to initiate a chain reaction if more than one neutron per fission on average went on to cause further fission reactions.

A chain reaction can release a huge amount of energy. If it is allowed to run out of control this energy is released explosively—this is the principle behind an "atom bomb." However, if the chain reaction can be controlled at a constant power, it can be used to generate electricity. This is the principle behind thermal nuclear power stations. The main problem facing both approaches is that 99.3% of natural uranium is uranium-238, a neutron absorber. This reduces the average number of neutrons per fission that can initiate further fission and effectively stops the reaction.

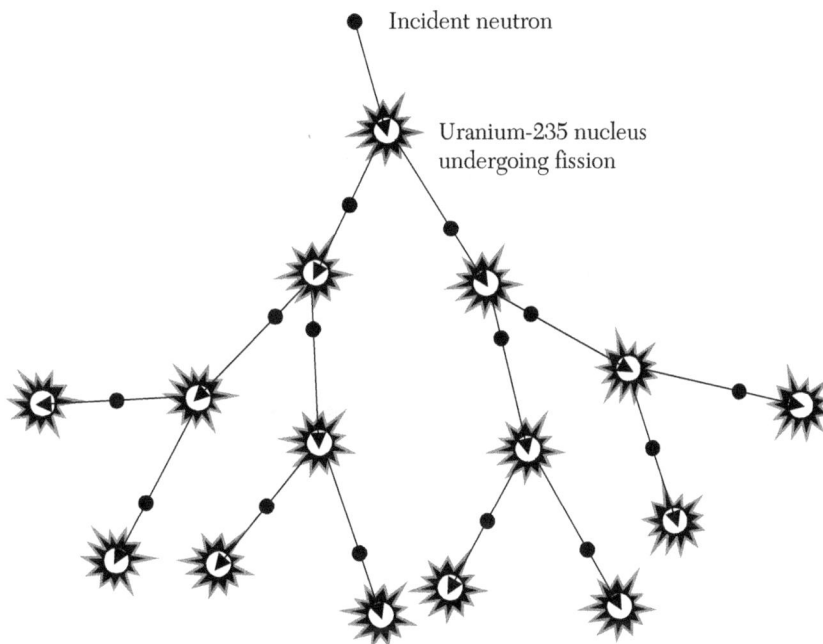

Incident neutron

Uranium-235 nucleus
undergoing fission

One way around this problem is to use enriched uranium, i.e., uranium with a higher content of uranium-235, but this is difficult to obtain in large quantities because uranium-235 and uranium-238 are isotopes of the same chemical element and so have the same chemical properties. Various separation techniques have been used but the most successful involves centrifuging uranium hexafluoride, a gaseous uranium compound, to increase the concentration of uranium-235. Since this method can be used to produce fuel for nuclear reactors (about 4% enrichment) *and* for weapons (about 80% enrichment) it is very difficult to distinguish between the peaceful production of enriched uranium and its production for atomic bombs.

26.3.2 Principle of the Atomic Bomb

The first atomic bombs were developed in the USA during World War II. In August 1945, atomic bombs were dropped on the Japanese cities of Hiroshima and Nagasaki. The Hiroshima bomb used fission of uranium-235 and the Nagasaki bomb used fission of plutonium-239.

To get an uncontrolled chain reaction a minimum or critical amount of fissionable material had to be produced. This is because the surface area to volume ratio decreases with the volume of material. The rate at which neutrons are produced depends on the volume of the sample, whereas the number lost from the surface depends on its surface area, so decreasing this ratio tips the balance toward the chain reaction. When a critical assembly is produced, the reaction proceeds very rapidly and there is a huge explosion.

In both the Hiroshima and Nagasaki bombs, sub-critical amounts of fissionable material were made critical when the bomb detonated. However, the techniques used were quite different. The Hiroshima bomb used a gun design where a sub critical amount of uranium-235 was fired (by conventional explosives) into another subcritical amount so that the combination became critical and exploded.

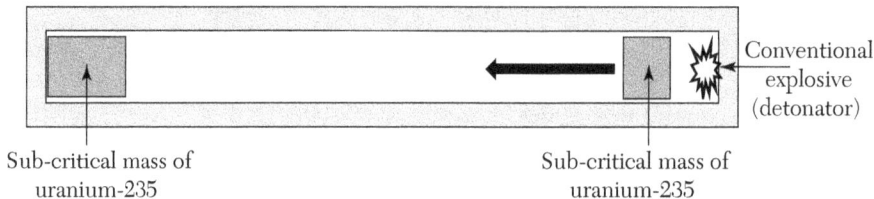

Sub-critical mass of uranium-235

Sub-critical mass of uranium-235

Conventional explosive (detonator)

In the Nagasaki bomb a spherical lump of plutonium-239 was surrounded by shaped conventional explosives. When it detonated shock waves compressed the sphere so that its surface area reduced and it became critical. It too exploded. The energy released by each bomb was equivalent to the explosion of about 10,000 tons of TNT (conventional high explosive).

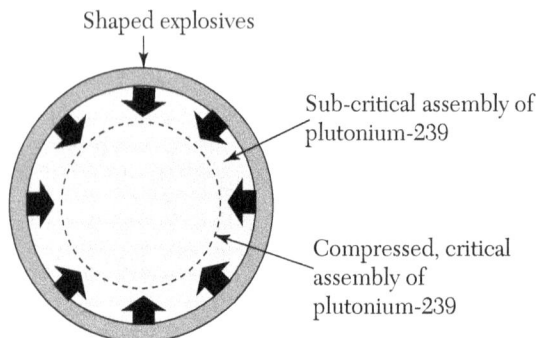

Shaped explosives

Sub-critical assembly of plutonium-239

Compressed, critical assembly of plutonium-239

While atom bombs are incredibly powerful weapons they release far less energy than a hydrogen bomb, which is based on nuclear fusion reactions. In order to initiate the intense conditions of temperature and pressure under which nuclear fusion reactions take place, a fission weapon is detonated first, for this reason such weapons are called **thermonuclear devices**.

26.3.3 Nuclear Reactors

In the core of a nuclear reactor the chain reaction is allowed to grow steadily until it reaches a constant power level and then it is kept at that level. Under these conditions exactly one neutron per fission goes on to initiate further fission reactions. These reactions heat the core and energy is extracted by pumping a coolant, usually water, through the core. The heated water is used to generate steam to drive turbo generators. In many respects this is the same principle as in a fossil-fueled power station. The main difference is that the source of the energy is nuclear rather than chemical.

To achieve a stable critical assembly a **moderator** is used. This is a material that slows down the neutrons emitted when a uranium-235 nucleus undergoes fission. The fast neutrons emitted by fission have a chance of being absorbed by a uranium-238 nucleus or initiating fission in a uranium-235 nucleus. Slowing them down increases both probabilities but has a much greater effect on the probability of initiating fission in uranium-235 so it allows a chain reaction to proceed at lower levels of enrichment than are needed for the spontaneous chain reaction in a bomb. The moderator contains nuclei of relatively light elements (e.g., carbon, water or heavy water) so that fast neutrons make a number of collisions with these nuclei and transfer energy and momentum to them, slowing down as they do so. Eventually the mean kinetic energy of the neutron is comparable to the thermal kinetic energy of particles in the moderator, so they are called **thermal neutrons**. Fuel rods are surrounded by moderating material:

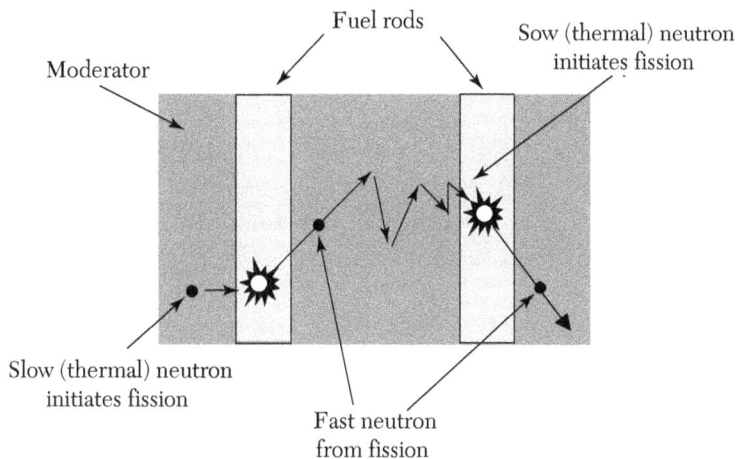

The reaction is controlled using neutron-absorbing material such as boron or cadmium in control rods that can be raised or lowered inside the core of the reactor. In an emergency, the control rods are dropped into the core and the chain reaction stops.

Coolant is pumped through the core of the reactor. This extracts energy to generate electricity but also stops the reactor core overheating and melting ("melt down"). Failure of the coolant system activates fail-safe systems that immediately lower the control rods and switch off the reactor.

The diagram shows a simplified arrangement for a pressurized water reactor. This is the most common type of nuclear reactor and is in use in many countries including the USA, France, Russia, Japan, and China.

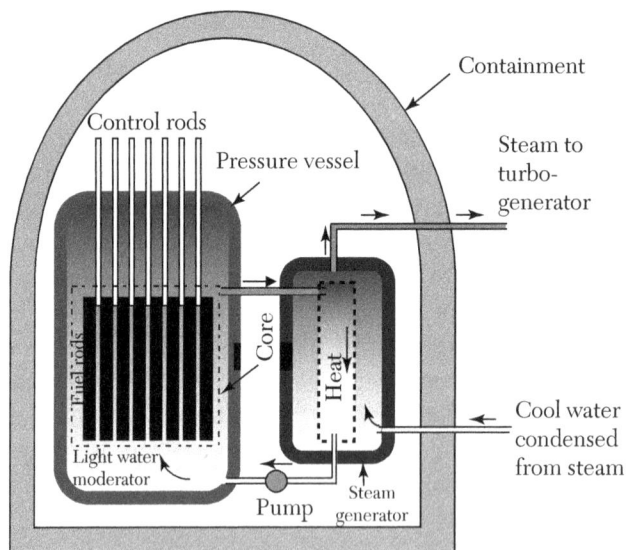

There are two cooling circuits. In the primary circuit water is pumped up through the core and then returns to the core via a heat exchanger. In the heat exchanger energy is transferred to water in the secondary circuit. This generates steam that is used to drive a turbo generator to generate electricity. The steam is then condensed and returned to the heat exchanger. Cold water from a lake or river is needed to operate the condenser and large cooling towers are used to cool this water once it has returned from the condenser.

26.3.4 Plutonium

Plutonium is a fissile material that can be used in bombs and reactors. However, it does not occur naturally on Earth in any significant quantities but it is created as a by-product of nuclear fission reactions in a reactor core. Neutrons that are absorbed by uranium-238 create an unstable and short-lived nuclide, uranium-239 that undergoes a beta-minus decay to form neptunium-239. This is also unstable with a relatively short half-life and it undergoes a second beta-minus decay to form plutonium-239:

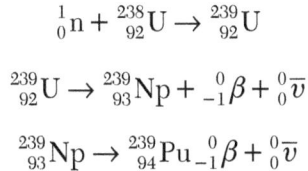

$$\frac{1}{0}n + \frac{238}{92}U \rightarrow \frac{239}{92}U$$

$$\frac{239}{92}U \rightarrow \frac{239}{93}Np + \frac{0}{-1}\beta + \frac{0}{0}\bar{v}$$

$$\frac{239}{93}Np \rightarrow \frac{239}{94}Pu \frac{0}{-1}\beta + \frac{0}{0}\bar{v}$$

Plutonium can be harvested from spent fuel rods. This is called "reprocessing."

26.3.5 Nuclear Fusion

Nuclear fusion is the combination of two light nuclei to form a more massive nucleus with the release of a great deal of energy. Here is an example of a nuclear fusion reaction which combines two isotopes of hydrogen (deuterium and tritium) to form helium-4 and release a neutron. This reaction might be used in future fusion reactors:

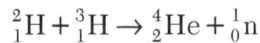

$$\frac{2}{1}H + \frac{3}{1}H \rightarrow \frac{4}{2}He + \frac{1}{0}n$$

The relevant atomic masses are:

Atom	Mass
$\frac{1}{0}n$	1.008664 u
$\frac{2}{1}H$	2.014102 u
$\frac{3}{1}H$	3.0160492 u
$\frac{4}{2}He$	4.002603 u

We need to work with nuclear masses, but since there are equal numbers of electrons to subtract from each side of the equation we can, once again, simply use the atomic masses.

Mass defect = $(2.014102 + 3.0160492)$ u $- (4.002603 + 1.008664)$ u
= 0.0189 u
The energy released is $E = 932.9 \times 0.0189 = 17.6$ MeV

This is about 3.5 MeV/nucleon compared to about 0.76 MeV/nucleon from nuclear fission (combustion releases less than 1 eV per nucleon!).

For nuclear fusion to take place the reacting nuclei must come close enough (a few times 10^{-15} m) for the short range strong nuclear force to bind them together. However, all nuclei are positively charged and repel one another. To approach close enough for fusion to take place they must have a very large kinetic energy. This can be achieved by accelerating the nuclei and then crashing them together in a device such as the Large Hadron Collider (LHC) at CERN or by confining the reactants and heating them to extreme temperatures.

Three situations that involve nuclear fusion reactions are:

- Nucleosynthesis—the formation of heavy nuclei from light nuclei in the cores of stars.

- Thermonuclear weapons—the nuclear fusion of isotopes of hydrogen in a bomb.

- Fusion reactors—commercial reactors designed to produce electrical energy from nuclear fusion.

26.3.6   Nucleosynthesis

Soon after the Big Bang the early universe consisted mainly of hydrogen with some helium and trace amounts of other nuclei. Nuclei of all the heavier elements were formed (and are still being formed) by nuclear fusion reactions taking place in stars. Nuclei of elements up to iron-56 are formed in the cores of stars during most of their "normal life" (when they are on the "main sequence"). Nuclei beyond iron-56 are formed when very massive stars explode at the end of their lives (forming supernovae).

Stars form when clouds of gas and dust collapse under their own gravitational forces. As gravitational potential energy falls the gas and dust heats up and when the temperature and pressure at the core of the collapsing mass become high enough nuclear fusion reactions begin. At this point a proto star is formed. The higher the mass of the star the more extreme the core conditions and elements higher up the periodic table can form. This process of nucleosynthesis stops at iron-56 because this is the most stable nuclide. Nuclides lighter than this are formed by exothermic reactions whereas those beyond iron-56 are only formed in endothermic fusion reactions and so need an external source of energy. The energy released in the core by exothermic nuclear fusion reactions generates an outward radiation pressure that supports the star against further gravitational collapse during most of its "life."

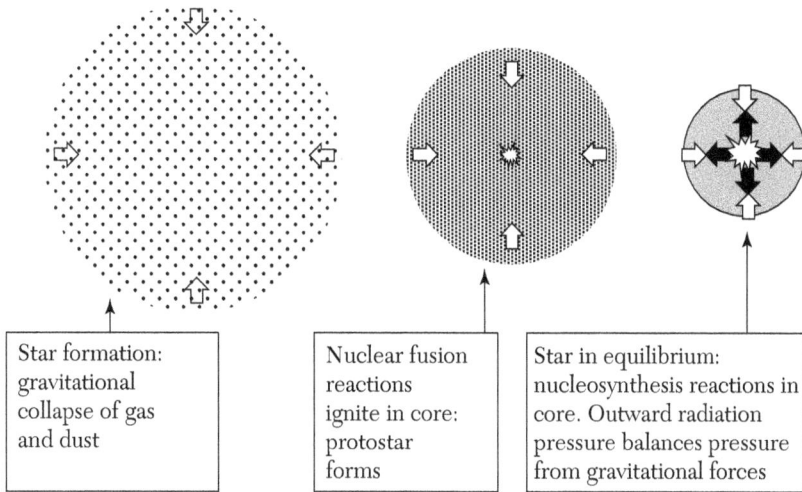

| Star formation: gravitational collapse of gas and dust | Nuclear fusion reactions ignite in core: protostar forms | Star in equilibrium: nucleosynthesis reactions in core. Outward radiation pressure balances pressure from gravitational forces |

As fuel for the fusion reactions in the core runs out gravitational forces cause the core to collapse. What happens next depends on the mass of the star (see Section 28.1.1) but when stars of mass greater than about 10 times the mass of the Sun collapse they undergo a sequence of fusion reactions and create all the nuclides up to iron-56 and then explode in a supernova. Some of the energy released in the explosion creates the heavier nuclei up to uranium and the explosion distributes them throughout space.

Our Sun is a medium-sized star and will spend almost all of its life synthesizing helium from the hydrogen in its core. The net effect is to convert four protons into a helium nucleus but the probability of this happening in one step by a fortunate collision of four particles with enough energy to get close enough to fuse is effectively zero. The main process by which helium is created is called the proton-proton cyclethat proceeds in three steps.

- Step 1: Two protons collide to form a deuteron, a positron (anti-electron) and a neutrino:

$$\,_1^1\text{H} + \,_1^1\text{H} \rightarrow \,_1^2\text{H} + \,_1^0\overline{e} + \,_0^0\nu$$

- Step 2 (twice): A proton collides with a deuteron to form a nucleus of helium-3 and emit a gamma-ray:

$$\,_1^1\text{H} + \,_1^2\text{H} \rightarrow \,_2^3\text{He} + \,_0^0\gamma$$

- Step 3: Two helium-3 nuclei collide to form a helium-4 nucleus and release two protons:

$$\,_2^3\text{He} + \,_2^3\text{He} \rightarrow \,_2^4\text{He} + 2\,_1^1H$$

This process releases about 26 MeV per helium-4 nucleus produced. The overall reaction for the proton-proton cycle is then:

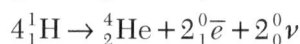

$$4\,^1_1\text{H} \rightarrow\, ^4_2\text{He} + 2\,^0_1\overline{e} + 2\,^0_0\nu$$

The two positrons created in the core almost immediately annihilate with electrons creating high-energy gamma-rays that contribute to the outward radiation pressure that supports the star. The neutrinos are very weakly interacting and pass through the outer layers of the Sun and into space. The flux of solar neutrinos detected on Earth gives astronomers a way to monitor the fusion processes going on in the Sun's core. Astronomers estimate that there is enough hydrogen left in the core for the Sun, which was formed about 5 billion years ago, to continue to shine for another 5 billion years.

26.3.7 Thermonuclear Weapons

The need to form a critical mass and for a significant amount of the fissile material in that mass to fission in a chain reaction limits the maximum yield of an atom bomb (fission weapon). A thermonuclear weapon uses a fission explosion to create the extreme temperatures and pressures under which isotopes of hydrogen can fuse. This makes the yield of a fusion weapon (hydrogen bomb) effectively unlimited. The most powerful thermonuclear bomb ever exploded was the RDS-220 or "Tsar bomb" detonated by Russia in 1961. It released an energy equivalent to about 50 million tons of TNT, which is about 5000 times more energy than the atom bomb dropped on Hiroshima.

The Teller-Ulam design of a hydrogen bomb, as shown below, has two distinct parts: a primary device rather like the implosion weapon dropped

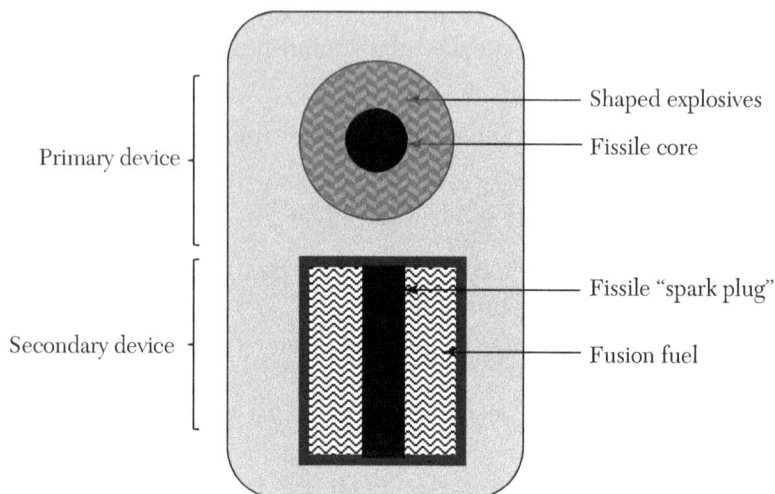

on Nagasaki, and a secondary device containing the fusion fuel that is imploded by a focused shock wave from detonation of the primary device. A fissile "spark plug" runs through the center of the second device and both enhances the reaction and propagates the shock wave that compresses the fusion fuel. The implosion creates the extreme conditions needed for the fuel (ultimately deuterium and tritium) to undergo nuclear fusion reactions and release a huge amount of energy.

26.3.8 Fusion Reactors

All existing commercial nuclear reactors use nuclear fission reactions in uranium or plutonium. While this is a well-established technology there are significant disadvantages to the use of nuclear fission to generate electricity:

- Fission reactors produce high level nuclear waste consisting of a complex cocktail of radioactive nuclides. Proposed strategies for the long term safe storage of nuclear waste (where these exist) are controversial.

- The fuel for fission reactors, uranium and plutonium, is limited, and sources of fuel are not accessible to all nations equally.

- Equipment needed to enrich uranium for reactors can also be used to enrich it to weapons grade so it is difficult to distinguish peaceful from military nuclear programs.

- When a reactor fails, e.g., in a melt-down of the core, there is the potential for massive environmental damage (e.g., Chernobyl, Fukushima) as radioisotopes escape from the reactor.

- Nuclear fission reactors are complex and the cost of decommissioning reactors at the end of their working life adds a significant amount to the cost of the electricity they generate.

If a reactor could be constructed that would generate electricity commercially from nuclear fusion reactions it would avoid most of the problems associated with nuclear fission.

- Fusion reactions involve light nuclei and produce very little radioactive waste.

- The fuel for fusion reactors, isotopes of hydrogen, are plentiful and spread all around the world.

- Nuclear fusion reactors work in a very different way to thermonuclear weapons so it would be easier to distinguish peaceful from military projects.

- When fusion reactions fail the reactor simply switches off, there is no core and no possibility of a meltdown.

- While the cost of a commercial fusion reactor is likely to be high the decommissioning costs should be significantly less than for a fission reactor because there is no highly radioactive core or spent fuel rods to deal with.

The challenge of creating a commercial fusion reactor is very great and stems from the need to create and maintain the extreme conditions under which fusion reactions can take place. The two main approaches are:

- Tokamaks—reactors that support a plasma (a gas of ionized particles) in a toroidal magnetic field and then pump it to high temperature to initiate the fusion reactions.

- Inertial confinement—focusing intense lasers onto a pellet of fusion fuel so that it implodes, reaches extreme temperatures and pressures and fuses.

Both methods have achieved some success in producing fusion reactions but have a long way to go before they can be used in a commercial reactor. The largest current project is the International Thermonuclear Experimental Reactor that is being built in France by an international collaboration of countries representing over half the world's population. Its aim is to test the feasibility of magnetic fusion reactors (Tokamaks).

The tokamak method is likely to use the fusion of deuterium (about 0.015% of all hydrogen atoms on Earth are deuterium) and tritium, which can be created from lithium (a common element in the Earth's crust).

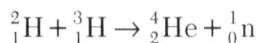

$$_{1}^{2}\text{H} + _{1}^{3}\text{H} \rightarrow _{2}^{4}\text{He} + _{0}^{1}\text{n}$$

The reactor would be surrounded by a lithium blanket. Once self-sustaining fusion reactions are taking place in the plasma a constant flux of neutrons can be absorbed in the blanket where they convert lithium to tritium, effectively creating more fuel for the reactor, e.g.,

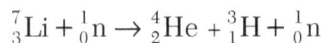

$$_{3}^{7}\text{Li} + _{0}^{1}\text{n} \rightarrow _{2}^{4}\text{He} + _{1}^{3}\text{H} + _{0}^{1}\text{n}$$

The blanket would also heat up and a suitable coolant (water) could be pumped through the lithium blanket and used to raise steam to drive turbo-generators.

26.4   Particle Physics

Throughout the twentieth century, experiments with cosmic rays and in nuclear physics revealed the existence of many new subatomic particles, with a range of different properties. As more and more were discovered, patterns and relationships between them were identified and physicists suspected that there might be something like the periodic table for these particles. They were right, and, as with the periodic table, the patterns were present because of an underlying structure to the particles.

The different types of atom are all constructed from just three types of particle: protons, neutrons and electrons. It turns out that all of the subatomic particles can be explained in terms of two simple classes of particle: the quarks and the leptons. The grand scheme that was constructed to explain particle physics is called the Standard Model. Here we will simply describe the main characteristics of each type of particle and the structure of the model.

26.4.1   Leptons

The anti-neutrino, emitted in beta-minus decay, and the neutrino, emitted in beta-plus decay, are closely related to the electron and positron that are also emitted in those decays. These are all **leptons**, particles that interact by the weak nuclear force. Surprisingly it turns out that there are also heavier versions of these particles so that there are three generations of leptons:

Generation	Particle/anti-particle	Particle/anti-particle
1st	Electron and positron	Electron-neutrino and anti-electron neutrino
2nd	Muon and anti-muon	Muon-neutrino and anti-muon-neutrino
3rd	Tau and anti-tau	Tau-neutrino and anti-tau neutrino

The muon and tau are effectively more massive versions of the electron and they tend to undergo decays that eventually produce electrons. The neutrinos are all neutral and have very low rest mass. Beams of neutrinos of any one type begin to oscillate between the different "flavors" of neutrino so that soon the beam contains equal numbers of electron-, muon-, and tau-neutrinos. The discovery that this occurs solved the so-called "solar neutrino problem" where only about 1/3 of the electron-neutrinos emitted by nuclear fusion reactions in the Sun were detected here on Earth. The other 2/3 had oscillated to muons or tausen route.

26.4.2   Hadrons and Quarks

Hadrons are strongly interacting particles like protons and neutrons but they are not fundamental, they consist of combinations of smaller particles called quarks. Most hadrons contain quark pairs (mesons) or quark triplets (baryons) and the quarks, like the leptons, come in three generations, each with a pair of different "flavors."

Generation	Particle/anti-particle	Particle/anti-particle
1st	Up and anti-up	Down and anti-down
2nd	Strange and anti-strange	Charm and anti-charm
3rd	Bottom and anti-bottom	Top and anti-top

The up, charm, and top quarks have a charge of $+2/3$ e.

The down, strange, and bottom quarks have a charge of $-1/3$ e.

Mass increases as we go down the table so that hadrons formed from bottom quarks tend to be very heavy (and highly unstable).

Theory suggests that quarks cannot exist as free particles and we do not see fractional charges in nature. Quarks bind together with the "color force" which is the origin of the strong nuclear force, and they do so into triplets (baryons) or pairs (mesons) with integer or no charge.

▪ Protons consist of two up and one down quark: uud. Charge $= (2/3 + 2/3 - 1/3)$ e $= +$e

▪ Neutrons consist of one up and two down quarks: udd. Charge $= (2/3 - 1/3 - 1/3)$ e $= 0$

While quarks interact by the strong force they are also affected by the weak nuclear force and this can cause them to change flavor. For example, when a beta-minus decay occurs and a neutron changes to a proton one down quark inside the neutron changes to an up quark (this is a weak interaction) creating an electron and anti-neutrino in the process.

Mesons are particles made from a quark and antiquark pair. For example the positive pi-meson $\pi+$ consists of an up quark and an anti-down quark: $u\bar{d}$. Virtual pi-mesons are exchanged between protons and neutrons to bind them together in nuclei.

26.4.3   Fundamental Interactions

There are four fundamental interactions in nature:

▪ Gravitation—infinite range; described by Einstein's general theory of relativity.

▪ Electromagnetism—infinite range; described by Maxwell's equations and quantum electrodynamics(QED).

- The weak nuclear force—short range; described by electroweak theory.

- The strong nuclear force (color force)—short range; described by quantum chromodynamics.

Three of these have been described using quantum theory and two of them, the weak nuclear force and electromagnetism, have been unified into a single theory, electroweak theory, so that they are seen as manifestations of the same underlying interaction. The color interaction is described by a similar quantum mechanism so it too is expected to unify with the electroweak interaction to form a single super interaction at very high energies.

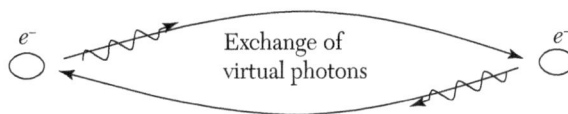
Exchange of virtual photons

The underlying process that explains how these three quantum forces work is based on the **exchange of virtual particles**. A virtual particle can be created by "borrowing" energy from the universe for a short time and then "paying it back" when the particle disappears. This is possible because of the energy-time Uncertainty principle in quantum theory. For example, the electromagnetic repulsion between two electrons comes about as a result of an exchange of virtual photons.

The exchange particles for electromagnetism are photons, for the weak force they are $W+$, $W-$, and Z^0 particles and for the strong force they are different kinds of gluons. Richard Feynman developed a pictorial way to represent interactions. His method was useful because it provided a link to the mathematical methods needed to solve problems in QED. The diagrams are known as **Feynman diagrams** and the diagram on the next page shows two of the many ways a pair of electrons might interact:

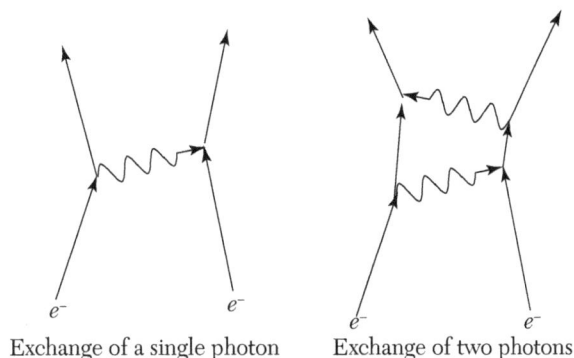

Exchange of a single photon   Exchange of two photons

In QED all possible ways in which an interaction could take place contribute to the probability of how it does take place, and Feynman diagrams provide a way to identify and order the different possibilities.

Gravity is still described by Einstein's general theory of relativity. This is an elegant and highly successful mathematical theory but it is based on continuous variations in the geometry of space-time and so far, no one has succeeded in finding an acceptable theory of quantum gravity. This is one of the most important outstanding questions in physics—how to connect general relativity and quantum theory.

26.4.4 Conservation Laws

We are familiar with conservation of momentum, mass energy, and charge but there are several other important conservation laws in particle physics, in particular:

- Conservation of baryon number: the total number of baryons in the universe cannot change. Each baryon, e.g., proton or neutron, has baryon number $B = 1$. Anti-baryons have number $B = -1$. Mesons have baryon number zero. Effectively quarks have a baryon number $+1/3$ and anti-quarks have baryon number $-1/3$.

- Conservation of lepton number: the total number of leptons in the universe cannot change. Each lepton, e.g., electron, neutrino, tau, has lepton number $L = 1$. Anti-leptons have lepton number $L = -1$.

(According to the Standard Model, lepton numbers in each generation are separately conserved and neutrinos are massless. However, recent discoveries have shown that neutrinos have a very small mass and can oscillate between the three flavors. While this does not affect conservation of total lepton number it does undermine the conservation of lepton number by flavor.)

26.4.5 Standard Model

The fact that everything can in principle be explained by a few leptons, quarks and force carriers is amazing, as is the achievement of physicists in discovering that this is the case.

The only particle that has been added to the Standard Model in recent years is the Higgs boson, discovered at the LHC in 2012. This particle is a quantum of the Higgs field that fills the Universe. The interaction of each particle with the Higgs field is responsible for setting the mass of the particle.

All of the particles in the table on the next page have anti-particles (or are their own anti-particle).

The table below summarizes all the particles in the Standard Model.

QUARKS			FORCE CARRIERS
1st generation	**2nd generation**	**3rd generation**	
UP Charge 2/3 e Spin ½	**CHARM** Charge 2/3 e Spin ½	**TOP** Charge 2/3 e Spin ½	**GLUON** Charge 0 Spin 1
DOWN Charge −1/3 e Spin ½	**STRANGE** Charge −1/3 e Spin ½	**BOTTOM** Charge −1/3 e Spin ½	**PHOTON** Charge 0 Spin 1
Increasing mass			**Z BOSON** Charge 0 Spin 1
ELECTRON Charge −e Spin ½	**MUON** Charge −e Spin ½	**TAU** Charge −e Spin ½	**W BOSON** Charge ±e Spin 1
ELECTRON-NEUTRINO Charge 0 Spin ½	**MUON-NEUTRINO** Charge 0 Spin ½	**TAU-NEUTRINO** Charge 0 Spin ½	
LEPTONS			

HIGGS BOSON
Charge 0
Spin 0

The fact that there are three generations of quarks and three generations of leptons suggest that there is an underlying symmetry linking the quarks and the leptons. This has led to several hypotheses about new particles and mechanisms for changing leptons to quarks and vice versa, but so far there has been no experimental evidence to support these ideas. While the Standard Model is incredibly impressive, it is unlikely to be the last word on particle physics; there are too many arbitrary constants that have to be put into the model to make it work.

26.4.6 Dark Matter and Dark Energy

Despite the success of physicists in constructing the standard model, discoveries in cosmology have raised the strong possibility that there are as yet undiscovered forms of matter. Careful measurements of the rotation rates of galaxies suggest that there is nowhere near enough visible matter in these galaxies (i.e., stars) to provide the gravitational force needed to maintain the rotation. In other words, the total gravitational force from visible matter inside the galaxy is much less than the required centripetal force for the outer parts to rotate as they do. This was first pointed out by Fritz Zwicky in 1933 when he tried to understand the motion of the Coma cluster of galaxies.

These observations led physicists to suggest that there must be a lot of invisible or **dark matter** in the galaxies to provide the additional centripetal force. This idea has been supported by evidence from gravitational lensing—the deflection of light close to galaxies because their mass distorts the local space-time geometry. The mass required to account for the observed lensing effects is much greater than the mass of the visible matter in the galaxies. While a small proportion of the dark matter is probably cool baryonic matter (i.e., "ordinary matter") the rest is yet to be identified. Dark matter is thought to make up 23% of the mass of the universe whereas ordinary matter is thought to make up just 4.6%. The remaining 72% of the mass is thought to be **dark energy**.

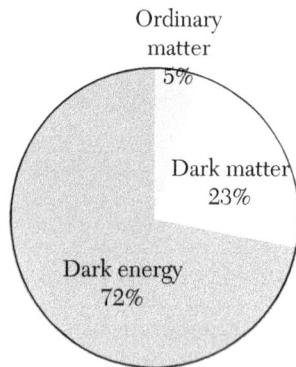

Dark energy is a relatively new idea put forward to explain an unexpected discovery in cosmology. In 1998, Saul Perlmutter and others measured the red-shifts of distant type-1a supernovae and came to the conclusion that the expansion of the universe is accelerating. This was a surprise because gravitational attraction between the galaxies would be expected tom slow the expansion. Other observations seemed to support the conclusion that the expansion rate is increasing so theoreticians needed to come up with a new model to explain this. Their idea is that an unknown form of energy fills the universe and creates an outward pressure, rather like a negative gravitational force, causing space-time to expand at an accelerating rate. The energy density is such that the mass equivalent of dark energy in the universe accounts for 72% of the total mass!

26.5 Exercises

1. (a) Explain what is meant by nuclear B.E..

(b) Sketch a graph to show how nuclear B.E. per nucleon varies with nucleon number and use it to explain:

(i) Why iron-56 is regarded as the most stable nucleus.

(ii) How nuclear fission reaction of some heavy nuclei can release a large amount of energy.

(iii) How nuclear fusion of some light nuclei can release a large amount of energy.

(c) Suggest why a nucleus of an element with atomic number 120 is likely to be highly unstable.

2. Use data from the table below to calculate the B.E. and B.E. per nucleon for iron-56 (Z = 26).

Particle	Mass
Electron	0.000549 u
Proton	1.007276 u
Neutron	1.008665 u
Iron-56 atom	55.934934 u

3. Uranium-235 (atomic number 92) is unstable and decays to thorium-231 by emitting an alpha particle of energy 4.77 MeV.

(a) Write down a nuclear transformation equation for the decay.

(b) Calculate the kinetic energy of the thorium-231 nucleus as it recoils.

(c) Calculate the mass defect in kilograms for the decay.

4. Here is data for the beta-minus decay of the rare isotope carbon-16.

Particle	Mass
$_{-1}^{0}\beta$	0.000549 u
$_{6}^{16}C$ (atom)	16.01470 u
$_{7}^{16}N$ (atom)	16.006103 u

(a) Write down a balanced nuclear equation for this decay.

(b) Calculate the maximum kinetic energy of an emitted beta-particle in this decay.

(c) Explain why this is a maximum-value and beta-minus particles from different decaying carbon-16 nuclei will have a range of kinetic energies.

(d) Carbon-16 has a half-life of just 0.74 s. Explain why it is not surprising that this nucleus is highly unstable.

5. Nitrogen (atomic number 7) has 7 isotopes from nitrogen-12 to nitrogen-18. 99.63% of all nitrogen is the stable isotope nitrogen-14 and the rest is almost entirely nitrogen-15, which is also stable. Here is some data that will be useful when answering the questions that follow.

Atom	Mass
Electron	0.000549 u
$^{12}_{7}N$	12.01864 u
$^{13}_{7}N$	13.005738u
$^{14}_{7}N$	14.003074 u
$^{15}_{7}N$	15.000108 u
$^{16}_{7}N$	16.006103 u
$^{17}_{7}N$	17.00845 u
$^{18}_{7}N$	18.0142 u
$^{13}_{6}C$	13.003354 u
$^{13}_{8}O$	13.0248 u
Proton	1.007276 u
Neutron	1.008665 u

Nitrogen-12 and nitrogen-13 decay by beta-plus emission while the other unstable isotopes of nitrogen decay by beta-minus emission.
(a) Write down nuclear transformation equations for the beta-plus and hypothetical beta-minus decay of nitrogen-13.
(b) Calculate mass defects for each of the reactions in (a) and use the values to explain why nitrogen-13 cannot decay by beta-minus emission.
(c) Calculate the B.E. per nucleon for nitrogen-14 and nitrogen-15. Both isotopes are stable but 99.63% of all nitrogen is nitrogen-14. Comment on this in the light of your calculation.
(d) Suggest, with reasons, how the half-lives of nitrogen-16, nitrogen-17 and nitrogen-18 are likely to compare.

6. Beta-decay is a weak interaction in which the flavor of a quark inside a baryon changes causing a change in the type of baryon in the nucleus. In beta-minus decay a neutron changes to a proton and in beta-plus decay a proton changes to a neutron.

(a) Sketch a graph of proton number against neutron number for the stable nuclides and use it to explain where on the chart beta-plus or beta-minus emitting nuclides are likely to be found.

(b) (i) Write down the nuclear equation for the beta-minus decay of carbon-14 to nitrogen-14.

(ii) Write down an equation for the decay of a neutron to a proton inside the carbon-14 nucleus in the decay in (b)(i).

(iii) Write down an equation for the change of flavor of a quark in the neutron in the decay in (b)(ii).

(c) Discuss whether a quark changes flavor when electron capture occurs.

CHAPTER 27

QUANTUM THEORY

27.1   Problems in Classical Physics

Classical physics is based on Newtonian mechanics (and gravitation) and Maxwell's electromagnetism. By the end of the nineteenth century these two theories had been incredibly successful in explaining diverse phenomena from the laws of thermodynamics and the propagation of radio waves to the paths of planets in their orbits. However, a few problems were beginning to emerge that resisted an explanation using classical physics. Here are three examples:

Black Body Radiation Spectrum

The shape of the black body radiation spectrum was well known, but all attempts to use classical physics to derive a formula for the spectrum failed. In fact, they agreed with the spectrum at long wavelengths (low frequencies) and but diverged drastically from it at short wavelengths (high frequencies), predicting an infinite amount of high frequency radiation from a hot body. This was clearly wrong. It was called the "**ultraviolet catastrophe**."

Heat capacities of gases

According to classical equipartition theory (see Section 9.3.5) thermal energy at temperature T should be distributed equally amongst all the degrees of freedom available to the particles of the gas such that each degree of freedom gains an energy of $\frac{1}{2}kT$. For a monatomic gas there are just three degrees of freedom corresponding to translations in the x-,y-, and

Power radiated per
unit wavelength

Prediction of classical
theory: ultraviolet
catastrophe

Black body radiation
spectrum

Wavelength

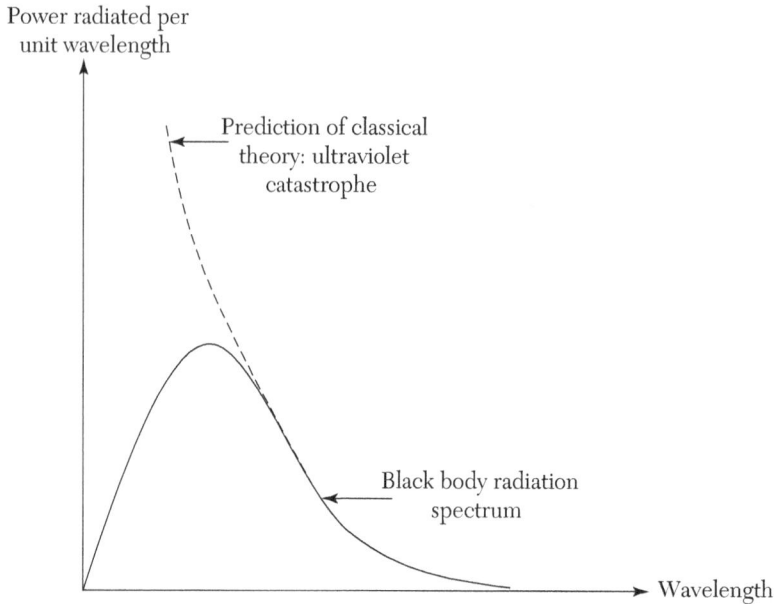

z-directions, so the energy per molecule is $3/2\ kT$. For more complex molecules, vibrational and rotational degrees of freedom can also be excited and these should, according to classical theory, get $\tfrac{1}{2}\,kT$ as well. The greater the energy per molecule at a particular temperature the higher the heat capacity of the gas, so measurements of heat capacity allowed physicists to test the predictions of classical equipartition theory. They found that while the predictions were usually good at high temperatures they sometimes broke down at lower temperatures, as if some degrees of freedom were not contributing to the heat capacity. This was unexpected and could not be explained using the classical theory.

Photoelectric effect

Heinrich Hertz discovered the photoelectric effect in 1887. He noticed that when ultraviolet light was shone onto a pair of electrodes it was easier to form sparks between them. Ultraviolet light seemed to be able to knock electrons out of the surface of the metal electrode. Further investigation showed that the effect depended on the frequency of the absorbed radiation but not on its intensity. This was a complete surprise. According to classical theory ejecting an electron from the surface of a metal should depend on the energy of the incoming radiation and not on its frequency.

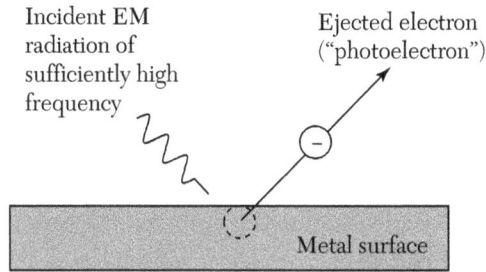

27.1.1   Planck and Black Body Radiation Spectrum

In 1900 Max Planck managed to derive the correct equation for the black body radiation spectrum, but to do so he had to make a revolutionary assumption. He assumed that the atomic oscillators vibrating in the black body can only have discrete amounts of energy and that the energy of any particular oscillator is a multiple of a smallest quantity or **quantum of energy** equal to a constant multiplied by its frequency. The energy of an atomic oscillator is:

$$E = nhf$$

n is an integer, f is the frequency of the atomic oscillator and h is the Planck constant.

$$h = 6.62607004 \times 10^{-34}\,\text{Js}$$

This was a radical suggestion. In classical physics energy is a continuous variable so atomic oscillators at any frequency could have any energy. By quantizing energy in this way Planck showed that the quantum of energy for a high frequency oscillator is much larger than for a low frequency oscillator.

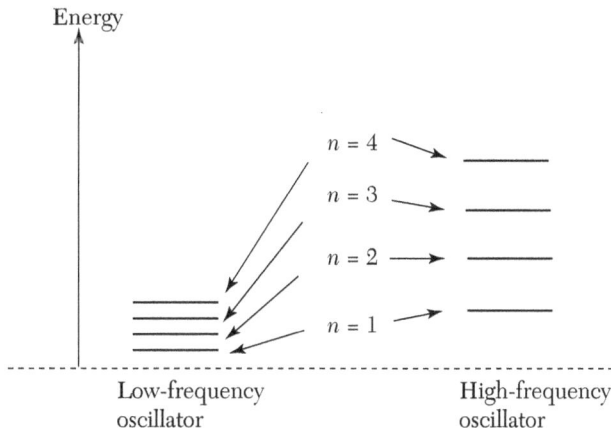

This makes it far less likely that a high frequency oscillator will be excited because it needs a large energy to start vibrating. The effect is to suppress high frequency vibrations and dramatically reduce the high-frequency electromagnetic radiation they emit, thus preventing the ultraviolet catastrophe. At low frequencies, the allowed energies are very close together and behave more like the classical continuum of allowed energies. That is why the long wavelength, low-frequency part of the curve *can* be explained classically.

Planck's quantization of energy was the first of several ad hoc quantizations discovered in the early part of the twentieth century. This was the beginning of quantum theory, but at that time no one understood why quantization worked!

27.1.2 Explaining Heat Capacities

Planck's idea, that the energy of an atomic oscillator is quantized, can also be used to explain why, at low temperatures, some modes of molecular rotation or vibration are suppressed. The argument is the same as for high-frequency atomic vibrations in a black body. The energy of a vibrational mode is quantized and if the mean thermal energy, ½ kT, is less than the minimum energy needed to excite a mode of vibration, hf, then those modes will not be excited and the heat capacity will be lower than expected. At high temperature ½ $kT > hf$ so the modes are excited and do contribute to the heat capacity. Measurements of heat capacity as a function of temperature shows that it increases toward the value predicted by equipartition as temperature increases.

27.1.3 Explaining the Photoelectric Effect

Einsteinrealized that the photoelectric effect could be explained by assuming that electromagnetic radiation can only exchange energy with matter in discrete amounts or quanta. These quanta or "photons" have an energy:

$$E = nhf$$

where f is the frequency of the radiation and h is the Planck constant.

There is a minimum energy, Φ, needed to remove an electron from the surface of a metal. This is called the "work function"for the metal and it depends on the element used, being lower for more reactive metals. Einstein assumed that when radiation is absorbed by matter one photon transfers all its energy to one electron. The photon energy must be greater than the work function for the electron to be ejected (some electrons will require more energy than Φ because they are not at the surface of the metal). For electrons to be ejected from the surface:

$$hf \geq \Phi$$

This explains why there is a threshold frequency f_0 below which photoelectric emission does not occur:

$$hf_0 = \Phi$$

This was another radical departure from classical physics where light was considered part of the electromagnetic spectrum and was assumed to be a continuous wave that can take any energy. Einstein's theory treated electromagnetic radiation as if it consisted of discrete packets of energy and transferred that energy discretely too. The photon theory treated light more like a particle model than a wave model.

27.1.4 Characteristics of Photoelectric Emission

To understand the significance of Einstein's photon theory we must first review the characteristics of the photoelectric effect. One way to do this is to use light sources of various frequencies to illuminate the negatively charged plate of a gold-leaf electroscope. If the leaf falls the light source is ejecting electrons. Typical experiments that can be carried out quite simply in a laboratory are shown below.

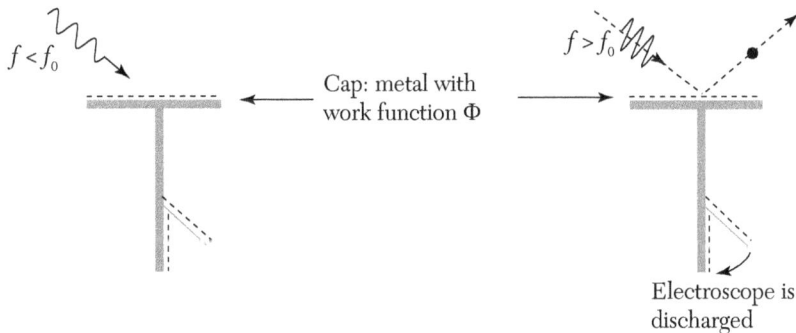

$f < f_0$

$f > f_0$

Cap: metal with work function Φ

Electroscope is discharged

- If the metal is zinc, then visible light does not discharge the electroscope even if it is very intense but ultraviolet radiation will discharge it even if its intensity is low.

- Increasing the intensity of the ultraviolet radiation discharges the electroscope more rapidly.

- When ultraviolet radiation is used, the leaf begins to fall as soon as it is illuminated: there is no delay.

▪ When ultraviolet radiation is used, the kinetic energy of the ejected electrons has a continuous spectrum up to a maximum value and this maximum kinetic energy increases if higher frequency ultraviolet radiation is used.

These observations cannot be explained using a classical wave model of light.

▪ According to the wave model, energy is spread across the wave front and delivered continuously to the metal surface. If this was the case, then energy would need to accumulate close to a surface electron before it was ejected. For typical light intensities used in these experiments this would take a long time, but no delay is observed. Electrons are emitted as soon as light of high enough frequency strikes the surface. The wave model cannot explain how this energy suddenly gets concentrated into a few different places to eject the electrons.

▪ In the wave model the parameter that determines how much energy is delivered is the intensity of the radiation and not the frequency, so the wave model cannot account for the threshold frequency or the increase of kinetic energy with frequency and not intensity.

Einstein's photon model can explain *all* the observations. Photons are distributed randomly in the arriving radiation but transfer all their energy to individual electrons. We have already seen that this explains the existence of a threshold frequency f_0. It also explains the immediate ejection of electrons: as soon as the first photon strikes the surface it will eject a photon (if $f > f_0$). Increasing the intensity of the radiation increases the number of photons per second arriving at the metal surface so the electroscope discharges more rapidly (again only if $f > f_0$). Finally, increasing the frequency above the threshold frequency gives photons more than enough energy to eject an electron from the surface so the excess energy gives the electron kinetic energy and the maximum kinetic energy will be:

$$KE_{max} = hf - \Phi = hf - hf_0$$

So the maximum kinetic energy is directly proportional to $(f - f_0)$ as observed.

27.1.5 Measuring the Planck Constant

The Planck constant can be determined using a photocell. This is an electrical component that uses the photoelectric effect to produce an output voltage when light is shone onto it. The symbol for a photocell is shown below. It consists of a metal emitter and a collector. When light strikes the emitter electrons are ejected and travel across to the collector. If the photocell is connected into a circuit it provides an emf that can transfer energy to other components.

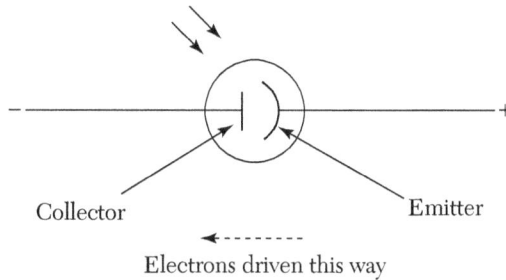

Collector   Emitter

Electrons driven this way

The circuit shown on the next page can be used to find the maximum kinetic energy of the emitted electrons by applying an opposing voltage to the cell and increasing this until the current in the circuit is reduced to zero. Measurements of this "stopping voltage" for incident light with a range of different frequencies can be used to find the Planck constant and the work function of the emitter.

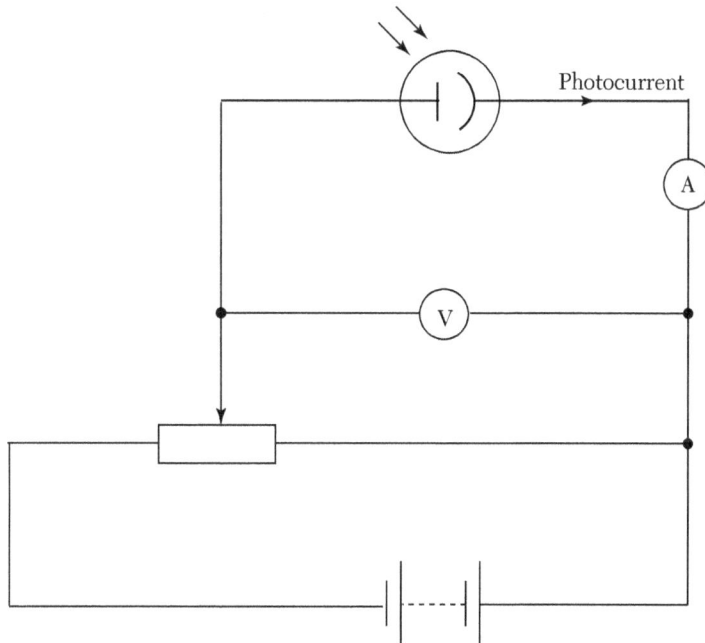

Photocurrent

A

V

By making the emitter positive with respect to the collector electrons must do work to move between them. When the voltage between the two electrodes is V the work that must be done is eV. Increasing V increases the work that must be done and the photocurrent falls to zero when the work is equal to the maximum kinetic energy. The voltage at which this occurs is called the stopping voltage V_S:

$$eV_S = KE_{max} = hf - \Phi$$

If the stopping voltage is measured for light sources with a range of frequencies (e.g., by using different colored filters in front of a white light source) then a graph of V_S against f can be used to find both the Planck constant and the work function.

$$V_S = \left(\frac{h}{e}\right)f - \left(\frac{\Phi}{e}\right)$$

Comparing this with:

$$y = mx + c$$

the gradient is h/e, the intercept on the voltage axis is $-\Phi/e$, and the intercept on the frequency axis is f_0:

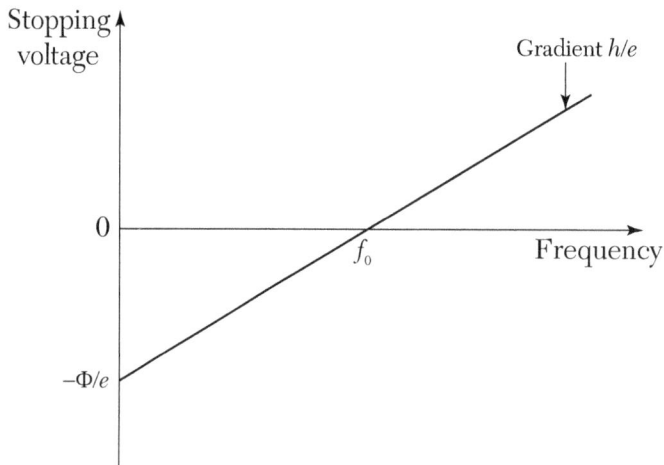

Robert Millikan used a similar method to make the first measurement of the Planck constant in 1916.

27.2 Matter Waves

Einstein's photon theory showed that a particle model can be used to explain some aspects of the behavior of light. However, superposition effects such as interference and diffraction can only be explained using a wave model. Neither model completely describes the nature of light and physicists sometimes say that light exhibits "wave–particle duality." In 1919 Louis de Broglie suggested that if radiation exhibits wave–particle duality then perhaps matter does too and he proposed an equation that could link the wave and particle models together. This is known as the de Broglie relation.

27.2.1 de Broglie Relation

de Broglie showed that the characteristic wave-like property of light, its wavelength, is linked to a characteristic particle-like property of the photon, its momentum. Here is an argument that shows how they are linked.

Photon energy:

$$E = hf$$

Mass equivalent to this energy:

$$m = \frac{E}{c^2}$$

Momentum associated with this mass:

$$p = mc = \frac{E}{c} = \frac{hf}{c} = \frac{h}{\lambda}$$

This is the de Broglie relation, the link between wavelength and momentum. While we have derived it for the photon, de Broglie's hypothesis was that it also applied to matter. This would imply that a moving particle has an associated wavelength given by:

$$\lambda = \frac{h}{p}$$

At velocities small compared to the speed of light this becomes:

$$\lambda = \frac{h}{mv}$$

The faster the particle moves the shorter its de Broglie wavelength.

27.2.2 Electron Diffraction

If de Broglie's hypothesis is correct then matter ought to exhibit wave-like properties such as interference and diffraction just like light. The question is, on what scale? If a person walking through a doorway behaves like a wave, shouldn't they diffract like light going through a narrow slit? A rough calculation shows that even if they do, the diffraction effects would be undetectable. Recall that the first minimum of a single slit diffraction pattern occurs at an angle whose sine is λ/b where λ is the wavelength and b is the width of the slit. For a 70 kg human walking at 1 ms^{-1} this is:

$$\sin\theta = \frac{6.6 \times 10^{-34}}{70 \times 1} \sim 10^{-35}$$

This is unbelievably tiny so the diffraction pattern would effectively have zero width and so would not affect the classical expectation that the person continues through the doorway with no diffraction at all.

However, if an electron is travelling at 5% of the speed of light its de Broglie wavelength is about 5×10^{-11} m. This is comparable to the spacing of atoms in a crystal lattice so we might expect to detect interference and diffraction effects when electron beams accelerated to comparable speeds are directed at crystals. This would only require an accelerating voltage of about 600 V.

The first physicists to detect electron diffraction were Davisson and Germer in 1925 when they fired electron beams at a crystalline sample of nickel. They noticed that the electrons were scattered into a pattern of maxima and minima similar to the one that would be obtained if X-rays were passed through the structure. Furthermore, the wavelength of X-rays needed to obtain the same pattern corresponded to the de Broglie wavelength of the electrons.

At almost the same time G.P. Thomson, the son of J.J. Thomson (who had shown that the electron can be treated as a particle), produced electron diffraction rings by passing a beam of electrons through a thin slice of graphite. The experimental arrangement is shown below.

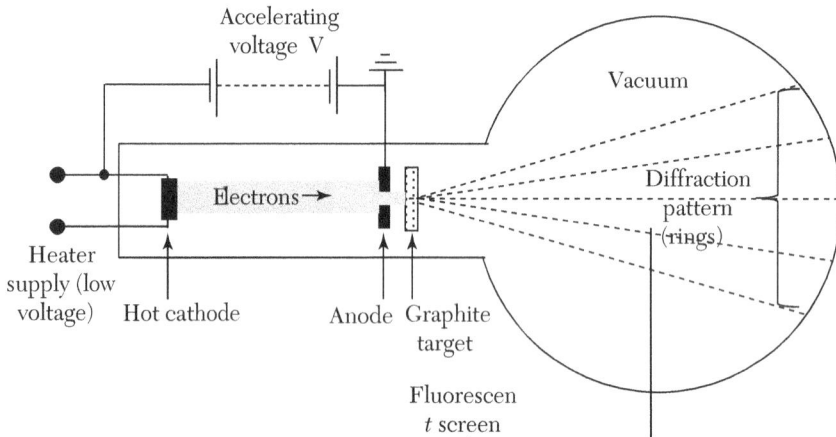

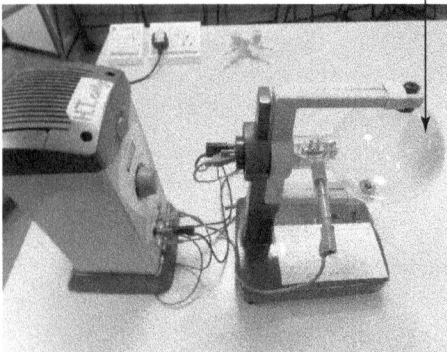

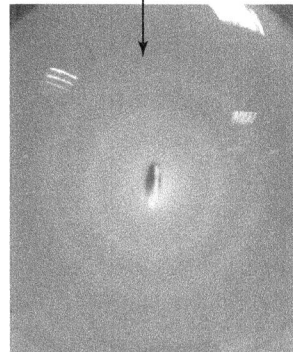

Graphite consists of planes of atoms arranged in a hexagonal pattern. Rows of atoms within these planes act like the lines in a diffraction grating. For planes separated by a distance d there will be diffraction maxima at angles given by:

$$\sin\theta = \frac{n\lambda}{d}$$

The fact that there are many such planes in all orientations relative to the incident beam results in diffraction maxima being formed around the surface of a cone with half angle θ for each set of planes and each order of diffraction. Where these cones of diffracted electrons hit the end of the vacuum tube they form a pattern of concentric rings.

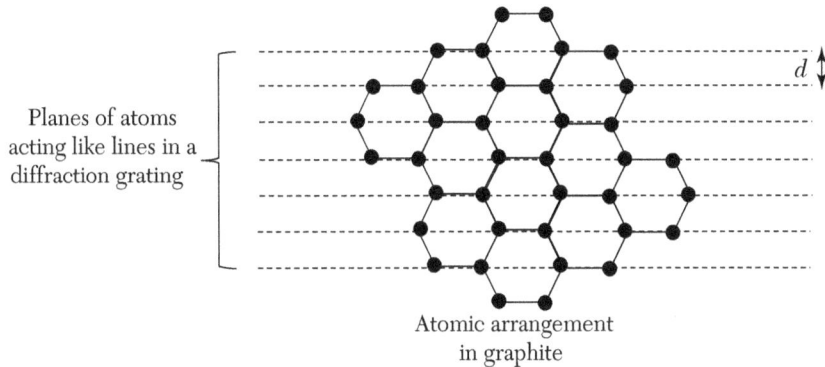

Planes of atoms acting like lines in a diffraction grating

d

Atomic arrangement
in graphite

The radius of diffraction rings can be used to find the spacing of atomic planes in the crystalline structure. Changing the accelerating voltage changes the wavelength of the electrons and the radius of each ring. Higher voltage gives the electrons greater momentum and smaller de Broglie wavelength so the rings get smaller:

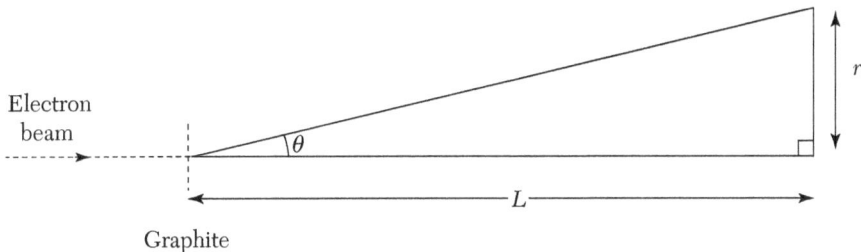

Electron beam

θ

r

L

Graphite

$$\sin\theta = \frac{\lambda}{d} = \frac{h}{dmv} = \frac{h}{d\sqrt{2meV}}$$

For small angles, $\sin\theta \sim \theta \sim \dfrac{r}{L}$, so that:

$$r \sim \frac{hL}{d\sqrt{2meV}}$$

where d is the spacing of a particular set of atomic planes and V is the accelerating voltage for the electrons. The radius of a diffraction ring is approximately proportional to the reciprocal of the square root of the accelerating voltage.

27.2.3 Compton Effect

In 1923, Arthur Compton showed that when X-rays are scattered by electrons the process behaves just like a collision between two particles. This was an important result because the wavelength shift in the scattered X-rays could not be explained using a wave model for the radiation. In Compton's experiment the X-ray photons had much greater energy than the ionization energy of the atom so the electrons behaved like free particles.

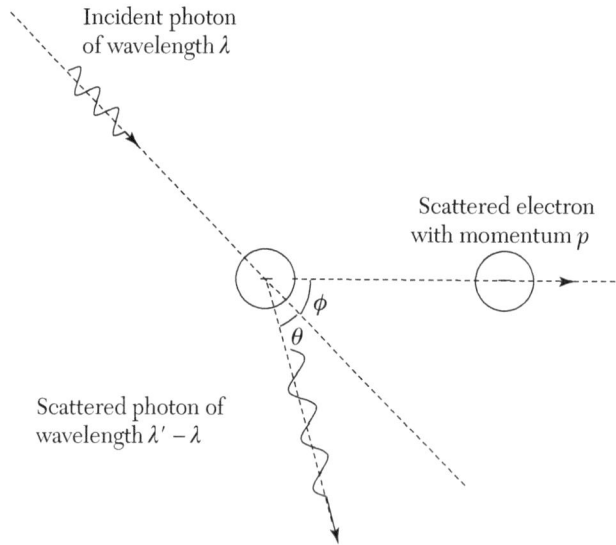

Incident photon
of wavelength λ

Scattered electron
with momentum p

ϕ

θ

Scattered photon of
wavelength $\lambda' - \lambda$

Compton treated this as a collision between particles and used the equations for conservation of energy and momentum to find the relationship between the X-ray scattering angle and the change in wavelength of the X-rays:

$$\left(\lambda' - \lambda\right) = \frac{h}{m_e c}(1 - \cos\theta)$$

This was verified experimentally and showed once again that the particle model must be used for some interactions between radiation and matter.

27.3 Wave–Particle Duality

The de Broglie equation applies to both electromagnetic radiation and matter—sometimes a wave model is needed and sometimes a particle model is needed. Neither model gives a complete description of the underlying phenomenon, but there is a deeper problem because the two models seem to contradict one another.

We have already seen that the wave model of light, with energy spread continuously across the wave front, cannot be used to explain photoelectric emission. However, if instead of directing the light onto a metal plate we had passed it through a double slit apparatus we would have to assume that the energy is spread continuously to explain the resulting interference pattern.

	EM radiation (e.g., light)	**Matter (e.g., electrons)**
Evidence for the wave model	Young's double slit experiment	Electron diffraction
	Diffraction patterns	Electron standing waves in atoms
Evidence for the particle model	Photoelectric effect	Discrete nature of electric charge
	Compton effect	Momentum of individual electrons

How can these two apparently irreconcilable models be related?

27.3.1 Young's Double Slit Experiment Revisited

The double slit experiment, first used to support the wave model of light, is an ideal example to use when trying to reconcile the wave and particle models.

In the double slit experiment a monochromatic light source is directed at a double slit and an interference pattern consisting of regularly spaced maxima and minima appears on a screen. The maxima occur in positions where waves arriving from each slit are in phase and interfere constructively. The minima occur where the waves arrive in antiphase (π phase difference) and interfere destructively. The resultant intensity at any point on the screen is calculated by adding the phasors from each slit and then squaring the resultant amplitude. This approach, the wave model, works (up to a point). It explains the intensity variation across the screen.

If we now assume that the light is emitted and absorbed as individual photons we run into difficulty. Consider, for example, a minimum position. When both slits are open the minimum is dark, so no photons arrive at this position. If we cover either one of the slits, then light does arrive at this position. So we have a problem. It seems that opening the second slit and allowing light to reach that position from either slit results in less light arriving …. how can identical photons cancel one another out? How can energy disappear?

The diagrams on the next page illustrate this. In the top diagram, only the top slit is open and N_A photons reach point P. In the middle diagram,

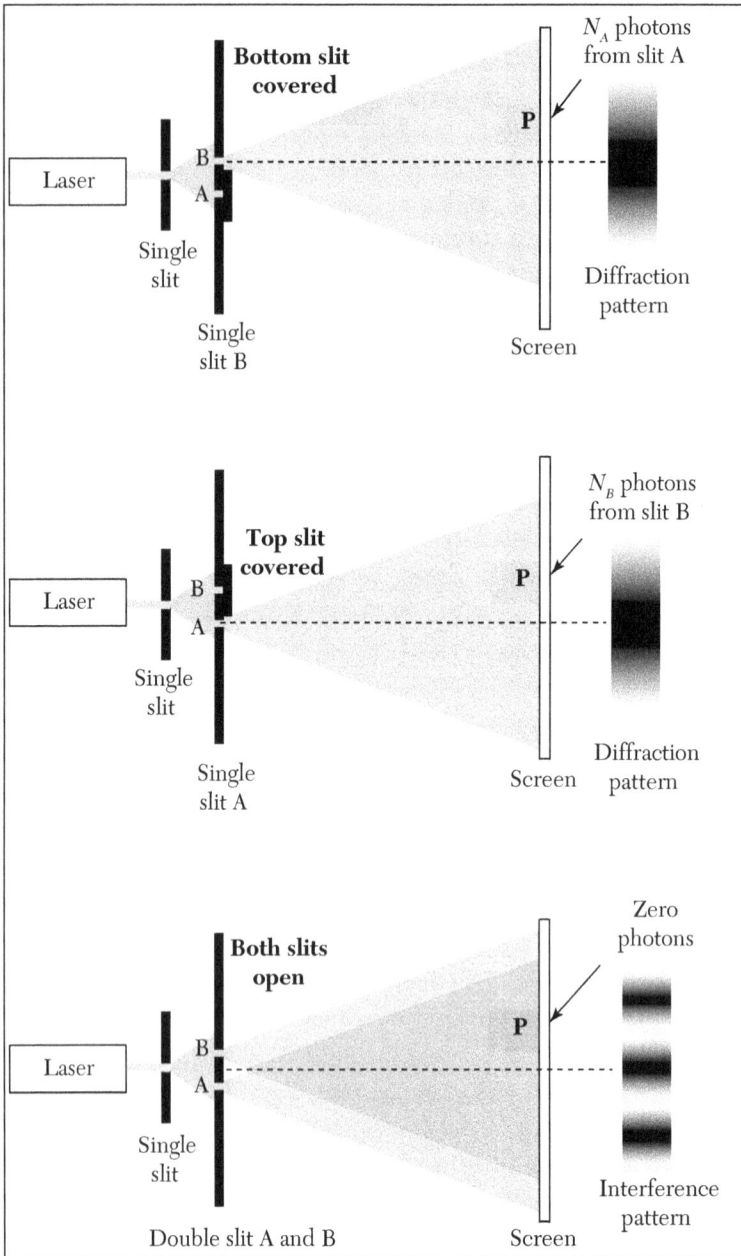

only the bottom slit is open and N_B photons reach point P. With both slits open one might expect $(N_A + N_B)$ photons to reach point P but in fact zero photons are detected at P.

It seems that opening the second slit affects where photons passing through the first slit can go. On the face of it, this is bizarre!

One, unlikely, possibility is that when both slits are open photons from *A* and *B* somehow interact on their way to the screen and this prevents them reaching *P*. To rule this out the experiment has been repeated using a filter in front of the source to reduce the intensity so far that only one photon at a time is interacting with the apparatus. Now photons arrive one at a time at the screen and cannot interact with one another en route. What happens?

At first the photons seem to be arriving completely randomly but after a short while it becomes apparent that the same patterns as before are produced with single or double slits. If we insist that the photons behave like particles then each photon can only pass through one of the slits. If it passes through slit *A* when *B* is closed it can reach *P*. If it passes through slit *B* when *A* is closed it can also reach *P*. However, if it passes through slit *A* when slit *B* is open it cannot reach *P*! Similarly, if it passes through slit *B* when *A* is open it cannot reach *P*! To maintain the particle model we would need to assume that when the photon passes through one of the slits its future path is affected by the state of the slit through which it did not pass.

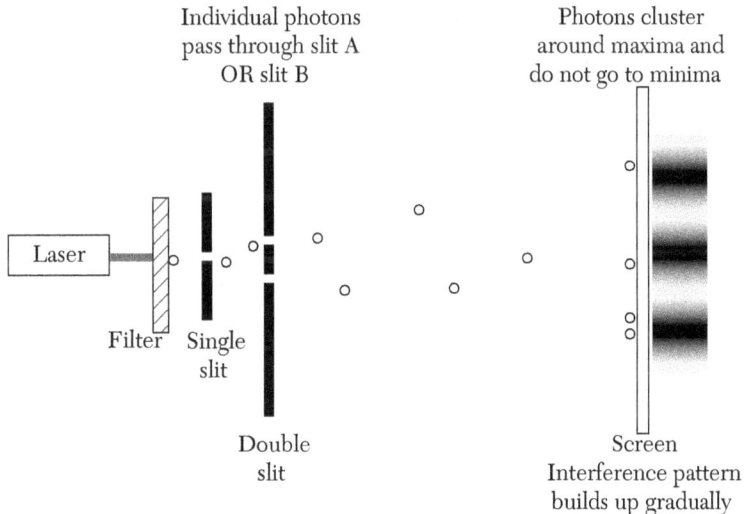

A single photon passing through a double slit apparatus cannot reach any of the minimum positions, even though it could reach all of them if it passed through either slit when the other one is closed! This shows that photons do not interfere with each other but every photon interferes with itself. Where does this leave us?

- The wave model can be used to explain the intensity distribution in superposition effects.

- The particle model can be used to explain the discrete emission and absorption of radiation.

A similar experiment can be carried out with electron beams. The results are exactly the same. Wave–particle duality affects matter and radiation in the same way.

27.3.2 Interpreting Wave–Particle Duality

Einstein suggested that the solution to the problem of wave–particle duality might be to treat the waves as waves of *probability* so that the higher the intensity of the radiation the higher the probability of finding a photon (or electron) there. Max Born developed this idea into the "**statistical interpretation**" of quantum mechanics and this became a cornerstone of the most popular interpretation of the theory—the **Copenhagen Interpretation** (see Section 27.5.1).

When light or electrons pass through a double slit apparatus each photon or electron is associated with a wave (the "wave function") whose intensity is directly proportional to the probability of finding a photon or an electron at each point in space. When an observation is made, for example a photon is detected on the screen, the wave function "collapses" to zero everywhere except at the position where the photon or electron is detected. This interpretation is radical in several respects.

- Prior to the observation the wave function changes in a continuous way but when the observation is made there is a discontinuous collapse so that the probability changes suddenly even at great distances from the observation. There is no known physical process that can explain the **collapse of the wave function**. This is sometimes called "**the measurement problem**."

- The idea that distant parts of the wave function can change when an observation is made shows that quantum theory is **non-local**.

- The wave function is the most complete description of the system but it deals only with probability. This means that even if we could know the initial conditions of a system with absolute precision (e.g., how a particular electron is approaching a double slit apparatus) we can only make statistical predictions about the future state of the system (e.g., where the electron will hit the screen). Quantum theory is **indeterministic**—the future of the universe is *not* uniquely determined by its present state. This is a complete departure from the determinism of classical physics.

27.3.3 Schrodinger Equation

Waves are solutions to differential equations. The one-dimensional wave equation is:

$$\frac{\partial^2 y}{\partial x^2} = \frac{1}{v^2}\frac{\partial^2 y}{\partial t^2}$$

where y is the wave disturbance and v is the speed of the wave in x-direction. A solution to this is:

$$y = A\cos(\omega t - kx)$$

where $\omega = 2\pi f$ and $k = 2\pi/\lambda$.

A similar pair of equations for the electric field (E) and magnetic field (B) in a vacuum can be derived from Maxwell's equations. These form the electromagnetic wave equations (in one dimension):

$$\frac{\partial^2 E}{\partial x^2} = \varepsilon_0 \mu_0 \frac{\partial^2 E}{\partial t^2}$$

$$\frac{\partial^2 B}{\partial x^2} = \varepsilon_0 \mu_0 \frac{\partial^2 E}{\partial t^2}$$

ε_0 is the permittivity of free space and μ_0 is the permeability of free space.

Comparing the electromagnetic equations with the original one-dimensional wave equation we can see that the speed of electromagnetic waves is:

$$c = \frac{1}{\sqrt{\varepsilon_0 \mu_0}}$$

The electromagnetic wave equations provide a way to derive the wave function for light, but is there a corresponding wave equation for the de Broglie waves of an electron? Erwin Schrodinger thought that there should be and in 1925 he found it. This is the most important equation in quantum theory: **the Schrodinger equation**. However, unlike the equations above, the Schrodinger equation deals with complex quantities (combinations of real and imaginary numbers). The details of how the Schrodinger equation was derived are beyond the scope of this book but here is the time-dependent equation:

$$\frac{ih}{2\pi}\frac{\partial \Psi}{\partial t} = \frac{-h^2}{8\pi^2 m}\frac{\partial^2 \Psi}{\partial x^2} + V\psi$$

where ψ is the wave function, i is the square root of minus one, and m is the mass of the electron. V is the potential energy that might vary with position and time. If the electron is moving freely in space then $V = 0$.

It is possible to find wave-like solutions to this equation. These represent the de Broglie waves of the electron. Once Schrodinger had published his equation physicists could use it to solve a vast range of problems, and Schrodinger himself showed how it could explain the energy level structure and spectrum of the hydrogen atom. In Schrodinger's atomic model the electron orbitals are three-dimensional standing wave solutions to the equation.

According to the Copenhagen Interpretation, the square of the magnitude of the wave function at each point in space is equal to the probability per unit volume of finding the electron at that point. Since the wave function is itself a complex quantity it is not directly observable, and its magnitude is found by multiplying it by its own complex conjugate:

Probability of finding electron in small region $\delta x \delta y \delta z = \Psi(x, y, z)\,\Psi^*(x, y, z)$

if this is integrated over all of space it must equal 1: the electron will be found somewhere in the universe!

27.4   Quantum Atom

The Rutherford nuclear model of the atom consists of electrons in orbit around a positively charged nucleus. Such a model would allow the electrons to orbit at any distance from the nucleus and to have any energy. However, there is a serious problem with this model. According to Maxwell's equations, charged particles radiate energy when they are accelerated and according to Newtonian mechanics a particle moving in orbital motion has a centripetal acceleration. Orbiting electrons ought to radiate energy and fall rapidly into the nucleus. Classical physics leads us to the conclusion that Rutherford's nuclear atom is unstable.

It was also known that excited atoms emit a line spectrum, i.e., the radiation emitted by isolated atoms contains a set of discrete frequencies. If the photon model is considered this suggests that the emission of radiation involves discrete energy jumps $\Delta E = hf$ for each spectral line. This further suggests that the electrons inside an atom cannot have any value of energy but can only exist at certain discrete energy levels.

27.4.1 Bohr's Model of the Hydrogen Atom

The fact that atoms exist and are stable shows that something is missing from Rutherford's nuclear atom. Niels Bohr realized that quantization might solve this problem and at the same time give a way to calculate the frequencies present in atomic line spectra.

Bohr assumed that:

- The electron can move in one of several discrete circular orbits and that each orbit has a specific energy (energy levels)

- The angular momentum of the electron is quantized in integer units of $h/2\pi$.

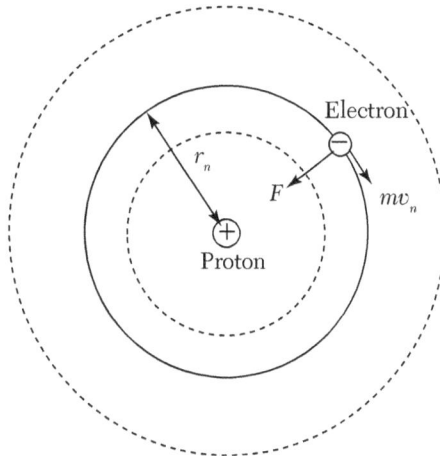

The diagram above shows an electron in the nth orbit inside the hydrogen atom. The dotted lines indicate other quantized orbits. The energy of the nth orbit is derived below.

Angular momentum:

$$mv_n r_n = \frac{nh}{2\pi}$$

Centripetal force:

$$\frac{mv_n^2}{r_n} = \frac{e^2}{4\pi\varepsilon_0 r_n^2}$$

Kinetic energy:

$$KE = \frac{1}{2}mv_n^2 = \frac{e^2}{8\pi\varepsilon_0 r_n^2}$$

Potential energy:

$$\text{PE} = -\frac{e^2}{4\pi\varepsilon_0 r_n}$$

Energy:

$$E_n = \text{KE} + \text{PE} = \frac{1}{2}mv_n{}^2 - \frac{e^2}{4\pi\varepsilon_0 r_n} = -\frac{e^2}{8\pi\varepsilon_0 r_n}$$

Bohr used these equations to eliminate r_n from the energy equation and to express the energy of an electron in the nth level in terms of the quantum number n.

$$E_n = \frac{-me^4}{8\varepsilon_0^2 h^2}\left(\frac{1}{n^2}\right) = \frac{-13.6\,\text{eV}}{n^2}$$

This is a very important result as it provides an accurate method to calculate the ionization energy of hydrogen and the frequencies of the lines in the hydrogen line emission spectrum. The number n is called the **principal quantum number** and $n = 1$ represents the lowest allowed energy state for an electron in the hydrogen atom: the **ground state**. All the energies are negative because the electron is bound to the nucleus.

The diagram below is an energy level diagram for the hydrogen atom based on the Bohr model.

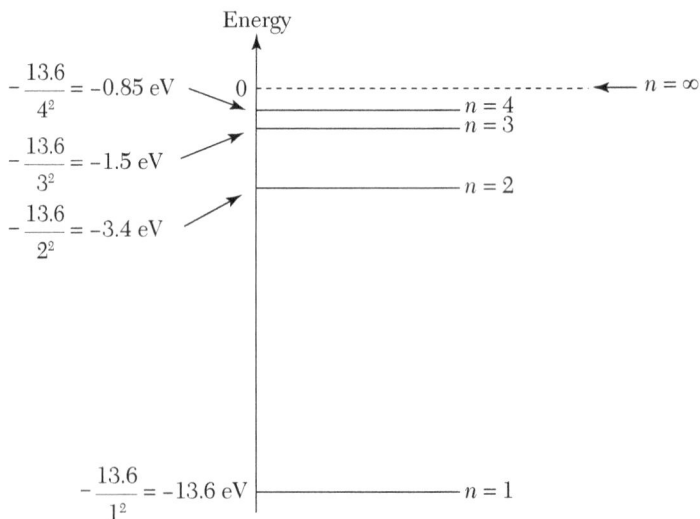

As n increases the energy levels get closer and closer together becoming a continuum of states as n approaches infinity. The energy needed to remove an electron from the ground state of the hydrogen atom (its first ionization energy) is 13.6 eV, the energy needed to move from $n = 1$ to $n = \infty$ where the energy would be zero (a free electron).

27.4.2 Explaining the Hydrogen Line Spectrum

It had been well known since the end of the nineteenth century that the wavelengths of the lines in the hydrogen spectrum fall into several series which are linked by simple mathematical formulae. The spectral series with most lines in the visible part of the spectrum is called the Balmer series after the Swiss school master who first worked out the formula that linked individual lines in the series. Balmer's formula (in a form later adopted by Johannes Rydberg) is:

$$\frac{1}{\lambda_n} = R\left(\frac{1}{2^2} - \frac{1}{n^2}\right)$$

R is the Rydberg constant. $R = 1.097373157 \times 10^7$ m^{-1}. n is an integer greater than 2.

Rydberg generalized this formula for all the series in the hydrogen spectrum:

$$\frac{1}{\lambda_n} = R\left(\frac{1}{m^2} - \frac{1}{n^2}\right)$$

where m and n are integers with $n > m$.

Bohr was able to derive this formula from his atomic model and to find an expression for the Rydberg constant. He assumed that the spectral lines correspond to quantum jumps made by electrons moving from excited state to less excited states so that the Rydberg formula corresponds to a quantum jump from the nth energy level to the mth energy level when $(n > m)$. When the electron loses an energy ΔE_{mn} it emits a photon of frequency f_{mn} given by $\Delta E_{mn} = f_{mn}$. The diagram below shows how a photon is emitted when an electron jumps from $n = 4$ to $m = 2$.

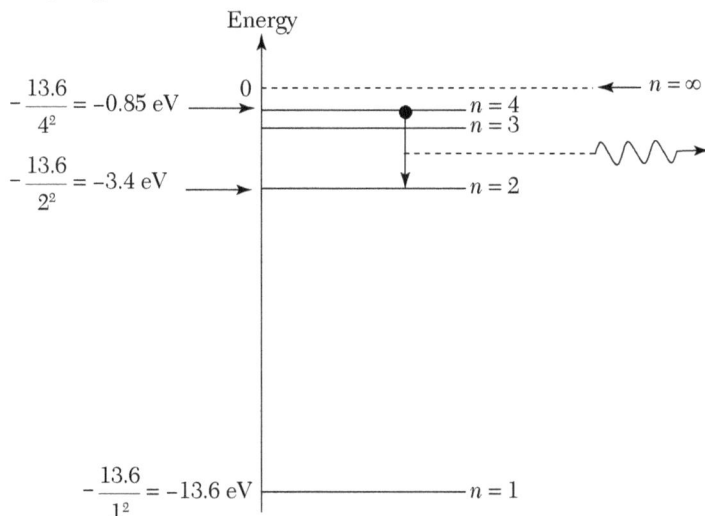

For the transition above (from $n = 4$ to $n = 2$):

$$\Delta E_{24} = E_4 - E_2 = \frac{-me^4}{8\varepsilon_0^2 h^2}\left(\frac{1}{4^2}\right) - \frac{-me^4}{8\varepsilon_0^2 h^2}\left(\frac{1}{2^2}\right) = \frac{me^4}{8\varepsilon_0^2 h^2}\left(\left(\frac{1}{2^2}\right) - \left(\frac{1}{4^2}\right)\right)$$

$$= \frac{3me^4}{128\varepsilon_0^2 h^2}$$

$$hf_{24} = \frac{3me^4}{128\varepsilon_0^2 h^2}$$

$$f_{24} = \frac{3me^4}{128\varepsilon_0^2 h^3}$$

$$\lambda_{24} = \frac{c}{f_{24}} = \frac{128c\varepsilon_0^2 h^3}{3me^4} = 4.9 \times 10^{-7}\,\text{m}$$

The Rydberg formula is derived in a similar way but using a transition from n to m.

$$\Delta E_{mn} = E_n - E_m = \frac{-me^4}{8\varepsilon_0^2 h^2}\left(\frac{1}{m^2}\right) - \frac{-me^4}{8\varepsilon_0^2 h^2}\left(\frac{1}{n^2}\right)$$

$$= \frac{me^4}{8\varepsilon_0^2 h^2}\left(\left(\frac{1}{n^2}\right) - \left(\frac{1}{m^2}\right)\right)$$

$$\Delta E_{mn} = hf_{mn} = \frac{hc}{\lambda_{mn}}$$

$$\frac{1}{\lambda_{mn}} = \left(\frac{me^4}{8c\varepsilon_0^2 h^3}\right)\left(\left(\frac{1}{n^2}\right) - \left(\frac{1}{m^2}\right)\right)$$

The Rydberg constant is:

$$R = \left(\frac{me^4}{8c\varepsilon_0^2 h^3}\right)$$

27.4.3 Electron Waves in Atoms

The success of the Bohr model was impressive but it was still an arbitrary quantization rule. However, an alternative way to think about the quantization rule gives us a clue to the real reason that atoms have energy levels. Bohr's quantization of angular momentum is:

$$mv_n r_n = \frac{nh}{2\pi}$$

which can be rearranged to give:

$$2\pi r_n = n\left(\frac{h}{mv_n}\right)$$

The de Broglie relation can be used to introduce a wavelength:

$$\lambda_n = \left(\frac{h}{mv_n}\right)$$

$$2\pi r_n = n\lambda_n$$

The left-hand side of this equation is equal to the circumference of the circular orbit. The right-hand side of the equation is an integer number of de Broglie waves. Bohr's quantum condition is equivalent to saying that energy levels can only occur when the circumference is an integer number of electron wavelengths. This guarantees only a discrete set of energy levels in the atom in a similar way that a stretched string can vibrate in only a discrete set of frequencies (the fundamental and harmonics).

The energy levels of an atom correspond to standing wave patterns for the de Broglie electron waves. The lowest or ground state is when the circumference of the orbit corresponds to a single electron wavelength.

27.4.4 Schrodinger Atom

While Bohr's model was surprisingly successful in explaining the energy levels and spectrum of the hydrogen atom it was quite obviously not the final word on the atom. In Bohr's model the electron is restricted to a planar circular orbit whereas an electron in a real atom could be anywhere in the three-dimensional space surrounding the nucleus and the orbits could be different shapes. Bohr's model was adapted to include elliptical orbits but the solution for a three-dimensional atom was found by Schrodinger using the Schrodinger equation.

The mathematical derivation of formulae for electron orbitals is beyond the scope of this book but it is interesting to outline the method and to describe some of the important results. First of all Schrodinger needed to put a potential energy term V into his equation. This is simply the electrostatic potential energy of the electron in the field of the nucleus:

$$PE = \frac{e^2}{4\pi\varepsilon_0 r}$$

Then he needed to solve the equation in spherical polar coordinates (r, θ, ϕ). These coordinates are used because of the symmetry of the problem—solving a spherically symmetric problem in Cartesian coordinates is much harder and the solutions are unlikely to be easy to interpret!

The equation separates into three separate equations in r, θ and ϕ. This results in three separate quantum numbers, one associated with each of the coordinates.

n = principle quantum number—(linked to r) this determines the energy of the electron as before

l = orbital quantum number—(linked to θ) this determines the angular momentum and shape of the orbit

m = magnetic quantum number—(linked to ϕ) this determines the orientation of the orbit.

There is another property of electrons that is not included in the Schrodinger equation, its spin. Electrons have spin angular momentum of magnitude $h/4\pi$. This is usually referred to as spin-½ because the unit of angular momentum used in atomic physics is $h/2\pi$ (or 'h-bar). Spin is a quantum mechanical property and when it is measured along any particular axis a spin-½ particle is found to have either +½ or −½ a unit of spin. This introduces another quantum number:

s = spin quantum number—if this is +1 the spin is "up" and if it is −1 the spin is "down" relative to the axis of measurement.

Pulling all of this together we can see that there will be four integer quantum numbers associated with each allowed state. The equation also constrains the values of these quantum numbers so that:

For any n: l can have any value from 0 to $(n - 1)$

For any l: m can take integer values from $-l$ to $+l$.

s can only be +1 or −1.

The three-dimensional shapes of the $n = 1$ (s) and $n = 2$ (2s and 2p) orbitals are shown approximately (and not to scale) below. The s-orbitals (with $l = 0$) are spherical. The p-orbitals ($l = 1$) form three dumbbell shaped orbitals with different orientations ($m = -1, 0, +1$). Two electrons can occupy each orbital. The shapes represent probability distributions for the electrons.

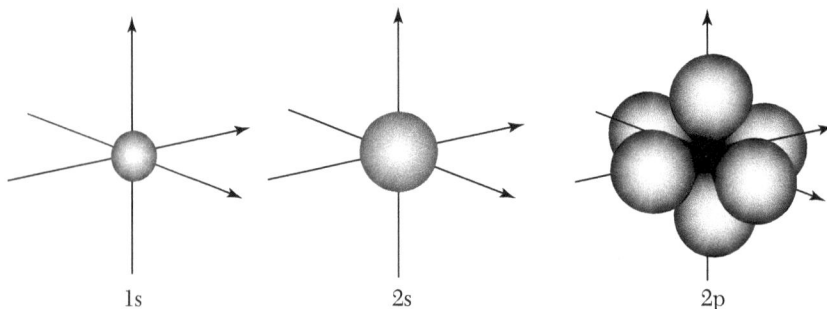

1s   2s   2p

Wolfgang Pauli realized that no two electrons in the same atom can have the same set of quantum numbers. This is an example of the **Pauli exclusion principle** and the consequence is incredibly important. If it were not the case then all the electrons in a multi-electron atom would fall into the lowest energy level and different elements would have very similar chemical properties. The Exclusion principle prevents this and ensures that the electrons must fill up each energy level in turn. In a very general sense this is what gives us chemistry. It certainly accounts for the Periodic Table. For example:

When $n = 1$: $l = 0$, $m = 0$ and $s = \pm 1$. Two electrons can occupy these states. These are the 1s states in the atom and correspond to a spherically symmetric wave function. Hydrogen, with one electron has a half-filled 1s shell. This can be represented by $1s^1$.

The first number is the energy level or principal quantum number, the letter refers to the type of orbital (related to the value of l) and the superscript is the number of electrons in the shell.

This accounts for its reactivity—it can complete the shell by reacting with other elements with one or more outer electrons.

Helium, has two electrons, so the 1s shell is full and the atom is very stable. Its electronic configuration is $1s^2$.

Lithium has three electrons so $n = 1$, $l = 0$, $m = 0$ and $s = \pm 1$; states are filled by two electrons so the 1s shell is full and the third electron has to go into one of the eight available states with $n = 2$ (the 2s shell). The lowest energy state is the $n = 2$, $l = 0$, $m = 0$, $s = 1$ (or −1) state. This leaves the 2s shell partially filled and lithium is again reactive. Its electronic configuration is:

$$1s^2 2s^1$$

This filling of energy shells continues as we move to larger and larger atoms. The periodicity of the periodic Table comes about as successive shells fill up and electrons start to fill the next shell.

The fifth element in the Periodic table is Boron. This must accommodate five electrons so both the 1s and 2s shells are full and the next electron must have an orbital quantum number $l = +1$. This is called a p-shell and has a different shape to the spherical s-shells. There are three p-shells ($m = -1$, 0 and +1) in different orientations and each is shaped like a dumbbell. Two electrons can go into each shell (with $s = +1$ or $s = -1$). The electronic configuration for Boron is:

$$1s^2 2s^2 2p^1$$

The table on the next page shows the order of the first few electronic energy levels and the corresponding electronic configurations. Just a few examples will show how the electronic configuration, determined by solutions to the Schrodinger equation, affects chemical properties.

▪ Atoms with filled shells tend to be very stable—e.g., helium, neon.

▪ Atoms with a single electron in an otherwise empty shell tend to be very reactive because they easily lose that electron to an atom that needs one or more electrons to complete its own outer shell—e.g., sodium.

▪ Atoms with an almost full shell are also very reactive, easily gaining electrons from other atoms that have only one or a few electrons in their outer shell—e.g., fluorine.

▪ Carbon has 2 electrons in the $2p$ shell, so to complete this shell carbon must gain four electrons. This makes it "4-valent." Carbohydrates (compounds of carbon and hydrogen) are the most important class of compounds for living things and illustrate how valency is linked to electronic configuration. In methane, CH_4, each hydrogen atom shares one electron so that the carbon $2p$ shell is completed and the hydrogen 1s shell is completed, so both achieve more stable (lower energy configurations) and four covalent bonds are formed.

	n	l	m	s	Configuration	Element (ground state)
1s shell	1	0	0	+1	$1s^1$	Hydrogen
				−1	$1s^2$	Helium
2s shell	2	0	0	+1	$1s^2 2s^1$	Lithium
				−1	$1s^2 2s^2$	Beryllium
2p shell		1	−1	+1	$1s^2 2s^2 2p^1$	Boron
				−1	$1s^2 2s^2 2p^2$	Carbon
			0	+1	$1s^2 2s^2 2p^3$	Nitrogen
				−1	$1s^2 2s^2 2p^4$	Oxygen
			+1	+1	$1s^2 2s^2 2p^5$	Fluorine
				−1	$1s^2 2s^2 2p^6$	Neon
3s shell	3	0	0	+1	$1s^2 2s^2 2p^6 3s^1$	Sodium
				−1	$1s^2 2s^2 2p^6 3s^2$	Magnesium
					...	

This gives a very simple explanation of key chemical properties. More detailed analysis of chemical bonding requires a solution of the Schrodinger equation for electrons in the field of both atoms. The Schrodinger equation is the fundamental equation in chemistry.

27.5   Interpretations of Quantum Theory

While relativity challenges our preconceived ideas about space and time, quantum theory challenges our ideas about the nature of reality. Wave–particle duality shows that it is impossible to capture all the features of a quantum object in a visualizable classical model and any such attempt leads to contradictions. However, the Schrodinger equation and other developments in quantum theory provide a powerful set of mathematical equations that allow us to make predictions about the physical world to an incredible degree of precision. While some physicists think we should just accept and use the equation and give up on trying to find any underlying meaning to them others have felt that it is important to interpret the wave function and seek a deeper understanding of what is actually going on in a quantum process. This has led to several different "interpretations" of quantum theory but there is still a lack of consensus. Here we will give a brief description of three alternative approaches to the theory:

- The Copenhagen Interpretation which developed around Niels Bohr and Werner Heisenberg in Copenhagen (where Bohr worked).

- The sum-over-histories approach developed by Richard Feynman.

- The many-worlds theory developed by Hugh Everett III.

27.5.1   Copenhagen Interpretation

The Copenhagen interpretation has been the "establishment" interpretation of quantum theory for nearly a century despite having obvious shortcomings. The central idea is the wave function (ψ). This is a solution of the Schrodinger equation that represents all it is possible to know about a physical system. For example, the wave function for light leaving a source might be a series of waves spreading out in all directions. But these waves are not directly observable, they are related to the probability of observing a photon at each point in space. The Schrodinger equation allows us to calculate how the waves change as they spread out and interact with the apparatus. For example, when the waves pass through a double slit arrangement they diffract at each slit and then superpose, producing an interference pattern consisting of regularly spaced maxima and minima. However, this pattern is only related indirectly to the light that is detected on the screen (or by any other kind of detector). The "intensity"($|\psi|^2$) of the interference pattern represents a probability distribution for the arrival of photons on the screen. High intensity corresponds to high probability and low intensity corresponds to low probability. The pattern applies to each individual photon, so if only one photon interacts with the apparatus the same probability distribution is used to work out where it is likely to be detected.

Now we reach the most controversial aspect of the Copenhagen interpretation, the act of observation or measurement. Up to this point the wave function has evolved continuously according to the Schrodinger equation. At the moment of observation or measurement the probability distribution, which spreads across the entire screen in the double slit experiment, suddenly and discontinuously changes. It becomes instantly zero everywhere except at the point where the photon is observed. This is called the "collapse of the wave function." This is not explained by the Schrodinger equation and there is no agreement on any physical mechanism by which it occurs. This is called the "measurement problem"and is the main reason many physicists think that the Copenhagen Interpretation is unacceptable (even if it does give us a useful way to describe quantum processes).

The diagram below gives a simplified explanation of the double slit experiment using the Copenhagen Interpretation.

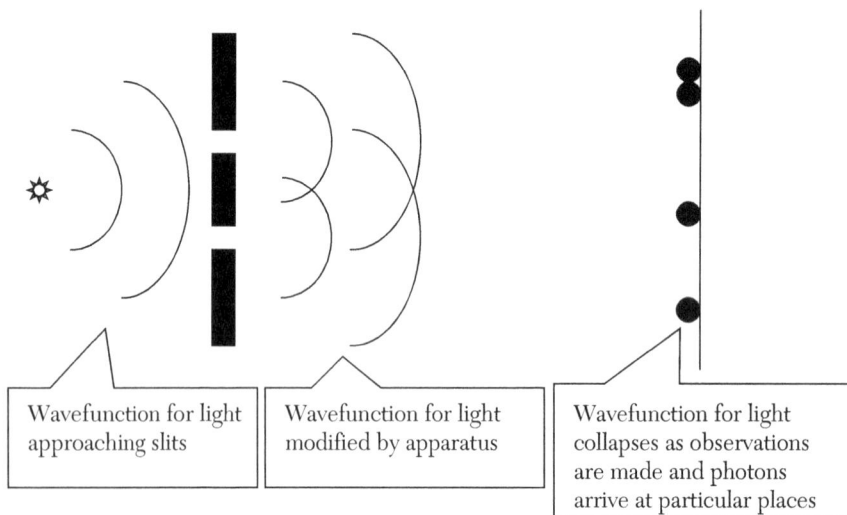

| Wavefunction for light approaching slits | Wavefunction for light modified by apparatus | Wavefunction for light collapses as observations are made and photons arrive at particular places |

Prior to observation the photon could be anywhere within the probability distribution. It is said to be in a "superposition of states." After the observation, the photon has arrived at a particular place on the screen.

Here is a summary of the main features of the Copenhagen Interpretation.

▨ The wave function ψ (a solution of the Schrodinger equation) gives the most complete description of a quantum state (this is unobservable).

▨ Schrodinger's equation governs the behavior of the wave function and its interaction with the apparatus.

▨ The evolution of the wave function is continuous and deterministic (i.e., if we know its initial state we can calculate, with certainty, its future state.

▨ $|\psi|^2$ gives probability of finding photon at each point in space (Born's statistical interpretation).

▨ Before an observation or measurement photons are in a superposition of states.

▨ Observation or measurement results in the discontinuous collapse of the wave function.

▨ We can only make statistical predictions about the outcome of wave function collapse so quantum theory is indeterministic.

When we use high intensity light (large numbers of photons) the quantum distribution is identical to that expected from classical physics. This is not really surprising, if there is a large amount of energy the discrete nature of individual quanta is hard to observe. The idea that quantum predictions merge smoothly into classical predictions at high energies is called the correspondence principle.

27.5.2 Heisenberg's Uncertainty (Indeterminacy) Principle

In classical physics it is, in principle, possible to measure the position and momentum of a particle as precisely as our instruments allow. It is assumed that a particle possesses a definite position and momentum at all times even if we do not choose to measure these quantities. The forces between the particles are also determined by other properties such as mass and charge so that it should be possible, in principle, to predict future positions and momenta by carrying out suitable calculations using the initial values. More importantly, classical physics is **deterministic**—the future is entirely determined by the present state whether or not we have this information. This is *not* the case with quantum theory.

Consider trying to determine the exact position of an electron at some moment. One way of doing this would be to direct single electrons at a narrow hole. If an electron gets through the hole and reaches the other side then it must have been localized in the region of the hole as it passed through, so we know its location at that moment with an uncertainty about equal to the diameter of the hole. However, we must take into account the wave nature of the electron. As it passes through the hole its wave function diffracts and spreads out in all directions. This is equivalent to giving the electron a random sideways momentum as it passes through the hole. As the hole is made smaller the diffraction effects increase and the random changes of momentum are likely to be larger. Making a precise measurement of the position of the electron in the plane of the hole introduces an uncertainty in the momentum of the electron in that plane.

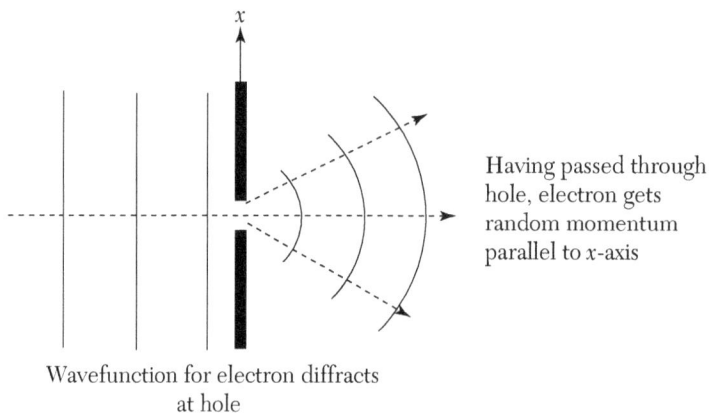

Having passed through hole, electron gets random momentum parallel to x-axis

Wavefunction for electron diffracts at hole

If the hole diameter is Δx the angular spread of the first diffraction maximum (which is where most electrons will be found after passing through the hole) is given by:

$$\sin\theta = \pm\frac{1.22\lambda}{\Delta x}$$

For small angles this is approximately:

$$\theta \sim \pm\frac{\lambda}{\Delta x}$$

The de Broglie wavelength links wavelength to momentum, so:

$$\lambda = \frac{h}{p}$$

$$\theta \sim \pm\frac{h}{p\Delta x}$$

where p is the original momentum of the electrons (perpendicular to x). After the electron passes through the hole its momentum also has an x-component Δp perpendicular to its original momentum:

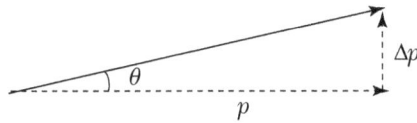

For small angles:

$$\theta \sim \frac{\Delta p}{p}$$

Combining the last two equations:

$$\frac{\Delta p}{p} \sim \pm \frac{h}{p\Delta x}$$

$$\Delta p \sim \frac{h}{\Delta x}$$

$$\Delta p \Delta x \sim h$$

While this is a rough derivation, it does suggest that the uncertainty in x-position and the uncertainty in x-momentum are inversely proportional and that their product is of the order of the Planck constant. The more precisely we measure the position of the electron (smaller hole) the greater the uncertainty in its momentum (greater spread). But this is not just about our ability to make precise measurements. The wave function contains complete information about the electron, so the properties of position and momentum are *indeterminate*—the electron does not possess independent properties of position and momentum at each moment.

This discovery, made by Werner Heisenberg in 1927, is one of the defining features of quantum theory and is known as the uncertainty principle or indeterminacy principle. Its more formal statement is in terms of an inequality:

$$\Delta x \Delta p \geq \frac{h}{4\pi}$$

If particles do not possess defined properties such as position and momentum then it is impossible to predict the future in detail based on the total information about the state of the present. This makes quantum theory **indeterminate** and leaves the future open. It also undermines our classical ideas about the nature of reality—how can an electron be regarded as a real particle if it does not possess definite values for momentum and position?

27.5.3 Sum-Over-Histories Approach

In 1980, Freeman Dyson described the sum-over-histories approach: *Thirty-one years ago, Dick Feynman told me about his "sum over histories" version of quantum mechanics. "The electron does anything it likes," he said. "It just goes in any direction at any speed, ….however it likes, and then you add up the amplitudes and it gives you the wave function." I said to him, "You're crazy." But he wasn't.*

Feynman embraced the idea that photons or (electrons or any other quantum object) are free to get from one point to another by any route at all, and assumed that these alternative "histories" contribute in some way to what the electron or photon actually does. This seems to be what is happening in the double slit experiment because the slit through which the photon does not pass does affect where it can hit the screen. To do this he associated a rotating phasor with each photon. Different routes have different lengths and take different times to complete so the phasors from different routes will have a range of phases (i.e., they point in a range of different directions) when they arrive at any point on the screen. The phase can be worked out if we know the wavelength associated with the photon (e.g., a path of length 72.5 wavelengths introduces a phase difference π). The next step is to add up the phasors at each point on the screen. The square of the resultant phasor amplitude is proportional to the probability that the photon is found at that point. The diagram below shows three phasors for three of the many possible routes a photon could take from one point, A, to another, B.

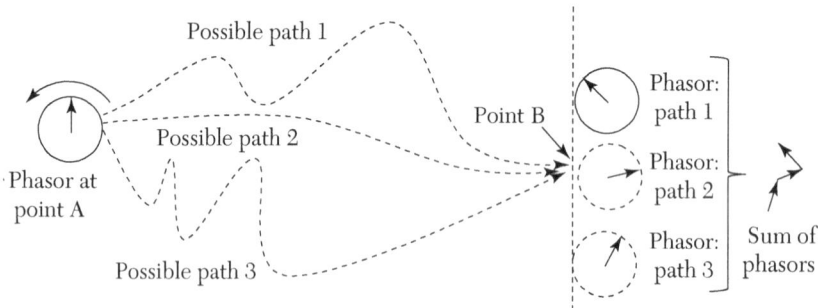

This can be summed up in a series of rules.

- **Rule 1:** photons/electrons "explore all paths"

- **Rule 2:** each path contributes a phasor

- **Rule 3:** the phasors add like vectors at the detector

- **Rule 4:** the square of the resultant phasor is proportional to the probability of finding the electron/photon at that point.

Considering double slits, the photon can reach the detector via either slit, so each possible route contributes a phasor. The paths are different lengths so there is a phase difference.

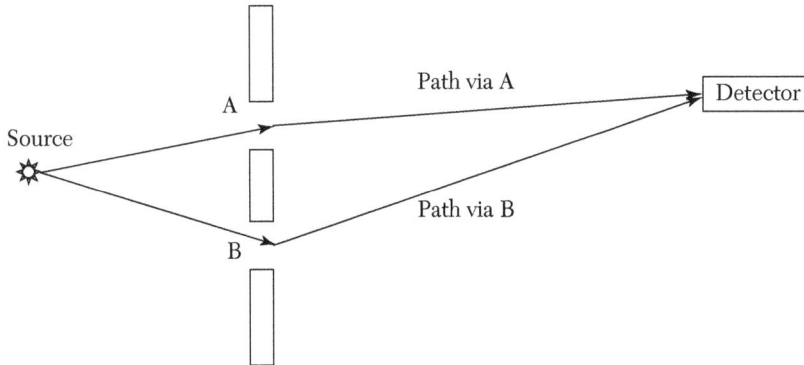

Path B is longer than path A, so the phasor has rotated more times and there is a phase difference between phasors via A and via B.

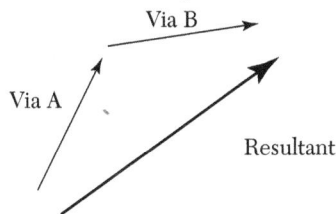

The length of the resultant will vary as the detector is moved to different positions. If the two phasors arrive in phase they reinforce so the probability at that point is a maximum value. At positions where the phasors arrive in antiphase (π phase difference) they undergo destructive interference and cancel out. The probability at that point is zero so no photons actually arrive there.

This "sum-over-histories"approach is particularly helpful in particle physics where all possible mechanisms for a particular interaction contribute a phasor and the sum of all the phasor amplitudes (squared) gives the probability of the process. As an interpretation of quantum theory it suggests that beneath what we regard as the real world of actual events there are potential events (possible paths) that may not actually be where the photon is found but which nonetheless contribute to what it can do.

27.5.4 Many-Worlds Theory

The many-worlds theory was first suggested in 1957 by Hugh Everett III and it was an attempt to solve the measurement problem. Everett thought that the Schrodinger equation should give a complete explanation of what happens in a quantum system and felt that the measurement problem only arises if we treat the system we are measuring as a quantum system and the apparatus we use to measure it (or even the observers who observe it) as classical systems. He argued that all observations and measurements were interactions between quantum systems so both the observed and the observer (the measured system and the measuring apparatus) must be described by the Schrodinger equation.

The wave function, a solution to the Schrodinger equation, represents a **superposition of states**. According to the Copenhagen Interpretation this superposition collapses into a particular state when an observation is made. Everett argued that if we really take the Schrodinger equation seriously we should assume the there is a wave function describing both the observed system and the observer and that this wave function contains a superposition of all possible states of both. For example, if an atom of a radioactive element has a 50% chance of decaying in 1 hour then the wave function of the system consists of two parts, one representing the undecayed atom and the other representing the decayed atom and the emitted radiation. During 1 hour the amplitude of the first part gets weaker and the second part gets stronger until at the end of the hour both parts have equal strength. This represents an equal chance of having decayed or not having decayed. Now consider an observer who can detect whether the atom has decayed or not. According to Everett the observer is also described by a wave function and this contains a superposition of two states, one in which the observer detects a decay and the other in which he does not. As time goes on the amplitude of the first part grows and the amplitude of the second part falls until after 1 hour they have equal amplitudes. No collapse has occurred but the description of the universe now contains two observers, one having observed the decay, the other having observed no decay. The single world has split into two.

According to Everett's model the wave function for the whole universe contains a superposition of all possibilities. We only think that the wave function collapses because we, as individuals, only experience one path through this ever-splitting universe even though copies of ourselves occur in many of the other parallel worlds. The attraction of the many-worlds theory is that it solves (or at least removes) the measurement problem. Its drawback is the weird multiplicity of universes it imagines.

- Every measurement causes the world to split into multiple copies of itself, each copy containing one possible outcome of the measurement AND a copy of the measuring device giving that outcome.

▣ This solves the **measurement problem**—No "collapse of the wave function"

▣ But it suggests that world is continually splitting into multiple copies of itself—**many-worlds**.

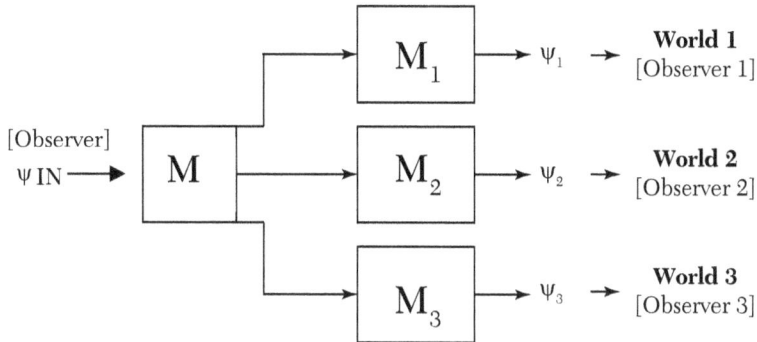

M is a measurement and M_1, M_2 and M_3 are three different possible results of that measurement that exist as super positions in the wave function.

In the many-worlds interpretation everything that is not impossible actually occurs in some world—if life is highly improbable but not impossible it must exist in part of the multiverse!

27.5.5 Schrodinger's Cat

Erwin Schrodinger realized that quantum theory challenged our conventional view of how the universe functions. He tried to draw attention to its counter intuitive nature using a thought experiment in which quantum effects on the atomic scale are amplified up to affect objects on an everyday scale. This thought experiment is known as Schrodinger's cat.

A cat is placed in a box along with a mechanism that is designed to break a vial of poison if a radioactive atom decays. If this does happen the poison will kill the cat. The box is closed and left for a time so that the probability that the atom has decayed and the poison has been released to kill the cat is exactly ½ . The box is then opened. The cat is either dead or alive but what was its state prior to opening the box?

Classically the answer to this question seems very obvious—the cat was either alive or dead before we opened the box, we just do not know which state it was in. Quantum theory gives a different and rather disturbing answer. The atom is described by a wave function that has two parts— one describing its undecayed state and the other describing its decayed state. Prior to opening the box and making an observation the most we can deduce about the system is that the atom is in a superposition of the decayed and undecayed states. But there is also a wave function for the state of the detector and the vial of poison. These too must be in a superposition of states. So must the cat! It is not dead or alive prior to opening the box but in a superposition of those states! And we can go further—the experimenter who opens the box can also be described by a wave function. Until we interact with him and ask him what has happened to the cat we must describe him, the cat, the detector and the atom by a wave function that includes a superposition of both alternatives....

If we stick to the Copenhagen Interpretation each wave function is collapsed by successive observations. This is where the many-worlds theory offers a way out of these endless wave function collapses. The cat is both dead and alive but not in the same world. In one world the atom decayed, the poison was released and the cat died. In that world the experimenter opened the box to find the dead cat. But in another world the atom did not decay, the poison remained trapped in the vial and when the experimenter opened the box the cat was alive and well.

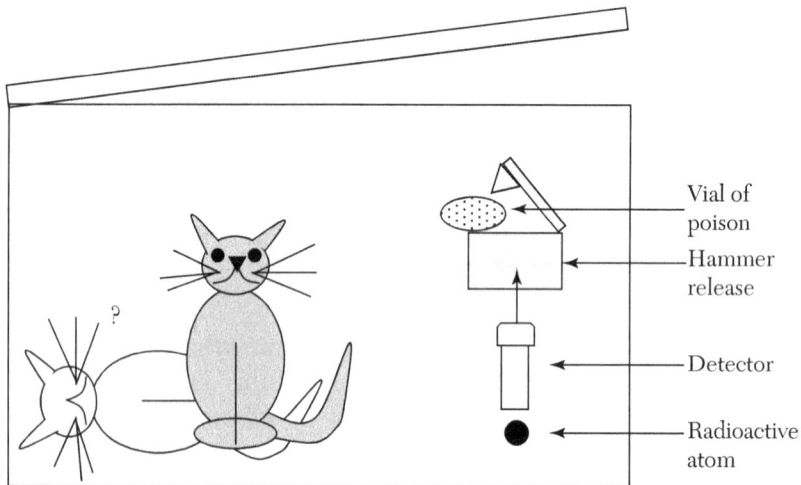

The arguments about how to interpret quantum theory go on!

27.6 Exercises

1. (a) Calculate the de Broglie wavelength of an electron that has been accelerated through a potential difference of 500 V.

(b) (i) Calculate the energy of an X-ray photon of wavelength 2.0×10^{-10} m. Express your answer in both J and eV.

    (ii) Calculate the de Broglie wavelength of an electron that has the same energy as the photon in (b)(i).

(c) The first ionization energy of most atoms is around 10 electron volts. Explain why visible light is not ionizing but X-rays are.

2. (a) Explain what is meant by each of the following:

    (i) the photoelectric effect,

    (ii) threshold frequency,

    (iii) work function

(b) Describe the dependence of the photoelectric effect on:

    (i) frequency of radiation, (ii) intensity of radiation

(c) Give one aspect of the photoelectric effect that cannot be satisfactorily explained using a wave model of light.

(d) Describe Einstein's photon theory and explain how it can account for the frequency and intensity dependence of the photoelectric effect.

3. (a) Why will weak red light emit no electrons from the surface of zinc?

(b) Why doesn't increasing the intensity of red light lead to the emission of electrons from the surface?

(c) Why can UV light of wavelength 200 nm emit electrons from zinc?

(d) What changes if the intensity of the UV light falling on the zinc is increased? Why?

(e) Blue light of wavelength 450 nm can emit electrons from potassium. Explain why this can occur for potassium but not zinc.

(f) How would electrons emitted from potassium by violet light differ from those emitted by blue light? Explain.

Work functions: zinc: 4.3 eV; potassium: 2.3 eV

4. (a) Calculate the maximum kinetic energy for electrons emitted from the surface of a metal of work function 2.0 eV by light of wavelength 450 nm.

(b) Calculate the maximum velocity of these electrons.

(c) Calculate the stopping voltage for these electrons.

(d) State and explain what happens if the intensity of the light is increased.

5. The energy levels of the hydrogen atom are given by the equation:

$$E_n = \frac{-13.6\,\text{eV}}{n^2}$$

(a) How much energy is needed to ionize a hydrogen atom from its excited $n = 2$ state?

(b) What is the wavelength of the photon emitted by a hydrogen atom when it decays from the $n = 4$ state to the $n = 2$ state? What part of the EM spectrum does this radiation belong to?

(c) An electron with kinetic energy 11 eV scatters from a hydrogen atom in its ground state. Describe the states of the hydrogen atom and electron after the collision if:
(i) It is an elastic collision. (ii) It is an inelastic collision.

6. (a) Estimate the number of photons of visible light emitted by a 100 W filament lamp per second. Assume the light efficiency is 10% and take photons of visible light to have a wavelength around 500 nm.

(b) Why don't we notice the quantum nature of light in everyday life?

7. (a) Derive an algebraic expression for the recoil force on a torch if it emits a beam of light of optical power P.

(b) Solar sails could be used to reflect light from the Sun and generate a thrust to propel a spacecraft away from the Sun. Estimate the minimum area of solar sail necessary to generate a thrust of 1.0 N at a distance of 1.5×10^{11} m from the Sun (1 AU). The Sun's luminosity is 3.8×10^{26} W.

8. (a) Summarize the main features of the Copenhagen Interpretation.

(b) Use the Copenhagen Interpretation to explain the formation of an interference pattern when monochromatic light is shone through double slits.

(c) Explain what is meant by "the measurement problem."

(d) Why is the "measurement problem" a problem for the Copenhagen Interpretation?

(e) How does the many-worlds interpretation "solve" the measurement problem?

9. (a) Describe the Schrodinger's cat thought experiment.

(b) At the end of the experiment a box is opened and the experimenter finds a cat that is either dead or alive. How does the description of the state of the cat inside the box, *before it is opened*, differ in classical physics and in quantum physics (according to the Copenhagen interpretation)?

(c) What happens, according to the Copenhagen interpretation, when the box is opened and the experimenter looks inside?

(d) Suggest a reason why this thought experiment is so shocking.

(e) How does the many-worlds theory account for what happens in this thought experiment?

10. The energy levels of the hydrogen atom are given by the equation:

$$E_n = \frac{-13.6\,\text{eV}}{n^2}$$

(a) Derive a similar equation for the singly ionized atom of helium $(Z = 2)$.

(b) Describe how you would expect the line emission spectrum from singly ionized helium to compare with that of hydrogen.

(c) Explain why we cannot use the same simple method to derive the energy levels for the second electron in the neutral helium atom.

28

ASTROPHYSICS

28.0   Physics Astrophysics and Cosmology

Much of astrophysics differs from other branches of the subject because we cannot set up experimental stars and vary parameters to see what happens. However, there are many billions of stars in space and we can test our theories by observing their radiation using telescopes here on Earth. The first telescopes used for this purpose were optical but in the second half of the twentieth century other parts of the electromagnetic spectrum were also used; first radio but now infrared, ultraviolet, X-ray and even gamma-ray telescopes have been built and used. The Earth's atmosphere is almost transparent to visible light and some radio wavelengths but much of the rest of the spectrum is strongly absorbed, so telescopes working at other wavelengths are usually mounted on satellites orbiting beyond the atmosphere.

Different wavelengths provide information about different processes going on in the source stars. It is usually the case that the higher the frequency of the radiation the more energetic the underlying process that emitted it. In recent years, the detection of gravitational waves (predicted in Einstein's general theory of relativity) has opened the possibility of gravitational wave telescopes in the future. These would provide evidence about some of the most dramatic astronomical events, e.g., the collapse of giant stars collisions of black holes.

Here we will focus on some of the physics that is important in astrophysics and include aspects of cosmology, the study of the entire universe.

28.1 Stars

Before the start of the twentieth century, physicists were unable to explain how a star like the Sun could continue to radiate energy for billions of years. No known chemical or gravitational process could provide a large enough source of energy. This was not a major problem until the theories of long-term geological processes and evolution by natural selection became established. Both required the Earth to have existed, and had a source of energy, for billions of years.

The problem of the Sun's energy source was solved by the discovery of Einstein's mass–energy equation and the process of nuclear fusion (see Section 26.3.4). Stars fuse light nuclei into heavy nuclei in their cores releasing a huge amount of energy. Radiation from the core supports the star against gravitational collapse while energy is transferred to its surface where it radiates out into space.

28.1.1 Mass

The most important parameter when modelling a star is its mass. The greater the mass of the star the higher the temperature at its core. This can allow fusion reactions to proceed faster and nucleosynthesis (see Section 26.3.5) to produce heavier elements. Increasing mass rapidly increases the reaction rate in the core so that more massive stars use up their fuel relatively more quickly than less massive stars and reach the end of their lives earlier.

Mass determines the fate of a star.

- Low mass stars are stable for a long time but their cores are not hot enough to create nuclei beyond carbon. When they run out of fuel they swell to become red giant stars before shedding their outer layers of gas and forming a planetary nebula. This exposes the white-hot core. This final state is called a white dwarf star. Fusion reactions have now ceased so the white dwarf cools down over a long period of time (of the order of a billion years) eventually becoming a dense black dwarf star.

- High mass stars have shorter lives but the core becomes hot enough for nuclear fusion to create iron. Iron is the most stable nuclide so at that point fusion reactions stop suddenly and the star undergoes a violent collapse and explosion called a supernova. This can increase the luminosity of a star by a factor of around 10^{10} for a short period (days or weeks). The supernova explosion has two effects—some of the energy creates heavier nuclei than iron and the process blasts these out into space, where they can become part of the raw material for second and third generation stars to form. Our own Solar System must have formed from supernovae remnants because it contains significant amounts of the heavy nuclides (e.g., uranium in the Earth's crust).

Mass also determines what happens to the core left after the supernova explosion.

- The core is so massive (typically 2–3 times the mass of the Sun) that the forces that prevent collapse of ordinary matter—i.e., forces that stop atoms being crushed (called electron degeneracy pressure)— are overcome by gravity. This effectively forces orbital electrons and nuclear protons to combine to form neutrons. When this happens the core radius decreases enormously (to about 10 km!) and the density of the core, which is now almost entirely made of neutrons, increases spectacularly to around 10^{17} kgm^{-3}. One centimeter cubed of this neutron star material would have a mass of one hundred million tons! The core is now called a neutron star. This collapse causes the rotation rate to increase too so neutron stars spin rapidly, sometimes completing a revolution in milliseconds). Neutron stars form from the collapse of stars with initial masses in the approximate range 10–30 times the mass of the Sun.

- Another effect of the collapse is to intensify the magnetic field of the star. This has the effect of directing a beam of radio waves out along the magnetic axis of the star. Since this axis can be in a different place to the rotation axis the beam sweeps around like the light from a lighthouse. If the Earth happens to be struck by this beam we receive regular pulses. When these pulses were first discovered they were so regular that astronomers thought they might be alien radio signals. They are called **pulsars**.

- If the mass of the core is greater than about 5 times the mass of the Sun then the collapse to a neutron star is not the end of the story. The gravitational forces are so strong that the neutrons themselves are crushed and at the present time we know of no physical force that prevents collapse to a point or singularity. A black hole is formed. The reason for the name is that at a certain distance from the central singularity the escape velocity is c, the speed of light. Since this is a universal speed limit no material or information from points closer to the singularity can reach the outside world. The surface at which this occurs is called the event horizon of the black hole (see Section 23.2.5).

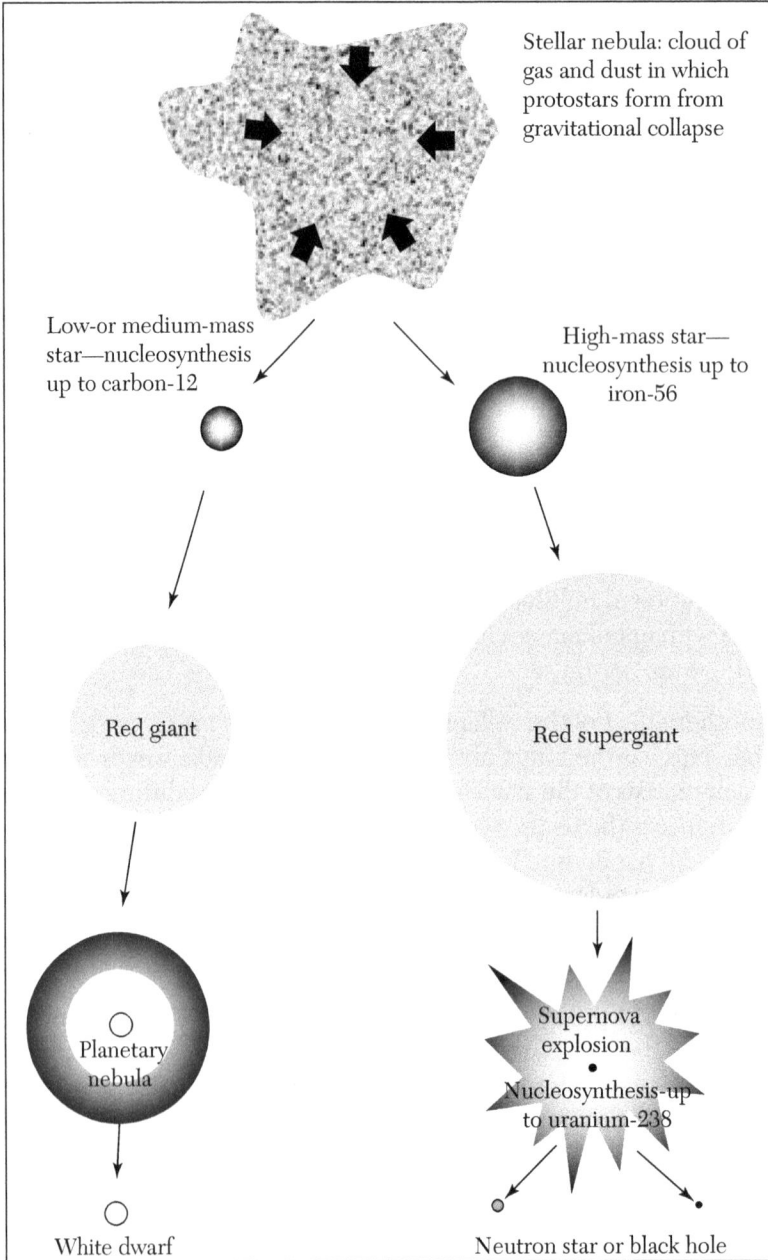

Stellar nebula: cloud of gas and dust in which protostars form from gravitational collapse

Low-or medium-mass star—nucleosynthesis up to carbon-12

High-mass star— nucleosynthesis up to iron-56

Red giant

Red supergiant

Planetary nebula

Supernova explosion

Nucleosynthesis-up to uranium-238

White dwarf

Neutron star or black hole

28.1.2 Stars as Black Bodies

Energy released by nuclear fusion reactions in the core of a star raises its surface temperature so that it emits a spectrum of electromagnetic radiation into space like a black body. The total power radiated by a star is called its luminosity L (measured in watts). However, whilst the black body model is fine for the overall shape of the spectrum, there is a great deal of fine detail too. This results in a complex pattern of dark absorption lines corresponding to elements in the outer layers of the star that have absorbed radiation at particular frequencies. Analysis of stellar spectra can tell us a great deal about the nature of the star, its surface temperature and composition, as well as allowing us to classify stars in a useful way. In addition to this, motion of the star relative to the Earth results in a Doppler shift of these absorption lines and by measuring this shift we can determine the relative velocity of the star.

The peak wavelength in the black body radiation spectrum can be measured using a telescope attached to a spectrometer. Wien law (see Section 8.5) can then be used to find the surface temperature of the star.

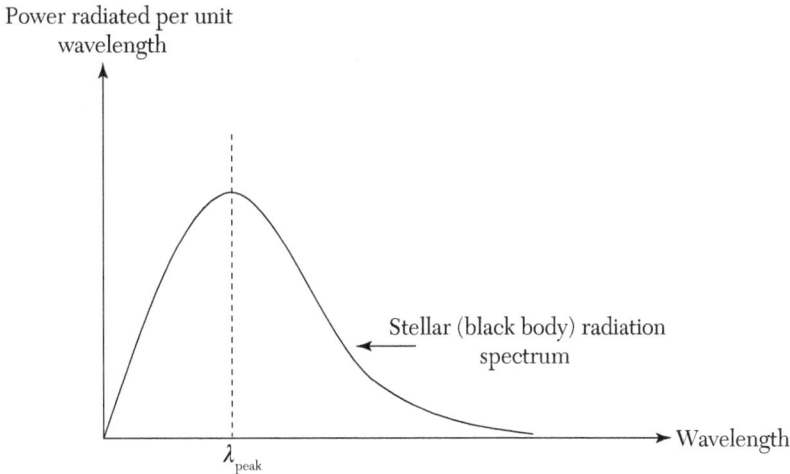

$$\lambda_{max} T = 2.9 \times 010^{-3} \ mK$$

$$T = \frac{2.90 \, 10^{-3}}{\lambda_{max}}$$

If the luminosity is also known it is possible to use Stefan law to calculate the radius of the star:

$$L = \sigma A T^4 = 4\pi r^2 \sigma T^4$$

$$r = \sqrt{\frac{L}{4\pi\sigma T^4}}$$

28.1.3 Stellar Spectra and the Hertzsprung-Russell Diagram

Astronomers classify stars according to their spectral type. This also corresponds to their surface temperature because as the temperature rises different types of spectral lines appear in the spectrum. The details of this do not need to concern us here but what is interesting is that when luminosity is plotted against temperature (spectral type) for all of the observable stars a clear pattern emerges. This was first done by Hertzsprung and Russell and the plot is called a Hertzsprung-Russell (HR) diagram. Note that the *x*-axis points in the direction of *decreasing* temperature.

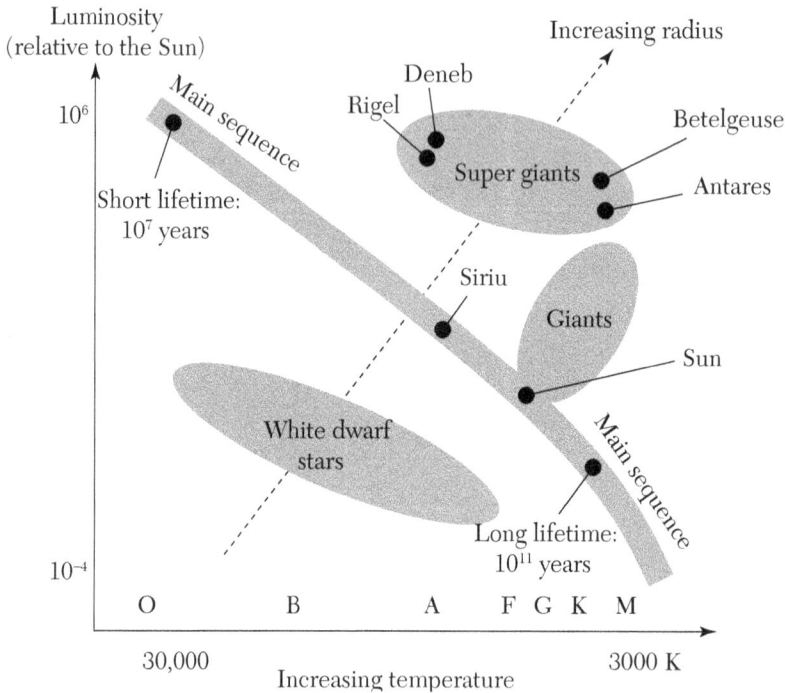

The diagram above shows (in a very simplified form) the main regions of the HR diagram. The letters refer to spectral classes used in astronomy. The Sun is in class G and has a surface temperature of about 5800 K.

A diagonal band runs from large luminous hot blue-white stars at top left to small dim red stars at bottom right. This is called the **main sequence** and stars spend most of their lives on this band. At the end of their lives, when nuclear fusion fuel in their core runs out, they move off the band as they become red giants or super giants and eventually white dwarf stars, neutron stars or black holes. These final two star types do not appear on the HR diagram because their luminosity and spectrum is not measured directly (and luminosity is very low).

28.2   Distances

One of the greatest challenges for astronomers and cosmologists is to find ways to determine accurate distances to the objects they observe. Ancient Greek astronomers managed to find ingenious methods to estimate the size of the Earth and the distances to the Moon and Sun but modern space exploration has provided accurate methods for surveying our immediate surroundings in space. Distances to objects beyond the solar system are determined by several different overlapping methods and these regions of overlap can be used to calibrate one technique against another.

28.2.1   Trigonometric Parallax

The Earth's orbital motion causes the apparent positions of relatively nearby stars to shift against the background of very distant ("fixed") stars. This parallax shift can be used to measure the distance to the nearby star.

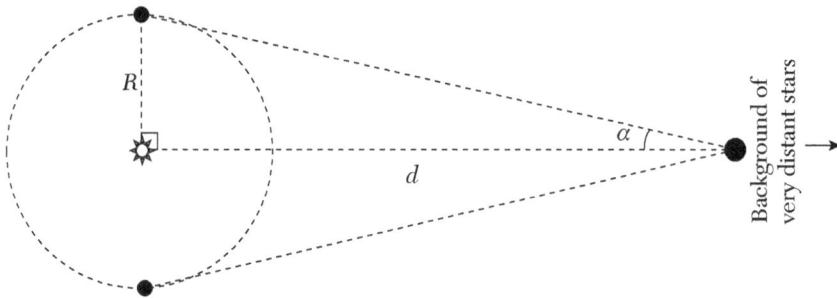

Telescopes can be used to measure the parallax angle α. This is half of the total angular shift in the star's position during a 6-month period (as the earth completes half of an orbit).

$$\frac{R}{d} = \tan \alpha$$

Parallax angles are tiny so we can use the small angle approximation and replace $\tan(\alpha)$ with α (in radians).

$$d = \frac{R}{\alpha}$$

where R is the radius of the Earth's orbit. This is known very accurately from laser ranging within the solar system and trigonometry. If R is measured in meters and α in radians then d will also be in meters. Astronomers often use different (non S.I. units):

1 Astronomical unit (AU) = 149,597,870,700 m

They also measure the parallax angle in seconds of arc where:

$$1 \text{ second of arc} = 1/3600 \text{ degree}$$

When these units are substituted into the equation for distance above, the result is in parsecs (pc) so that a star with a parallax angle of 0.1 seconds of arc is at a distance of 10 pc.

$$1 \text{ parsec (pc)} = 3.0857 \times 10^{16} \text{ m} = 2.26156 \text{ light years}$$

The parallax method using Earth-based telescopes is limited to about 100 pc because as distance increases the parallax angles soon become too small to be resolved. However, space telescopes (e.g., the Hipparcos satellite) can extend this method to about 1000 pc.

28.2.2 Inverse-Square Law and Cepheid Variables

Telescopes can be used to measure the intensity (or flux) I of stellar radiation as it reaches the Earth. If we also know the luminosity L of the star being observed then the inverse-square law can be used to calculate its distance d.

$$I = \frac{L}{d^2}$$

$$d = \sqrt{\frac{L}{I}}$$

In 1912, Henrietta Leavitt discovered a relationship between the period of a certain type of variable star, a Cepheid variable, and its luminosity. This period-luminosity rule provides a way to discover the luminosity of Cepheid variables by measuring their period from the Earth. This is easily done simply by monitoring the flux from the star over a period of time.

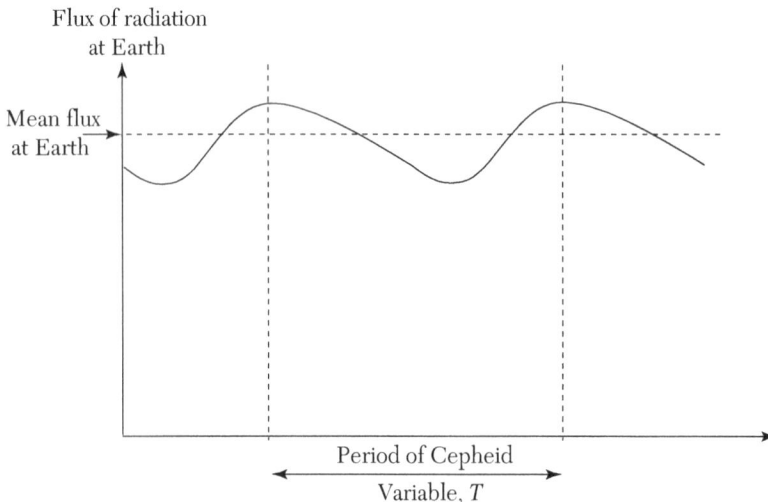

Cepheid variables are very luminous so they can be detected out to a very great distance. This gave astronomers a method to extend distance measurements from our own galaxy to other quite distant galaxies.

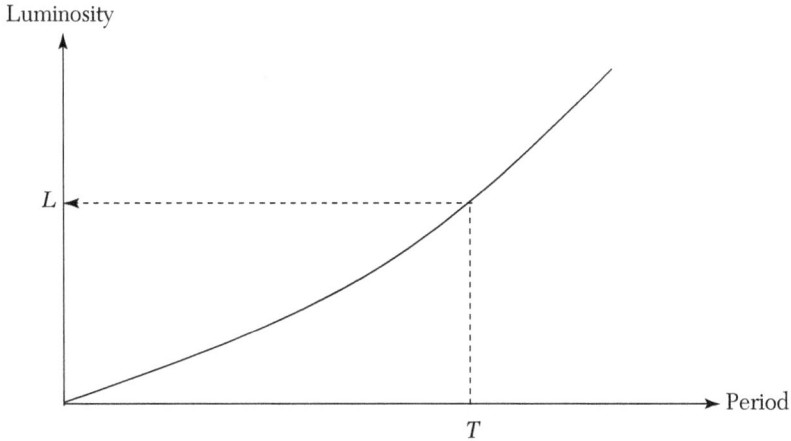

The way to determine distance using Cepheid variables is summarized below:

- Monitor the flux from a Cepheid variable.

- Measure its period T and mean intensity I.

- Use the period-luminosity relation to find the luminosity L.

- Use the inverse-square law to find the distance d.

This method can be calibrated against the parallax method using nearby Cepheids.

Hubble used this distance method with groups of Cepheids in nebulae and showed that the nebulae were actually separate galaxies outside the Milky Way.

28.2.3 Hubble Law

In the 1920s, Edwin Hubble and Vesto Slipher measured the spectra and distances for many observable galaxies. When a galaxy is in motion relative to the Earth the spectrum measured on Earth is shifted relative to the spectrum of the same elements from a stationary source. This is a Doppler effect (see Section 14.4.1) and can be used to calculate the velocity of the source, in this case the velocity of the galaxy. The wavelength shift $\delta\lambda$ is given by:

$$\delta\lambda = \left(\lambda - \lambda_0\right)$$

Where λ_0 is the wavelength from a stationary source and λ is the wavelength from the moving source.

Astronomers usually work with the fractional shift, $z = \dfrac{\delta\lambda}{\lambda_0}$. If this is positive the wavelengths are increased and it is called a "red-shift." If it is negative the wavelengths are decreased and it is called a "blue-shift."

For velocities small compared to the speed of light the relationship between the fractional shift in wavelength $\dfrac{\delta\lambda}{\lambda_0}$ and the velocity is given by:

$$z = \frac{\delta\lambda}{\lambda_0} = \frac{v}{c}$$

To their surprise they discovered that:

- $z = \dfrac{\delta\lambda}{\lambda_0}$ is positive for all distant galaxies, i.e., light from all distant galaxies is red-shifted;

- the red-shift is directly proportional to the distance of the galaxy:

$$z = \frac{\delta\lambda}{\lambda_0} = \frac{v}{c} \propto d$$

Hubble realized that:

- Distant galaxies are moving away from us—**the universe is expanding**. (A few nearby galaxies are actually moving toward us but that is because their own proper motion is greater than the global motion due to expansion.)

- The recession velocity of a distant galaxy is directly proportional to its distance:

$$v \propto d$$
$$v = H_0 d$$

This final equation is the **Hubble law**.

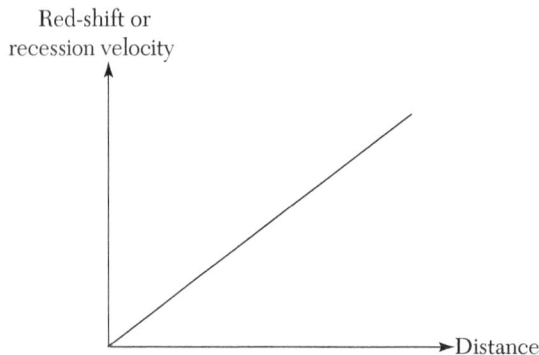

Having established Hubble law, red-shifts can be used to determine the distances to distant galaxies. However, for very distant objects the source must be extremely bright otherwise the flux of radiation reaching the Earth is too weak for measurements to be made. Fortunately, there are extremely bright objects that can act as "**standard candles**"for these measurements. One such object is a type 1a supernova. Astronomers understand the physics of these stars and can predict their luminosity, which is great enough for them to be seen in the most distant galaxies. At these distances the recession velocities are a significant fraction of the speed of light so relativistic effects must also be taken into account.

28.3   Cosmology

Cosmology is the science of the universe as a whole, dealing with its origin, nature, evolution and end. All of the evidence that we have suggests that the laws of physics we have discovered from our own planet operate throughout the universe, so we use these to try to understand it. Whilst cosmology deals with physics on the largest scale it is intimately linked to physics on the smallest scales and discoveries in particle physics and cosmology are often linked. The enormous energies present soon after the Big Bang are reproduced in particle physics experiments such as the Large Hadron Collider at CERN.

28.3.1   Origin and Age of the Universe

Hubble law shows that the universe is expanding, so it is tempting to think that if all the distant galaxies are moving away from us (as they are) then we must be at the center of this expansion. However, this is not how a cosmologist would view the expansion. An observer on any of the galaxies would see all the other galaxies moving away from him in the same way that we do. There is no center to the expanding universe, all points are equivalent. A way to understand this is to imagine that the universe is two-dimensional and shaped like the surface of a balloon with the galaxies as spots on the surface. As the universe expands the balloon gets larger and all the spots move apart. Those initially close together move apart slowly, those that started farther apart separate more rapidly, exactly as the Hubble law describes. An observer placed on any one of the spots could discover this law and would see the universe expanding in exactly the same way—there would be no center to this surface. A section of the expanding surface is shown below as it doubles in size. The black dots represent three galaxies—A, B, and C. Distance AC is double the distance AB.

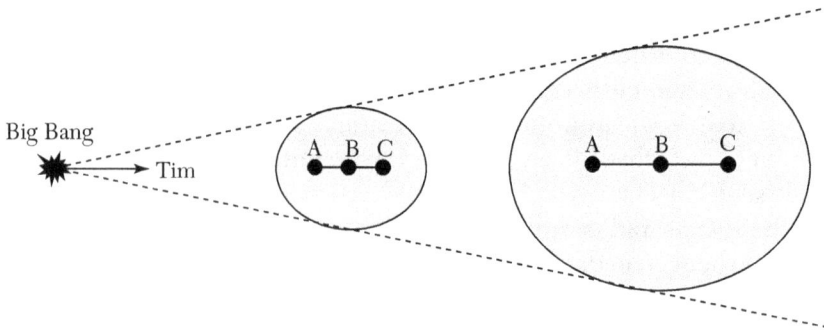

As the scale doubles, AB also doubles. An observer on A sees B recede at some velocity v. AC also doubles, so the same observer (on A) also sees C recede. However, the recession velocity of C is double that of A because AC = 2AB in both diagrams. This confirms the Hubble law—a galaxy at double the distance has double the recession velocity, $v \mu d$. The same argument would hold equally well for observers at B or at C.

If the universe is expanding now, it must have been much smaller in the past. Stephen Hawking and Roger Penrose showed that it must have begun as a point or singularity of infinite density that exploded and has been expanding ever since. This initial explosion is called the Big Bang.

We can use Hubble law to estimate how much time has passed since the Big Bang. This is an estimate of the age of the universe. The method assumes that the galaxies have always had their present relative velocities. We can calculate how long they have been separating by dividing their current separation by their current recession velocity:

$$T = \frac{d}{v} = \frac{1}{H_0}$$

This is the **Hubble time**. It does not take into account the variation in galactic velocities caused by gravitational forces but does give a good order of magnitude for the age of the universe. More sophisticated methods using evidence from the cosmic background radiation (measured by the Wilkinson Microwave Anisotropy probe [WMAP] in 2012) have given a much more precise estimate of the age of the universe.

Age of universe = $13.772 \pm 0.059 \times 109$ years

So far we have described the red-shifts as if they are caused by the motions of galaxies through space. However, we have already seen that the concept of absolute space had to be abandoned and only relative motions are significant. Einstein's general theory of relativity goes further showing that space has geometrical properties that can be changed by the presence of matter or energy (space-time curvature). This theory provides a different interpretation of red-shifts and the Big Bang and expansion. According to general relativity space itself is expanding (rather like the surface of the balloon in the analogy above). The red-shifts are therefore a result of the stretching of electromagnetic waves as they cross the expanding space between galaxies. Hubble law is consistent with this approach (as can be seen from the example above). One new consequence of Einstein's approach is that there will be "horizons" in the universe. Whilst no object can move through space faster than the speed of light it is possible for the space between two galaxies to expand so fast that light cannot travel between them. When this happens the galaxies effectively disappear over a cosmic horizon.

28.3.2   Evidence for the Big Bang

The Big Bang model of the origin of the universe is supported by several strong strands of evidence:

- The red-shifts of distant galaxies are consistent with the universe having evolved from a tiny dense point.

- The universe is filled with cosmic microwave background radiation. This has a blackbody radiation spectrum corresponding to a temperature of about 2.7 K. This was predicted from the model and detected, accidentally, by Penzias and Wilson in 1964. Soon after the Big Bang the universe was filled with high-intensity gamma-radiation that was in thermal equilibrium at a very high temperature. As the universe expanded it cooled and the radiation was red-shifted to longer wavelengths forming the microwave background that we detect today. The fact that the spectrum is a blackbody spectrum confirms that the radiation was in (almost) thermal equilibrium and the fact that it is uniform from all parts of the sky confirms that it filled the universe.

- Tiny fluctuations in the microwave background radiation (detected and measured by the COBE and WMAP satellites) are of the correct magnitude to account for galaxy formation. If the current microwave background radiation had been perfectly uniform, then the early universe would not have contained enough concentrations of matter for galaxies to form.

▪ The ratios of light nuclides (e.g., hydrogen, helium, lithium) throughout the universe are consistent with the amounts expected to have been formed by nuclear fusion reactions as the universe expanded. There was a brief period after the Big Bang when the universe was hot and dense enough for some nucleosynthesis to take place but this soon stopped as the universe expanded and cooled. All the heavier elements were synthesized by nuclear fusion reactions in stars.

▪ The Big Bang and expanding universe model is consistent with Einstein's general theory of relativity. In fact his theory requires a Universe that either expands or contracts.

28.4 Exercises

1. (a) The surface temperature of the Sun is about 5800 K. Sketch a graph to show how the intensity of radiation varies with wavelength for electromagnetic waves leaving the surface and calculate the wavelength at which the peak of this distribution occurs.

(b) The luminous flux from the Sun is about 1400 Wm^{-2} at the radius of the Earth's orbit. Calculate the luminous flux at the orbit of Mars (about 1.5 times further away than the Earth).

(c) The luminosity of the Sun is about 4×10^{26} W with a surface temperature of about 5800 K. Calculate its radius.

(d) Toward the end of its life the Sun will become a white dwarf star with a surface temperature of about 10^5 K. How will this affect the spectrum of radiation it emits? Support your answer with a relevant calculation.
Wien constant = 2.90×10^{-3}mK,
Stefan constant = 5.67×10^{-8} Wm^{-2}K^{-4}

2. The peak wavelength in the spectrum of light emitted by the super-giant star Betelgeuse is 830 nm and its luminosity is 3.5×10^{31} W

(a) Calculate its surface temperature.(b) Calculate its radius.
Wien constant = 2.9×10^{-3}mK, Stefan constant = 5.67×10^{-8} Wm^{-2}K^{-4}

3. A distant galaxy has a red-shift of 0.05. The Hubble constant is about 2.2×10^{-18} s^{-1} in S.I. units.

(a) Calculate the velocity of the galaxy relative to the Earth and state the direction in which it is moving.

(b) Calculate the distance of the galaxy.

(c) Use the Hubble constant to estimate the age of the universe in years and explain why the actual age is likely to differ from this value.

4. (a) Describe a parallax method for measuring the distance to a relatively nearby star.

   (b) Show that 1 parsec is about 3×10^{16} m and about 3.26 light years.

   (c) Explain why stellar parallax cannot be used to measure the distance to very distant stars.

$$1 \text{ AU} = 1.496 \times 10^{11} \text{ m}$$

   (d) How far away (in pc) is a star with parallax 0.052 second of arc?

5. (a) Two similar galaxies are observed. The brightest blue stars in galaxy A have an apparent brightness 10,000 times greater than those in galaxy B. Galaxy A is 107 light years away. How far away is galaxy B?

   (b) Two type Ia supernovae are observed one week apart. The first is 100 Mpc away but its apparent brightness is only 0.070 of the apparent brightness of the second. How far away is the second supernova?

6. (a) Explain what is meant by a "standard candle" in astronomy.

   (b) Explain how a Cepheid variable can be used as a standard candle to measure distances to galaxies beyond the Milky Way.

7. (a) A rocket is traveling away from the Earth at a velocity of 9.0×10^6 ms^{-1} when it transmits a signal to the Earth on a carrier frequency of 10.000×10^{10} Hz. To what frequency should the receiver on Earth be tuned?

   (b) A distant galaxy has a red-shift of 0.030. Aliens on a planet in this galaxy transmit a signal to Earth on a carrier frequency of 10.000×10^{10} Hz. To what frequency should the receiver on Earth be tuned?

   (c) Compare your answers to (a) and (b) and discuss whether or not the shift in frequency has the same physical cause.

   (d) Estimate the time taken for the signal in (b) to reach the Earth.

$$H_0 = 2.2 \times 10^{-18} \text{ s}^{-1}$$

8. (a) Explain how observations of galactic spectra led to the ideas of the expanding universe, Hubble law, and the Big Bang.

   (b) State and explain two other pieces of evidence for the Big Bang as the origin of the universe.

CHAPTER 29

MEDICAL PHYSICS

29.1 Ultrasound

29.1.1 Overview of Ultrasound

Ultrasound (sonography) uses high-frequency sound waves to form images of structures inside the human body and is particularly suited to imaging soft tissues. Typical frequencies are in the range 1–20 MHz with corresponding wavelengths from 2–0.1 mm. Distances to tissue boundaries are computed using the time for reflected pulses to return:

$$\text{Distance to boundary} = \frac{1}{2} \times \text{speed of ultrasound in tissue}$$

$$\times \text{time for pulse to return}$$

The speed of ultrasound in the body is about 1550 ms^{-1}. The higher frequency waves provide higher resolution but are absorbed more strongly so do not penetrate so far into the body.

Ultrasound transducers use piezoelectric crystals to generate and detect the waves. When an alternating voltage is applied across the crystal it vibrates and when the crystal experiences an alternating stress (as it absorbs an ultrasound wave) it generates an alternating voltage.

There are several different types of ultrasound scan. An A-scan ("amplitude mode") is the simplest procedure, using a single transducer to detect echoes. As the scanner is moved along a line the depth is computed and displayed on a screen. A B-scan ("brightness mode") uses a linear array of transducers and creates a 2D image as the scanner is moved. It is also possible to ultrasound pulses to compute the velocities of tissue boundaries in order to create a moving image, this is M-mode ("motion mode"). When ultrasound reflects from a moving object it is Doppler shifted. Doppler mode ultrasound uses these shifts to measure and display blood flow rates.

Studies into the safety of ultrasound techniques have shown that they present very low risk with no confirmed evidence that normal ultrasound scans cause any significant damage to humans. This makes them preferable to X-ray techniques for many diagnostic uses, including prenatal scanning.

29.1.2   Ultrasound and the Eye

A-scans are used to measure the structure and dimensions of the human eye to ensure that the correct lens implant is used following surgery for the removal of a cataract. The diagram below shows how an A-scan result relates to the eye. The speed of the ultrasound waves in the eye is 1550 ms^{-1}. The dotted lines on the upper diagram correspond to tissue boundaries

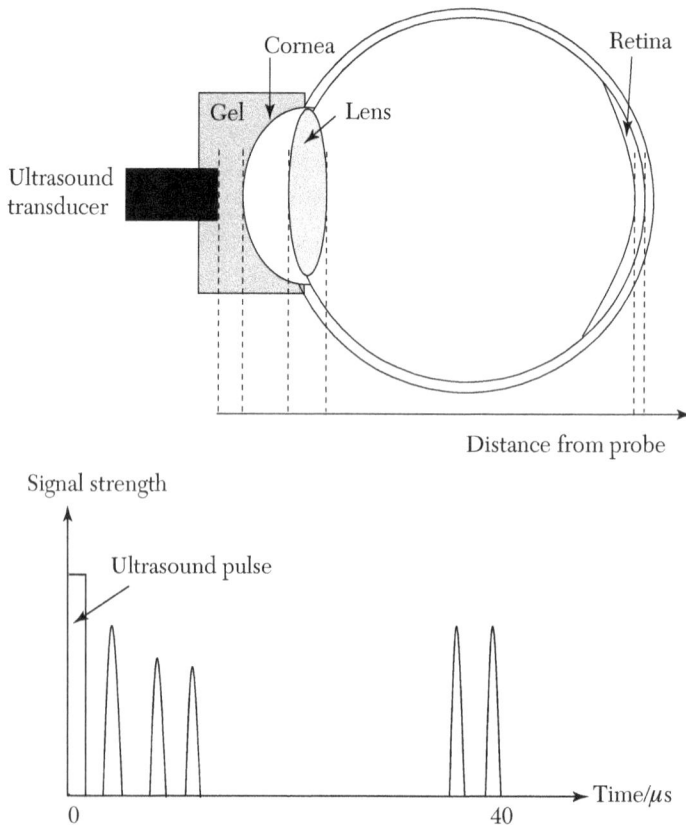

responsible for the peaks in the detected signal. The gel between the transducer and the cornea reduces the amount of the incident signal that is reflected by the outer surface and does not enter the eye.

B-scans of the eye are used to diagnose problems such as a detached retina, glaucoma, or cataracts or to monitor the shape and size of a tumour. In the B-scan the incident ultrasound is moved back and forth to scan slices of the eye. These can then be combined to form an image. The eye is quite small so ultrasound does not have to penetrate far and higher frequencies can be used. Recently frequencies up to 50 MHz have been used to provide extremely high resolution images of structures at the front of the eye. At these frequencies penetration is just a few millimeters.

B-scans in prenatal scanning have to penetrate farther into the body so these are limited to lower frequencies. This reduces their resolution but the structures being imaged tend to be larger so this is not a major problem.

29.1.3 Doppler Ultrasound for Blood Flow Measurements

Ultrasound is directed into a blood vessel and reflects from the moving cells. The reflected waves are Doppler shifted and the frequency difference between the incident and reflected waves is proportional to the speed of the flow. The diagram below illustrates how this is carried out:

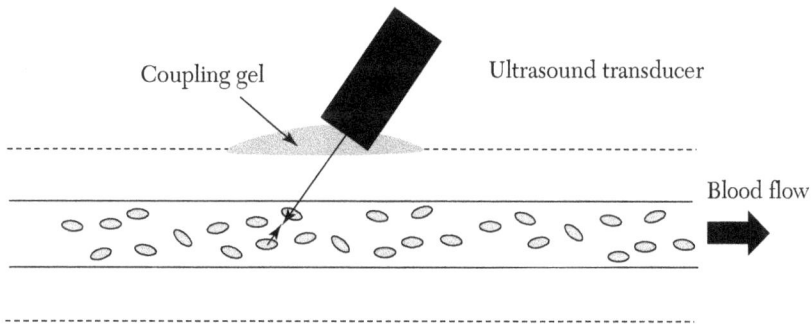

As the blood flow pulses the frequency difference shows a series of peaks:

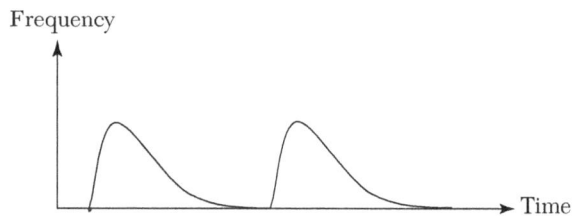

29.1.4 Using Ultrasound to Break Kidney Stones

Kidney stones can be painful and might get stuck in the tubes connecting your kidney to your bladder. Small stones are usually passed out of the body in urine but larger ones can be located using ultrasound and then broken into small pieces by high intensity ultrasound pulses. This technique is called Extracorporeal Shock Wave Lithotripsy.

29.2 X-rays

29.2.1 Overview of Medical X-Rays

X-rays were discovered in 1895 by Wilhelm Roentgen and he won the very first Nobel Prize for physics in 1901. X-rays are a form of high-frequency, short-wavelength electromagnetic radiation emitted when electrons moving at high speed crash into a target and stop suddenly. The radiation is ionizing and highly penetrating and Roentgen took the first X-ray photograph, of his wife's hand, soon after his discovery.

X-rays are now used routinely in medicine to create images of the inside of the human body. In conventional X-ray radiography the X-rays pass through the body and are absorbed to differing extents by the tissues through which they pass, creating a shadow image that can be captured on film or by arrays of detectors. A more sophisticated technique, called computed tomography (CT), involves rotating the X-ray source and detectors around the body to create a 3D image.

X-rays are ionizing radiation so they can damage tissues and doctors must always balance risk against benefit when deciding whether to use them. The risks depend on the wavelength and intensity of the X-rays, the

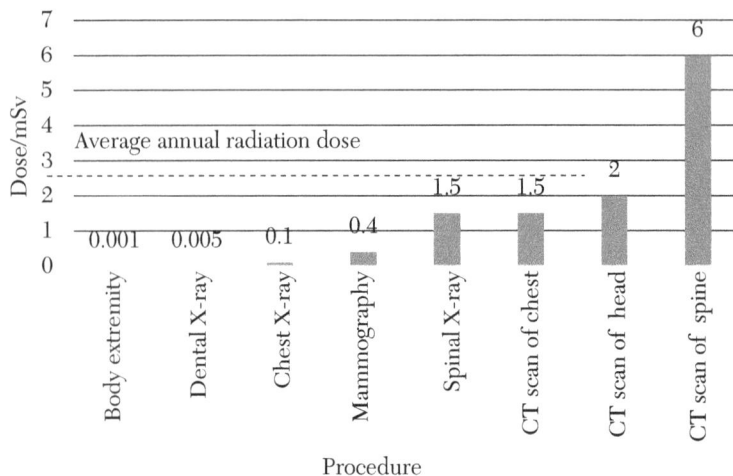

duration of the procedure and the tissues being exposed. To assess the risk, the X-ray dose is compared to the annual radiation dose from natural background sources. The chart below shows typical doses from different types of X-ray procedure.

29.2.2 Generating X-Rays

To generate X-rays an electron beam is accelerated through a potential difference of between 30 and 150 kV and directed onto a target cathode made of tungsten. As the electrons stop some of their kinetic energy is transferred to X-ray photons. The radiation is called "bremsstrahlung," which means "braking radiation." The rest of the incident energy (99% or more) is transferred to heat in the target. Removal of heat from the target is a serious problem in X-ray tube design. Modern CT machines are rated at up to 100 kW and the focal spot on the anode can reach temperatures in excess of 2000°C. A high melting point target, e.g., tungsten, is mounted in a material with high thermal conductivity, e.g. copper and a coolant is pumped through the copper. The shape of the anode is designed to produce a narrow beam of X-rays from a fairly broad beam of electrons.

There are two main adjustments to the tube—the electron current and the accelerating voltage. Increasing the current increases the number of electrons per second striking the target and this increases the number of X-ray photons emitted per second. Increasing the voltage increases the

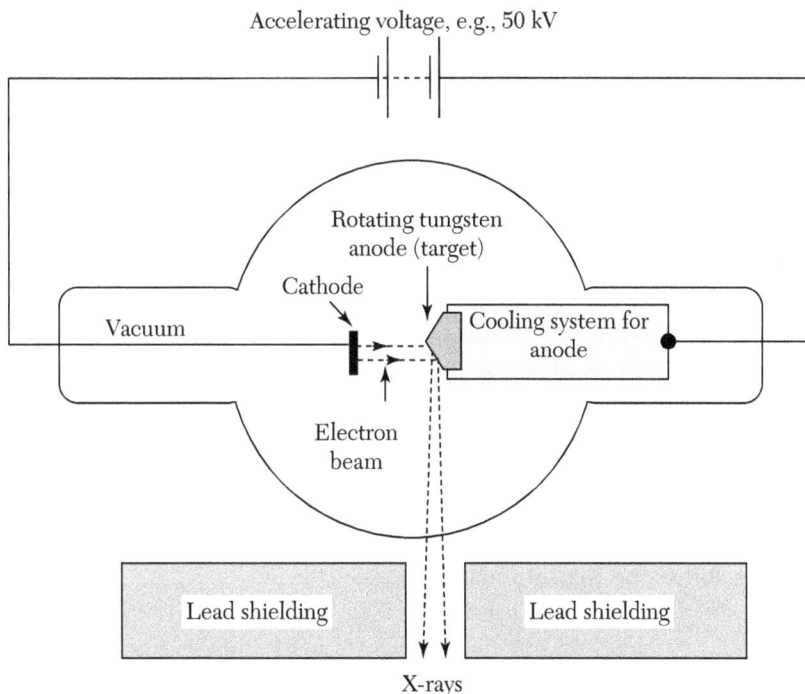

energy of the electrons and the maximum frequency of the X-ray photons. It also increases the number of photons emitted.

The spectrum of X-rays produced is continuous, but also contains some sharp emission lines that are characteristic of the target element. These lines are created when electrons strike atoms in the target and eject an electron from an inner orbits (e.g. K-shell or L-shell). The vacancy is then filled by electrons from higher orbits cascading down and emitting photons as they do so.

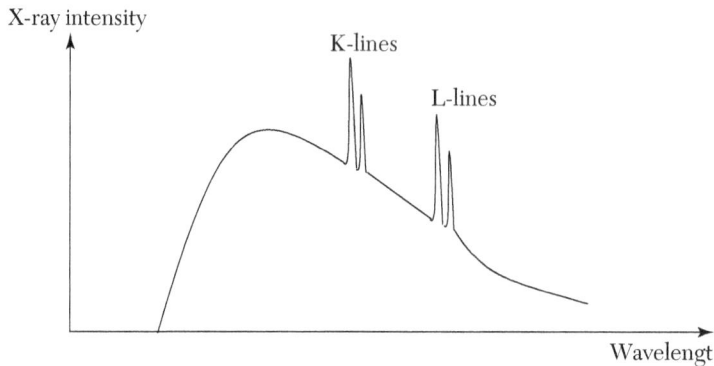

The largest energy jumps are for electrons dropping into the innermost shell, the K shell, and these correspond to the shortest wavelength spectral lines. The energy jumps into the K and L shells correspond to X-ray photon energies.

The short wavelength cut off corresponds to all the energy of one incident electron being transferred to a single X-ray photon. The rest of the continuous spectrum corresponds to more complex interactions and multiple collisions.

The higher frequency, shorter wavelength X-rays are the most penetrating. These are called "hard X-rays" and are the ones needed for image formation. Longer wavelength, "soft X-rays" are usually filtered out because they do not contribute to the image but do increase the radiation dose. The absorption of X-rays by tissues attenuates the beam. The amount of attenuation increases with the density of the tissue, so bones absorb X-rays more strongly than the surrounding soft tissues. This is what creates the contrast in an X-ray image and X-rays are particularly good for imaging bones and bone damage. Soft tissues do not create much contrast so often a contrast medium is injected prior to the X-ray. CT scans are better at imaging soft tissue than standard X-rays.

29.2.3 Attenuation of X-Rays in Matter

There are two mechanisms by which X-rays are absorbed in matter:

▦ The photoelectric effect—where an incident X-ray transfers all of its energy to an electron and this electron undergoes multiple collisions. The probability of photoelectric scattering drops rapidly with X-ray energy. The photoelectric effect is the mechanism that determines contrast in an X-ray image.

▦ Compton scattering—the incident X-ray transfers some energy to an electron but scatters off with lower energy (longer wavelength) and interacts with other electrons. The probability of Compton scattering falls only slowly with increasing X-ray energy. Compton scattering is the mechanism that determines the noise in an X-ray image.

The intensity of an X-ray beam falls off exponentially as it passes through matter according to the Beer-Lambert law:

$$I_{out} = I_{in}e^{-\mu x}$$

where μ is the linear absorption coefficient of the material (m^{-1}).

This has two important consequences. A certain minimum intensity is required to produce an image so the exposure time will depend on the size of the patient and larger patients will need longer exposures and will receive a larger dose. Also, since the intensity falls exponentially through the patient, tissues near the top surface, where the X-rays enter the body, receive a larger dose than tissues near the X-ray detector. Most of the dose is absorbed close to the skin at the top surface of the patient, and this is where there is most risk of tissue damage.

Different materials have different absorption coefficients and these differences are responsible for creating contrast in X-ray images. The differences become larger at lower X-ray energies so these are preferable for high contrast imaging. The drawback is that the low energy X-rays are absorbed more strongly than high energy X-rays so the dose increases.

Absorption coefficients for soft tissues do not vary much so contrast media are often used to create X-ray images of blood vessels, the fallopian tube, the urinary tract, the digestive system, etc. These are based on iodine or barium compounds that absorb X-rays strongly.

29.2.4 Creating X-Ray Images

A simple medical X-ray imaging machine uses a film placed underneath the patient to detect the X-rays.

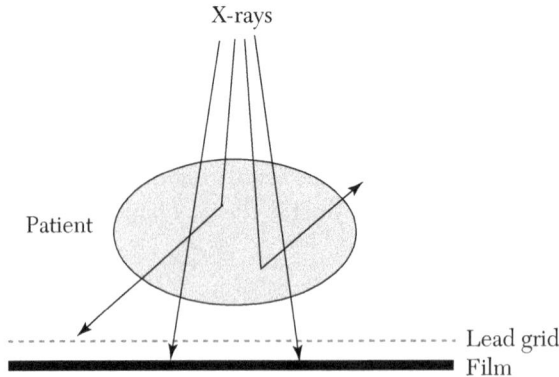

Some of the X-rays passing through the patient are scattered off-axis so a lead grid is slowly moved between the patient and the film during the exposure. This eliminates off-axis X-rays and increases the signal to noise ratio for the image.

An intensifying screen can be used to increase the number of light photons created from each X-ray photon. This consists of thin layers of fluorescent material placed in front of the film. The fluorescent layers absorb the X-rays and emit visible photons. The arrangement is housed in a cassette that is placed under the patient for exposure.

An X-ray filter (usually a metal plate) is placed between the X-ray source and the patient to filter out the low energy (long wavelength) X-ray photons. This reduces the total dose given to the patient but does not affect the intensity of the image because low energy photons would have been absorbed in the body.

An X-ray image intensifier can also be used to increase the brightness of the image. Incident X-rays strike a phosphor screen that converts the

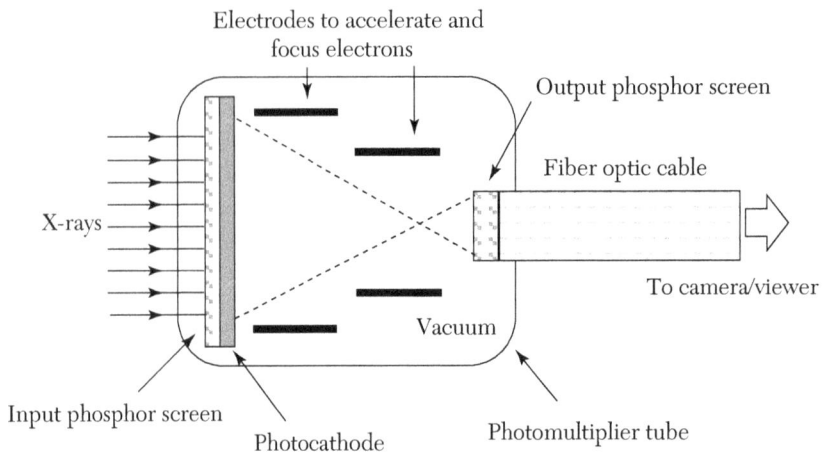

X-rays to photons of visible light. These strike a photocathode which emits electrons. The electrons are accelerated and focused in a **photomultiplier** tube and converted back to photons when they strike another phosphor screen.

CT scans rotate an X-ray source and collimator around the patient to create detailed images of slices of the body. An array of fixed detectors surround the patient.

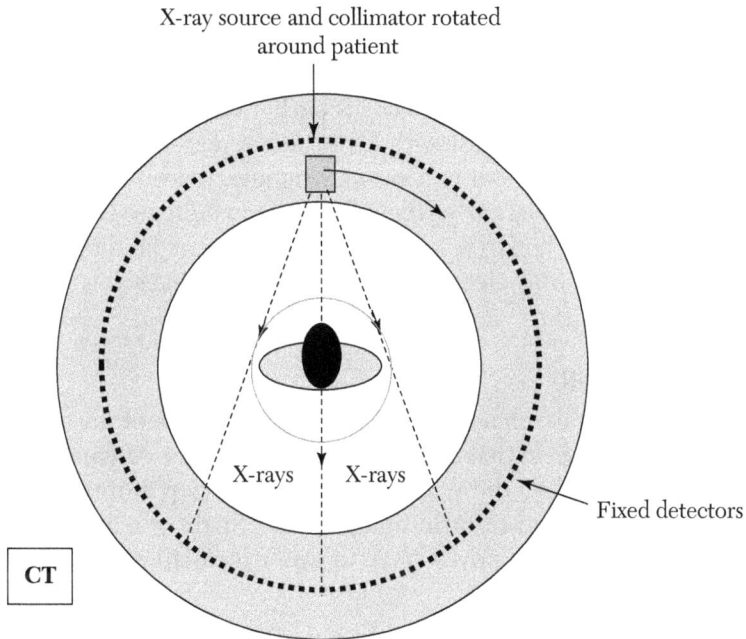

X-ray source and collimator rotated around patient

X-rays | X-rays

Fixed detectors

CT

Whereas 2D images are composed of 2-dimensional picture elements or "pixels,"3D images are composed of volume elements or "**voxels**." Each

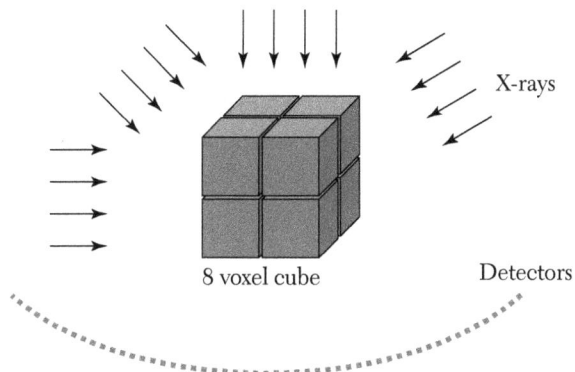

X-rays

8 voxel cube

Detectors

voxel is assigned a value that represents how strongly it attenuates the X-rays that pass through it. The voxel values are calculated by a computer algorithm based on the attenuation of X-rays passing through from different directions (hence the need to rotate the X-ray source around the patient).

The X-ray attenuation across different directions is used to determine the value for each individual voxel.

29.3 Magnetic Resonance Imaging (MRI)

29.3.1 Overview of MRI

Magnetic resonance imaging (MRI) is particularly useful for imaging soft tissues that have low contrast with X-ray techniques, e.g., brain imaging. It involves placing the patient in a strong magnetic field but does not involve the use of ionizing radiation, so the risks to the patient are very low and long exposures are possible. However, the strong magnetic field must be created using superconducting electromagnets and this makes this an expensive procedure.

29.3.2 Physics of MRI

Nuclei are positively charged and when the nucleus of an atom spins it creates a magnetic field that is like that of a bar magnet. Magnetic resonance techniques manipulate the nuclear spin of hydrogen atoms to generate a signal that can be used to form images. Since hydrogen atoms are spread throughout the human body, mainly in water molecules, they are ideal for imaging all parts of the body.

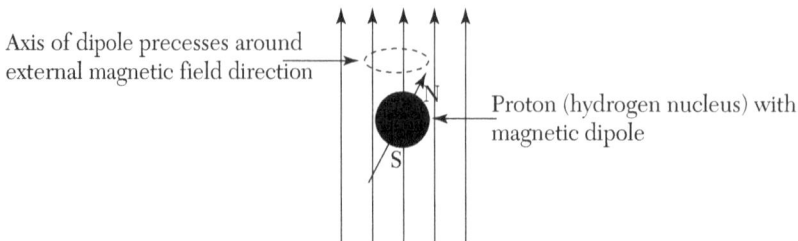

Axis of dipole precesses around external magnetic field direction

Proton (hydrogen nucleus) with magnetic dipole

When an external magnetic field is applied the spinning nuclei precess around it at a fixed frequency called the Larmor frequency:

$$f = \frac{\gamma B_0}{2\pi}$$

where γ is the gyromagnetic ratio.

The gyromagnetic ratio for a hydrogen nucleus (a proton) is 42.6 MHzT^{-1}.

Clinical MRI scanners use magnetic field strengths in the range 0.2–3.0 T and research scanners use up to 11 T. These give Larmor frequencies that correspond to radio waves. An additional weaker variable magnetic field is applied along the axis of the patient so that the Larmor frequency varies with position. Pulses of radio waves corresponding to the Larmor frequency across a particular slice of the body are then transmitted through the body. This disturbs the precessing nuclei in that slice so that they create a rotating magnetic field at the same frequency. It is this field that is detected using coils placed outside the body. The rate at which this field decays depends on the type of tissue surrounding the nuclei so can be used to distinguish different types of soft tissue and to achieve much higher contrast than X-ray techniques. By varying the frequency of the radio waves nuclei in different locations resonate and an image can be built up.

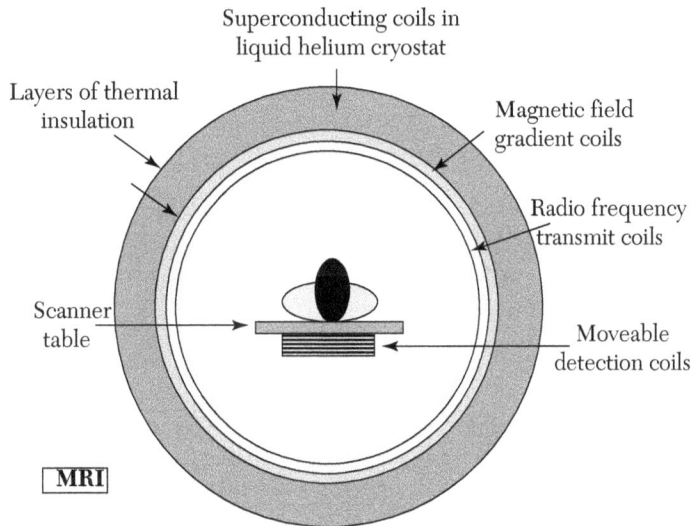

29.4 Radioactive Tracers

29.4.1 Overview of the Use of Radioactive Tracers

A radioactive tracer is an element or compound containing a radioactive atom that is introduced into the body so that the emitted radiation can be detected outside the body. If the tracer is linked to an element or compound used in a particular metabolic process the doctors can monitor the build-up and discharge of this compound in particular organs. The thyroid gland in a human uses iodine to produce hormones, so introducing some radioactive iodine as a tracer can help a doctor to diagnose a faulty thyroid gland. The meta-stable gamma-emitter, technetium-99m can be added to a number of important biological compounds and is used to diagnose a wide range of illnesses, including kidney problems.

Radioactive tracers are usually gamma-emitters because gamma-rays are the most penetrating and least ionizing form of radiation and can be detected, using a gamma-camera, outside the body. Sources with a half-life of a few hours are ideal because they remain in the body long enough for the procedure to be carried out but fall to safe levels relatively soon afterwards.

X-rays are used to image static structures inside the body but radioactive tracers are used mainly to monitor functions.

29.4.2 Gamma-Camera

Gamma-rays are detected when they fall onto a scintillator containing sodium iodide crystals that emit photons of visible light when they absorb gamma-ray photons. The scintillator is mounted in front of an array of photomultipliers that emit electrons and amplify the signal. A computer processes their electrical output to produce the final image. A lead grid is placed in front of the scintillator to reject off-axis gamma-rays and create a sharper image.

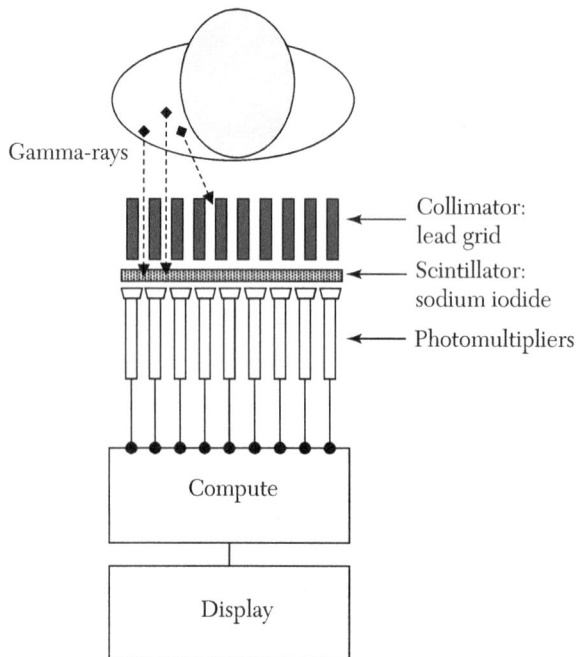

29.5 Positron Emission Tomography (PET scans)

Positron emission tomography (PET) detects gamma-rays emitted when a positron annihilates with an electron inside the body. The positrons result from the beta-plus decay of a radioactive tracer injected into the body prior to the scan. This technique is particularly useful for investigating how different organs are functioning or to monitor the extent, development and response to treatment of cancers. It can be combined with CT scans or MRI scans to produce detailed 3D images of the body.

29.5.1 Physics of PET Scans

The radioactive tracers used for PET scans are short half-life beta-plus emitters such as fluorine-18 (half-life about 110 minutes), carbon-11 (half-life about 20 minutes), or oxygen-15 (half-life about 2 minutes). These are attached to compounds that the body uses for particular biological pathways, e.g., glucose. The uptake of the compound can be then monitored from outside the body by detecting the pairs of 511 keV gamma-rays emitted as the beta-plus particles, positrons, annihilate with electrons in the surrounding tissue.

The equations below summarize the process:

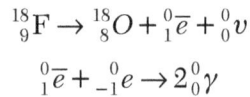

$$\,^{18}_{9}\text{F} \rightarrow \,^{18}_{8}O + \,^{0}_{1}\overline{e} + \,^{0}_{0}v$$

$$\,^{0}_{1}\overline{e} + \,^{0}_{-1}e \rightarrow 2\,^{0}_{0}\gamma$$

The positron emitted in this decay only travels a short distance (<1 mm) through the tissue before meeting an electron and annihilating. The annihilation of the electron-positron pair results in emission of a pair of gamma-rays that travel in opposite directions (a pair must be emitted in order to conserve linear momentum). The position of the annihilation along the line determined by the two gamma-rays is determined from the time delay in arrival at detectors on either side of the patient.

The detectors only respond to near simultaneous pairs of photon arrivals (within about 10 ns of each other) and then measure the small additional time delays for each pair.

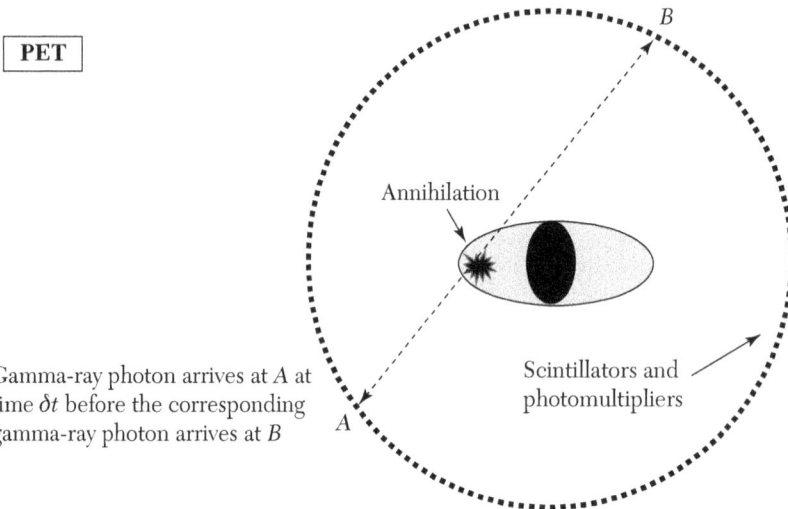

PET

Annihilation

Gamma-ray photon arrives at *A* at time δt before the corresponding gamma-ray photon arrives at *B*

Scintillators and photomultipliers

For a time delay δt the annihilation event must have been at a position that is $c\delta t/2$ closer to A than the center of the chord AB, i.e. a distance AB/2 $-c\delta t/2$ from A and AB/2 $+ c\delta t/2$ from B.

In practice the two gamma-rays emitted from an annihilation event are not emitted at exactly 180° so this introduces an uncertainty into the position of the chord. In addition to this the detector can only resolve events that are more than about 0.50 ns apart so this introduces an uncertainty into the positon along the chord. The image quality and resolution improves as more events are detected (signal to noise ratio falls) and resolutions of about 1-2 mm are possible with clinical scanners. This is not as good as a CT image but PET scans can be used to investigate a very wide range of metabolic pathways and when used alongside CT or MRI scans information about both structure and function can be combined.

The injection of a radioactive tracer means that the patient remains radioactive for a short time after the procedure. For a typical PET scan involving fluorine-18 the total activity injected is about 370 MBq. The patient will absorbs a radiation dose equivalent to that of a full body CT scan (about 7 mSv). If the PET scan is combined with a CT scan the total dose will be the sum of doses from the two procedures.

29.6   Exercises

1. (a) State and explain the risks associated with X-ray imaging of the human body.

When an X-ray tube is used to produce a photographic image of a patient an aluminum plate is placed between the patient and the X-ray tube.

(b) Explain how this can reduce the radiation dose without affecting the intensity of the X-ray image that is formed.

2. Some tissue injuries can be imaged equally effectively using ultrasound or an MRI scan.

(a) Compare the physical principles used in the two processes.
(b) If you were a doctor which would you recommend and why?

3. The diagram below shows a typical spectrum from a medical X-ray machine.

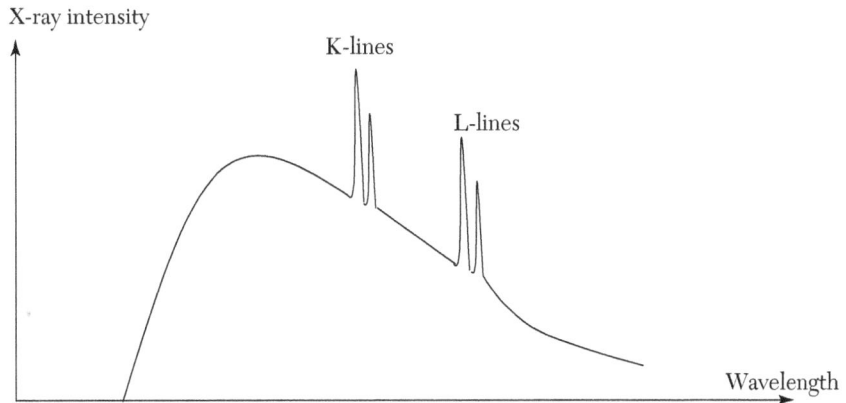

(a) Explain how the continuous spectrum arises and why it has a short wavelength cut-off.
(b) Calculate the cut-off wavelength for a 40 kV X-ray machine.
(c) Explain how the line spectra come about and account for the difference between the K and L lines.

4. State and explain four factors that must be considered when selecting a suitable radioisotope to use as a tracer in the human body.

5. (a) Explain why the annihilation of an electron and a positron during a PET scan is likely to result in two gamma-rays of the same wavelength emitted in opposite directions.

(b) Calculate the wavelength of these gamma-rays.

A

ESTIMATIONS AND FERMI QUESTIONS

A.0   Fermi and the Trinity Test

The first atomic bomb was tested at the Trinity site in New Mexico in 1945. The physicists and engineers involved in its design and construction placed bets on the energy that would be released by the explosion. Enrico Fermi, the Italian physicist who had designed and tested the first controlled nuclear chain reaction, observed the test from a safe distance of about 12 km. The explosion created a shock wave in the atmosphere that spread out in all directions. Fermi dropped several pieces of paper as the shock wave reached him and observed that they were displaced horizontally about 2.5 m by the disturbance. From this information (and his knowledge of physics!) he estimated the yield of the bomb to be about 10 kT (i.e., equivalent to the explosion of 10 kT of TNT, a conventional chemical high explosive). Later, more detailed calculations showed that the actual yield was closer to 20 kT.

It seems remarkable that Fermi was able to do this, based on such minimal information and a very simple experiment, but a good understanding of the underlying physics provides the tools to make useful estimates of a very wide range of physical quantities and the skills used in making these estimations can also be employed to non-physical problems. Such methods are invaluable because real-world problems are rarely fully defined and it is important to have a ball-park figure in mind before committing to a project or experiment. It is also useful to estimate expected results so that you can tell whether the actual results from an experiment or calculation are reasonable.

Fermi built up quite a reputation for being able to estimate values based on minimal information and these types of problem are often called "Fermi problems" for that reason.

I do not know what method Fermi used to make his estimation but there are usually many ways to solve a Fermi problem and here is a (very simple) approach to this one.

Let's assume that the energy released by the explosion pushes the atmosphere back so that a hemisphere of air with a radius equal to Fermi's distance from ground zero (the position of the explosion) is displaced outwards by 2.5 m. Work must be done against the atmospheric pressure further out, so the energy transfer can be calculated using force times distance:

$$\text{Surface area of hemisphere} = 2\pi r^2$$

$$\text{Outward force } F = 2\pi r^2 \times p$$

where p is the atmospheric pressure (about 10^5 Pa).

$$\text{Work done} = F\delta x$$

Where δx is the outward displacement of the air.

$$\text{Work done} = 2\pi \times (12{,}000)^2 \times 10^5 \times 2.5 = 2 \times 10^{14} \text{ J}$$

$$1 \text{ kT} = 4.2 \times 10^{12} \text{ J}$$

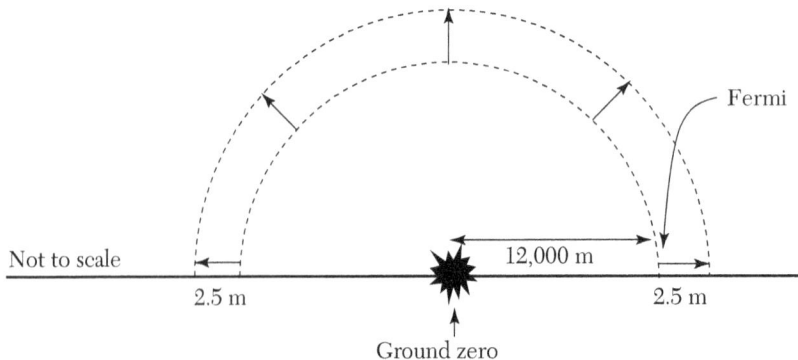

This gives a result of about 50 KT, much greater than Fermi's estimate and about 2.5 times the actual yield of the Trinity test. However, it is the correct order of magnitude, which is pleasing given the incredibly simple model used to make the estimate! Fermi would have used a more sophisticated model, taking into account the actual pressure differences in the shock wave and the proportion of the input energy that went into it. His estimate of

10 kT was impressive, but did not win the bet. Isodor Rabi was the winner with an estimated yield of 18 kT. We do not know how he did this (or maybe he just got lucky).

A.1 Making Estimations

A good estimation is usually based on a simplification of the relevant physics and reasonable estimates of the size of relevant parameters. In addition to this we usually need to make some assumptions to simplify the actual calculations. In the example above we reduced a complex process of energy transfer to the expansion of a gas at constant pressure so that we could use the equation $W = p\delta V = F\delta x$. We estimated the atmospheric pressure and we assumed that the pressure and displacement was constant over the hemisphere. We also assumed that all the energy released by the explosion was used to move this layer of air back. You will also notice that we tended to round values off quite severely, often to one significant figure (e.g., distance to ground zero and atmospheric pressure). This makes sense because we are not dealing with definite known values and we cannot justify great precision. Many estimates will produce a value that can only be quoted to one significant figure and that has an uncertainty of over 100%! In this section, we will illustrate how to make estimates by using a few examples.

A.1.1 How Many Air Molecules in the Earth's Atmosphere?

Method 1

Atmospheric pressure is due to weight of air in atmosphere, so we can use the equation for fluid pressure at depth: $p = \rho gh$ to work out the effective depth of the atmosphere if it was of uniform density.

$$\rho gh = 10^5$$

$$h = \frac{10^5}{1 \times 10} = 10^4 \text{ m}$$

This is much less than the radius of the Earth so the atmosphere can be treated as a thin layer (like a carpet with an area equal to the surface area of the Earth). The volume of this layer can then be used to find how many moles of gas are present and this can be used to find the number of molecules. The atmosphere is equivalent to a 10 km deep uniform layer. The fact that the density of the atmosphere varies with height does not affect this estimate—it would have the same weight and exert the same pressure at the surface for any pattern of density variation because it is only the total weight of all the molecules that is responsible for the surface pressure.

The surface area of the Earth is $A = 4\pi \times (6.4 \times 10^6)^2 = 5.15 \times 10^{14} \, m^2$

The volume of air at atmospheric temperature and pressure that would exert the same pressure at sea level as the Earth's atmosphere is:

$$V = 5.15 \times 10^{14} \times 10^4 \, m^3$$

1 mole of an ideal gas at atmospheric temperature and pressure occupies a volume of 24 liters or 0.024 m³, so the number of moles of air molecules in the atmosphere is:

$$n = \frac{5.15 \times 10^{14} \times 10^4 \, m^3}{0.024} = 2.1 \times 10^{20} \, moles$$

1 mole contains 6.02×10^{23} molecules so the total number of air molecules in the atmosphere is:

$$N = 2.1 \times 10^{20} \times 6.02 \times 10^{23} = 1.3 \times 10^{44} \text{ molecules}$$

There are about 10⁴⁴ air molecules in the Earth's atmosphere.

Method 2

This method has some similarities to the first but uses the total mass of the atmosphere and the mass of an "air molecule." The total weight of the atmosphere is equal to the atmospheric force exerted on the entire surface of the Earth:

$$F = pA = 10^5 \times 4 \times \pi \times (6.4 \times 10^6)^2 = 5.15 \times 10^{19} \, N$$

The mass of the atmosphere is:

$$m = \frac{F}{g} = 5.25 \times 10^{18} \, kg$$

Air consists mainly of oxygen (molar mass = 32g) and nitrogen (molar mass = 28 g) so an "air molecule" is taken to have a molar mass of 30 g.

Number of moles of air molecules in the atmosphere:

$$n = \frac{5.25 \times 10^{18}}{0.030} = 1.75 \times 10^{20} \, moles$$

The number of air molecules in the atmosphere:

$$N = 1.75 \times 10^{20} \times 6.02 \times 10^{23} = 1.05 \times 10^{44} \, molecules$$

There are about 1044 air molecules in the Earth's atmosphere.

This is (not surprisingly) consistent with our first method.

A.1.2   What is the Minimum Area for a Parachute?

First we must interpret the question. Why is there a minimum area? If the parachute is too small the terminal velocity of the parachutist will be too high and they will be injured or killed when they hit the ground. The first thing to do therefore is to estimate a maximum safe landing speed. It might be tempting to guess what this is but it is better to make a reasoned estimate based on sensible assumptions. One way to do this would be to think about the highest object you could jump, from and still expect to land safely—perhaps this is about 2 m high. This can then be used to calculate a maximum safe landing speed.

$$v^2 = u^2 + 2gh = 0 + 2 \times 9.8 \times 2 = 39.2$$

$$v = 6.3 \, \text{ms}^{-1}$$

So a maximum safe landing speed is about 6 ms^{-1}.

Now we need a physical model of the falling parachutist. At terminal velocity the weight of parachutist and parachute is equal and opposite to the total drag force acting upwards on the parachute. Where does this drag force come from? It comes from the collision of the parachute canopy with air molecules as it moves downwards—the canopy exerts a downward force on the molecules and the molecules, by Newton's third law, exert an equal upward force on the canopy. Using Newton's second law, the magnitude of these forces must equal the rate of change of momentum of the air molecules. This is what we will estimate next.

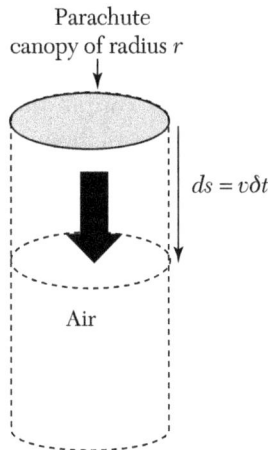

Parachute canopy of radius r

$ds = v\delta t$

Air

During a time δt the canopy moves down through a distance $v\delta t$ and sweeps through a volume $\pi r^2 v\delta t$ of air. The mass of air in this volume is equal to $\rho \pi r^2 v\delta t$ where ρ is the density of the air. If we assume that all of this air must be accelerated up to speed v as the canopy passes we can work out the rate of change of momentum.

$$\text{Rate of change of momentum} = \frac{\text{mass of air swept out in } \delta t \times \text{speed}}{\delta t}$$

$$= \frac{r2v^2t}{t} = r2v^2$$

This is equal to the drag force on the parachute and, at terminal velocity, the weight of the parachute and parachutist:

$$r2v^2 = mg$$

This equation can be rearranged to give an expression for r, the radius of the parachute:

$$r = \frac{mg}{2v^2}$$

Now we need to input some reasonable values: mass of parachutist and equipment, m = 100 kg; density of air, ρ = 1.2 kgm⁻³; maximum speed, v = 6 ms⁻¹

This gives a minimum safe radius of:

$$r = \frac{100 \times 9.8}{1.2 \times 2 \times 6^2} = 7.2\,\text{m}$$

This is probably an over estimate but is certainly of the correct order of magnitude. It corresponds to a minimum area of about 160 m².

The approach above was based on momentum but different estimates result if we consider kinetic energy instead. To make the estimate more realistic we should consider the drag coefficient for a parachute of a particular shape. Nonetheless, our simple method has produced a useful equation to begin an investigation of how parachute size and terminal velocity are related.

A.2 Useful Values

You will have noticed that values such as atmospheric pressure and the density of air at a.t.p. featured in the estimations above. As a physicist it is useful to have some values at your fingertips because they are important in so many different situations. Here are some numbers that are worth remembering (they are not precise—these are for use in estimates):

Speed of light : 3.00×10^8 ms⁻¹

Speed of sound (in air at sea level) : 330 ms⁻¹

Density of air at a.t.p.	:	1.2 kgm^{-3}
Density of water	:	1000 kgm^{-3}
Density of aluminum	:	2700 kgm^{-3}
Density of steel	:	8000 kgm^{-3}
Density of mercury	:	$13{,}600 \text{ kgm}^{-3}$
Specific heat capacity of water	:	$4200 \text{ Jkg}^{-1}{}^{\circ}\text{C}^{-1}$
Atmospheric pressure	:	10^5 Pa
Molar volume of an ideal gas at a.t.p.	:	0.024 m^3 (24 liters)
Wavelength of visible light	:	~500 nm
Wavelength range of visible light	:	400–700 nm
Audible frequency range for humans	:	20 Hz–20 kHz
Speed of ultrasound in water or tissue	:	$\sim1500 \text{ ms}^{-1}$
Size of an atom	:	10^{-10} m
Size of a nucleus	:	10^{-15} m
Avogadro number	:	$6 \times 10^{23} \text{mol}^{-1}$
Mass of a proton	:	$1.7 \times 10^{-27} \text{ kg}$
Mass of an electron	:	$9.1 \times 10^{-31} \text{ kg}$
Charge on an electron	:	$-1.6 \times 10^{-19} \text{ C}$
Radius of Earth	:	6400 km
Mass of Earth	:	$6 \times 10^{24} \text{ kg}$
Mass of Sun	:	$2 \times 10^{30} \text{ kg}$
Distance to Moon	:	400,000 km (~1.3 light seconds)
Distance to Sun	:	$1.5 \times 10^{11} \text{ m}$ (8 light minutes)
Age of the universe	:	13.7 billion years
Number of stars in the Milky Way	:	400 billion (4×10^{11})

You can add to this list. It is also helpful to know the values of several physical constants and be able to recall the important equations!

A.3   Fermi Questions

Whilst there is no clear distinction between a simple estimation and a fermi question the latter tend, on first reading at least, to seem rather abstract. Perhaps the most famous of these is:

"How many piano tuners in Chicago?"

This seems to have nothing to do with physics and yet questions like this have been used in selection interviews for places to read the subject at top Universities. Why? The reason is simple. A good physicist has to be able to use her knowledge and skills to solve problems in unfamiliar contexts. She must deconstruct the question and then tackle it logically, making reasonable estimates of quantities she will need to feed into a calculation. Look at the question above and ask yourself what information you might need to solve it. Here are a few suggestions:

- the population of Chicago (a)

- the fraction of the population that have a piano (b)

- the number of times per year on average each piano is tuned (c)

- the time taken to tune a piano (d)

- the average number of hours worked per day by a piano tuner (e)

- the number of days per year that a piano tuner works (f)

  there may well also be other things to consider or estimate, but nothing in this list should cause too much of a problem or give too wild a result. Once these values have been estimated they can be fed into a calculation like the one below:

- Number of pianos to be tuned in 1year:abc

- Number of pianos tuned by one piano tuner per year:fe/d

- Number of piano tuners in Chicago:abcd/fe

Try it! Which estimates are you most confident about and which have largest uncertainty? How would varying these parameters affect the final estimate? Can you put an upper and lower limit on the number of piano tuners in Chicago? Is there any way to test your estimate?

A.4   Drake Equation

In 1961, Dr. Frank Drake posed a Fermi question:

"How many advanced civilizations exist in our galaxy?"

Drake approached this problem in the same way that we approached the problem of piano tuners in Chicago. He considered different factors that might affect the number of advanced civilizations and then constructed a formula to determine that number. This formula is known as the **Drake equation**. However, the factors he identified are much harder to estimate than those in our problem so the Drake equation is used to discuss possibilities rather than to provide a ball park answer. Adjusting the factors can lead to a prediction that the galaxy is teeming with intelligent life or that we are alone in it!

As we discover more about the Universe we are gradually able to fine tune the values that we input to the equation. For example, one of the factors is the number of stars in our galaxy that have planets orbiting them. In Drake's time the only star known to have a planetary system was our Sun. Now we know that a very large number of extrasolar planets exist and it might be the case that most stars have planetary systems. We have also discovered some "Earth-like" planets orbiting other stars. This increases another factor in the equation making the likely number higher. However, the uncertainties in other factors, and especially in those linked to the emergence and evolution of life, are huge.

$$N = Rf_p\, n_e f_l f_i f_c\, L$$

where

- N = number of advanced civilizations whose EM communications are detectable

- R = rate of formation of suitable stars (stars such as our Sun)

- f_p = fraction of those stars with a planetary system

- n_e = number of Earth-like worlds per planetary system

- f_l = fraction of those Earth-like planets where life actually develops

- f_i = fraction of planets where life develops that evolve intelligent life

- f_c = fraction of planets where intelligent life develops that develop detectable communications

- L = "lifetime" of communicating civilizations

Some of these factors are reasonably well known—e.g., R is thought to be about 10 new stars per year in the Milky Way and f_p is probably between 0.5 and 0.8. L is usually assumed to be of the order of human written history, around 10,000 years, but that's a rather arbitrary choice. The others are harder to pin down. Make estimates and see what these imply for N. This should give you a feel for how this equation can promote arguments between astrobiologists!

A.5 Estimates and Fermi Questions

Here are some problems to try. There are no definite correct answers but they are not guesses either! Start from what you know or can reasonably estimate and then use physical and logical principles to work your way to an answer. Once you have done this ask yourself whether it seems reasonable.

1. How many atoms in a golf ball?

2. How far can a car go before it rubs off a layer of rubber one molecule thick?

3. How many molecules from Isaac Newton's last breath do we inhale when we breathe in?

4. What area of solar panels is needed to supply all of Britain's energy requirements?

5. How many trucks would be needed to cart away all the rock in Mount Everest?

6. How long would it take to fill an Olympic swimming pool from a kitchen tap?

7. How many photons enter your eye per second on a sunny day?

8. How many times more expensive is electrical energy from an AA battery than from the mains electrical supply?

9. If all the atoms in an elephant were put in a line, how long would the line be?

10. How many photons are emitted by a camera flashgun?

11. What is the speed of the wing tip of a bee?

12. How many leaves on a mature oak tree?

13. What is the drag force on a large truck travelling at speed on a motorway?

14. What is the power output of a racing cyclist?

15. What is the greatest distance at which the human eye can resolve car headlamps?

16. Will relativistic effects be important for an electron in the ground state of a hydrogen atom?

17. What is the spring constant of a car's suspension system?

18. How many hairs on your head?

19. How much energy can be supplied by a car battery?

20. What is the temperature at the center of the Sun?

B

EXPERIMENTAL INVESTIGATIONS

B.0   Introduction: Nature of Science

The defining characteristic of science is that it is based on evidence and that evidence comes from experiment. You might have a wonderful theory about the nature of the universe, but if it is not testable it is not a scientific theory. If it is a scientific theory then it can be used to make predictions that can be tested by experiment. If repeated experiments agree with these predictions then your theory is a good one. However, if the results of experiments do not agree with your theory, and the experiments have been carried out carefully and by reputable scientists and repeated by different groups, then the theory might have to be abandoned or modified. One interesting consequence of this definition of science is that even if we have a brilliant theory that agrees with all the evidence we have ever collected, we still do not know that it is the final theory—there might be other tests that we have not thought about or conditions we could not create in our tests and the theory might then fail. For example, the theory that atoms are the smallest units of matter is a good one but failed when J.J.Thompson showed that all atoms contain electrons. A good theory is a survivor but we cannot ever be absolutely certain it will not fail and be replaced by a better or more comprehensive theory at a later time.

The philosopher Karl Popper suggested that the criterion that defines science is that it is in principle "**falsifiable**." When Einstein's general theory of relativity suggested that the presence of matter should curve space and deflect light, the astronomer Arthur Eddington set out to detect and measure this deflection. His measurements, during a total eclipse of the Sun in 1919, showed that light *is* deflected by matter and the measured deflection was consistent with Einstein's predictions. This propelled Einstein to international fame and helped secure the general theory of relativity. Had Eddington failed to detect the deflection, or shown it to be much larger or smaller than predicted, then Einstein's theory would have been thrown into doubt. Of course, the experiment too would have been critically assessed to make sure it was capable of detecting the predicted effects and it would have been repeated, but a continued failure to detect the deflection of starlight by matter would have undermined general relativity.

The success of general relativity was at the expense of the existing theory, Newton's theory of gravity, so in a sense Eddington's experiment falsified Newton's theory. Does that mean that, from 1919 on we have abandoned Newtonian gravitation? If so at least one chapter of this book is redundant! No: whilst we know that the Newtonian theory is not the most comprehensive theory of gravity, it is much simpler to use than Einstein's theory and it works very well for calculating planetary orbits or spacecraft trajectories. However, if we want to explain gravitational lenses or the behavior of space and time close to a black hole, we must use Einstein's full theory. On a fundamental level our best theory of gravity is Einstein's but on a practical level Newton's theory is precise enough for the majority of applications. The reason for this is that Einstein's equations reduce to Newton's when the gravitational field is weak and the curvature of space-time is small. So, continuing to use Newton's equations does not contradict our knowledge that general relativity gives a more complete description of space and time.

B.1 Carrying Out an Experiment

An experiment is an attempt to gather evidence that can be used to answer a question. There are several reasons why you might do this:

- To investigate the relationship between different parameters—e.g., how does mass and resultant force affect the acceleration of a dynamics vehicle?

- To make a careful measurement of a physical quantity—e.g., what is the resistivity of nichrome?

▨ To determine the parameters that affect an important quantity—e.g., how does changing the length, mass, and amplitude of a simple pendulum affect its period?

▨ To test a hypothesis—e.g., does painting a coffee cup silver keep coffee hot for a longer time?

To carry out and write up an experiment effectively you will need to:

▨ Have a clear aim.

▨ Consider relevant underlying physical principles (this might involve research).

▨ Identify relevant parameters and decide which will be varied and which will be held constant.

▨ Select appropriate measuring instruments bearing in mind their range and precision.

▨ Plan an experimental procedure.

▨ Carry out a risk assessment.

▨ Decide how to record and analyze the data including uncertainties.

▨ Draw a conclusion.

▨ Evaluate the experiment.

▨ List any references you have used.

B.1.1   Variables

To make your experiment a **fair test**, the parameter you are investigating must be affected only by the parameter you are varying. For example, if you are investigating factors affecting the acceleration of a dynamics vehicle you might vary its mass by adding loads to the vehicle. However, if you then pull it with different forces you have varied two parameters, both of which affect its acceleration, so you will not be able to separate the effect of mass from the effect of resultant force. To make this fair you need to keep the resultant force constant while varying the mass and then carry out a separate experiment in which the mass is kept constant and the resultant force is varied. The three parameters, mass, resultant force and acceleration are examples of the three types of variable parameter in all experimental work:

Independent variable: the unique variable that we change (e.g., mass)

Dependent variable: the variable that we are investigating (e.g., acceleration)

Control variable: the variable we keep constant (e.g., resultant force) so that it will not affect the dependent variable

In most experiments there are many control variables that must be kept constant.

B.1.2 Selecting Measuring Equipment

It is important to select measuring equipment that can be used to provide good quality data and that is safe for the selected use. For example, measuring the temperature of a Bunsen flame with a mercury in glass thermometer is dangerous, the glass might melt and release the mercury, and impractical, the range of the thermometer does not go to high enough. A better choice would be a thermocouple or a pyrometer.

You need to consider the precision and range of each measuring instrument.

▨ Precision: smallest scale division

▨ Range: difference between smallest and largest value that can be measured

Even simple measurements of length require careful thought. Four common pieces of laboratory apparatus to measure length are tape measures, rulers, calipers and micrometer screw gauges. The table below shows typical values and uses for each of these.

Instrument	Range	Precision	Example of use
Tape measure	10.00 m	0.01 m (1 cm)	Displacement of a student in the lab
Ruler	1.000 m	0.001 m (1 mm)	Length of a simple pendulum
Vernier caliper	0.0200 m	0.0001 m (0.1 mm)	Thickness of small density blocks
Micrometer screw gauge	0.01000 m	0.00001 m (0.01 mm)	Wire diameter

Selecting appropriate instruments for electrical experiments is particularly important. If you connect a sensitive ammeter into a circuit which carries a large current you will damage the meter so it is important to consider the likely currents before switching on! It is also important to connect the apparatus correctly. An ideal ammeter has zero resistance so if you accidentally connect it in parallel with a component (as if it was a voltmeter) it will short the component and might damage the circuit and meter.

When you are writing up an experiment it is important to list or label the measuring equipment and include its range and precision.

You should also consider whether the measuring equipment itself will affect the measurement. For example, a real ammeter has a small resistance that will reduce the current in the circuit. If the ammeter resistance is comparable to that of the circuit this will be a significant effect. An ideal voltmeter should have infinite resistance but real voltmeters have a large finite resistance. This will only be a problem if the voltmeter is connected across a component which also has a comparably large resistance. A mechanical example is the use of tickertape—the tape drags as it moves through the ticker-timer and this affects the motion that is being measured.

B.1.3 Planning a Procedure

Once you have decided what you are going to measure and how you are going to measure it you must consider three things:

- The **range of measurements** for each variable. Try to make this as wide as possible because a limited range might not reveal the underlying pattern of behavior—all curves look straight if you only consider a small section!

- The **number and spacing of measurements** within the range. Aim for at least five well-spaced readings across the entire range of measurements. A larger number is usually better and if the behavior is non-linear it will be worth taking more readings where it is changing rapidly.

- **Repeating and averaging measurements**. If possible repeat each measurement several times and use an average value. This helps to reduce the effects of random errors and if one of your repeats is very different from the others it can be ignored—this is called an **anomalous** result.

Recording and analyzing data, plotting graphs and dealing with uncertainties is dealt with in detail in Chapters 2 and 3.

B.1.4 Risk Assessments

You should carry out a risk assessment before performing any experiment. Many experiments have low risk but you will only know this if you think about it! Some simple experiments can be dangerous, for example, overstretching a steel spring can result in it flying up and hitting you in the eye. This risk should be identified before attaching loads to the spring and suitable eye protection should be worn. If you are carrying out an experiment that involves the use of radioactive sources you will need to consider the likely total radiation dose you will receive an ensure that this is low enough to justify doing the experiment. There are three stages to a risk assessment:

- Identify the risks to you and others.

- Adapt the experimental procedure to minimize the risk—i.e., take precautions.

- Decide whether the risk is low enough to justify carrying out the experiment.

There are guidelines that must be followed when you work with hazardous chemicals, electricity, radioactive sources, vacuum containers, etc. Your teacher or supervisor should be consulted about these and if you are ever in doubt about the risk of an experimental procedure, do not proceed!

B.1.5 Writing Upon Experiment

An experimental account should address all the points noted in B.1 and could use these headings:

Aim: this can be a simple statement and should not run to more than two or three sentences. State what you are hoping to achieve—e.g., to determine how the current in a thermistor varies with temperature.

Theory: use what you know or have found out through research to propose a hypothesis—e.g., thermal energy can release additional charge carriers and the number is expected to depend on the Boltzmann factor $e^{-\frac{\Delta E}{kT}}$ so the current should be given by an equation of form: $I = Ce^{-\frac{\Delta E}{kT}}$.

Variables: state the independent and dependent variable and all of the important control variables, e.g., dependent variable is current, independent variable is temperature, control variable is potential difference across the thermistor.

Apparatus: you could give a list of instruments, or label them clearly on a diagram showing the experimental setup. Either way it is important to state (and if necessary justify) their range and precision, e.g., a suitable thermometer and ammeter. A voltmeter will also be needed to ensure that the potential difference is constant.

Procedure: a clearly labeled diagram showing the experimental arrangement can save a lot of writing and is easier to understand than a block of text. Take care over the diagram and if you draw it yourself use a ruler. Explain what you will measure and why and state the range and number of readings you intend to take, including any repeats. For an electrical experiment draw a circuit diagram and make sure the symbols you use are correct!

Risk assessment: once you have planned a procedure look at it critically. Assess the risks and plan precautions to minimize them. This might mean modifying your experiment or seeking advice from your teacher or supervisor. Do not carry out an experiment if you have any doubts about its safety. Include a summary of the risk assessment as part of the experimental account.

Results: tabulate your results including the uncertainty in each measurement. Take care over: table headings (include units) and significant figures (especially for any calculated values). If you repeat measurements and take an average include all the raw data in your table.

Analysis: this will often involve plotting a graph and taking a gradient (see Chapter 2). Take care over: axis labels; plotting points; drawing best fit and worst acceptable lines. Make sure everything is clearly labeled. The analysis should be based on the theory you have used to propose a hypothesis, e.g., for the current through a thermistor a graph of $\ln (I)$ against $(1/T)$ should be a straight line with a negative gradient.

Uncertainties

Include an analysis of uncertainties so that any value calculated from your data should be quoted to an appropriate degree of precision and be accompanied by its own uncertainty. For example, if you are carrying out an experiment to determine the specific heat capacity of water and come up with a value of 4000 $\text{Jkg}^{-1}\text{°C}^{-1}$ there is no way to know if your experiment was a good one even though your value is within 5% of the expected value (4200 $\text{Jkg}^{-1}\text{°C}^{-1}$). However, if your result is 4000 ± 400 $\text{Jkg}^{-1}\text{°C}^{-1}$ it is clear that the expected value is within the range of your experimental uncertainties and that your result has an uncertainty of 10%. Refining your methods should reduce the uncertainties but keep the expected value within the range of your results. If you end up with a result of 4000 ± 100 $\text{Jkg}^{-1}\text{°C}^{-1}$ this does not include the expected value so either you have underestimated the uncertainties or there is a systematic error in your measurements.

Conclusion

It is surprising how often students forget to state a conclusion to their experiments. This should be a simple statement of what you have achieved in your experiment and should relate back to the original aims.

Evaluation

Having carried out the experiment and drawn a conclusion you should consider:

▪ How strongly the conclusion is supported.

▪ How any calculated values compare with expected or known values.

▪ The significance of the uncertainties.

▪ Likely sources of error.

▪ Suitability of the apparatus used.

▪ How the experiment could be improved.

If possible make your evaluation quantitative by referring to the range of uncertainties in measurements and their impact on your final result.

Glossary

Include a list of technical terms you have used in your account along with brief explanations.

References

Include a list of references indicating where you have used each source. Provide enough information so that someone reading your report can easily find the information you used. This should include:

▪ the title of the work or article

▪ the author or authors

▪ publisher and publication date

▪ the page or pages used

▪ URL and date accessed (for websites)

B.2 Investigations

An investigation is a more open-ended project that involves a considerable amount of preliminary research and pilot experiments (to try things out and to explore the phenomena). These will help you to plan a sequence of experiments, but as you proceed you should be prepared to modify your plans in the light of results, or when you discover that a particular experiment does not work!

It takes time to get into an investigation, and you must not be easily deterred, especially at the start when there is a lot of uncertainty about how to begin. Researchers spend a lot of time getting nowhere, but without trying a range of different approaches you are unlikely to hit upon the method that actually works. You might also discover interesting and unexpected aspects of the problem along the way.

Science is a collaborative endeavor and carrying out an investigation usually requires you to discuss ideas with your teacher or supervisor and to think about the feedback they give you. You are also likely to need a fair amount of apparatus, and this could involve discussing your needs with laboratory technicians or equipment suppliers (in advance!) In both cases it is important to be as clear as you can about what you need and what you need it for—asking for a "block of wood" is pretty meaningless: what type of wood? What dimensions? What is it for?

Writing up an investigation can be a daunting task, especially if you leave it all until you have finished in the laboratory. It is worth using a laboratory notebook and it is essential to write up and process data from every experiment before moving on. A good investigation cannot be completely planned in advance, it evolves as you discover more about the problem you are tackling. There are many ways to write up an investigation but the written report could take the following form:

Aim: statement of what is to be investigated

Background physics: summary of research about the problem identifying aspects for investigation.

Pilot experiments: these should be used to explore the phenomena and test methods or instruments to see if they are suitable. These pilot experiments are not intended to produce precise data for analysis although they might suggest relationships to be investigated in more detail later.

Plan of investigation: this should outline a sequence of related experiments that can be used to collect good relevant data to move the investigation forward.

Experiments: each experiment should be written up fully (using the guidelines in Section B.1) including risk assessments and taking account of anything learnt in earlier experiments. The glossary and references can be left to the end of the complete report.

Conclusions: whilst each individual experiment might lead to its own conclusion this section takes an overview of the whole investigation and must be written at the end of the work. It should relate back to the original aims.

Evaluation: this covers the same points as an individual experimental evaluation but refers to the whole investigation. It might consider the relevant merit and contribution of different experiments.

Glossary: a list of technical terms you have used in your investigation along with brief explanations.

References: The point of a reference is so that someone reading your work can find the information easily without having to search through an entire book or article, so give full details. There are several standard ways to provide references; one of the most widely used is the Harvard system.

C

UNITS, CONSTANTS, AND EQUATIONS

C.1 S.I. Units

Base units

Quantity	Name	Symbol
		S.I. base unit
Length	Meter	m
Mass	Kilogram	kg
Time	Second	s
Electric current	Ampère	A
Thermodynamic temperature	Kelvin	K
Amount of substance	Mole	mol
Luminous intensity	Candela	cd

Derived units

Derived quantity	Name	Symbol	
	S.I derived units		S.I base units
Force	Newton	N	$kgms^{-2}$
Pressure	Pascal	Pa	$kgm^{-1}s^{-2}$
Energy	Joule	J	kgm^2s^{-2}
Power	Watt	W	kgm^2s^{-3}
Charge	Coulomb	C	As
Resistance	Ohm	Ω	$kgm^2s^{-2}A^{-2}$

Derived quantity	Name		Symbol	
	S.I derived units			**S.I base units**
Potential difference	Volt		V	$kgm^2s^{-2}A^{-1}$
Capacitance	Farad		F	$kg^{-1}m^{-2}s^3A^2$
Magnetic field strength	Tesla		T	$kgs^{-2}A^{-1}$
Magnetic flux	Weber		Wb	$kgm^2s^{-2}A^{-1}$
Inductance	Henry		H	$kgm^2s^{-1}A^{-2}$

C.2   Simple Approximate Combinations of Uncertainties

Combination	Rule	
Uncertainty in a sum: $y = a + b$	Add absolute uncertainties	$\delta y = \delta a + \delta b$
Uncertainty in a product: $y = ab$	Add fractional uncertainties	$\dfrac{\delta y}{y} = \dfrac{\delta a}{a} + \dfrac{\delta b}{b}$
Uncertainty in a quotient: $y = \dfrac{a}{b}$	Add fractional uncertainties	$\dfrac{\delta y}{y} = \dfrac{\delta a}{a} + \dfrac{\delta b}{b}$
Uncertainty in a power: $y = an$	Multiply fractional uncertainty by power	$\dfrac{\delta y}{y} = n\dfrac{\delta a}{a}$

C.3   Useful Derivatives

Function: y	Derivative (rate of change): dy/dx
Constant value, e.g. $y = 8$	0
Power law, e.g. $y = Ax^n$	$\dfrac{dy}{dx} = nAx^{n-1}$
Exponential function: $y = e^x$	$\dfrac{dy}{dx} = e^x$
Exponential relationship: e.g. $y = Ae^{bx}$	$\dfrac{dy}{dx} = bAe^{bx}$
Sine function: $y = \sin x$	$\dfrac{dy}{dx} = \cos x$
Cosine function: $y = \cos x$	$\dfrac{dy}{dx} = -\sin x$
Sinusoidal variation: e.g. $y = A \sin (bx)$	$\dfrac{dy}{dx} = bA\cos(bx)$
Cosinusoidal variation: $y = A \cos (bx)$	$\dfrac{dy}{dx} = -bA\sin(bx)$

C.4 Differential Equations

Topic	Differential equation	Solution	Conditions
Radioactive decay	$\dfrac{dN}{dt} = -\lambda t$	$N = N_0 e^{-\lambda t}$	$N = N_0$ at $t = 0$
Capacitor discharge	$\dfrac{dQ}{dt} = -\dfrac{Q}{RC}$	$Q = Q_0 e^{-\frac{t}{RC}}$	$Q = Q_0$ at $t = 0$
Newton's second law (constant acceleration)	$\dfrac{d^2x}{dt^2} = \dfrac{F}{m}$	$x = ut + \dfrac{1}{2}\left(\dfrac{F}{m}\right)t^2$	F, m, u constants $x = 0$ at $t = 0$
Simple harmonic motion	$\dfrac{d^2x}{dt^2} = -\omega^2 x$	$x = A\cos(\omega t + \phi)$	A, w, f constants $(w = 2\pi f)$

C.5 Differentials and Integrals

Context	Differential form	Integral form
Dynamics	$v = \dfrac{ds}{dt}$	$s = \int v\,dt$
Dynamics	$\dfrac{dv}{dt}$	$v = \int a\,dt$
Newton's laws	$F = \dfrac{d(mv)}{dt}$	$mv - mu = \int F\,dt$
Electric circuits	$I = \dfrac{dQ}{dt}$	$Q = \int I\,dt$
Radioactivity	$\dfrac{dN}{dt} = -N$	$\int \dfrac{dN}{N} = -\int dt$
Capacitors	$\dfrac{dQ}{dt} = -\dfrac{Q}{RC}$	$\int \dfrac{dQ}{Q} = -\int \dfrac{dt}{RC}$

C.6 Equations

Mechanics

Equations for constantly accelerated motion:

$$v = u + at$$

$$s = \frac{(u + v)t}{2}$$

$$s = ut + \frac{1}{2}at^2$$

$$s = vt - \frac{1}{2}at^2$$

$$v^2 = u^2 + 2as$$

Range of a projectile $\quad:\quad R = \dfrac{u^2 \sin 2\theta}{g}$

Coefficient of friction $\quad:$

$$F_{\text{limit}} = \mu N, \quad \mu_S = \tan \theta \ (\theta \text{ is limiting angle})$$

Weight $\quad:\quad W = mg$

Density $\quad:\quad \rho = \dfrac{m}{V}$

Pressure $\quad:\quad p = \dfrac{F}{A}$

Linear momentum $\quad:\quad p = mv$

Newton's second law (constant mass) $\quad:\quad F = \dfrac{dp}{dt} \qquad a = \dfrac{F}{m}$

Impulse and change of momentum $\quad:\quad \displaystyle\int F dt = \int dp = p$

$$Ft = (mv - mu) \quad \text{(constant mass)}$$

Rockets $\quad:\quad v_f = v_0 + u \ln\left(\dfrac{m_0}{m_f}\right)$

Work done $\quad:\quad W = F_s \cos \theta$

Gravitational potential energy $\quad:\quad \Delta \text{GPE} = mgh$

Kinetic energy $\quad:\quad \text{KE} = \frac{1}{2}mv^2$

Efficiency $\quad:\quad \eta = \dfrac{\text{useful output energy}}{\text{total input energy}} \times 100\%$

Power $\quad:\quad P = \dfrac{dE}{dt} \qquad P = F\dfrac{ds}{dt} = Fv$

Photon momentum $\quad:\quad p_{\text{photon}} = \dfrac{E}{c}$

Radiation pressure $\quad:\quad p = \dfrac{I}{c}$

Lagrangian $\quad:\quad L = T - V$

Euler–Lagrange equations $\quad:\quad \dfrac{\partial}{\partial t}\left(\dfrac{\partial L}{\partial \dot{q}}\right) = \dfrac{\partial L}{\partial q}$

Fluids

Fluid pressure : $p = \rho g h$

Reynold number : $R_e = \dfrac{v\rho L}{\eta}$

Equation of continuity : $\rho_1 A_1 v_1 = \rho_2 A_2 v_2$

Stoke law : $F = 6\pi h r v$

Turbulent drag : $F = \frac{1}{2} C_D \rho\, A v^2$

Bernoulli equation : $P_1 + \dfrac{1}{2}\rho v_1^2 + \rho g h_1 = P_2 + \dfrac{1}{2}\rho v_2^2 + \rho g h_2$

Poiseuille equation : $Q = \dfrac{\pi p a^4}{8\eta l}$

Materials

Hooke law : $F = ke$

Spring systems : $\dfrac{1}{k_{series}} = \dfrac{1}{k_1} + \dfrac{1}{k_2} + \dfrac{1}{k_3}\ldots$

$k_{parallel} = k_1 + k_2 + k_3 \ldots$

Strain energy (spring) : $EPE = \dfrac{1}{2}Fe = \dfrac{1}{2}ke^2$

Strain energy (wire) : $\dfrac{EPE}{V} = \dfrac{1}{2}\sigma\varepsilon$

Stress and strain : $\sigma = \dfrac{F}{A} \quad \varepsilon = \dfrac{e}{l_0}$

Young modulus : $E = \dfrac{\sigma}{\varepsilon}$

Thermodynamics

Temperature scales : $T = \theta + 273.15; \theta = T - 273.15 (\theta$ in °C, T in K)

Thermal conduction : $\dfrac{dQ}{dt} = -kA\dfrac{d\theta}{dx}$

Wien displacement law : $T = \text{constant} = 2.910^{-3}\,\text{mK}$

Stefan–Boltzmann law : $P = e\sigma A T^4$

Specific heat capacity : $c = \dfrac{\Delta E}{m\Delta\theta}$

Specific latent heat : $E = mL$

Ideal gas equations

Boyle law : $pV = \text{constant (constant mass and temperature)}$

Charles law : $V/T = \text{constant (constant mass and pressure)}$

Gay Lussac law : $p/T = \text{constant (constant mass and volume)}$

Ideal gas equation : $\dfrac{pV}{T} = \text{constant}; \quad pV = nRT$

Adiabatic compression : $pV^{\gamma} = \text{constant}$

Kinetic theory equation : $pV = \dfrac{1}{3}\text{Nmv}^2; \; p = \dfrac{1}{3}\rho v^2$

Mean molecular KE : $\dfrac{1}{2}mv^2 = \dfrac{3}{2}\dfrac{R}{N_A}T = \dfrac{3}{2}kT$

RMS molecular speed : $v_{\text{rms}} = \sqrt{v^2} = \sqrt{\dfrac{3kT}{m}}$

Internal energy : $U = \text{Total KE} = \dfrac{1}{2}N_A mv^2 = \dfrac{3}{2}RT$

Heat capacities (ideal gas) : $c_V = \dfrac{3}{2}R \quad c_P = \dfrac{5}{2}R$

Adiabatic gas constant : $\gamma = \dfrac{c_P}{c_V}$

Speed of sound in gas : $c = \sqrt{\dfrac{\gamma p}{\rho}}$

Boltzmann factor : $f = e^{-\frac{\Delta E}{kT}}$

First law of thermodynamics : $\Delta U = Q - W$

Heat engine efficiency : $\dfrac{W}{Q_1} = 1 - \dfrac{Q_2}{Q_1} \leq 1 - \dfrac{T_2}{T_1}$

Reversible heat transfer : $\Delta S = \displaystyle\int_{\text{state A}}^{\text{state B}} \dfrac{dQ}{T}$

Entropy change : $S = k\ln(W)$

Thermodynamic temperature : $\dfrac{1}{T} = \dfrac{dS}{dQ}$

Refrigerator	:	$\mathrm{CoP}_{\text{refrigerator}} = \dfrac{Q_{\text{cold}}}{W} \leq \dfrac{1}{\dfrac{T_{\text{hot}}}{T_{\text{cold}}} - 1}$
Heat pump	:	$\mathrm{CoP}_{\text{heat pump}} = \dfrac{Q_{\text{hot}}}{W} \leq \dfrac{1}{1 - \dfrac{T_{\text{cold}}}{T_{\text{hot}}}}$

Oscillations

Frequency	:	$f = \dfrac{1}{T}$
Angular frequency	:	$\omega = 2\pi f$
SHM	:	$a = \dfrac{d^2 x}{dt^2} = -\omega^2 x$
	:	$x = A\cos(\omega t + \delta)$
		$x = A\cos\omega t$
		$v = \dfrac{dx}{dt} = -\omega A \sin\omega t$
		$a = \dfrac{dv}{dt} = -\omega^2 A \cos\omega t = -\omega^2 x$
Mass-spring system	:	$f = \dfrac{1}{2\pi}\sqrt{\dfrac{k}{m}} \quad T = 2\pi\sqrt{\dfrac{m}{k}}$
Simple pendulum	:	$f = \dfrac{1}{2\pi}\sqrt{\dfrac{g}{l}} \quad T = 2\pi\sqrt{\dfrac{l}{g}}$
Total energy	:	$\mathrm{TE} = \dfrac{1}{2} m\omega^2 A^2$
Damped SHM	:	$x = Ae^{-\gamma t}\cos(\omega t)$
Resonance condition	:	$f_d = f_0$

Rotational dynamics

Angles in radians	:	$\theta = \dfrac{l}{r}$
Small angle approximations	:	As $\theta \to 0$: $\sin\theta \to \theta$, $\cos\theta \to 1$, $\tan\theta \to \theta$ (θ in radians)
Angular velocity	:	$\omega = \dfrac{v}{r}$
Tangential acceleration	:	$\alpha = \dfrac{a}{r}$
Centripetal acceleration	:	$a = \dfrac{v^2}{r} = r\omega^2$
Centripetal force	:	$F = ma = \dfrac{mv^2}{r} = mr^2$

Equations of motion for constant angular acceleration (by analogy):

$$v = u + at \qquad \rightarrow \qquad \omega_f = \omega_i + \alpha t$$

$$s = \frac{(u+v)t}{2} \qquad \rightarrow \qquad \theta = \frac{(\omega_i + \omega_f)t}{2}$$

$$s = ut + \frac{1}{2}at^2 \qquad \rightarrow \qquad \theta = \omega_i t + \frac{1}{2}\alpha t$$

$$s = vt - \frac{1}{2}at^2 \qquad \rightarrow \qquad \theta = \omega_f t - \frac{1}{2}\alpha t^2$$

$$v^2 = u^2 + 2as \qquad \rightarrow \qquad \omega_f^2 = \omega_i^2 + 2\alpha\theta$$

Moment of inertia	:	$I = \left(\sum_{i=1}^{i=N} m_i r_i^2 \right)$
Rotational KE	:	$\mathrm{RKE} = \frac{1}{2}I\omega^2$
Angular momentum	:	$L = I\omega$
Torque	:	$\Gamma = \frac{d(L)}{dt} = I\frac{d\omega}{dt} = I\alpha$

Moments of inertia for uniform objects:

Point mass	:	$I = mr^2$
Rod	:	$I_{\mathrm{end}} = \frac{1}{3}ml^2 \quad I_{\mathrm{CM}} = \frac{1}{12}ml^2$
Thin hoop	:	$I = mr^2$
Disc/cylinder	:	$I_{\mathrm{CM}} = \frac{1}{2}ma^2$ (radius = a)
Sphere	:	$I_{\mathrm{CM}} = \frac{2}{5}ma^2$ (radius = a)

Waves

Wave speed	:	$v \;\; - = f\lambda$
1D traveling wave	:	$y = A\cos(\omega t - kx)$
Intensity	:	$I \propto A^2$
Wave speed in medium	:	$v = \frac{c}{n}$
Refraction	:	$n_1 \sin\theta_1 = n_2 \sin\theta_2$
TIR	:	$\sin c = \frac{n_2}{n_1}$

Malus law (polarization)   : $I_{trans} = I_0 \cos^2\theta$

Brewster law   : $\tan\theta_B = \dfrac{n_2}{n_1}$

Speed of light   : $c = \sqrt{\dfrac{1}{\varepsilon_0\mu_0}}$ (vacuum)

$v = \sqrt{\dfrac{1}{\varepsilon_0\varepsilon_r\mu_0\mu_r}}$ (medium)

Refractive index   : $n = \sqrt{\varepsilon_r\mu_r}$

Power of a lens   : $P = \dfrac{1}{f}$

Linear magnification   : $m = \dfrac{h_i}{h_o} = \dfrac{v}{u}$

Lens equation   : $\dfrac{1}{f} = \dfrac{1}{u} + \dfrac{1}{v}$

Astronomical telescope   : $M = \dfrac{\beta}{\alpha} = \dfrac{f_o}{f_e}$

Doppler shift (light)   : $\Delta\lambda = \pm\dfrac{v}{c}\lambda_0$

Red shift   : $z = \dfrac{\lambda' - \lambda_0}{\lambda_0} = \dfrac{v}{c}$

Hubble law   : $v = H_0 d$

Phase difference   : $\Delta\phi = \dfrac{2\pi\Delta x}{\lambda}$

Young's double slit   : $\lambda = \dfrac{sy}{d}$

Diffraction grating maxima   : $n\lambda = d\,\sin\theta$

Single slit minima   : $\sin\theta = \dfrac{1.22}{D}$

Rayleigh criterion   : $\theta \geq \dfrac{1.22}{D}$ (for resolution)

Waves on a string   : $v = \sqrt{\dfrac{T}{\mu}}$

Speed of sound   : $v = \sqrt{\dfrac{\gamma RT}{M}}$ (ideal gas); $v = \sqrt{\dfrac{E}{\rho}}$ (solid)

Decibel scale	:	intensity level $(B) = \log_{10}\left(\dfrac{I}{I_0}\right)$
Acoustic impedance	:	$Z = (\text{speed of sound}) \times (\text{density})$
Reflection	:	$\dfrac{I_r}{I_0} = \left(\dfrac{Z_2 - Z_1}{Z_2 + Z_1}\right)^2$

Electricity

Electric current	:	$I = \dfrac{dQ}{dt}; I = \dfrac{Q}{t}$; (constant current)
Coulomb law	:	$F = \dfrac{Q_1 Q_2}{4\pi\varepsilon_0 r^2}$
Electric field strength	:	$E = \dfrac{F}{q}; E = -\dfrac{dV}{dx}$
E-field strength (point charge)	:	$E = \dfrac{F}{q} = \dfrac{Q}{4\pi\varepsilon_0 r^2}$
Electric potential	:	$V = \dfrac{\text{EPE}}{Q}$
Potential difference	:	$\Delta V = \dfrac{W}{Q}$
Electric potential (point charge)	:	$E = \dfrac{Q}{4\pi\varepsilon_0 r}$
Gauss theorem	:	$\displaystyle\int_{\text{surface}} E.dS = \dfrac{\sum_{i=1}^{i=N} Q_i}{\varepsilon_0}$
Electric current	:	$I = nAve$
Kirchhoff second law	:	$\displaystyle\sum_{\substack{\text{closed}\\\text{loop}}} \text{emfs} = \sum_{\substack{\text{closed}\\\text{loop}}} \text{p.d.s}$
Resistance (Ohm's law)	:	$R = \dfrac{V}{I}$
Resistors in series	:	$R_{\text{series}} = \displaystyle\sum_{i=1}^{i=N} R_i$
Resistors in parallel	:	$\dfrac{1}{R_{\text{parallel}}} = \displaystyle\sum_{i=1}^{i=N} \dfrac{1}{R_i}$

Resistivity	:	$\rho = \dfrac{RA}{l}$
Electrical energy	:	$E = VIt$
Real cell	:	$V = E - Ir$
Capacitance	:	$C = \dfrac{Q}{V}$
Energy stored (capacitor)	:	$E = \dfrac{Q^2}{2C} = \dfrac{CV^2}{2} = \dfrac{QV}{2}$
Parallel plate capacitance	:	$C = \dfrac{\varepsilon_0 \varepsilon_r A}{d}$
Capacitor discharge	:	$Q(t) = Q_0 e^{-\frac{t}{RC}} I(t)$
		$= I_0 e^{-\frac{t}{RC}} V(t) = V_0 e^{-\frac{t}{RC}}$
Capacitor charging	:	$Q = Q_F \left(1 - e^{-\frac{t}{RC}}\right)$
		$V = V_S \left(1 - e^{-\frac{t}{RC}}\right)$
		$I(t) = I_0 e^{-\frac{t}{RC}}$
Time constant	:	$\tau = RC$
Capacitors in series	:	$\dfrac{1}{C_{\text{series}}} = \sum\limits_{i=1}^{i=n} \dfrac{1}{C_i}$
Capacitors in parallel	:	$C_{\text{para}} = \sum\limits_{i=1}^{i=n} C_i$
Capacitance (charged sphere)	:	$C_{\text{sphere}} = \dfrac{Q}{V} = 4\pi\varepsilon_0 a$

Magnetism

Magnetic force	:	$F = BIl \sin\theta$ (on current)
		$f = Bqv \sin\theta$ (on moving charge)
Lorentz force	:	$f = q\underline{E} + q\underline{v}{\wedge}B$
Radius of curvature in B-field	:	$r = \dfrac{mv}{Bq}$
Biot–Savart law	:	$\delta B = \dfrac{\mu_0 I \sin\theta}{4\pi x^2} \delta l$

Magnetic field strength : $B = \dfrac{\mu_0 NI}{2r}$ (narrow coil)

$B = \dfrac{\mu_0 I}{2\pi r}$ (long straight current-carrying wire)

$B = \dfrac{\mu_0 \mu_r NI}{l}$ (long solenoid at center)

$B = \dfrac{\mu_0 \mu_r NI}{2l}$ (long solenoid at end)

Ampère theorem : $\displaystyle\int_{\substack{\text{closed}\\ \text{loop}}} B.dl = \sum_{\substack{\text{enclosed}\\ \text{by loop}}} \mu_0 I$

Torque on coil in B-field : $\Gamma = NBIA$

Magnetic flux : $\Phi = \displaystyle\int_{\text{surface}} B\cos dA$ (flux-linkage = NF)

Faraday law : $E = -\dfrac{d(N\Phi)}{dt}$

Self-inductance : $E = -L\dfrac{dI}{dt}$

Mutual inductance : $E_2 = M\dfrac{dI_1}{dt}; E_1 = M\dfrac{dI_2}{dt}$

Energy stored in inductor : $W = \dfrac{1}{2}LI^2$

Ideal transformer : $\dfrac{I_2}{I_1} = \dfrac{V_1}{V_2} = \dfrac{N_1}{N_2}$

A.C. circuits

Alternating current and voltage : $V = V_0 \sin\omega t \quad I = I_0 \sin\omega t$

RMS values : $I_{\text{rms}} = \dfrac{I_0}{\sqrt{2}}; V_{\text{rms}} = \dfrac{V_0}{\sqrt{2}}$

A.C. power : $P_{AC} = I_{\text{rms}} V_{\text{rms}}$

Reactance (capacitor) : $X_C = \dfrac{V_0}{I_0} = \dfrac{1}{\omega C} = \dfrac{1}{2\pi f C}$

Reactance (inductor) : $X_L = \dfrac{V_0}{I_0} = \omega L = 2\pi f L$

Impedance : $Z = \dfrac{V_0}{I_0}$

$$Z = \frac{V_S}{I} = \sqrt{R^2 + X_L^2 - X_C^2} = \sqrt{R^2 + (\omega L)^2 - \left(\frac{1}{\omega C}\right)^2} \text{ (RCL series circuit)}$$

Phase angle : $\tan\phi = \dfrac{V_L - V_C}{V_R} = \dfrac{X_L - X_C}{R}$

(RCL series circuit)

Resonant frequency : $f_0 = \dfrac{1}{2\pi}\sqrt{\dfrac{1}{LC}}$
(RCL circuit)

Gravitational fields

Newton's law of gravitation : $F = -\dfrac{Gm_1 m_2}{r^2}$

Gravitational field strength : $g = \dfrac{\text{gravitational force}}{\text{mass}} = \dfrac{F}{m}$

$g = \dfrac{F}{m} = -\dfrac{GM}{r^2}$ (point mass or uniform spherical mass)

Gravitational potential : $V_G = \dfrac{\text{GPE}}{m}$

$V_G(r) = -\dfrac{GM}{r}$ (point mass or uniform spherical mass)

$$\Delta\text{GPE} = mgh \qquad (h \gg r)$$

Escape velocity : $v_{esc} = \sqrt{\dfrac{2GM}{r}}$

Scwarzschild radius : $R_S = \dfrac{2GM}{c^2}$

Kepler's third law : $\dfrac{r^3}{T^2} = \dfrac{GM}{4\pi^2}$

Tidal forces : $\Delta F = \dfrac{-2GMmr}{R^3}$

Gravitational time dilation : $T' = T\left(1 + \dfrac{\Delta V}{c^2}\right)$ (weak fields)

Special relativity

Gamma-factor : $\gamma = \dfrac{1}{\sqrt{1 - \dfrac{v^2}{c^2}}}$

Time dilation : $T' = \gamma T$

Length contraction $\quad:\quad L' = \dfrac{L}{\gamma}$

Lorentz transformations $\quad:$

$x' = \gamma(x - vt) \qquad\qquad z' \quad z$

$y' = y \qquad\qquad\qquad t' = \gamma\left(t - \dfrac{vx}{c^2}\right)$

Inverse Lorentz transformations :

$x = \gamma(x' + vt') \qquad\qquad z = z'$

$y = y' \qquad\qquad\qquad t = \gamma\left(t' + \dfrac{vx'}{c^2}\right)$

Velocity addition $\quad:\quad w = \dfrac{(u + v)}{\left(1 + \dfrac{uv}{c^2}\right)}$

Mass increase $\quad:\quad m = \gamma m_0$

Mass and energy $\quad:\quad \Delta E = c^2 \Delta m; E^2 - p^2 c^2 = \left(m_0 c^2\right)^2$

Atomic and nuclear physics

Inverse-square law for intensity $\;:\quad I = \dfrac{R}{4\pi r^2}$

Attenuation of gamma-rays $\quad:\quad I = I_0 e^{-\mu x}$

Half-thickness $\quad:\quad x_{1/2} = \dfrac{\ln 2}{\mu}$

Range of radiation $\quad:\quad R = \dfrac{\sigma}{\rho}$

Radioactive decay $\quad:\quad N = N_0 e^{-\lambda t}$

Activity $\quad:\quad A = -\dfrac{dN}{dt} = -\lambda N_0 e^{-\lambda t} = -\lambda N$

Half-life $\quad:\quad t_{1/2} = \dfrac{\ln 2}{\lambda}$

Nuclear binding energy $\quad:\quad \text{B.E.} = c^2 \Delta m$

Binding energy per nucleon $\quad:\quad \text{B.E.}/A = c^2 \Delta m /A$

Quantum physics

Photon energy $\quad:\quad E = hf$

Photoelectric effect $\quad:\quad \text{KE}_{\text{max}} = hf - \Phi = hf - hf_0$

$\qquad\qquad\qquad\qquad eV_S = \text{KE}_{\text{max}} = hf - \Phi$

de Broglie relation $\qquad$: $\quad \lambda = \dfrac{h}{p}$

Compton Effect $\qquad$: $\quad (\lambda' - \lambda) = \dfrac{h}{m_e c}(1 - \cos\theta)$

Hydrogen atom energy levels $\quad$: $\quad E_n = \dfrac{-me^4}{8\varepsilon_0^{\,2}h^2}\left(\dfrac{1}{n^2}\right) = \dfrac{-13.6\,\text{eV}}{n^2}$

Balmer series $\qquad$: $\quad \dfrac{1}{\lambda_n} = R\left(\dfrac{1}{2^2} - \dfrac{1}{n^2}\right)$

Rydberg formula $\qquad$: $\quad \dfrac{1}{\lambda_n} = R\left(\dfrac{1}{m^2} - \dfrac{1}{n^2}\right)$

Rydberg constant $\qquad$: $\quad R = \left(\dfrac{me^4}{8c\varepsilon_0^{\,2}h^3}\right)$

Heisenberg indeterminacy relation $\qquad$: $\quad \Delta x \Delta p \geq \dfrac{h}{4\pi}$

Astrophysics

Radius of a star $\qquad$: $\quad r = \sqrt{\left(\dfrac{L}{4\pi\sigma T^4}\right)}$

Hubble law $\qquad$: $\quad v = H_0 d$

Hubble time $\qquad$: $\quad T_H = \dfrac{1}{H_0}$

Medical physics

Beer–Lambert law $\qquad$: $\quad I_{\text{out}} = I_{\text{in}}e^{-\mu x}$

Larmor resonant frequency $\qquad$: $\quad f = \dfrac{\gamma B_0}{2\pi}$

C.7 Constants

Speed of light in a vacuum	c	$3.00 \times 10^8\,\text{ms}^{-1}$	$299\ 792\ 458\ \text{ms}^{-1}$
Electronic charge (magnitude)	e	$1.60 \times 10^{-19}\,\text{C}$	$1.60217662 \times 10\text{-}19\,\text{C}$
Planck constant	h	$6.63 \times 10^{-19}\,\text{Js}$	$6.62607004 \times 10^{-34}\,\text{Js}$
Gravitational constant	G	$6.67 \times 10^{-11}\,\text{Nm}^2\text{kg}^{-2}$	$6.67408 \times 10^{-11}\,\text{Nm}^2\text{kg}^{-2}$
Avogadro constant	L	$6.02 \times 10^{23}\,\text{mol}^{-1}$	$6.02214086 \times 10^{23}\,\text{mol}^{-1}$
Boltzmann constant	k	$1.38 \times 10^{-23}\,\text{JK}^{-1}$	$1.38064852 \times 10^{-23}\,\text{JK}^{-1}$
Molar gas constant	R	$8.31\ \text{J mol}^{-1}$	$8.3144598\ \text{mol}^{-1}$

Unified atomic mass unit	u	1.66×10^{-27} kg	$1.660539040 \times 10^{-27}$ kg
Permeability of free space	μ_0	$4\pi \times 10^{-7}$ Hm^{-1}	$1.25663706 \times 10^{-6}$ Hm^{-1}
Permittivity of free space	ε_0	8.85×10^{-12} Fm^{-1}	$8.85418782 \times 10^{-12}$ Fm^{-1}
Stefan constant	σ	5.67×10^{-8} Wm^{-2}K^{-4}	5.670367×10^{-8} Wm^{-2}K^{-4}
Wien constant	W	2.90×10^{-3} mK	2.8977685×10^{-3} mK
Proton mass	m_p	1.67×10^{-27} kg	$1.6726219 \times 10^{-27}$ kg
Electron mass	m_e	9.11×10^{-31} kg	$9.10938356 \times 10^{-31}$ kg
Neutron mass	m_n	1.67×10^{-27} kg	$1.674927471 \times 10^{-27}$ kg
Rydberg constant	R	1.10×10^7 m^{-1}	$1.0\,973\,731.568\,508 \times 10^7$ m^{-1}
Standard acceleration of gravity	g	9.81 ms^{-2}	$9.806\,65$ ms^{-2} (defined)
Standard atmospheric pressure	atm	1.01×10^5 Pa	$101\,325$ Pa (defined)
Molar mass of carbon-12	$M(^{12}\text{C})$	1.2×10^{-2} kg·mol^{-1}	1.2×10^{-2} kg·mol^{-1} (defined)

SOLUTIONS TO EXERCISES

Descriptive answers are not usually included and can be found by referring to the relevant chapter.

Chapter 1   The Language of Physics

1. (a) 0.267 kg; (b) 25 000 000 mm; (c) 5000 000 cm³; (d) 22.2 ms⁻¹; (e) 0.0045 m²

2. (a) 1.44×10^8 km; (b) 9.47×10^{15} m; (c) 1.27 light seconds

3. (a) 2.00; (b) 0.00209; (c) 0.00950; (d) 3.14

4. (a) Nm²kg⁻² or m³kg⁻¹s⁻² (in base units)

5. (a) 0.012, 0.016, 0.050 (b) 1.2 %, 1.6 %, 5.0 % (c) 5300 ± 200 cm² (d) $105\,000 \pm 8000$ cm³

6. (a) 0.234 ± 0.051 kg; (b) time period $\left(2\dfrac{\delta T}{T} > \dfrac{\delta k}{k} \right)$.

7. (a) Systematic errors affect all readings in the same way (e.g. constant addition or subtraction) and can be corrected for if the error is known (e.g. subtracting a zero error). Random errors cannot be predicted and affect each data point independently.

(b) Repeat reading several times and use an average value (neglecting obvious anomalies). (c) 0.32 mm

8. (a) 5.5×10^3; (b) 7×10^{-10}; (c) 1.2×10^{24}

9. (a) 9.0×10^{15}; (b) 9.0×10^{-3}; (c) 2.0×10^{8}; (d) 2.0×10^{16}

10. (a) 3.3×10^{7} s or 1 year 20.6 days; (b) 4.0×10^{9} s or 126 years 275.8 days.
(c) 2.8×10^{13} s or 900 000 years;
(d) 9.5×10^{20} s or 3.0×10^{13} years (>> age of Universe!)

Chapter 2   Representing and Analyzing Data

1. (a) 6; (b) − 3; (c) 1.748; (d) − 0.488; (e) 0

2. (a) 10,000; (b) 501.2; (c) 1.122; (d) 0.001995; (e) 0.8913

3. (a) $1/v$ on y-axis and $1/u$ on x-axis; (b) y-intercept is $1/f$

4. (a) Plot p against $1/V$. If pV = constant graph will be a straight line
through the origin. (b) −

5. (a) Plot log (T) against log (r). This should give a straight line with
gradient equal to n and $n = 3/2$; (b) 225 days

6. Plot ln (I) against t. to get a straight line with gradient equal to $- 1/T$.
$RC \sim 44 - 45$ s

Chapter 3   Capturing, Displaying, and Analyzing Motion

1.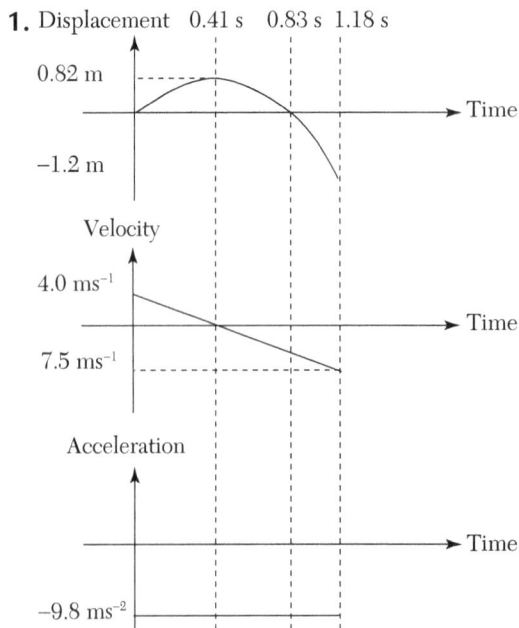

2. (a) 8.3 ms⁻¹; (b) 1.67 ms⁻²

3. (a) 5.0 s; (b) 2.4 ms⁻²

4. Car's speed is 13.9 ms⁻¹(a) 27.8 m; (b) 36.1 m

5. 23.7 m

6. 0.067 ms⁻²

7. Car B wins by 0.64 s

8. (a) 14.3 m; (b) 61.1 m

Chapter 4 Forces and Equilibrium

1. All of them.

2. (a) 2.83 N at 45° to left of vertical; (b) 5.10 N at 32.1° left of vertical

3. 6.2 N

4. 0.75 N

5. (a) F_A = 1225 N; F_B = 1715 N; (b) 2940 N

6. (a) μ_S = 0.466; (b) μ_K = 0.443

(c)

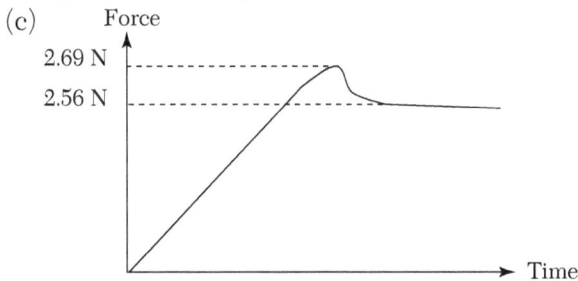

Chapter 5 Newtonian Mechanics

1. Tension $T = mg + ma$. If the stone is lifted gradually the acceleration is small and $T \sim mg$. If the string is jerked the acceleration is large so ma is large and the tension can exceed the breaking force for the thread.

2. (a) Passengers continue at constant velocity until acted upon by a resultant force (N1). When the bus slows down no additional force acts on the passengers so they continue moving forward relative to the decelerating bus and this makes them think they have been thrown forwards even though no forward force acts on them.

(b) There is an interaction between the bullet and gun when the propellant explodes. The forward force on the bullet is equal to the backward force on the gun (N3).

(c) To accelerate upwards there must be a resultant upward force on you (N2). This arises because the contact force from the floor of the lift is greater than your weight: $F_C - mg = ma$. You do not feel your weight but you do feel the contact force and it is this that makes you feel "heavier."

(d) These are opposite ends of the same interaction so they must be equal (N3).

3. (a) 0.50 ms^{-2} to the right; (b) $a = 19.4$ ms^{-2} to the right

4. (a) 30 000 kgms^{-1}; (b) 0.16 kgms^{-1}

5. 1.3 ms^{-1} to the right

6. (a) $v_1 = 0.594$ ms^{-1}; $v_2 = 0.316$ ms^{-1}; (b) Initial KE = 0.32 mJ and final KE = 0.23 mJ so the collision was inelastic. 094 mJ has been transferred to other forms in the collision.

7. (a) 2500 N; (b)(i) Power needed to overcome drag is unchanged but additional power must be supplied to increase the GPE of the car as it rises; (ii) 111 kW

8. (a) $s = uT + \dfrac{mu^2}{2B}$

(b)

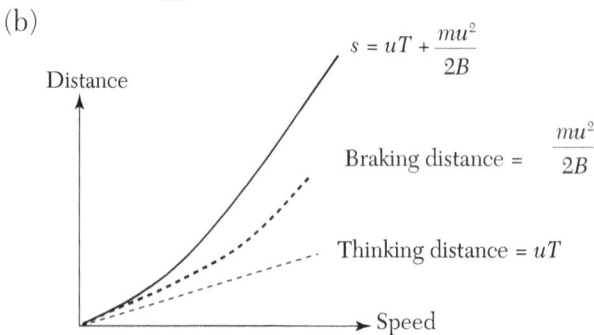

(c) Braking distance depends on u^2, so halving u reduces this by a factor of 4.

Chapter 6 Fluids

1. (a) Δp (due to 10 m of seawater) = 101 kPa ~ 1 atmosphere;
(b) 111 MPa; (c) Actual pressure > 111 MPa because there is more

mass in each unit volume so the weight of water creating the pressure is greater than if the water was incompressible.

2. (a) $\Delta p = \rho gx$; (b) $F = \frac{1}{2}\rho glh^2$; (c) Integrate moments on a narrow strip of width δx and equate the result to the moment of the total force at height a above the base. Show that $a = h/3$.

3. (a) –; (b) The constant is dimensionless so changing its value does not affect the balance of dimensions in the equation;(c) The Reynold number will be too high; the flow will change from laminar to turbulent.

4. (a) Sensible estimates must be made for each quantity in the expression for Reynold number. E.g. $v \sim 30$ ms^{-1}, $\rho \sim 1.2$ kgm^{-3}, $L \sim 2$ m, $\eta \sim 2 \times 10^{-5}$ Pas giving $R_e \sim 4 \times 10^6$. This is very high and implies turbulent flow.

(b) (i) Smooth surfaces, streamlined shape, reduced cross-sectional area; (ii) 765 N; (iii) $\times$ 8; (iv) 68 ms^{-1}; (v) Engine efficiency likely to fall and other frictional forces will increase.

5. (a) ~ 0.08 m^3; (b) ~ 800 N; (c) ~ 1 N; (d) about 0.1 % so probably not noticeable.

6. (a) 6.83×10^{-5} N; (b) 1.12×10^{-5} N

(c) Free body diagram has two upward forces: buoyancy B and drag $D(v)$ and one downward force, weight mg. The resultant force $\Sigma F = mg - (B + D(v))$ so the downward acceleration is $a = g - (B + D(v))/m$. The weight and buoyancy forces are constant but D increases with speed v. When the ball bearing is released $D = 0$ so the initial acceleration is $a(0) = g - B/m$. As D increases the resultant force and acceleration decrease until $mg - (B + D(v)) = a = 0$, so the ball bearing then falls at a constant terminal velocity.

(d) 1.4 Pas assuming Stoke law is valid—i.e., ball bearing falling slowly in the center of a tube of diameter >> diameter of ball.

(e) Droplets were very small. Stoke law assumes that fluid is a continuum. For tiny droplets we need to take into account the particle nature of the air—this affects average viscosity.

7. (a) $\rho_{air} << \rho_{water}$; (b) 27 ms^{-1}; (c) ~ 1600; (d) Assumption of laminar flow is dubious. If flow is turbulent result is not valid.

8. $p_Y = \dfrac{16p_X + p_Z}{17}$

9. (a) Pressure at top of straw is reduced as you suck, so it is less than atmospheric pressure. Pressure in straw at level of external fluid is atmospheric. Pressure difference creates an upward force on fluid inside straw; (b) Maximum pressure difference is 1 atmosphere so $\rho g h_{max} \sim 10^5$ Pa. For water $h_{max} \sim 10$ m.

10. E.g. by balancing units in Stoke law equation.

11. 183 ms^{-1}

12. (a) 1370 Pa; (b) 103370 Pa; (c) Mercury has a higher density so the displacements are smaller for the same pressure difference. This increases range but reduces precision.

13. At higher altitude the weight of air above each square meter of surface is less so pressure is lower.

14.

	Area / cm^2	Speed / cms^{-1}	Volume flow rate / cm^{3}s^{-1}
Aorta	3.0	30	90
Arteries	100	**0.9**	**90**
Capillaries	900	0.10	**90**
Veins	**200**	0.45	**90**
Vena cava	18	**5.0**	90

(b) The largest Reynold number is ~ 1000 in the aorta. This is close to the limit for laminar flow so some turbulence could occur, but elsewhere the flow will be laminar.

Chapter 7   Mechanical Properties

1. (a) 2000 kgm^{-3}; (b) fractional uncertainty = 0.1, absolute uncertainty = ±320 kgm^{-3}

2. (a) 167 Nm^{-1}; (b) 2.7 J; (c) 0.12 m

3. (a) –; (b) 11 N; (c) 0.30 Ncm^{-1} or 30 Nm^{-1}; (d) 33.2 cm; (e) 28.4 cm; (f) $e \mu F$ (Hooke law obeyed); (g) 98.8 cm; (h) It has passed its elastic limit and permanent plastic deformation has occurred; (i) Less stiff—increases in extension become greater for the same increases in load.

4. (a) 1.1 mm; (b) 8.3 J

5. Loading/unloading cycle shows hysteresis so stretching forces are always greater than recompression forces at the same extension. More work is done to stretch it than it does when it recompresses. The difference is transferred to heat.

6. A: stiff, strong, brittle.

    B: less stiff and strong than A, small amount of plastic deformation before fracture.

    C: less stiff and strong than B, a significant amount of plastic deformation before fracture.

    D: Non-linear behavior. Not stiff. Tough and has a low yield stress.

Chapter 8   Thermal Physics

1. (a) ~ 6 kW; (b) Other heat losses—e.g. through ground, draughts, etc…

2. 20.2 °C

3. Cu: 24.4 J°C^{-1} mol^{-1}, Al: 24.2 J°C^{-1} mol^{-1}, Fe: 25.0 J°C^{-1} mol^{-1}, Hg: 28.1 J°C^{-1} mol^{-1}

4. 5.3 MJ

5. Molten iron cooling / state change, liquid to solid while latent heat is dissipated at constant temperature / solid iron cooling

Chapter 9   Gases

1. (a) 4.4×10^5 Pa, 15°C

    (b) No change. Work is done on the gas but this is equal to the heat transferred to the surroundings.

    (c) $\Delta U = 0$, $Q = -W$; (d) 1152 K (879°C)

    (e) Same number of molecules occupy larger volume so collision rates are lower. To exert the same pressure the particles must move faster (have more KE) so the gas must be heated and its temperature must rise.

2. (a) 0.018 moles

    (b) External forces cause compression/expansion and do work on the tire and air, heating it up.

    (c) 32°C

    (d) Assumes constant volume. In practice, volume might increase slightly.

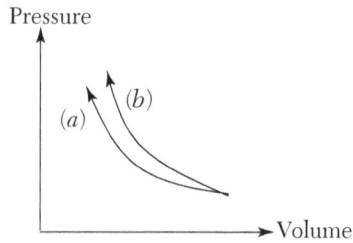

3. (a),(b)

(c) Slow compression: temperature is constant so U is constant. Heat flow out of system is equal to work done on the system. Fast compression: heat flow is minimal so work done on system increases its internal energy, temperature, and pressure.

4. (a) 500 ms^{-1}; (b) Same mean KE at same temperature but molecular masses are different: more massive molecules (oxygen) have lower rms speed.

5. (a) Molecular volume can no longer be ignored.(b) Collisions become inelastic.(c) Interactions can no longer be ignored (bonds will form).

6. –

7. (a) 0; (b) 519.5 ms^{-1}; (c)) 519.9 ms^{-1}; (d) 7.3 × 10^{-21} J

8. Internal energy U depends only on absolute temperature T. In an isothermal change U is constant and so is T. Work done to compress the gas equals the heat that flows out of the gas to the surroundings.

9. (a) –; (b)(i) $\Delta U = 0$; (ii) $W = H$

10. (a) –; (b)(i) exponential increase; (ii) exponential decay; (iii) exponential decay; (c) Increasing T increases the B.F. so a larger fraction of collisions exceeds the activation energy and the reaction rate increases.

11. 2.5 ×

12. 1.3 × 10^{-5}

13. (a) As T increases B.F. increases so a larger fraction of the electrons have enough energy to jump to the conduction band.

(b) $I = GV = \text{const.} \times V \times e^{-\frac{\Delta E}{kT}}$ so a graph of ln I against $1/T$ has gradient $-\Delta E/k$

14. (a) –; (b) 0.67; (c) 0.174 gs^{-1}

Chapter 10   Statistical Thermodynamics and the Second Law

1. –

2. (a) –; (b) Newton's laws are reversible.(c) Entropy increases so the future is distinct from the past as long as the universe started in a low entropy state.(d) Cosmological expansion (but this is also dependent on the universe starting in a low entropy state).

3. (a) It was very low.(b) If all possible initial conditions are considered equally likely the one in which the universe actually began was one of very low probability (i.e. a microstate belonging to a very small number of microstates).

4. –; 5. –; 6. –

Chapter 11   Oscillations

1. (a) AC, (b) B, (c) B, (d) C, (e) B, (f) C.

2. (a) Position of zero resultant force; (b) $F\mu - x$; (c) 0.50 Hz; (d) 6.0 cm; (e) $y = 6.0 \cos(\pi t)$ (+ve to right); (f) –; (g) $v_{max} = 0.19$ ms^{-1}, KE$_{max}$ = 0.014 J; (h) No change; (i) Increasing mass reduces acceleration at each point increasing T; increasing k increases acceleration at each point decreasing T; (j) Work done against frictional forces. Total energy and amplitude decay; (k) E.g. look for constant ratio: 5.4/6.0 ~ 4.9/5.4 ~ 4.4/4.9 etc. This supports exponential decay; (l) 2.1 cm

3. (a) Differentiate twice; (b)(i) 2.5; (ii) 0.20 s; (iii) 78.5 ms^{-1}; (iv) 2470 ms^{-2}; (c)(i) $1540 \sin^2(10\pi t)$; (ii) –; (iii) –

4. $f > \dfrac{1}{2\pi}\sqrt{\dfrac{g}{A}}$

5. (a) 9.93×10^{14} Hz; (b) 3.02×10^{-7} m, UV

6. (a) 0.635 m; (b) 0.5 % increase; (c) Mass-spring period unchanged, pendulum period about 3.9 s

Chapter 12   Rotational Dynamics

1. (a) –; (b) Force and velocity are perpendicular so $F.\delta s = 0$.

2. (a) $\dfrac{5}{2}mrg$

(b) N

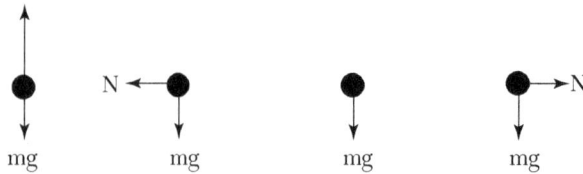

mg   mg   mg   mg

(c) No. Ball will have tangential acceleration as well as centripetal acceleration and resultant force is parallel to resultant acceleration.

(d) $v = \sqrt{rg}$

3. Velocities shown on diagram. Accelerations all equal to v^2/r and directed toward center.

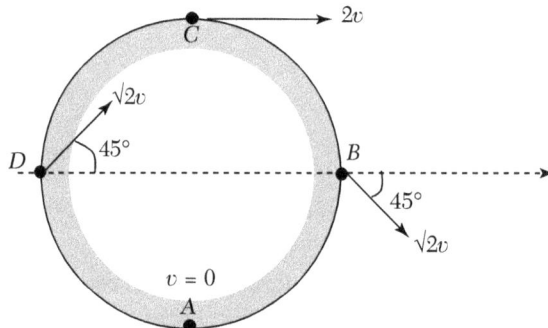

4. (a) 1.82 s; (b) 0.55 Hz; (c) 3.5 rad s^{-1}; (d) 7.8×10^{-3} Js; (e) 0.054 Nm; (f) 0.013 J

5. –

6. (a) 4.7×10^9 kgm^2; (b) Astronaut experiences an inward contact force which maintains his circular motion about the centerof rotation. This feels like the reaction to a gravitational field of strength $g = rw^2$; (c) 0.22 rad s^{-1}; (d) 94 MJ

7. (a) 0.63 Js; (b) 26.2 s

8. No external resultant torque. Angular momentum is conserved. $L = I\omega$. I is reduced so ω increases. Work must be done by the children as they use forces to move inwards. This increases RKE.

9. 8.3×10^{28} Nm

Chapter 13   Waves

1. (a) amplitude = 25 m frequency = 4.8 Hz, wavelength = 3.14 m; (b) 15 ms^{-1}; (c) In the negative x-direction.

2. (a) One cycle with period 20s and amplitude 1.2 m; (b) f = 0.10 Hz, λ = 12 m; (c) 90°, $(\pi/2)$

3. –

4. (a) **Row 1:** air or vacuum / 2.0 × 10^8 ms^{-1} / 25°; **Row 2:** 3.0 × 10^8 ms^{-1} / 3.0 × 10^8 ms^{-1} / 75°; **Row 3:** 2.3 × 10^8 ms^{-1} / 1.2 × 10^8 ms^{-1} / 38°; **Row 4:** diamond / glass / 1.2 × 10^8 ms^{-1}
   (b) –
   (c) Lower critical angle, so light reflects internally to a greater extent and returns to observer by many paths.

5. HINT: Exploit the symmetry.

6. (a) $I_T = \frac{1}{2} I_0 \cos\theta \cos(90° - \theta) = \frac{1}{4} I_0 \sin 2\theta$ maximum when $2\theta = 90°$, i.e., $\theta = 45°$; (b) 0.22

Chapter 14   Light

1. (a) 2.57 s; (b) to about 0.1 ns; (c) diffraction.

2. (a) –; (b) 4 times
   (c) Image distance and magnification increase as u approaches f. When $u = f$ the image is at infinity. For $u > f$ a real image is formed on the far side of the lens, inverted, in the space between the eye and the lens (if the eye is far enough away).

3. **1st row:** 20 cm, 2, virtual erect; **2nd row:** infinity, infinity, not defined; **3rd row:** 60 cm, 2, real inverted; **4th row:** 40 cm, 1, real inverted; **5th row:** 20 cm, 0, not defined

4. **1st row:** 6.7 cm, 0.67, virtual erect; **2nd row:** 10 cm, o,50, virtual erect; **3rd row:** 12 cm, 0,40, virtual erect; **4th row:** 20 cm, 0, not defined

5. (a) 30 times; (b) 20 times and length increased by 2.0 cm.

6. (a) 12.5 ms^{-1}; (b) 3.4 × 10^{24} m

7. E.g. by measuring the Doppler shift of 21 cm hydrogen lines from spiral arms and calculating their velocities relative to earth. From these 9 and some geometry) the angular velocity of the galaxy can be calculated.

Chapter 15   Superposition

1. (a) 0 (b) π (c) $\lambda/2$; (d) Amplitude $\times$ 2, intensity $\times$ 4; (e) 0;
(f) Amplitude $\times$ 1/2 , intensity $\times$ ¼ (g) 0.026 m, 13.1 kHz; (h) doubles;
(i) halves; (j) the phase difference is between $\pi/2$ and π

2. (a) –; (b) 9.8 mm; (c)(i) maxima closer together; (ii) minima not
completely dark; (iii) no effect; (iv) maxima farther apart; (v) maxima
farther apart; (vi) clarity of fringes varies—maximum at 0 and π, but no
fringes at $\pi/2$ and $3\pi/2$ when polarizations are perpendicular.

3. Same fringe separation but maxima 9/4 times brighter.

4. (a) 8.28°, 8.97°; (b) 1.44°; (c) 6 orders
(d) 1st order diffraction minimum at 25.6° and third order grating
maxima at 25.6° and 27.9° so maxima absent or very dim here.

5. –

6. (a) 0.15°; (b) –; (c) –; (d) Minima closer to centerof pattern.

7. –

8. (a) 1.2×10^{-4} rad; (b) about 8 m; (c) 2.4×10^{-4} light years; (d) Limited
by other factors—e.g. retinal cell density/sensitivity; optical quality of
cornea and lens; (e) Minimum angle resolved is inversely proportional to
diameter and $d_{scope} >> d_{eye}$

9. (a) Assume: separation ~ 2 m, d_{max}~ 20 km (in practice less than this
because of other factors); (b) Needs objective diameter ~ 30 m, so no.

10. (a) 8.82×10^{-5} kgm^{-1}; (b) 543 ms^{-1}; (c) 362 Hz; (d) 362 Hz, 724 Hz,
1086 Hz

Chapter 16   Sound

1. –

2. (a) 30 dB; (b) doubles to 60 Db

3. (a) 0.283 m; (b) –; (c) h_1 = 37.9 cm, h_2 = 23.8 cm, h_3 = 9.6 cm

4. (a) Otherwise most of the incident ultrasound would reflect off the
surface of the skin. Gel matches the impedance of tissue so most is
transmitted; (b) 31 %

Chapter 17   Electric Charge and Electric Fields

1. (a) 14 400 C; (b) 3.0×10^{-3} C; (c) 0.20 A

2. –

3. (a) **Field strengths:** A: 9.0×10^{15} Vm^{-1} to left; B: 4.0×10^{15} Vm^{-1} to right; C: 8.4×10^{14} Vm^{-1} to left; D: 4.0×10^{15} Vm^{-1} to right; E: 8.38×10^{14} Vm^{-1} to left; **Electric potentials:** A: 0V; B: 9.0×10^{9} V; C: 0V; D: -9.0×10^{9} V; E: 0V

4. (a) 8.19×10^{-8} N; (b) 2.2×10^{6} ms^{-1}, 2.2×10^{-18} J;(c) -4.3×10^{-18} J; (d) -2.2×10^{-18} J; (e) $+2.2 \times 10^{-18}$ J

5. (a) –
   (b) $E = 62500$ Vm^{-1} at 4.0 cm and 8300 Vm^{-1} at 8.0 cm; $V = 2500$ V at 4.0 cm and 910 V at 8.0 cm.

6. Field lines originate on charge and meet surface at 90°. Equipotentials perpendicular to field lines.

7. (a) Gaussian surface inside sphere encloses zero charge so net flux is also zero. By symmetry zero everywhere inside.
   (b) Use surface of small flat cylinder embedded in conductor with ends inside and outside surface of conductor. All field lines must pass through external surface and must be perpendicular to conductor so $EA = Q/\varepsilon_0$ and $E = \sigma/\varepsilon_0$ where A is the surface of the conductor inside the cylinder.
   (c) Gaussian surface enclosing empty space encloses zero charge so net flux through surface is zero and flux entering volume must equal flux leaving it.

8. (a) –; (b) –; (c) 250 000 Vm^{-1}; (d) 8.0×10^{-14} N; (e) No change (uniform field); (f) $a_e = 60\,000\ a_{ion}$

9. (a) 2.3×10^{7} ms^{-1}; (b) parabolic downward; (c) 2.2×10^{-9} s; (d) 7.7×10^{6} ms^{-1}; (e) 18.5°; (f) 2.7×10^{-7} J

Chapter 18   D.C. Circuits

1. (a) –; (b) –

2. 7.0×10^{-5} ms^{-1}

3. (a) 0.26 A; (b) 3.3×10^{5} Am^{-2}; (c) 17 Ω^{-1}m^{-1}; (d) 0.026 Ω^{-1}

4. (a) B lights normally, A does not light; (b) A lights almost normally and B lights just below normal brightness.

5. (a) –; (b) $V = \dfrac{ER}{(R+r)}$; (c) Plot 1/V against 1/R: gradient = r/E and intercept = 1/E; (d) –

6. (a) copper; (b) 0.62 Ω, 0.39 Ω, 2.22 Ω; (c) 44%, 54%, 91%; (d) 0.43 Ω; (e) 4.6 A

7. (a) 4.0 V, 0.020 A; (b) 3.87 V, 0.019 A

8. (a) in series: R, 2R, 3R; in parallel: R/2, R/3; series+parallel: 2R/3, 3R/2

9. (**Hint:** redraw equivalent circuit as 3 + 6 + 3 groups of parallel resistors) 6R/5

10. (a) Dark 5.9 V, Light 0.55 V; (b) For high values the range of voltages becomes smaller. This makes the device less sensitive but more linear.

11. ammeter 1: 0.060 A, ammeter 2: 0.135 A, ammeter 3: 0.075 A; voltmeter: 3.0 V

Chapter 19   Capacitors

1. (a) 1.32 mC; (b) 0.00396 J; (c) charge doubles, energy quadruples

2. (a) 0.103 s; (b) ~ 0.5 s; (c) no change, time constant is the same; (d) 30.7 kΩ, 330 μA

3. (a) exponential decay from I_0 = V/R; (b) V_1 = constant, V_2 exponential decay, V_3 growing from zero at decaying rate toward V.

4. (a) 0.00040 C, 0.0016 J; (b) Q_1 = 0.00013 C, Q_2 = 0.00027 C; (c) 0.00053 J energy dissipated as heat in connecting wires as charge is redistributed; (d) Q_1 = 0.00040 C, Q_2 = 0.00080 C, Q_{tot} = 0.00120 C; (e) 0.0048 J, cell does extra work to charge both capacitors to same voltage

5. Series: C, C/2, C/3; parallel: 2C, 3C; series+parallel: 2C/3, 3C/2

6. Charge is halved; current forced back through cell as plates separate; energy halved; voltage constant.

7. 0.0011 C, 2.34 V

Chapter 20   Magnetic Fields

1. Temperature in core is above Curie temperature so permanent magnetization not possible.

2. (a) Circular filed lines around each wire: direction from RH rule. Field of either wire intersects other current at 90° creating a "motor effect" force: directions from FLHR. Attraction; (b) 1.78×10^{-6} Nm^{-1}; (c) Currents on opposite sides of coil have a repulsive interaction producing an outward (explosive) force. This will be large if current and number of turns is large.

3. (a) –; (b) $r_1 = \dfrac{m_1 v}{Bq}$; $r_2 = \dfrac{m_2 v}{Bq}$; (c) separation $= \dfrac{2(m_1 - m_2)v}{Bq}$

4. (a) –; (b) Distance of line of action of force from axis varies as coil turns. Max. at 0°, min. at 90°; (c) 0.0312 Nm, 0.027 Nm, 0.022 Nm, 0.016 Nm. 0 Nm

Chapter 21   Electromagnetic Induction

1. (a) Flux in coil is changing. By Faraday law this results in an induced emf.

(b)(i) Larger deflection—greater $\dfrac{dN\Phi}{dt}$; (ii) negative deflection—sign of $\dfrac{dN\Phi}{dt}$ has changed; (iii) No deflection $\dfrac{dN\Phi}{dt} = 0$

2. (a) 2.0×10^{-6} Wb; (b) 4.0×10^{-4} Wb; (c) –;(d) 1.3×10^{-4}A; (e) 6.7×10^{-5} C; (f) current increased $\times 5$ but charge is the same

3. Field of falling magnet cuts through conducting copper walls. Rate of change of flux-linkage in copper creates eddy currents. These create a magnetic field that acts back on the falling magnet opposing its motion (Lenz law). Magnet accelerates until upward magnetic field balances weight and then falls at a terminal velocity.

4. (a) Wings are conductors cutting through the magnetic field. Rate of cutting flux is equal to induced emf; (b) 0.40 V; (c) No. Return path would have same emf so sum of emfs around circuit would be zero. OR: closed circuit encloses constant flux so no change and no emf around loop.

5. (a) 100 turns; (b) –; (c) core forms a magnetic circuit so that the changing flux in the primary coil also links the secondary coil;

(d) laminations interrupt eddy currents reducing the heat dissipated in the core. (e) Eddy currents / ohmic heating in coils / magnetic losses caused by constant magnetization and demagnetization.

6. (a) 0.0017 H, 0.50 Ω; (b) 0.034 J; (c) When switch is opened I falls to zero very rapidly so $\dfrac{dN\Phi}{dt}$ is large and there is a large induced emf. This appears across separating switch contacts. Air breaks down and a spark is formed.

7. (a) 90.5 V; (b) –; (c) 1.81 A; (d) 164 W; (e) answer (b) is unchanged, (c) and (d) are reduced

Chapter 22   Alternating Current

1. (a) D.C. batteries, compact, portable, no need to rectify but expensive, limited life, disposal issues; A.C. supply: can be stepped up or down with transformers, cheap at point of use, easy to generate

(b) Can be stepped up at generator and transmitted at high voltage and low current reducing heat loss from transmission lines.

2. (a) 28.3 V; (b) 20 V; (c) May be shorter with A.C.> because of continual heating/cooling stress

3. Resistor: 100 Ω at all frequencies. Capacitor: infinity / 16 Ω / 1.6 Ω / 0.16 Ω. Inductor: 0.063 Ω / 0.63 Ω / 6.3 Ω

4. (a) 46.4 Ω; (b) 0.35 A; (c) 14 V; (d) 2.6 W; (e) current and voltage have a 90° phase difference so the integral, over one cycle, of $P = IV$ is zero; (f) voltage leads current by 62°

5. (a) infinity / 12.1 Ω / 10.0 Ω / 10.3 Ω / 32.9 Ω; (b) 4800 Hz; (c) current rises to peak of 1.2 A at f = 4800 Hz; (d) From low frequency limit with voltage lagging current by 90° through 0 at 4800 Hz to voltage leading current by 90° at high frequency limit; (e) Circuit is purely resistive at resonance so power dissipated is maximum. Energy is dissipated as heat in thee resistor but the capacitor and inductor alternately store and release energy back to the circuit, they do not dissipate energy.

Chapter 23   Gravitational Fields

1. (a) 9.8 N; (b) 2.72×10^{-3} N; (c) 3600 and 3602—i.e. the ratios are the same so it is consistent.
(d) –; (e) –; (f) –

2. (a) $V_P = \dfrac{-GM}{R}$; (b) $V_Q = \dfrac{-GM}{4R}$; (c) $W = \dfrac{3GMm}{4R}$; (d) $v = \sqrt{\dfrac{GM}{2R}}$;

(e) $v_{esc} = \sqrt{\dfrac{2GM}{R}}$; (f) $v_{esc} = \sqrt{\dfrac{GM}{2R}}$; (g) $\dfrac{GMm}{8R}$

3. (a) $g = \dfrac{4}{3}G\pi\rho r$; (b) g is greater at poles than equator; (c) about 3 %

4. 3.44×10^8 m from Earth

5. (a) –; (b) –; (c) $\dfrac{r^3}{T^2} = \dfrac{GM}{4\pi^2}$ so constant depends on central mass;

(d) 4.32×10^7 m; (e) orbit is centered on center of Earth so anything other than an equatorial orbit would vary in latitude and not be stationary with respect to the surface. (f) 1.4 h

6. 1.9×10^{27} kg

7. More than 6 times as far

8. –;

9. –

Chapter 24   Special Relativity

1. (a) –; (b) –; (c) –

2. (a) 87.2 m; (b) 5680 m; (c)(i) 22.2 y; (ii) 9.69 y; (d) 8.72 ly; (e) Earth twin 65.4, traveler 40.4

3. (a) –; (b) 0.14 c / 0.42 c / 0.75 c / 0.98 c; (c) – ; (d) if v << c then $\gamma \sim 1$ and relativistic effects are negligible; (e) –; (f) 1.2 GV

4. (a) Traveling clock slows—shows less elapsed time; (b) 18 ns

5. (a) –; (b) –; (c) –; (d) $\dfrac{l_0}{\gamma}$; (e) B sees a synchronization error on A's clocks

with X started before Y so that A's measurement of time is incorrect. B also sees both of A's clocks running slow. A sees B's clock running slow. Both disagree with the way measurements have been carried out in the other reference frame. However, both are in inertial reference frames so their measurements are valid within their own frame.

Chapter 25   Atomic Structure and Radioactivity

1. (a) –; (b) –; (c) –; (d) –; (e) 4.6×10^{-14} m; (f) scattering is purely electrostatic / Coulomb law is valid

2. Can control energy using an accelerator, Electrons do not feel the strong nuclear force but alpha particle do so for close approach electrons are only scattered by electrostatic forces.

3. (a)

Element	Symbol	Atomic number	Nucleon number	Protons	Electrons	Neutrons
Hydrogen	$^{1}_{1}\text{H}$	1	1	1	1	0
Carbon-12	$^{12}_{6}\text{C}$	6	12	6	6	6
Carbon-13	$^{13}_{6}\text{C}$	6	13	6	6	7
Carbon-14	$^{14}_{6}\text{C}$	6	14	6	6	8
Oxygen-16	$^{16}_{8}\text{O}$	8	16	8	8	8
Iron-56	$^{56}_{26}\text{Fe}$	26	56	26	26	30
Gold-	$^{157}_{79}\text{Au}$	79	197	79	79	118
Uranium-235	$^{235}_{92}\text{U}$	92	235	92	92	143
Uranium-238	$^{235}_{92}\text{U}$	92	238	92	92	146

(b) E.g., carbon isotopes: same atomic number but different mass number

(c) Electrostatic repulsion between protons is a cumulative long range repulsion. Strong nuclear interaction between neutrons and protons is short range. For a large nucleus more neutrons are needed to stabilize the nucleus against the Coulomb force.

4. (a) 20 cpm; (b) radioactive decay is a random process so it fluctuates; (c) 8 h; (d) alpha and gamma

5. (a) $1.22 \times 10^{-4}\, y^{-1}$ or $3.85 \times 10^{-12}\, s^{-1}$; (b) 22 800 y assuming no contamination with "younger" carbon; (c) 19,000 years

6. 2

7.

Time / Years	P	Q
0	N	0
1000	N/2	N/2
2000	N/4	3N/4
3000	N/8	7N/8
5000	N/32	31N/32

Q rises from zero to asymptote at N, P falls exponentially from N to zero

$P + Q = N$ at all points sop graph lines cross at 1000 years.

8. (a) –; (b) 3.43 mm; (c) 23 cm

9. –

10. (a) 8 alpha and 6 beta

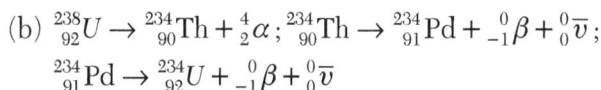

(b) $^{238}_{92}U \rightarrow {}^{234}_{90}Th + {}^{4}_{2}\alpha$; $^{234}_{90}Th \rightarrow {}^{234}_{91}Pd + {}^{0}_{-1}\beta + {}^{0}_{0}\overline{v}$;

$^{234}_{91}Pd \rightarrow {}^{234}_{92}U + {}^{0}_{-1}\beta + {}^{0}_{0}\overline{v}$

(c) Abundances in approximate ratio of half-lives

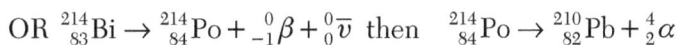

(d) $^{214}_{83}Bi \rightarrow {}^{210}_{81}Tl + {}^{4}_{2}\alpha$ then $^{210}_{81}Tl \rightarrow {}^{210}_{82}Pb + {}^{0}_{-1}\beta + {}^{0}_{0}\overline{v}$

OR $^{214}_{83}Bi \rightarrow {}^{214}_{84}Po + {}^{0}_{-1}\beta + {}^{0}_{0}\overline{v}$ then $^{214}_{84}Po \rightarrow {}^{210}_{82}Pb + {}^{4}_{2}\alpha$

11. Half-life approx. 185 s, decay constant approx. $3.7 \times 10^{-3}\ s^{-1}$

Chapter 26   Nuclear Physics

1. –

2. B.E. = 470 MeV, B.E./A = 8.3 MeV/nucleon

3. (a) $^{235}_{92}U \rightarrow {}^{231}_{90}Th + {}^{4}_{2}\alpha$; (b) Use conservation of momentum:

KE_{Th} = 81 keV; (c) 0.0052 u

4. (a) $^{16}_{6}C \rightarrow {}^{16}_{7}N + {}^{0}_{-1}\beta + {}^{0}_{0}\overline{v}$

(b) 8.0 MeV; (c) Energy is shared randomly with anti-neutrino

(d) Neutron rich nucleus. Large mass defect for decay. High probability of decay and therefore short half-life.

5. (a) $^{13}_{7}N \rightarrow {}^{13}_{6}C + {}^{0}_{+1}\beta + {}^{0}_{0}\overline{v}$ (1); $^{13}_{7}N \rightarrow {}^{13}_{8}O + {}^{0}_{-1}\beta + {}^{0}_{0}\overline{v}$ (2)

(b) For (1) Δm = 0.0024 u, for (2) $\Delta m = -0.019$ u, so mass of products in (2) is greater than mass of original nucleus so that reaction cannot proceed spontaneously.

(c) 7.5 MeV / nucleon, 7.7 MeV / nucleon—both are stable and nitrogen-15 might be expected to be more abundant. In fact nitrogen-14 is much more abundant because the odd-odd configuration is much more stable than an odd-even configuration of nucleons, allowing protons and neutrons to pair up.

(d) $t_{1/2}(C-16) > t_{1/2}(C-17) > t_{1/2}(C-18)$. Neutron excess destabilizes nucleus because neutrons are themselves unstable.

6. (a) –; (b)(i) $^{14}_{6}C \rightarrow {}^{14}_{7}N + {}^{0}_{-1}\beta + {}^{0}_{0}\overline{v}$; (ii) $^{1}_{0}n \rightarrow {}^{1}_{1}p + {}^{0}_{-1}\beta + {}^{0}_{0}\overline{v}$;

(iii) $^{1/3}_{-2/3}d \rightarrow {}^{1/3}_{+1/3}u + {}^{0}_{-1}\beta + {}^{0}_{0}\overline{v}$

(c) Yes—proton (uud) changes to a neutron (udd) so an up quark changes to a down quark.

Chapter 27 Quantum Theory

1. (a) 5.5×10^{-11} m; (b)(i) 6.2 keV; (ii) 1.6×10^{-11} m

(c) Visible photon energy < 10 eV, X-ray photon energy > 10 eV

2. –

3. (a) photon energy < work function

(b) photons only transfer energy to single electrons (at the intensities used)

(c) UV photon energy > 4.3 eV

(d) more photons per second arrive so more electrons per second emitted

(e) work function zinc > blue photon energy > work function of potassium

(f) Max. KE increases ($= hf - \Phi$)

4. (a) 0.76 eV; (b) 5.2×10^5 ms^{-1}; (c) 0.76 eV

(d) No change in max. KE or stopping voltage but more electrons are emitted per second so photocurrent increases.

5. (a) 3.4 eV; (b) 4.9×10^{-7} m, visible;

(c) (i) Elastic—scattered electron has 11 eV of KE.
(ii) Inelastic: electron excites atom to $n = 2$ level. Atom absorbs 10.2 eV and electron is scattered with 0.8 eV of KE.

6. (a) 2.5×10^{19} s^{-1}; (b) Photon energy is so small that changes in intensity are effectively continuous.

7. (a) $F = \dfrac{IA}{c}$; (b) Approx. 100 000 m^2

8. –;

9. –

10. $E_n = -\dfrac{54\,\text{eV}}{n^2}$

Chapter 28   Astrophysics

1. (a) 5.0×10^{-7} m; (b) 620 Wm^{-2}; (c) 7.0×10^8 m; (d) Peak of curve shifts to shorter wavelengths. 2.90×10^{-8} m (UV)

2. (a) 3500 K; (b) 5.7×10^{11} m

3. (a) 0.05 c moving away; (b) 7.2 light years

4. (a) –; (b) 3.09×10^{16} m, 3.26 light years; (c) parallax angle becomes too small to measure; (d) 19.2 parsecs

5. (a) 10700 light years; (b) 26 Mpc

6. –

7. (a) 9.7×10^{10} Hz; (b) 9.7×10^{10} Hz

(c) Shift for rocket is caused by Doppler Effect due to relative motion, shift for galaxy is caused by cosmological expansion of space.
(d) 4.3×10^8 years

8. –

Chapter 29   Medical Physics

1. –; 2. –

2. (a) –; (b) 3.1×10^{-11} m; (c) –

3. 4. –

4. (a) Photons have momentum. In the CM frame the electron and positron have net zero momentum sop if only a single photon was emitted this would violate the law of conservation of momentum. In the CM frame a pair of identical photons are emitted in opposite directions.

5. 2.4×10^{-12} m

GLOSSARY

Absolute space: proposed by Newton as a background against which all physical processes take place. All observers, regardless of their own motion, would agree on the separation of events in absolute space. Einstein's special theory of relativity showed that this cannot be the case.

Absolute time: proposed by Newton as a universal time in which all observer, regardless of their own motion, would agree on time intervals and the rate of flow of time. Einstein's special theory of relativity showed that this cannot be the case.

Absolute uncertainty: likely range of values represented by a measurement—e.g., a measurement of length might be 4.565 ± 0.002 m: an absolute uncertainty of ± 0.002 m (± 2 mm). The smaller the uncertainty the more precise the measurement.

Absolute zero: lower fixed point of the thermodynamic temperature scale. According to kinetic theory, molecular motion would stop at this temperature.

Absorption lines: dark lines in a spectrum corresponding to wavelengths that are strongly absorbed by the medium through which radiation has passed.

Acoustic impedance: product of the speed of sound in a medium and the density of the medium.

Action-at-a-distance: Newton's original idea that gravitational forces act on distant objects instantaneously and with nothing acting as an intermediary.

Activation process: process that depends, on a molecular scale, on the molecules gaining an additional energy ΔE above the average energy at that temperature.

Activity: number of decays (disintegrations) per second inside a radioactive source.

Adiabatic change: change that takes place with no transfer of heat in or out of the system ($Q = 0$ in the first law of thermodynamics).

Alpha particle: helium nucleus. Emitted from certain unstable nuclei in radioactive decay.

Alternating current (A.C.): current alternates its direction of flow in the circuit and the polarity of the supply alternates in a periodic way.

Ampère theorem: theorem that relates the line integral of the magnetic field strength around a closed loop to the current passing through the area enclosed by the loop.

Amplitude: maximum disturbance from equilibrium in a wave or oscillation.

Angular momentum: sum of moments of linear momentum about an axis of rotation for all points in a rotating body, given by $L = I\omega$ and measured in Js.

Anode: positive electrical terminal.

Aphelion: point on an elliptical planetary orbit when the planet is farthest from the Sun. The equivalent point in the orbit of an Earth satellite is called apogee.

Archimedes principle: that the buoyancy force on an object placed into a fluid is equal to the weight of the fluid it displaces.

Arrow of time: distinction between the past and the future linked to an irreversible physical change, such as increasing entropy.

Atomic mass number (A): number of nucleons (protons + neutrons) in the nucleus. Also called nucleon number or just mass number.

Atomic number (Z): number of protons in the nucleus. Corresponds to the position of the element in the Periodic Table and the number of orbiting electrons in the neutral atom.

Attenuation: reduction of intensity with distance as a result of absorption in a medium.

Balmer formula: numerical relationship between visible wavelengths in the hydrogen atom line emission spectrum. This relationship could not be explained using classical physics.

Barometer: instrument used to measure atmospheric pressure.

Base unit: seven agreed independent units from which the S.I. system is constructed.

Battery: several cells connected together in series or parallel to form a power supply.

Beer–Lambert law: law of attenuation when radiation passes through an absorbing medium.

Bernoulli equation: equation describing conservation of energy in a flowing fluid. Can be used to relate pressures and flow rates in one part of the system to those in another.

Beta radiation: high-energy electron (β^-) or positron (β^+) emitted when an unstable nucleus decays.

Big Bang: origin of the universe exploding from a point about 13.7 billion years ago.

Biot–Savart law: equation used to find the magnetic field due to electric currents.

Black body radiation: spectrum of radiation emitted by an ideal radiator.

Black hole: object whose gravitational field is so strong that, at a certain distance from its center, the escape velocity is equal to the speed of light.

Bohr model: model of the hydrogen atom in which electrons move in circular orbits around the central nucleus but can only occupy orbits in which their angular momentum is an integer multiple of $h/2\pi$. This results in a discrete set of allowed energy levels.

Boltzmann factor: ratio of numbers of particles expected to have energy $E + \Delta E$ to the number having energy E in a system. This is equal to the probability that a particle can gain the extra energy ΔE for an activation process: $f = e^{-\frac{\Delta E}{kT}}$

Bourdon gauge: instrument used to measure pressure.

Boyle law: relationship between pressure and volume for a constant mass of an ideal gas at constant temperature: $pV = $ constant.

Bremsstrahlung: "braking radiation"—continuous spectrum of X-ray radiation emitted when an electron beam strikes a metal target and the electrons decelerate as they collide with atoms in target.

Brewster law: light incident on the surface of a transparent material at the Brewster angle, given by $\tan\theta_B = \dfrac{n_2}{n_1}$, splits into refracted and reflected beams that are polarized perpendicular to one another.

Brittle: material property when little or no plastic deformation occurs before fracture and fracture mechanism is by crack propagation.

Buoyancy: upward force on a body in a fluid caused by the pressure difference between the top and bottom of the body. Also called "upthrust." The size of the buoyancy force is given by Archimedes principle.

Capacitor: pair of conductors separated by an insulator. When there is a potential difference between the two conductors they store opposite charges of magnitude $Q = CV$, where C is the capacitance of the capacitor, measured in farads.

Cathode: negative electrical terminal.

Center of gravity: point where the resultant gravitational force on a body acts.

Centripetal acceleration: an object in uniform circular motion has an acceleration directed toward the center of the circle in which it moves.

Centripetal force: resultant force acting toward the center of the circle when an object is in uniform circular motion.

Cepheid variable: type of variable star with a well understood intensity-luminosity relationship so that its absolute luminosity can be inferred from its period of variation. Used as a standard candle for distance measurements.

Ceramics: solid non-metallic materials formed by high temperature firing, usually strong hard and brittle, e.g., pottery or brick.

Charge carriers: particles within an electrical conductor that are responsible for the electric current (e.g., electrons in most metals).

Charles law: relationship between volume and temperature for a constant mass of an ideal gas at constant pressure: $\dfrac{V}{T} = \text{constant}$

Classical physics: usually refers to Newtonian mechanics (including the law of gravitation) and Maxwell's laws of electromagnetism. It corresponds to the known physics prior to 1900 and excludes: relativity, quantum theory, and particle physics.

Closed system: in mechanics, a system with no external forces acting upon it. In thermodynamics, a system which does not exchange heat with its surroundings.

Coefficients of friction: ratio of frictional force to normal contact force when the surfaces just begin to slip (static coefficient of friction) or when they are sliding over one another (dynamic coefficient of friction).

Coherent sources: monochromatic sources that remain in phase or maintain a constant phase difference.

Commutator: rotating switch in a D.C. motor that ensures the current direction changes every half rotation so that the turning effect on the coil is always in the same direction.

Composite materials: combinations of two or more different materials designed to take advantage of the desirable properties of each individual component.

Compound pendulum: pendulum in which the mass is spread out rather than concentrated at one point (i.e., in the pendulum bob).

Compton Effect: scattering of X-rays from atomic electrons which can be analyzed as if it was a collision between particles. The scattered X-ray has a change of wavelength related to the angle of scatter.

Conservative field: if a particle is moved around a closed loop in a conservative field there is no net change in energy. The electric and gravitational fields are both conservative fields.

Continuity (equation of): equation stating that the mass flow rate of a fluid at one point in a pipe is equal to the mass flow rate at another: $\rho_1 A_1 v_1 = \rho_2 A_2 v_2$. If the fluid is regarded as incompressible this reduces to: $A_1 v_1 = A_2 v_2$.

Control variables: variables that must be kept constant during an experiment so that any change in the dependent variable is only caused by a change in the independent variable. All control variables must be kept constant for the experiment to be a fair test of the relationship between the independent and dependent variables.

Convection: transfer of heat in a fluid as a result of bulk movement of particles in convection currents. These arise as a result of density changes in the fluid when it is heated.

Conventional current: direction of current flow defined to be from the positive terminal of the supply to the negative terminal of the supply in the external field (i.e., the direction in which a positive charge carrier would move) regardless of the sign(s) of the charge carriers that make up the current.

Copenhagen Interpretation: mainstream interpretation of quantum mechanics in which the state of a system is described by a wave function and there is a collapse of the wave function when an observation or measurement is made.

Copernican model: heliocentric (Sun at the center) model of the solar system

Cosmic microwave background radiation: black body radiation corresponding to a temperature of about 2.7 K present throughout the universe. One of the key pieces of evidence supporting the idea of a Big bang and the expanding universe.

Cosmology: the study of the origin nature and end of the entire universe.

Coulomb law: force law between electric charges: $F = \dfrac{Q_1 Q_2}{4\pi\varepsilon_0 r^2}$.

Couple: moment of a pair of forces of equal magnitude, acting in opposite directions through different points in the same body: couple = magnitude of one force × distance between lines of action of the two forces.

Creep: gradually increasing strain with constant stress.

Crystalline materials: having long range geometric microstructure, e.g., metals.

CT (computed tomography): medical imaging technique using X-rays to create detailed images of slices of the body.

Curie temperature: temperature above which thermal motions prevent the permanent magnetization of a ferromagnetic material.

Damping force: force opposing motion causing the oscillator to do work as it oscillates and resulting a decay of amplitude and energy.

De Broglie relation: fundamental relationship between the wave and particle models of radiation and matter: $\text{wavelength} = \dfrac{\text{Planck constant}}{\text{momentum}}$.

Decibel scale: logarithmic scale of relative intensity used to compare intensity levels with the level at the threshold of human hearing (I_0): $\text{intensity level (dB)} = 10\log_{10}\left(\dfrac{I}{I_0}\right)$

Degeneracy pressure: the Pauli Exclusion Principle prevents fermions (half-integer spin particles) from existing in the same set of quantum states, so that when a gas of fermions is compressed the lower energy states become filled and it exerts an outward pressure.

Degree of freedom: in kinetic theory, the different modes in which a particle can absorb energy, e.g., translation, rotation, vibration.

Dependent variable: the variable you are investigating to find out how it depends on an independent variable.

Derived unit: unit built up from more than one base unit, e.g., ms^{-1} for velocity. Some derived units have their own name, e.g., the joule, for energy. The joule is equivalent to kgm^{2}s^{-2} in base units.

Deterministic: where having complete knowledge of the present state of a system is sufficient to make a complete prediction of its future state. Newtonian mechanics is a deterministic system.

Dielectric material: insulating material that is polarized by an external electric field. When a dielectric material is placed between the plates of a capacitor increases its capacitance.

Diesel cycle: idealized cycle for a diesel internal combustion engine.

Diffraction grating: an optical component consisting of a large number of narrow parallel slits—used in spectroscopy to analyze light.

Diffraction: spreading of a wave into a region of geometric shadow after passing through an aperture or past an object or edge.

Dimension: the nature of a physical parameter independent of its quantity, e.g., the dimension of time but not the actual duration of a particular time.

Dipole (electric): two charges of opposite sign and equal magnitude separated by a distance—e.g., in a polar molecule.

Dipole (magnetic): object (e.g., a bar magnet) with a south magnetic pole at one end and a north magnetic pole at the other end.

Direct current (D.C.): current that always flows in the same direction around a circuit. The polarity of a D.C. supply is constant.

Dispersion: when refractive index depends on wavelength (e.g., for different wavelengths of visible light in glass) the amount of refraction will also be wavelength dependent (e.g., when a triangular prism spreads white light into a spectrum of colors).

Doppler effect: shift in observed wavelength and frequency of a wave as a result of relative motion between the source and the observer.

Drag coefficient: constant in the drag equation related to the shape and nature of the surface of a body moving through a fluid.

Drift velocity: mean velocity of charge carriers in a current-carrying conductor. Usually much lower than the random thermal velocities of the charge carriers.

Ductile: property of a material that can be drawn out into wires—depends on plastic deformation.

Dynamic (kinetic) friction: frictional force between two surfaces sliding over one another.

Earth (electrical): connection to the planet Earth taken to be at a constant potential of 0V.

Eddy currents: current loops inside a conductor when a changing magnetic field passes through it. These cause energy losses in the ferromagnetic core of a transformer and are important in electromagnetic damping systems.

Elastic limit: up to the elastic limit extensions/strains are reversible when the force/stress is removed. Beyond the elastic limit they are irreversible and permanent plastic deformation occurs.

Electric field strength: force per unit charge acting on a charged particle at a point in an electric field. Also equal to the potential gradient at each point.

Electric potential energy: energy a charged object has as a result of its position within an electric field. A property of the object placed at that point in the field.

Electric potential gradient: equal to the electric field strength at a point in the field.

Electric potential: electric potential energy per unit charge at a point in an electric field. Property of the field at that point. The zero of electric potential is defined to be at infinity.

Electromagnet: usually a coil that creates a magnetic field through its center when current flows.

Electromagnetic induction: process where changing flux-linkage through a coil or conductor induces an emf in the coil or conductor. The essential principle behind transformers and generators.

Electron-capture: process in some proton-rich nuclei whereby an inner electron combines with a nuclear proton to create a neutron and emit a neutrino.

Electrostatics: study of electric fields and forces from stationary arrangements of charges.

EMF: energy transferred from other forms to electrical energy per unit charge passing through a power supply, measured in volts.

Energy availability: extent to which energy within a system can be harnessed to do useful work.

Equilibrium: situation in which the resultant force and resultant moment on a body are both zero.

Equipartition of energy: hypothesis that when energy is supplied to a thermodynamic system each degree of freedom gets an average energy of $\frac{1}{2}kT$.

Equipotential surfaces: surfaces perpendicular to electric field lines. No work is done on or by a charged particle when it moves from one point on an equipotential surface to another.

Equivalence principle: Einstein's thesis that the laws of physics in a freely falling reference frame are indistinguishable from the laws of physics in a region of uniform gravitational field.

Error bars: drawn as a vertical or horizontal bar on either side of each plotted point to indicate the range of uncertainty in that point. Error bars can then be used to find the worst acceptable lines and the range in gradient and intercept.

Escape velocity: minimum velocity that will allow an object to escape from a point in a gravitational field to infinity (neglecting effects of non-gravitational forces such as atmospheric friction).

Event horizon: surface surrounding a black hole such that no matter or radiation can escape from within this surface and no event occurring inside this surface can have an effect on an observer in the outside universe.

Exponential change: growth or decay that changes by a constant proportion in a constant time—e.g., activity of a radioactive source has a constant half-life. Described mathematically by an equation of the form: $y = Ae^{\pm ax}$.

Extension: difference between unstretched length and stretched length, e.g., of a wire.

Faraday cage: a conducting box (often a metallic mesh box) enclosing a region of space and preventing the transmission of electromagnetic waves into or out of the box.

Faraday law: fundamental law of electromagnetic induction equating the induced emf to rate of change of flux-linkage.

Ferromagnetic material: e.g., iron or nickel-containing atoms which are themselves magnetic dipoles and which can be aligned with an external field and remain aligned when the field is removed.

Fleming's left-hand rule: used to work out the direction of the motor effect force.

Forced oscillator: an oscillator coupled to and driven by an external oscillating force.

Fractional uncertainty: ratio of absolute uncertainty δx to measured value of x: $\frac{\delta x}{x}$.

Free fall: motion of an object when it falls solely under the influence of gravitational forces.

Freebody diagram: diagram representing a body as a single object and showing all of the forces that act on it.

Frequency: number of complete oscillations per second given by $f = \frac{1}{T}$, measured in Hz (1 Hz = 1 cycle per second).

Galilean relativity: idea that the laws of mechanics are the same in all uniformly moving (inertial) reference frames.

Galvanometer: meter used to detect and measure small currents.

Gamma-factor: relativistic factor, $\gamma = \dfrac{1}{\sqrt{\left(1 - \dfrac{v^2}{c^2}\right)}}$, that occurs in the calculation of many relativistic effects (e.g., time dilation, length contraction, mass increase with velocity) and that can be used to gauge the significance of relativistic effects.

Gauss law: mathematical equation connecting the flux of a field through a closed surface to the number of sources enclosed by the surface, e.g., for an electric field: the flux of electric field through the surface is equal to the total charge enclosed divided by the permittivity of free space.

Gay Lussac law: relationship between pressure and temperature for a constant mass of an ideal gas at constant volume: $\frac{p}{T} = \text{constant}$

Geiger counter: instrument used to detect and measure ionizing radiation from radioactive sources.

General relativity: Einstein's theory of gravity as a distortion of space–time geometry rather than a field of force.

Geodesic: shortest path through curved space–time. The path followed by a freely moving body or a light ray. (A geodesic on the surface of the Earth would be a great circle route.)

Geostationary satellite: satellite in an equatorial orbit placed at such a distance that its period of orbit is equal to the Earth's period of rotation on its axis (23 hours and 56 minutes). Geostationary (or geosynchronous) satellites remain above the same point on the Earth's surface and are used for global communications.

Glasses: similar to ceramics but they have a completely amorphous microstructure and result from a rapidly cooled melt

Graham law of diffusion: rate of diffusion in a gas is inversely proportional to the square root of the molecular mass.

Gravitational field strength: gravitational force per unit mass at a point in the field. A property of the field.

Gravitational potential energy: energy an object has because of its position in a gravitational field. A property of the body placed in the field.

Gravitational potential: gravitational potential energy per unit mass. A property of the field at each point in space. The zero of potential is defined to be at infinity.

Gravitational time dilation: effect of gravitational fields on the rate at which time passes—a clock placed in a stronger gravitational time ticks more slowly than one in a weaker field.

Gravitational waves: periodic disturbances of space–time geometry that travel outwards from their source (e.g., a binary star system or colliding black holes) at the speed of light. The gravitational equivalent of electromagnetic waves.

Ground state: lowest allowed energy state for an electron in an atom ($n = 1$ state). Corresponds, in the Bohr model, to the state in which the circumference of the orbit is exactly one wavelength.

Half-life: time taken for the activity of a radioactive source to halve or for the number of unstable nuclei in the source to halve.

Hard/soft magnetic material: difficult/easy to magnetize and demagnetize.

Hard: resists indentation and scratching

Heat capacity: energy required to raise the temperature of an object by 1 K.

Heat death: idea that, as entropy continues to increase, the far future of the universe will be characterized by an equilibrium state in which energy availability has fallen to zero.

Heat engine: engine designed to extract useful work from a heat reservoir.

Heat pumps: system in which work is used to pump heat from a heat reservoir at lower temperature to a heat reservoir at a higher temperature.

Heating: energy transfer as a result of a temperature difference.

Heisenberg's uncertainty principle: principle that sets a limit on how much can be known about certain pairs of variables, e.g., position and momentum or energy and time. For example, the more precisely we determine the location of an electron the smaller the uncertainty in its position but the larger the uncertainty in its momentum.

Hertzsprung–Russell diagram: chart displaying luminosity against spectral type (or inverse surface temperature) of stars, revealing several distinct groups or bands including: the main sequence, red giants, and white dwarf stars.

Homogeneous equation: an equation in which quantities, units, and dimensions balance.

Hubble law: relationship between speed of recession and distance for galaxies $v = H_0 d$.

Hubble time: reciprocal of the Hubble constant, a rough approximation of the age of the universe.

Hydrostatic pressure: pressure due to a stationary fluid.

Hysteresis: e.g., when the force to load a sample is different from force as it is unloaded, a cycle of loading and unloading results in a closed loop on a graph of force against extension. This is a hysteresis loop. The area of the loop is related to the energy dissipated by the sample during the process.

Ideal fluid: incompressible inviscid fluid.

Ideal gas equation: equation of state for an ideal gas, incorporating all three gas laws: $pV = nRT$.

Ideal gas: theoretical model of a gas whose equation of state is $pV = nRT$.

Impedance: ratio of the peak voltage to the peak current in an A.C. system even though these values occur at different times. Resistance and reactance are special cases of impedance when the phase difference between voltage and current is 0 or $\pi/2$ respectively. Measured in ohms.

Impulse: integral of force and time equal to the change of momentum. When force is constant impulse is $Ft = mv - mu$.

Independent variable: the variable you vary to find its effect on the dependent variable. Sometimes called the "manipulated variable."

Indeterministic: where having complete knowledge of the present state of a system is insufficient to make a complete prediction of its future state. Quantum theory is an indeterministic system.

Indicator diagrams: plot of pressure against volume for a cyclic process in a heat engine.

Inertial force: an apparent force that is a result of applying Newton's laws to an accelerating reference frame as if it is not accelerating. For example, thinking that a force throws you forward when the train in which you are travelling suddenly slows down. No physical force pushes you forward.

Inertial reference frame: an unaccelerated reference frame, one that is at rest or moving at a constant velocity.

Intensity: energy per unit area per second in a wave front, measured in Wm^{-2} and proportional to the amplitude-squared.

Interaction: all forces arise from interactions. In nature there are four fundamental interactions: gravitational, electromagnetic, and the strong and weak nuclear forces.

Interferometer: instrument in which light rays travelling along two perpendicular paths are brought back together and then allowed to superpose and interfere in order to measure small differences in the optical paths (e.g., to detect the distortions caused by gravitational waves passing through the apparatus).

Internal energy: sum of random thermal kinetic energies and potential energies of all particles in the body.

Internal resistance: resistance inside a cell or battery that dissipates energy when current is drawn from the cell and results in a lost voltage so that the terminal voltage is less than the emf of the supply.

Invariant: a quantity that is the same for all inertial observers (e.g., the 4D interval between events).

Inviscid fluid: fluid with zero viscosity.

Ionizing radiation: radiation capable of ionizing atoms, e.g., in the electromagnetic spectrum, short-wavelength UV, X-ray, and gamma-rays. Alpha and beta emissions from unstable nuclei are also forms of ionizing radiation.

Isobaric changes: changes that take place at constant pressure.

Isochoric changes: changes that take place at constant volume.

Isothermal changes: changes that take place at constant temperature.

Kelvin scale: thermodynamic scale based on the absolute zero of temperature and the triple point of water.

Kepler laws: three laws of planetary motion based on elliptical orbits with the Sun at the center of the system.

Kinetic theory: particle model of matter used with Newton's laws to derive the ideal gas equation.

Kirchhoff's second law: statement of energy conservation for an electric circuit stating that the sum of emfs must equal the sum of potential differences around any closed loop in an electric circuit.

Larmor frequency: e.g., precession frequency for the axis of a magnetic dipole rotating about the direction of an applied external magnetic field during an MRI scan.

Law of Dulong and Petit: postulate that the heat capacity of all metals is $3R$.

Length contraction: the observed reduction of lengths in a reference frame that is moving relative to the observer.

Lenz law: states that the direction of an induced emf is such as to oppose the change that caused it. This ensures that energy is conserved.

Lepton: fundamental particle related to the electron. There are six different leptons, each with its own antilepton.

Linear absorption coefficient: constant that relates the attenuation of light in an absorbing medium to the properties of the medium.

Linear momentum: the product of mass and velocity, a vector quantity with unit kgms^{-1}.

Local force: idea that forces do not come from a distance but are the result of a local field that acts directly on a particle. This also explains why effects take time to propagate from one place to another as changes spread out through the field at a fixed speed (e.g., electromagnetic waves).

Logarithm: power of some base number that represents a quantity— e.g., the logarithm of 1000 to base 10 is 3 because 10 to the power 3 is equal to 1000.

Logarithmic scale: a scale that increases by a constant multiple, e.g., 1, 10, 100, 1000…. Useful for displaying data with a very wide range of values onto a single graph or chart.

Longitudinal wave: wave in which the vibration direction is parallel to the direction in which the wave transfers energy (e.g., sound/ultrasound) creating regions of compression and rarefaction in the medium.

Lorentz transformation: series of equations that transform space and time coordinates of an event from one inertial reference frame to another (in agreement with the principle of relativity).

Luminiferousether: hypothetical medium supporting electromagnetic fields and through which electromagnetic waves travel at the speed of light. Einstein's special theory of relativity and the Michelson–Morley experiment both showed that this cannot be the case.

Luminosity: total power radiated from a star.

Magnetic field strength: equivalent to magnetic flux density. Measured in tesla (T) or webers per square meter(Wbm^{-2}).

Magnetic field: field created by moving charges (e.g., in an electric current) that exerts forces on other moving charges (or currents).

Magnetic flux linkage: the magnetic flux through a coil multiplied by the number of turns in the coil. Measured in webers (Wb).

Magnetic flux: integral of the perpendicular component of magnetic field strength and area. Measured in Wb.

Magnetic resonance imaging (MRI): medical imaging technique that detects the radio waves emitted when nuclear magnetic dipoles in hydrogen atoms inside the body align with an external field.

Main sequence: diagonal band on the HR diagram, running from top left (high luminosity and high temperature) to bottom right (low luminosity and low temperature). Most stars will spend most of their lives in this band.

Malleable: property of a material that can be beaten out into sheets—depends on plastic deformation.

Manometer: U-tube containing a liquid, used to measure pressure differences.

Many-worlds theory: interpretation of quantum theory put forward by Hugh Everett III in which the wave function never collapses and each separate possibility is realized in a separate world.

Mass spectrometer: instrument used to find the relative masses and abundances of isotopes in a sample.

Mass: fundamental property of matter that determines its inertia (response to a resultant force) and its effect on and response to gravitational fields.

Maxwell distribution: probability distribution for molecular speeds or kinetic energies within a gas.

Maxwell's equations: fundamental equations of electromagnetism describing how electric and magnetic fields are related to each other and to charges and how electromagnetic waves propagate.

Michelson–Morley experiment: an attempt to measure the effect of the Earth's motion through the postulated luminiferousether on the speed of light relative to the Earth and thereby infer the velocity of the Earth relative to the ether. No effect was detected.

Moment of inertia: property of a body that resists changes in angular velocity, given by $\sum_{i=1}^{i=N} m_i r_i^2$ and measured in kgm². Its value depends on the axis about which the body rotates.

Moment: turning effect of a force in Nm. Also called a torque.

Monatomic gas: gas consisting of individual atoms acting as particles with no internal degrees of freedom.

Monochromatic: light consisting of a single wavelength (single "color").

Monoenergetic: particles having a single energy.

Motor effect: force on a current-carrying conductor when placed into a magnetic field such that there is a component of the field perpendicular to the current.

Mutual inductance: when two coils are close together, a changing current in either coil creates a changing magnetic field that affects the other coil and induces an emf in it. The strength of this effect is measured by the mutual inductance of the system of two coils in henries (H).

Natural frequency: frequency of free oscillations when the oscillator is displaced and released.

Neutral point: point in space where the fields caused by two or more sources cancels out.

Neutron star: fate of a heavy star that has formed a planetary nebula. Its core continues to collapse beyond the white dwarf stage until it is prevented from further collapse by neutron degeneracy pressure.

Newton's law of gravitation: the gravitational force between two point masses is proportional to the product of the masses and the inverse-square of their separation: $F = \dfrac{Gm_1 m_2}{r^2}$.

Newton's laws of motion: three fundamental laws of mechanics related to the effects of resultant forces and the nature of interactions.

Nucleon: particle found in the nucleus—a proton or a neutron.

Null result: when an expected effect is absent even though the method and precision should have detected it. The Michelson–Morley experiment is the most famous example of a null result.

Optical Fiber: narrow transparent fiber along which light or infrared radiation can be transmitted because it repeatedly undergoes total internal reflection at the boundary. Used to transmit information, e.g., for computer networks, telephone systems, and cable TV.

Oscillation: periodic motion about an equilibrium position, such as the vibration of a mass on a spring or the swing of a simple pendulum.

Otto Cycle: idealized cycle for a petrol internal combustion engine.

Pair annihilation: conversion of the mass of a matter particle and its corresponding anti-particle into energy in the form of gamma-rays.

Parsec: unit of distance in astronomy equal to about 3.26 light years.

Percentage uncertainty: ratio of absolute uncertainty δx to measured value x expressed as a percentage: $\dfrac{\delta x}{x} 100\%$

Perihelion: point on an elliptical planetary orbit when the planet is closest to the Sun. The equivalent point in the orbit of an Earth satellite is called perigee.

Permanent magnet: magnetic material in which the atoms are themselves magnetic dipoles. If these are aligned (e.g., in a ferromagnetic material such as iron) the sample becomes a magnetic dipole.

Permeability: property of a medium related to its ability to support a magnetic field

Permittivity: property of a medium related to its ability to support an electric field.

Phase velocity: velocity at which a point of constant phase in a wave moves—e.g., the velocity of a wave crest.

Phase: position within a cycle of oscillation on a scale of 0 to 360° or 0 to 2π radians.

Phasor: rotating vector used to represent an oscillation or point on a wave. Length corresponds to amplitude, angle corresponds to phase, and its rotation frequency is the same as the oscillation or wave it represents.

Photoelectric effect: ability of light, above a certain threshold frequency, to eject electrons from a metal surface.

Photomultiplier tube: very sensitive light detector that can respond to single photons by massively amplifying the number of electrons emitted by each photon that is absorbed.

Photon: quantum of electromagnetic radiation.

Piezoelectric crystal: type of crystal used in ultrasound transmitters and detectors. When the crystal is stressed it generates a voltage and when a voltage is applied to it there is a corresponding strain.

Planck constant (h): fundamental constant in quantum theory.

Poincaré recurrence: idea that given enough time a system will return to all of its possible macroscopic states an unlimited number of times.

Polar orbit: satellite orbit passing over the poles of the Earth. These orbits are usually relatively low and with periods of a few hours, so the satellite will pass over every part of the Earth's surface as it completes several orbits.

Polarization: selection of a particular vibration direction for a transverse wave—e.g., vertical plane polarized waves or horizontally plane polarized waves.

Polymers: materials consisting of long chain hydrocarbon molecules which tend to align with an applied stress, e.g., rubber or polythene.

Positron emission tomography (PET): medical imaging technique using a beta-plus emitter as a tracer. Positrons emitted from the tracer annihilate with electrons in the body sending out a pair of gamma-rays of the same energy moving in opposite directions.

Potential difference: difference in electric potential between two points in space (or in an electric circuit).

Potential divider: arrangement of two resistors across a power supply so that the voltage across one of them is a fraction of the supply voltage determined by the resistance ratio.

Power: rate of transfer of energy, measured in watts (W). $1W = 1Js^{-1}$

Precession: for example, when the magnetic dipole axis of a nucleus rotates about the direction of an applied magnetic field.

Principle of moments: condition for equilibrium of moments acting on the same body; sum of clockwise moments must be equal to the sum of counterclockwise moments about any point.

Progressive/traveling wave: wave that transmits energy from a source to an absorber.

Pulsar: rapidly rotating neutron star whose intense magnetic field results in two jets of radiation emitted in opposite directions. If the Earth lies in the path of one of these jets then regular pulses of radiation will be detected.

Quantum of energy: smallest discrete unit of energy transfer—e.g., for the emission or absorption of electromagnetic radiation at frequency f the minimum transfer is one photon with energy $E = hf$.

Quark: fundamental particle found in all hadrons (baryons and mesons), e.g., protons and neutrons. There are six different types of quark, each with its own anti-quark.

Radian: unit for measurements of angle based on the geometry of the circle. One radian is the angle subtended by an arc of length equal to one radius of the circle. There are 2π radians in a complete circle so 2π radians $= 360°$.

Radioactive emission: ionizing radiation emitted when an unstable nucleus decays.

Radiological dating: use of known radioactive half-lives to work out the age of archaeological or geological samples.

Random error: error making measured values larger or smaller than true values by an unpredictable amount—e.g., in repeated measurements of the time period of a pendulum. Significance of random errors can be reduced by repeating measurements and using an average value.

Ray: line perpendicular to a wave front in the direction of energy transfer.

Rayleigh criterion: rule for comparing the diffraction limit to the resolving power of optical instruments.

Reactance: A.C. impedance of a component (capacitor or inductor) where the current and voltage vary with a phase difference of $\pi/2$. Reactance is equal to the ratio of the peak voltage to the peak current in the component even though these occur at different times. Measured in ohms.

Red giant star: star approaching the end of its life. As fuel for nuclear fusion reactions in its core begins to run out it swells up and its surface cools until it is a huge red giant star.

Red-shift: ratio of increase in wavelength to original wavelength, $z = \dfrac{\lambda - \lambda_0}{\lambda_0}$, for waves reaching an observer from a source that is moving away from the observer.

Refraction: change in direction of a wave when it crosses a boundary between two media in which the wave speed is different.

Refractive index: the absolute refractive index of a transparent material is the ratio of the speed of light in a vacuum (c) to the speed of light in the medium (v): $n = \dfrac{c}{v}$.

Refrigerator: system in which work is used to extract heat from an object at low temperature and dump it into a heat reservoir at a higher temperature (e.g., the environment).

Relativity of simultaneity: two separate events that are simultaneous for one observer can occur at different times for an observer in a different inertial reference frame.

Resistance: ratio of potential difference across an electrical conductor to the current in it: $R = \dfrac{V}{I}$. Resistance is a property of a particular component. Measured in ohms.

Resistivity: property of a material equal to the resistance across opposite faces of a uniform cube of the material with sides of 1m. measured in ohm-meters.

Resolving power: minimum angular separation of object points that results in separate image points in an optical instrument.

Resonance: strong response of an oscillatory system when it is driven at its natural frequency.

Resultant force: vector sum of all forces acting on the same body.

Reynolds number Re: dimensionless number that relates inertial forces to viscous forces. For large Reynold number the flow is turbulent, for small Reynold number the flow is laminar.

RMS values: root mean square value of a quantity, particularly important in A.C. The rms value for quantities that vary sinusoidally is $\dfrac{1}{\sqrt{2}}$ times the peak value. For example, $V_{rms} = \dfrac{V_{peak}}{\sqrt{2}}$

Rotational kinetic energy: energy of a rotating body given by RKE = $\frac{1}{2} I\omega^2$

Rutherford scattering experiment: in which alpha particles were fired at thin gold foil. Analysis of the scattering data led to the discovery of the atomic nucleus.

S.I.:   Système international d'unités, an agreed international system of units including m, kg, s.

Scalar field: region of space in which each point is associated with a quantity having magnitude only—e.g., gravitational potential or electrical potential.

Scalar: physical quantity with magnitude only, e.g., mass, temperature, energy.

Schrödinger's cat: thought experiment in which a microscopic quantum effect, the decay of a radioactive atom, is linked to a macroscopic event, the death of a cat. Prior to opening the box to see if the cat has survived it exists in a superposition of states: both dead and alive.

Schrödinger's equation: fundamental equation in quantum theory. Solutions to the Schrodinger equation are wave functions.

Schwarzschild radius RS: surface surrounding a spherical black hole at which the escape velocity is equal to the speed of light.

Scientific notation: method of expressing quantities in terms of a power of 10 multiplied by a number between 1 and 10, e.g., $c = 3.00 \times 10^8$ ms^{-1}

Scintillator: material that emits light when it absorbs X-rays.

Self-inductance: when the current in a coil changes the magnetic field also changes and affects the coil, inducing a (back) emf in the coil, opposing the change. The strength of this effect is measured by the inductance of the coil in henries (H).

Semiconductors: materials such as silicon and germanium that have a small density of charge carriers at room temperature (compared to a metal) but whose conductivity increases with temperature as more charge carriers are freed.

Simple harmonic motion: oscillatory motion in which the acceleration is directly proportional to displacement from equilibrium and directed back toward equilibrium.

Simple pendulum: point mass suspended from a light inextensible string.

Snell law: law of refraction stating that the ratio of the sine of the incident angle to the normal to the sine of the refracted angle to the normal at the boundary of two media is constant.

Solenoid: long coil, usually used to create an electromagnet.

Sonography: use of ultrasound to scan the body.

Space–time curvature: disturbance of the geometry of space and time as a result of the presence of matter or energy.

Specific heat capacity: energy required to raise the temperature of 1 kg of an object by 1 K.

Specific latent heat: energy required to change the state of 1 kg of a material from solid to liquid or from liquid to gas with no change in temperature (i.e., at its melting or boiling point).

Spectrometer: an instrument used to analyze the spectrum of a light source.

Spectroscopy: analysis of light which involves spreading the light into a spectrum and determining the wavelengths present and their relative intensities.

Spectrum: range of wavelengths which might be continuous (e.g., the colors of the rainbow) or might consist of lines or bands (e.g., atomic or molecular emission spectra).

Speed of light (c): speed of electromagnetic waves in a vacuum. Fundamental maximum speed at which matter and information can be transmitted from one place to another.

Spring constant: ratio of force to extension in Hooke law, a measure of the stiffness of a spring.

Stability: the extent to which a system can return to equilibrium after being displaced from it.

Standard candle: astronomical object whose absolute luminosity is known (e.g., Cepheid variable, type 1a supernova) so that it can be used to determine distances.

Standard Model: current best model of all particles and forces in the universe.

Standing/stationary wave: localized wave disturbance consisting of a stationary pattern of nodes (positions of minimum disturbance) and antinodes (positions of maximum disturbance).

Static friction: frictional force between two surfaces in contact with each other and at rest.

Stefan–Boltzmann law: relationship between the power per unit area emitted by a radiator and its temperature.

Stiff: material that has a large stress to strain ratio (large Young modulus)—i.e., hard to stretch.

Strain energy: energy stored because of deformation, e.g., in a stretched spring.

Strong: large breaking force (for a sample) or large breaking stress (UTS) for a material.

Sum-over-histories: approach to quantum theory suggested by Richard Feynman in which all possible paths contribute a phasor and the square of the resultant phasor at each point represents the probability of the process taking place.

Supernova: explosion of a massive star at the end of its life.

Superposition: when two or more waves are present at the same point in space the resultant disturbance is the vector sum of the disturbances due to each wave.

Symmetry principle: when an operation carried out on a system leaves it unchanged.

Systematic error: measurement error that affects all measurements in the same way—e.g., making them all too large or too small by the same quantity or proportion. If the error is known it can be corrected for (e.g., by subtracting a constant value from each measurement).

Tensile strain: ratio of extension to original length. Dimensionless.

Tensile stress: ratio of axial force applied to cross-sectional area of sample perpendicular to the force, measured in Nm^{-2}.

Thermal conduction: transfer of heat as a result of particle to particle interactions.

Thermal equilibrium: when two objects are at the same temperature and, if placed in thermal contact, there is no net transfer of heat between them.

Thermal radiation: emission of electromagnetic radiation with a spectrum that depends on the temperature of the emitting body.

Thought experiment: an imagined experiment used to explore the implications of theory, e.g., Schrodinger's cat or the twin paradox.

Tidal forces: differential forces arising because of the difference in gravitational force across the diameter of an orbiting body. Tidal forces tend to distort the body along and perpendicular to the line joining it to the body around which it orbits.

Time dilation: the observed slowing of time in a reference frame that is moving relative to the observer.

Time period: time for one complete cycle of oscillation.

Torricelli vacuum: space above the mercury inside a mercury barometer containing very low pressure mercury vapor.

Total internal reflection: when a wave strikes the boundary between a medium of higher refractive index and one of lower refractive index above a certain critical angle and all of the wave energy is reflected back into the first medium.

Tough: undergoes a considerable amount of plastic deformation and absorbs a lot of energy before fracture.

Transformer: electrical device usually consisting of a primary coil and a secondary coil wound onto the same ferromagnetic core. An A.C. voltage in the primary can be stepped up or down according to the transformer equation: $\dfrac{V_2}{V_1} = \dfrac{N_2}{N_1}$.

Transverse wave: wave in which the vibration direction is perpendicular to the direction in which the wave transfers energy (e.g., all electromagnetic waves).

Trigonometric parallax: method for determining the distance to a star by measuring the change in its apparent direction (parallax) angle as the Earth orbits the Sun. The more distant the star, the smaller the parallax angle. A method of triangulation with the Earth's orbital diameter as baseline.

Triple point of water: unique temperature at which ice, water and water vapor are in equilibrium.

Twin paradox: thought experiment in which two twins separate on a particular date with one taking a high-speed return journey to a star the other remaining behind on Earth. According to special relativity the twins should have different ages when they reunite—but which twin should be younger?

Ultrasound: sound waves at a higher frequency than the upper limit of human hearing, usually taken to be $f > 20$ kHz.

Ultraviolet catastrophe: the failure of classical theory, at short wavelengths, to derive an expression for the black-body radiation spectrum. The best attempts predicted an ever-increasing amount of radiation at the UV end of the spectrum.

Uniform gravitational field: field of constant strength and direction throughout a region of space. The gravitational field near the surface of the Earth is approximately uniform over distances that are small compared to the Earth's radius.

Universal constant of gravitation: "big G," the constant in Newton's law of gravitation.

Vector field: region of space in which each point is associated with a quantity having magnitude and direction—e.g., gravitational field strength or electric field strength.

Vector: physical quantity with magnitude and direction, e.g., force, momentum, velocity.

Velocity selector: region of space in which a magnetic field and an electric field act at right angles to one another and when charged particles are fired through this region only those with a unique velocity remain undeflected. Used in a mass spectrometer to ensure that all of the ions enter the device at the same velocity.

Viscosity: measure of a fluid's resistance to flow. Units of viscosity are Pas.

Voxel: a volume element in a 3D image that is assigned a value determined by how strongly it absorbs X-rays during a CT scan.

Wave function (Ψ): mathematical expression that describes the state of a physical particle or system. The "intensity" of the wave function at a point in space ($|\Psi|^2$) is proportional to probability, e.g., for a photon $|\Psi|^2\delta V$ represents the probability of finding the photon in a volume of space δV.

Wave fronts: lines of constant phase (usually crests and/or troughs) representing wave motion. Wavefronts are perpendicular to rays.

Wavelength: shortest distance between two points oscillating in phase in a wave.

Wave–particle duality: loose description given to the observation that a particle model and a wave model are needed to explain different aspects of the behavior of matter and radiation, but neither model can give a complete explanation of all aspects of behavior.

Weight: gravitational force acting on a body.

White dwarf star: fate of a medium mass star (like our Sun) after it has swollen to become a red giant and its outer layers have drifted off into space. A white-hot core remains, and electron degeneracy pressure prevents further collapse.

Wien displacement law: relationship between wavelength at the peak of the black body radiation spectrum and the temperature of the radiating body: λT = constant.

Work: energy transfer as a result of movement in the direction of an applied force.

World line: path through space–time (line on a space–time diagram).

Yield stress: stress at which a material starts to deform plastically.

Young modulus: measure of the stiffness of a material, equal to the ratio of stress to strain when the material obeys Hooke law (linear part of graph of stress against strain).

Young's double slit experiment: famous experiment in which monochromatic light is passed through a pair of narrow parallel slits and forms an interference pattern. This originally provided the first measurement of the wavelength of light (thereby supporting a wave model for light) and is now often used to explore ideas about the interpretation of quantum theory.

Zero error: non-zero reading on an instrument when it should read zero (e.g., when a micrometer screw gauge is closed but the scale reads 0.02 mm). Must be subtracted from readings when using the instrument.

*I*NDEX